Virus Diseases of Ornamental Plants

S. K. Raj • Rajarshi Kumar Gaur • Zhimin Yin
Editors

Virus Diseases of Ornamental Plants

Characterization, Identification, Diagnosis and Management

Springer

Editors
S. K. Raj
Plant Molecular Virology Laboratory
CSIR-National Botanical Research Institute
Lucknow, Uttar Pradesh, India

Rajarshi Kumar Gaur
Department of Biotechnology
D.D.U. Gorakhpur University
Gorakhpur, India

Zhimin Yin
Plant Breeding and Acclimation Institute – National Research Institute (IHAR-PIB)
Młochów Research Center
Młochów, Poland

ISBN 978-981-16-3918-0   ISBN 978-981-16-3919-7 (eBook)
https://doi.org/10.1007/978-981-16-3919-7

This Springer imprint is published by the registered company Springer Nature Singapore Pte Ltd.
The registered company address is: 152 Beach Road, #21-01/04 Gateway East, Singapore 189721, Singapore

Contents

About the Editors

S. K. Raj did his Ph.D. degree in Botany (Plant Virology) from CSIR-National Botanical Research Institute (NBRI), Lucknow (registered at Kanpur University, Kanpur), in 1990 on the topic "Characterization of a virus causing ring spot disease in Brinjal (*Solanum melongena* L.) and its management." He started his carrier as Scientist B at Molecular Virology Laboratory, CSIR-NBRI, Lucknow, India, in 1986 and continued as Chief Scientist/Sr. Deputy Director/Prof. AcSIR till his superannuation on 31 July 2015. He has been a visiting scientist at the Institute of Genetics and Crop Research, Gatersleben, Germany, in 1991 (INSA-DFG fellowship) with Prof. R. M. Leiser; and at Scottish Crop Research Institute, Dundee, UK, in 1997 (INSA-Royal Society fellowship) and 2000 (DBT fellowship) with Prof. Micheal Taliansky to carry out research on molecular biology of plant viruses with special reference to develop viral resistance in plants through pathogen-derived resistance strategies utilizing viral genes. He had been the principal investigator of many projects funded by the Department of Biotechnology (Govt. of India), U.P. Council of Science and Technology (UP-CST), and Department of Science and Technology (Govt. of India). Out of his research work, he has published more than 211 research papers in national and international journals and guided twelve Ph.D. students. He has been honored by Vigyan Ratan Award 2010–2011 from Uttar Pradesh Science and Technology (UPCS&T), Lucknow, U.P. Government, India, for his scientific contributions in the area of Plant Virology and Biotechnology. He has been conferred Plant Pathology Leadership Award 2012 and honored and elected as the Fellow of Indian Virological Society by the Indian Virological Society, New Delhi, India, in 2013. At present he has been actively engaged in science by contributing his services to national research institutes, scientific organizations, and universities in India.

Rajarshi Kumar Gaur earned his PhD in 2005 and is now Professor in the Department of Biotechnology at Deen Dayal Upadhyaya Gorakhpur University, Gorakhpur, Uttar Pradesh, India. His Ph.D was on molecular characterization of sugarcane viruses, viz., mosaic, streak mosaic, and yellow luteovirus. He received MASHAV fellowship of Israel government for his postdoctoral studies and joined the Volcani Centre, Israel, and BenGurion University, Negev, Israel. In 2007 he received the Visiting Scientist Fellowship from Swedish Institute Fellowship,

Sweden, to work in the Umeå University, Umeå, Sweden. He received Post-Doc Fellowship from ICGEB, Italy, in 2008. He has made significant contributions on sugarcane viruses and published 130 national/international papers, authored 17 edited books, and presented about 50 papers in national and international conferences. He has visited several laboratories of USA, Canada, New Zealand, UK, Thailand, Sweden, and Italy. Currently, he is handling many national and international grants and international collaborative projects on plant viruses and disease management.

Zhimin Yin received a PhD in agricultural sciences with specialization in horticulture in 2002 at the Department of Plant Genetics, Breeding and Biotechnology, Faculty of Horticulture and Landscape Architecture, Warsaw Agricultural University—SGGW and now works on plant virology at the Plant Breeding and Acclimatization Institute—National Research Institute (IHAR-PIB), Młochów Research Center, Poland. He has worked on 2 international and 7 national projects, published 33 research articles, 5 book chapters, 20 conference papers, and gave 42 oral presentations in national/international conferences, invited lectures, and seminars. He was awarded for his PhD study and scientific achievements by SGGW-Poland in 2001, 2002, and 2007, Ministry of Education of China in 2004, Polish Academy of Sciences in 2007, and IHAR-PIB in 2016 and 2018. He has made scientific visits to SASA UK and ECU USA and was the Polish representative of the PVYwide organization.

Tospovirus Diseases of Ornamental Plants: Characterization, Identification, Diagnosis and Management

1

G. Karthikeyan, K. Nagendran, Shweta Kumari, R. Priyanka, C. Senthilraja, and Betsy D. Haokip

Abstract

Ornamental industry is one of the commercial sectors contributing maximum for the growth of agriculture. Ornamental crops are grown for its aesthetic value and are categorized as flowering, foliage, shrubs and trees. Due to their perennial nature and availability throughout the year, they are very prone for the infection of several viruses. The *Orthotospovirus* is one of the most devastating viruses of ornamental plants and flower crops worldwide. It is one among the virus genera infecting ornamental crops across 46 countries belonging to the six continents. The diseases caused by orthotospoviruses (order *Bunyavirales*; family Tospoviridae; genus *Orthotospovirus*) are emerging as a significant limiting factor for the sustainable production of ornamental crops around the globe. These orthotospoviruses are known to be transmitted through thrips vector in a persistent and propagative manner. Till date 18 species of orthotospoviruses are reported to be infecting more than 600 plant species of ornamental crops belongs to all categories. Asian continent is found to show maximum diversity harboring nine different orthotospovirus species. Symptoms of orthotospovirus infections are highly variable with the host and the virus species interactions. In this chapter information on symptomatology, transmission, distribution, diagnosis, host-pathogen interactions and management of tospoviruses on ornamental plants are consolidated and presented in an organized manner.

G. Karthikeyan (✉) · R. Priyanka · C. Senthilraja · B. D. Haokip
Department of Plant Pathology, Tamil Nadu Agricultural University, Coimbatore, India
e-mail: gandhikarthikeyan@tnau.ac.in

K. Nagendran · S. Kumari
Division of Crop Protection, ICAR-Indian Institute of Vegetable Research, Varanasi, India

S. K. Raj et al. (eds.), *Virus Diseases of Ornamental Plants*,
https://doi.org/10.1007/978-981-16-3919-7_1

Keywords

Tospoviruses · Thrips · Epidemiology · Host-pathogen interaction · Diagnosis · Management

1.1 Introduction

Ornamental plants are grown for their aesthetic value. Maintaining visual appearance of ornamental plants is the major component of ornamental production facilities. There are different groups of ornamental plants, and they are classified based on their life span as annual, biennial and perennials. Similarly, based on their aesthetic part of the plant, they are grouped as flower crops, foliage/shrubs and trees. In recent days, ornamental industry has attained remarkable growth and has developed into commercial branch of modern agriculture. Economy of ornamental industry depends in sale of potted plants, tree nursery production, seed material for flower bulbs and other propagation materials and their maintenance. The Netherlands, the USA and Japan are the pioneer countries of ornamental production (Kazinczi et al. 2007).

The ornamental plants are affected by several biotic and abiotic stresses. Among them the diseases caused by viruses impose a serious threat to the ornamental industry. Tospoviruses are the plant-infecting viruses belonging to the genus *Orthotospovirus* under the family *Bunyaviridae* and are thrips transmissible. These viruses cause significant quality and quantity losses in horticultural crops including ornamentals in many parts of the world (Pappu et al. 2009). INSV infection causes crop losses in flower production of *Dendranthema* spp. (51%) and *Sinningia speciosa* (100%) (Matteoni and Allen 1989). The utilization of molecular tools enabled the characterization of tospoviruses in to 30 species grouped under five different serogroups (EPPO 2020; Chu et al. 2001). Ubiquitous nature of tospoviruses adds every day new hosts to their host range list. Due to the thrips transmissible nature, wide and overlapping host ranges of thrips and tospoviruses, lack of effective and economical management options and lack of resistant sources in hosts contribute to the repeated occurrence of tospoviruses in diverse crops both under open and protected cropping conditions (Daughtrey et al. 1997; Brown et al. 2005; Gent et al. 2006). Several reviews and book chapters extensively organized the various aspects of tospoviruses and its thrips vector in recent years (Whitfield et al. 2005; Kazinczi et al. 2007; Pappu et al. 2009; Turina et al. 2016). In this chapter, we summarized the current status of tospovirus epidemics in ornamental crops.

1.2 Host Range of Tospoviruses on Ornamental Crops

The economic impact of tospoviruses are increasing every day by causing huge losses in terms of quality and quantity in a wide variety of ornamental crops. The host range of tospoviruses is highly variable between the species. In global scenario, around 204 plant species were reported to be infected by 18 different tospovirus species (Table 1.1). Tospoviruses are infecting all kind of ornamentals such as flowering, foliage/shrubs, grasses and trees of annuals, biennials and perennials. Nearly eight different tospovirus species are affecting the cultivation of chrysanthemum in several countries. Tomato spotted wilt virus (TSWV) and impatiens necrotic spot virus (INSV) are the two major tospoviruses threatening the ornamental crop cultivation. Both of them were known to infect around 600 species of ornamental plants (Daughtrey et al. 1997; Ullman 1998). Alstroemeria necrotic streak *virus* (ANSV), alstroemeria yellow spot *virus* (AYSV), calla lily chlorotic spot *virus* (CCSV) and ligustrum necrotic ringspot *virus* (LNRV) are the tospovirus species having very narrow host range and documented only on ornamental crop plants. The chrysanthemum stem necrosis *virus* (CSNV) is found to infect ornamental crops such as chrysanthemum, *Callistephus* spp., *Eustoma grandiflorum*, *Gerbera* and lisianthus. The INSV is having intermediate host range than the TSWV and commonly infecting annuals and perennial ornamentals. Iran and the USA are the two countries found to have wider host range for the tospovirus infection among ornamental crops. Many tospovirus species are known to infect weed species very often which are playing a major epidemiological role in disease dissemination. Tree ornamentals such as *Cycas* sp., *Ficus benjamina*, *Schefflera*, etc. are also infected by tospoviruses.

1.3 Symptomatology of Tospoviruses on Ornamental Crops

In ornamentals, tospovirus infection produces symptoms similar to the other biotic and abiotic stresses. Tospovirus symptoms on floral crops include concentric ring spots; line or zonal patterns, malformation and necrosis of apical shoots and flowers, necrosis of leaf petiole, reduced growth, etc. were reported (Fig. 1.1) (Kazinczi et al. 2007). Chrysanthemum plants are very often observed with chlorotic and necrotic spots on leaves and stems, stunted growth and death of young plants upon infection of TSWV and INSV (Cho et al. 1989). TSWV produces chlorotic or black ringspot and line pattern of leaves, malformed leaves and terminal bud necrosis on impatiens; chlorotic and necrotic ringspot of leaves, stem necrosis and terminal bud necrosis on *Eustoma*; mild mosaic and ringspot of leaves on *Zinnia*; and reddish concentric brown rings on the leaves and malformed flowers on *Gloxinia* (Antignus et al. 1997). TSWV infection reduces the floral fresh weight, number of petals, number of flowers per plant and inhibiting adventitious sprouting of buds in perennials (Whitfield et al. 2003). CSNV infection causes mild or severe streaks on stem, wilting of leaves and stems, chlorotic and necrotic spots and rings on leaves, necrosis of the stem and floral receptacles, necrotic lesions surrounded by yellow halo on leaves and distinct dark stem lesions on chrysanthemum (Verhoeven et al. 1996; Duarte et al. 1995; Mumford et al. 2003). Persley et al. (2006) observed ringspots, line patterns, chlorotic blotches and necrotic etching on *Hoya* spp. upon capsicum chlorosis virus (CaCV) infection. CCSV caused chlorotic spots on their leaves and stems of *Zantedeschia* species in Taiwan (Lin et al. 2006). IYSV infection in plants of *Alstroemeria* spp. show necrotic streak symptoms (Okuda et al. 2005).

Table 1.1 Distribution and host range of tospoviruses infecting ornamental crops

Continent	Country	Number of hosts	Hosts
1. Alstroemeria necrotic streak virus (ANSV)			
North America	USA	1	*Hoya carnosa*
South America	Colombia	1	*Alstroemeria* sp.
2. Alstroemeria yellow spot virus (AYSV)			
Europe	Netherlands	1	*Alstroemeria* sp.
3. Calla lily chlorotic spot virus (CCSV)			
Asia	China	1	Spider lily
Asia	Taiwan	1	Calla lily
4. Capsicum chlorosis virus (CaCV)			
Asia	China	5	*Rieger begonias*, *Gloxinia*, orchids, African violet, *Crossandra*
Asia	India	1	*Hippeastrum*
Asia	Indonesia	1	Orchids
Asia	Iran	1	*Rudbeckia hirta*
Asia	Taiwan	4	Orchids, *Hippeastrum*, calla lily, *Scadoxus multiflorus*
Asia	Thailand	1	*Hymenocallis littoralis*
Australia	Australia	1	*Hoya* sp.
North America	USA	2	*Hoya* sp., *Gloxinia*
5. Chrysanthemum stem necrosis virus (CSNV)			
Asia	China	1	*Chrysanthemum*
Asia	Japan	3	*Chrysanthemum*, *Lisianthus*, *Eustoma grandiflorum*
Asia	South Korea	1	*Chrysanthemum*
Australia	New Zealand	1	*Chrysanthemum*
Europe	Belgium	1	*Chrysanthemum*
Europe	Netherlands	1	*Chrysanthemum*
Europe	Russia	1	*Chrysanthemum*
Europe	Slovenia	2	*Chrysanthemum*, *Gerbera*
Europe	UK	1	*Chrysanthemum*
South America	Brazil	4	*Chrysanthemum*, *Lisianthus*, *Callistephus* spp., *Eustoma grandiflorum*
6. Groundnut bud necrosis virus (GBNV)			
Asia	India	8	Orchids, *Phaius* sp., *Anthurium*, periwinkle, *Gerbera*, jasmine, *Dahlia*, *Chrysanthemum*
7. Groundnut ringspot orthotospovirus (GRSV)			
South America	Brazil	1	*Hippeastrum*
South America	Paraguay	1	*Petunia*
8. Hippeastrum chlorotic ringspot virus (HCRV)			
Asia	China	9	

(continued)

Table 1.1 (continued)

Continent	Country	Number of hosts	Hosts
			Hymenocallis littoralis, *Oxalis corniculata*, *Philodendron*, *Crinum asiaticum*, *Zephyranthes candida*, *Clivia miniata*, spider lily, *Ficus elastic*, *Hippeastrum*
9. Impatiens necrotic spot virus (INSV)			
Africa	Egypt	1	*Tropaeolum*
Asia	China	5	Begonia, orchids, spider lily, *Hymenocallis*, *Gentiana* spp.
Asia	Japan	10	*Anemone coronaria*, *Begonia*, *Chrysanthemum*, *Eustoma* sp., *Gentiana* sp., *Gerbera jamesonii*, *Pericallis*, *Exacum* spp., *Cyclamen*, *Eucharis* spp.
Asia	South Korea	2	*Hoya* sp., *Peperomia obtusifolia*
Asia	Iran	29	African violet, *Impatiens*, *Alstroemeria*, *Pelargonium* spp., *Cycas* spp., *Rosa* spp., *Scindapsus* spp., *Ficus* spp., *Gazania* spp., *Erica* spp., *Syngonium*, *Gladiolus* spp., *Anthurium*, *Codiaeum*, *Cheiranthus*, *Dieffenbachia*, *Philodendron* spp., *Zinnia*, *Althaea* spp., *Bougainvillea* spp., Chamomile, Marigold, *Dracaena* spp., *Pilea cadierei*, *Cissus* spp., *Spathiphyllum* spp., *Leucanthemum vulgare*, *Dahlia* spp., *Dianthus barbatus*
Asia	Israel	3	*Anemone*, *Impatiens*, *Torenia* spp.
Australia	New Zealand	9	*Begonia*, *Freesia refracta*, *Gardenia jasminoides*, *Gerbera jamesonii*, *Hibiscus rosa-sinensis*, *Impatiens*, *Primula*, *Ranunculus* sp., *Cyclamen*
Europe	Belgium	1	*Impatiens*
Europe	Bosnia and Herzegovina	1	*Begonia* spp.
Europe	Bulgaria	1	*Hydrangea* spp.
Europe	Czech Republic	11	*Columnea* sp., *Curcuma* sp., *Impatiens*, *Hippeastrum*, *Aeschynanthus* spp., *Anthurium*, *Philodendron* spp., *Osteospermum* spp., *Verbena* spp., *Saxifraga stolonifera*, *Cyrtomium falcatum*
Europe	Finland	1	*Lobelia* spp.
Europe	Germany	2	*Anemone*, *Impatiens*
Europe	Hungary	3	*Impatiens*, *Eustoma* sp., *Cyclamen*
Europe	Italy	22	*Antirrhinum*, *Scindapsus* spp., Periwinkle, calla lily, *Oncidium*, *Primula* spp., *Peperomia* spp., *Clarkia amoena*, *Anemone* spp., *Ruscus* spp., *Lobelia* spp., *Cyclamen*, *Cordyline*, *Ranunculus*, *Maranta* spp., *Episcia* spp., *Spathiphyllum* spp., *Isotoma axillaris*, *Limonium*, *Dahlia* spp., *Delphinium* spp., *Dianthus* spp.
Europe	Poland	1	*Impatiens*
Europe	Portugal	10	

(continued)

Table 1.1 (continued)

Continent	Country	Number of hosts	Hosts
			Pelargonium spp., *Gazania spp.*, *Gladiolus* spp., *Sinningia* spp., *Penstemon*, *Ruscus* spp., *Aphelandra*, *Helichrysum* spp., *Gerbera* spp., *Hydrangea* spp.
Europe	Russia	4	*Gloxinia*, *Impatiens*, *Ruellia*, *Streptocarpus*
Europe	Serbia	3	*Begonia* sp., *Impatiens*, *Tulip* sp.
Europe	Spain	1	*Asplenium*
Europe	The Netherlands	5	*Impatiens*, *Iris*, *Lysimachia*, *Nerine* spp., *Gentiana* spp.
Europe	UK	3	*Impatiens*, *Pericallis*, *Opuntia* spp.
North America	Canada	2	*Begonia*, *Impatiens*
North America	Costa Rica	6	*Antirrhinum majus*, *Hippeastrum* sp., *Impatiens*, *Iris* sp., orchids, *Plectranthus*
North America	Mexico	1	*Impatiens*
North America	USA	46	*Abelia* spp., *Antirrhinum*, Begonia, *Browallia speciosa*, *Calceolaria* spp., *Callistephus*, *Dianthus* sp., *Digitalis* sp., *Epidendrum* spp., *Eustoma* sp., *Gloxinia* spp., *Hosta* sp., *Hoya wayetii*, *Impatiens*, *Maranta leuconeura*, *Monarda didyma*, *Nemesia*, *Nipponanthemum*, *Ocimum basilicum*, *Lobelia* spp., orchids, snapdragon, *Schizanthus* spp., *Pelargonium* spp., periwinkle, *Pericallis*, *Primula* spp., *Sinningia* spp., *Penstemon*, *Exacum* spp., *Cyclamen*, *Weigela*, *Scaevola*, *Hydrangea* spp., *Mimulus* spp., *Oxydendrum arboreum*, *Ilex glabra*, *Photinia fraseri*, *Gaillardia* spp., *Diascia* spp., *Sempervivum* spp., *Dahlia* spp., *Franklinia* spp., *Halesia carolina*, *Rhaphiolepis indica*, *Calycanthus floridus*
10. Iris yellow spot virus (IYSV)			
Asia	Israel	4	*Lisianthus*, *Eustoma* sp., *Hippeastrum*, *Lilium* spp.
Asia	Japan	3	*Eustoma* sp., *Lisianthus*, *Alstroemeria*
Europe	Iran	5	*Chrysanthemum*, *Cycas* spp., *Pelargonium*, *Rosa* spp., *Scindapsus* spp.
Europe	Netherland	3	*Iris*, *Eustoma* sp., *Impatiens* spp.
Europe	Poland	1	*Chrysanthemum*
Europe	UK	2	*Lisianthus*, *Eustoma* sp.
North America	USA	1	*Eustoma* sp.
11. Lisianthus necrotic ringspot virus (LNRV)			
Asia	Japan	1	Lisianthus
12. Pepper chlorotic spot virus (PCSV)			
Asia	China	1	Erigeron
13. Tomato chlorotic spot virus (TCSV)			

(continued)

Table 1.1 (continued)

Continent	Country	Number of hosts	Hosts
North America	USA	4	Periwinkle, *Marsdenia floribunda*, *Hoya wayetii*, *Schlumbergera truncata*
South America	Argentina	1	*Eustoma*
South America	Brazil	5	*Bouvardia*, *Dieffenbachia*, *Mirabilis jalapa*, *Spilanthes oleracea*, *Eustoma*
14. Tomato spotted wilt virus (TSWV)			
Asia	China	10	*Argyranthemum*, *Bidens*, *Chrysanthemum*, *Dahlia*, *Geranium*, *Iris*, *Platycodon grandiflorus*, *Tropaeolum majus*, *Zinnia*, *Campanula* spp.
Africa	Egypt	1	*Chrysanthemum*
Asia	India	4	*Antirrhinum majus*, *Callistephus*, *Chrysanthemum*, marigold
Asia	Iran	30	*Althaea*, *Anthurium*, *Asplenium*, *Calendula* spp., *Canna indica*, *Codiaeum variegatum*, *Cupressus sempervirens*, *Cycas* spp., *Dahlia* spp., *Dieffenbachia amoena*, *Dracaena* spp., *Erica* spp., *Ficus benjamina*, *Gazania* spp., Jasmine, *Mirabilis jalapa*, orchids, *Pericallis*, *Philodendron* spp., *Pteris cretica*, *Rosa* spp., *Saintpaulia ionantha*, *Rudbeckia* spp., *Salvia* spp., *Schefflera*, *Scindapsus* spp., *Strelitzia reginae*, marigold, *Torenia fournieri*, *Zinnia*
Asia	Israel	2	*Eustoma*, *Pittosporum tobira*
Asia	Japan	11	*Arctotis*, *Callistephus*, *Chrysanthemum*, *Pericallis*, *Valeriana fauriei*, *Dahlia* spp., *Dimorphotheca sinuate*, *Eustoma*, *Felicia amelloides*, *Ranunculus* spp., *Verbena* spp.
Asia	Jordan	2	Gerbera, marigold
Asia	Saudi Arabia	1	Periwinkle
Asia	South Korea	11	*Angelica gigas*, *Anthurium*, *Impatiens*, *Brugmansia* sp., *Chrysanthemum*, *Eustoma*, *Gerbera*, *Hoya* sp., *Peperomia*, *Ranunculus*, *Nasturtium*
Africa	Zimbabwe	2	*Chrysanthemum*, *Gypsophila*
Australia	Australia	5	Spider lily, *Agapanthus*, *Gladiolus* spp., *Lupinus* spp., *Tanacetum* spp.
Australia	New Zealand	9	Chrysanthemum, *Cineraria* sp., *Dahlia*, *Hoya* sp., *Impatiens*, *Ocimum*, *Pinellia tripartita*, marigold, *Ornithogalum* spp.
Europe	Belgium	1	*Lobelia* spp.
Europe	Bosnia and Herzegovina	2	*Chrysanthemum*, *Gloxinia*
Europe	Bulgaria	14	*Leuzea carthamoides*, lisianthus, *Adenium*, *Althaea*, *Calendula* spp., *Helichrysum bracteatum*, *Hydrangea* spp., *Inula helenium*, *Physostegia* spp., *Salvia* spp., marigold, *Thymus vulgaris*, *Verbena* spp., *Zinnia*

(continued)

Table 1.1 (continued)

Continent	Country	Number of hosts	Hosts
Europe	Czech Republic	3	*Campanula* spp., *Columnea*, *Gazania* spp.
Europe	France	4	Dahlia, *Arum* spp., *Dimorphotheca sinuate*, *Salvia* spp.
Europe	Germany	1	Lysimachia
Europe	Greece	9	*Aristolochia clematitis*, Periwinkle, *Dimorphotheca sinuate*, *Fuchsia* spp., *Gazania* spp., *Justicia brandegeeana*, *Portulaca* spp., *Salvia* spp., marigold
Europe	Hungary	3	*Aristolochia clematitis*, *Echinacea purpurea*, *Heuchera sanguinea*
Europe	Italy	21	Chrysanthemum, *Coprosma repens*, *Gerbera*, *Iberis semperflorens*, *Anemone coronaria*, *Arctotis*, *Arum* spp., *Asclepias*, *Dimorphotheca sinuate*, *Euphorbia*, *Eritrea*, *Eustoma*, *Gazania* spp., *Limonium perezii*, *Nerium oleander*, orchids, *Polygala myrtifolia*, *Primula* spp., *Ranunculus* spp., *Ruscus racemosus*, *Verbena* spp.
Europe	Lithuania	2	*Dicentra Formosa*, *Lilium* spp.
Europe	Montenegro	6	*Aquilegia* sp., *Calceolaria* sp., *Chrysanthemum*, *Gerbera*, *Petunia*, *Primula* sp.
Europe	Poland	5	*Chrysanthemum*, *Ocimum*, *Dieffenbachia amoena*, *Hippeastrum* spp., *Hydrangea* spp.
Europe	Portugal	11	*Anthurium*, *Aphelandra*, periwinkle, *Fuchsia* spp., *Gazania* spp., *Gerbera*, *Gladiolus* spp., *Helichrysum bracteatum*, jasmine, *Penstemon* spp., *Sinningia speciosa*
Europe	Russia	1	*Streptocarpus*
Europe	Serbia	7	*Brugmansia* sp., *Chrysanthemum*, *Dahlia*, *Gerbera*, *Impatiens*, *Petunia*, *Sinningia speciosa*
Europe	Spain	1	*Ficus benjamina*
Europe	The Netherlands	11	*Adenium*, *Anthurium*, *Columnea*, *Dahlia* spp., *Hippeastrum* spp., *Impatiens* spp., *Iris*, *Lobelia* spp., *Lysimachia* spp., *Nerine* spp., *Schizanthus* spp.
Europe	Turkey	1	*Ranunculus*
Europe	UK	4	*Impatiens*, *Delphinium* spp., *Paeonia* spp., *Verbena* spp.
North America	Canada	7	Periwinkle, *Cyclamen persicum*, *Fuchsia* spp., *Gaillardia* spp., *Impatiens* spp., orchids, *Verbena* spp.
North America	Mexico	1	Zinnia
North America	USA	76	Orchids, *Chrysanthemum*, *Dahlia*, *Eustoma*, *Hoya* sp., *Lobelia* spp., *Matricaria*, *Pittosporum tobira*, *Stokesia*, *Tulbaghia violacea*, *Adenium*, *Anemone coronaria*, *Aeschynanthus*, *Aglaonema*, *Anthurium*, *Antirrhinum*, *Aphelandra*, *Arum* spp., Begonia, *Calceolaria* spp., *Callistephus chinensis*, periwinkle,

(continued)

Table 1.1 (continued)

Continent	Country	Number of hosts	Hosts
			Celosia cristata, *Centranthus ruber*, *Clarkia amoena*, *Cordyline terminalis*, *Cosmos* spp., *Cyclamen persicum*, *Delphinium* spp., *Digitalis* spp., *Dracaena* spp., *Eucharis grandiflora*, *Exacum* spp., *Gardenia jasminoides*, *Gerbera*, *Gladiolus* spp., *Gomphrena globosa*, *Helichrysum bracteatum*, *Hosta* spp., *Hydrangea* spp., *Impatiens* spp., *Lupinus* spp., *Melampodium divaricatum*, *Opuntia* spp., *Osteospermum*, *Pedilanthus*, *Penstemon* spp., *Peperomia* spp., *Pericallis*, *Petunia* spp., *Phlox* spp., *Pinus* spp., *Pittosporum tobira*, *Plectranthus australis*, *Portulaca grandiflora*, *Primula* spp., *Ranunculus* spp., *Rhododendron* spp., *Rohdea* spp., *Rudbeckia* spp., *Saintpaulia ionantha*, *Salvia* spp., *Schizanthus* spp., *Schlumbergera truncate*, *Sedum* spp., *Lychnis chalcedonica*, *Sinningia speciose*, *Stephanotis floribunda*, *Streptocarpus* spp., *Syngonium podophyllum*, *Tradescantia* spp., *Tropaeolum majus*, *Verbena* spp., *Veronica* spp., *Weigela florida*, *Zantedeschia* spp.
South America	Argentina	2	*Alstroemeria*, *Eustoma*
South America	Brazil	2	*Eucharis grandiflora*, *Campanula* spp.
South America	Ecuador	1	*Chrysanthemum*
South America	Venezuela	2	*Chrysanthemum*, *Gerbera*
15. Tomato yellow ring virus (TYRV)			
Asia	Iran	22	*Tropaeolum*, *Cineraria*, *Gazania*, *Chrysanthemum*, *Alstroemeria*, Anemone, *Antirrhinum* spp., *Althaea* spp., *Beaucarnea recurvata*. *Bougainvillea* spp., *Dahlia*, *Dieffenbachia*, *Dracaena*. *Ficus benjamina*, *Impatiens*, *Jasminum*, *Pelargonium*, *Pericallis hybrida*, *Rosa* spp., *Spathiphyllum*, *Syngonium*, marigold
16. Tomato zonate spot virus (TZSV)			
Asia	China	3	*Bidens*, *Crinum asiaticum*, *Iris*
17. Watermelon bud necrosis virus (WBNV)			
Asia	India	1	*Chrysanthemum*
18. Watermelon silver mottle virus (WSMoV)			
Asia	Taiwan	1	Calla lily

Source: NCBI database and Sastry et al. 2019

Fig. 1.1 Symptom variation of tospovirus infection on different ornamental crops

1.4 Transmission of Tospoviruses

Vectors play a major role in dissemination of tospoviruses and escalating them as devastating pathogen in the global status. The tospoviruses are transmitted by thrips (Thysanoptera: Thripidae) in a persistent, circulative and propagative manner (German et al. 1992). Among the 5500 known thrips species, nine species (*Frankliniella* sp., *Thrips* sp. and *Scirtothrips* sp.) were reported to be associated with tospoviruses (Mound 1998). All thrips species transmitting tospoviruses are polyphagous in nature and distributed worldwide causing the tospovirus a menace to ornamentals. Transmission of tospoviruses by thrips undergoes specific virus-vector interaction. One of the most interesting features is adults thrips were not able to acquire the virus in the thrips-tospovirus interaction. Tospoviruses acquired by first and second instar larval stage alone can able to transmit virus in their adult stage (Ullman et al. 1992; Sakimura 1963). Once thrips acquire the virus, it remains infectious throughout its life span because of its transstadial transmission nature (Wijkamp and Peters 1993; Wijkamp et al. 1995).

Frankliniella occidentalis (western flower thrips), a notorious insect pest, exists globally and is the most important vector of TSWV (Goldbach and Peters 1994). In petunia, *F. occidentalis* appeared to be the most effective in transmitting four tospoviruses tested (TSWV, INSV, tomato chlorotic spot *virus* (TCSV) and groundnut ringspot virus (GRSV)) in comparison to other thrips species through leaf disk assay. Also, one among four populations of *T. tabaci* was able to transmit TSWV in a low transmission efficiency (Wijkamp et al. 1995). Upon, ingestion of tospovirus by both adults and larvae of *F. occidentalis*, barrier in midgut prevents the virus movement out of the midgut of adults. Virions accumulated within epithelium gut of adults, whereas virions are seen in the midgut epithelia, fat body, brain, nerve cells, haemocoel and salivary glands of larvae (Ullman et al. 1992). Nagata and De Avila (2000) studied the transmission efficiency of CSNV by *F. schultzei* (78.1%), *F. occidentalis* (65.1%) and *T. tabaci* (0.0%). High CSNV titre was found in 75.9% of *F. schultzei* and 97.4% of *F. occidentalis* adults. Studies of Sharman et al. (2020) showed *Thrips palmi*, *F. schultzei* and *Microcephalothrips abdominalis* were transmitting CaCV, while no transmission was achieved with *F. occidentalis*. Medeiros et al. (2004) reported the activation of immune system in *F. occidentalis* upon acquiring TSWV. Beside thrips transmission, the dissemination of tospovirus also occurs through vegetatively propagated material such as tubers, corms, cuttings, etc. in many ornamental species (Wilson 2001; Adkins 2003). TSWV persists in flower bulbs and roots of lilies and dahlias (Kazinczi et al. 2007). Transmission of tospovirus through seeds does not occur among ornamentals (Mumford et al. 1996; Antignus et al. 1997). Major vectors of tospoviruses infecting ornamental crops are described in Table 1.2.

Table 1.2 Vectors of orthotospovirus species infecting ornamental crops

Virus	Vector
ANSV	*Frankliniella occidentalis*
AYSV	*Thrips tabaci*
CCSV	*T. palmi*
CaCV	*Ceratothripoides claratris, F. schultzei, T. palmi*
CSNV	*F. gemina, F. occidentalis, F. intonsa, F. schultzei*
GBNV	*F. schultzei, Scirtothrips dorsalis, T. palmi*
GRSV	*F. gemina, F. intonsa, F. occidentalis, F. schultzei*
HCRV	Unknown
INSV	*F. fusca, F. intonsa, F. occidentalis, F. schultzei*
IYSV	*F. fusca, T. tabaci*
LNRV	Unknown
PCSV	Unknown
TCSV	*F. intonosa, F. occidentalis, F. schultzei*
TSWV	*F. bispinosa, F. cephalica, F. fusca, F. gemina, F. intonsa, F. occidentalis, F. schultzei, T. setosus, T. tabaci*
TYRV	*Microcephalothrips abdominalis, T. tabaci*
TZSV	*F. occidentalis*
WBNV	*T. palmi*
WSMoV	*T. palmi*

Source: (EPPO 2020)

1.5 Occurrence and Distribution of Tospoviruses on Ornamental Crops

Tospoviruses are a highly cosmopolitan virus genus to have a worldwide distribution. The tospoviruses infecting ornamental crops are reported in 46 countries distributed representing all 6 continents (Fig. 1.2). TSWV and INSV are predominant tospovirus species infecting wide host range and are distributed across 38 and 26 countries on ornamental crops, respectively. TSWV is found to be prevalent among six continents and INSV in five continents and are ubiquitous infecting varieties of crops. INSV has become a serious pathogen on flower crops cultivated in greenhouse conditions in the USA (Daughtrey et al. 1997). Among the 18 ornamental-infecting tospoviruses, nine viruses, viz. CCNV, GBNV, HCRV, LNRV, PCSV, TYRV, TZSV, WBNV and WSMoV, seem to be restricted only to Asian countries. Totally 14 tospovirus species are infecting ornamental crops of Asian continent. Interestingly, GBNV is found on eight ornamental plants only in India, while HCRV infects nine plants only in China. In the Australian continent, four species (CaCV, CSNV, INSV and TSWV) are infecting different plant species such as *Hoya* sp., chrysanthemum, *Begonia*, *Cyclamen*, *Freesia*, *Gardenia*, *Gerbera*, *Hibiscus*, *Impatiens*, *Primula*, *Ranunculus*, spider lily, *Agapanthus*, *Gladiolus*, *Lupinus*, *Tanacetum*, *Cineraria*, *Dahlia*, *Ocimum*, *Pinellia*, marigold and *Ornithogalum*. Among the different countries, maximum of eight different tospovirus species are distributed across the China followed by six in the USA and five in Iran, Japan and the Netherlands (Table 1.1).

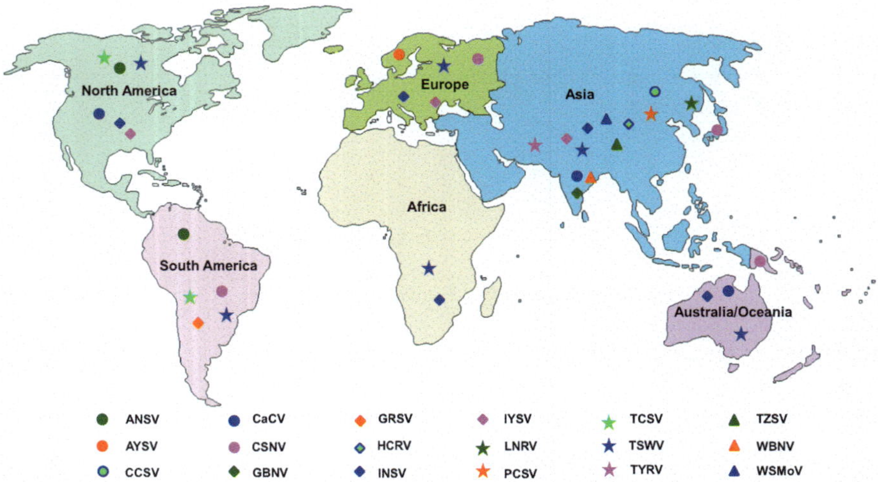

Fig. 1.2 Geographical distribution of tospoviruses species infecting ornamentals in different continents

1.6 Diagnostics of Tospoviruses

In this omics era, several sophisticated methods were utilized to detect tospoviruses. Exploiting viral properties such as morphology, transmission ability, serological property and nucleotide sequence identity is employed for tospovirus identification and characterization.

1.6.1 Visual Observation

Tospoviruses produces wide range of symptoms on ornamental plants upon infection. But some peculiar symptoms such as chlorotic or necrotic lesions, ringspots with concentric rings, green island mosaic on leaves, bronzing, chlorosis, tip necrosis and wilting are commonly seen across the hosts irrespective of tospovirus species infected. As they are infecting extensive hosts, symptom expression varies according to the hosts and virus interaction (Mumford et al. 1996). It is not feasible to identify the virus species based on the symptomatology, and therefore the serological and molecular tools are extensively used in the tospovirus detection.

1.6.2 Morphology

Use of electron microscope (EM) in the visualization of plant viruses was first demonstrated with tobacco mosaic virus (Kausche et al. 1939). EM has been extensively used to view the characteristic morphology of enveloped quasi-spherical particles of tospoviruses. Tospovirus-like particles of 90–120 nm in diameter were detected from the leaf extracts of chrysanthemum when negatively stained with 2% uranyl acetate (Duarte et al. 1995; Verhoeven et al. (1996). Gera et al. (1999) visualized INSV from the leaf-dip preparations prepared from leaf samples of *Anemone coronaria* plants through transmission electron microscopy (TEM). EM is used as a preliminary detection method of tospoviruses, and it does not satisfy their complete identification.

1.6.3 Bioassay

Mechanical inoculations on indicator plants such as *Nicotiana* sp., *Vigna unguiculata* (cowpea), etc. are used for the generic detection of tospoviruses. They produce characteristic local or systemic symptoms on indicator hosts. Inoculation of TSWV from chrysanthemum produced local chlorotic, concentric rings that later turned to necrotic spots 6 days post inoculation (DPI) on cowpea and pig weed (Renukadevi et al. 2015). Though it does not lead to identification at species level, it can be used for subsequent identification by other techniques.

1.6.4 Serological Methods

Serological method detects the presence of virus particle directly through the interaction between coat protein of virus (antigen) and the immunoproteins raised against the antigen in warm-blooded animals (antibody). Enzyme-linked immunosorbent assay (ELISA) is a very commonly used serological assay for the detection of tospovirus. Variants of ELISA such as double antibody sandwich [(DAS)-ELISA] and triple antibody sandwich [(TAS)-ELISA] are performed using antibody that has been raised against the specific tospovirus species. Based on the same principle, other methods such as dot immuno-binding assay, lateral flow assay, tissue immuno-binding assay, etc. are being carried out. As there are five sero-groups prevalent among the tospoviruses, cross reaction of particular antibody with other tospovirus members will occur commonly in the serological assays. Through this method, presence or absence of tospovirus can be ascertained, but the species level identity cannot be confirmed. To confirm the identity of tospovirus species, application of molecular method will be of more helpful. Shahraeen et al. (2002) utilized the TAS-ELISA assay to detect INSV among the various ornamental plants grown in commercial nurseries and field in Iran. Renukadevi et al. (2015) used DAS-ELISA technique to detect the TSWV in chrysanthemum.

1.6.5 DNA Technology

The development of PCR and nucleic acid-based assays such as RT-PCR and quantitative (q) RT-PCR provides a simple, quick and precise means for tospovirus detection. According to the ICTV nomenclature, species are defined on the coat protein (N) sequence with amino acid identity of >90% than that of previously described tospovirus species called as new species (Plyusnin et al. 2011). Therefore, sequencing of tospovirus genome is indispensable for its proper identification. For generic identification, conventional RT-PCR assays using universal degenerate primers are used. Several utilized degenerate primers for the preliminary detection of tospoviruses in ornamental plants (Chu et al. 2001; Renukadevi et al. 2015; Huang et al. 2018; Basavaraj et al. 2020). Naidu et al. (2007) documented the natural infection of impatiens necrotic spot virus on *Monarda didyma* in Washington State by molecular method. RT-PCR assay using species-specific primer followed by sequencing of amplified fragments allows confirmation of tospovirus at species level. Uga and Tsuda (2005) developed multiplex RT-PCR for the simultaneous detection of five tospoviruses (WSMoV, TSWV, INSV, MNSV and IYSV) in ornamental crops. Charoenvilaisiri et al. (2014) standardized a multiplex RT-PCR-ELISA to detect and differentiate four tospovirus species found in Thailand, viz. CaCV, MYSV, TNRV and WSMoV. Targeted viral genomes amplified and labelled with digoxigenin (DIG) in a single RT-PCR reaction followed by distinguishing four species by parallel hybridizations with species-specific biotinylated probes in streptavidin-coated microtiter wells were used in ELISA detection. Assay sensitivity was 10 to 1000-fold higher than conventional RT-PCR. Real-time RT-PCR assay based on TaqMan probe has been demonstrated for the detection and quantification of TSWV in infected plant samples. Assay detected TSWV in 500 fg of total RNA and sensitized TSWV to tenfold higher than ethidium bromide staining (Roberts et al. 2000).

1.7 Molecular Biology of Tospoviruses

Tospoviruses are having a tripartite genome of ssRNA molecules comprising a total of five open reading frames (ORFs) enveloped inside the quasi-spherical particles with a diameter of 80–120 nm. The three single-stranded RNA (ssRNA) segments are denoted as S, M and L RNAs with 2.9 kb, 5 kb and 8.9 kb in size, respectively (Fig. 1.3). The L RNA is of negative polarity and encodes a putative 331-kDa RNA polymerase that may be associated with the ribonucleoprotein complex. The other two genomic RNAs use an ambisense coding strategy. The M RNA encodes a nonstructural (NSm) protein (34 kDa) in the viral (v) sense and a protein (155 kDa) to serve as the precursor for the Gn (95 kDa) and Gc (58 kDa) glycoproteins in the viral complementary (vc). The NSm protein may be involved in cell-to-cell movement of nonenveloped ribonucleocapsid structures. Tubular structures are specifically formed when the NSm gene is expressed in both plant and insect cell systems. The Gn and Gc glycoproteins are believed to form spikes on the viral envelope which help in virus acquisition and transmission by thrips. The S RNA encodes in the 'v' sense a 52-kDa nonstructural (NSs) protein that forms filamentous inclusion bodies, and in the 'vc' sense, a 29-kDa nucleocapsid protein (NP) encapsidates viral RNAs (Adkins et al. 1996; de Haan et al. 1990, 1992; Kormelink et al. 1992; van Poelwijk et al. 1997; Chu and Yeh 1998; Law et al. 1992; Satyanarayana et al. 1996; Whitfield et al. 2005). RNA molecules are generally seen as pseudocircular because of highly conserved and complementary first eight nucleotides (5'-AGAGCAAU and 3'-UCUGCUUA) at the termini of all RNA segments. These sequences act as promotor region for viral RNA replication and protein transcription (Turina et al. 2016; Kormelink 2011).

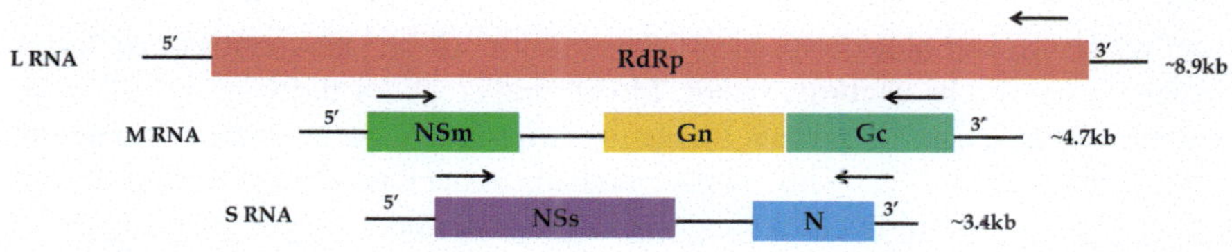

Fig. 1.3 Genome organization of tospovirus

1.8 Host-Pathogen Interaction

Host-pathogen interaction is a multifaceted process, mediated by the pathogen- and host-derived molecules which mainly include proteins, sugars and lipopolysaccharides. Plant viruses move cell to cell via plasmodesmata through the vascular system. Plant viruses encoding nonstructural proteins such as movement protein are specifically required for movement within their hosts. Movement proteins can able to increase the plasmodesmal size exclusion limit (SEL) to facilitate cell-to-cell movement of virions (Boevink and Oparka 2005). In the tospovirus infection cycle, only very few host factors involved were studied at various stages in plants. Interaction between DNA-J protein of host with NSM (movement protein) of TSWV suggested the movement of infectious ribonucleoproteins possible involvement of a mechanism that requires Hsp-70 (Soellick et al. 2000; von Bargen et al. 2001). Also, another chaperone protein (At-4/1) which shares homology with α-helical domains of ankyrin-, myosin- and kinesin-like proteins showed to be involved in both intra- and extracellular movement of TSWV virions (Paape et al. 2006; Morozov et al. 2014). Investigation on changes in the jasmonic acid (JA) and salicylic acid (SA) pathways upon TSWV inoculation through thrips revealed upregulation in the SA pathway in the infected plants than the healthy plants. Although feeding of thrips normally induces JA pathway, virus infection counteracts and suppresses the JA (antiherbivore response) pathway. These facilitate the thrips to get more attracted towards the infected plants rather than the healthy plants. This interaction supports the dissemination of the TSWV by thrips (Maris et al. 2004; Abe et al. 2012; Turina et al. 2016). Though few of the factors involved in tospovirus-host interaction were studied in model plants such tobacco and *Arabidopsis*, no factors were understood with the interaction of tospovirus-ornamental plants.

1.9 Management of Tospoviruses in Ornamental Crops

Early detection gives rise to early implementation of effective management strategies that can limit crop damage and economic loss. Several management strategies were demonstrated under open field and controlled conditions grown crops against tospoviruses across agricultural and horticultural crops. But information on tospovirus management in ornamental crops are limited. In general, management of tospoviruses can be achieved through cultural, biological, chemical, host plant resistance and biotechnological interventions. These interventions may target tospovirus as well as its thrips vector. Monitoring crops for thrips using yellow or blue sticky traps, checking every 7 days for thrips in foliage or flowers by tapping over a white surface or blowing lightly into buds or open flowers to draw thrips out for detection is helpful in controlling the thrips population. Proper maintenance of greenhouses plays an important role in the management strategy of tospoviruses in ornamental crops. All plant materials should be inspected before it is brought into the greenhouse and preferably kept in a separate section for a week or more before incorporating the new material into the production area, to assure that thrips are not introduced into the main production area. The greenhouse should be kept free of weeds inside and outside since both thrips and tospoviruses are easily harboured on greenhouse weed plants. Virus indicator plants can also be employed in the greenhouse. Growing crops with proper isolation distances from the TSWV source of inoculum provided there are no weed plant playing reservoir host in between will contain the disease spread (Coutts et al. 2004). Several cultural practices demonstrated to slow down the spread of tospoviruses are periodical removal of virus sources, avoidance of side-by-side and continuous plantings, planting of non-host barrier crops, growing under thrips proof net, mulching with black-silver polythene sheet, installation of yellow sticky traps, etc. under open field conditions (Jones 2004; Priyanka et al. 2019).

Use of insecticides in the management of thrips vector is the primary strategy for management of tospoviruses. Generally, the insecticides used in the thrips management can be grouped as broad-spectrum insecticides (pyrethroids, neonicotinoids, organophosphates and carbamates) and narrow-spectrum insecticides (pyridalyl and lufenuron) (Mouden et al. 2017). Regular use of broad-spectrum insecticides invites resistance development in thrips and detrimental to the natural enemies. Spinosad and spinetoram are being extensively used to manage thrips efficiently due to its tendency to be compatible with natural enemies (Gao et al. 2012; Li et al. 2016). However, the use of insecticides will ultimately lead the development of resistance in insect vectors. Therefore, it is necessary to use the insecticides accurate, precise and safe (He et al. 2020).

Biological means of tospovirus management can be achieved by targeting either the virus or the insect vector. Yoon et al. (2020) developed a novel approach in the management of TSWV in chrysanthemum by combining application of soil-dwelling predatory mite (*Stratiolaelaps scimitus*) @ 60/m^2 placed in soil along with foliar spray of four essential oil mixtures (cinnamon oil, cinnamon bark oil, oregano oil and thyme oil). These treatments were toxic to eggs, larvae and adults of *F. occidentalis*. Treatment has reduced the incidence of TSWV to 0.93% in comparison to chemical insecticides (32.05%) and untreated controls (84.85%) in chrysanthemum. *Amblyseius cucumeris*, another predatory mite, gave excellent control of *F. occidentalis* on *Impatiens* and reduced the spread and severity of TSWV under glasshouse conditions in the UK (Bennison et al. 2002). Another approach in management of tospoviruses in ornamental crops is development of resistant varieties. Some of the resistance genes such as *Tsw* and *Sw-5* were identified in solanaceous vegetables and are deploying resistance against TSWV (Turina et al. 2016). Unfortunately, resistance source identification against tospoviruses among the ornamental crops was in infant stage. Biotechnological interventions such RNAi, clustered regularly interspaced short palindromic repeats (CRISPR)-associated protein 9 (Cas9) and exogenous application of dsRNA to mediate resistance against tospovirus have been demonstrated by several workers (Tabein et al. 2020; Carbonell et al. 2019). Though these findings were demonstrated in other crop plants, these tospoviruses are also known to infect ornamental crops too. Therefore, these biotechnological tools are inviting major attention among the researchers to manage the tospoviruses.

1.10 Conclusion

The literature clearly indicates the prospective of tospovirus to cause damage to the ornamental crops worldwide. Though there are several reports for the infection of tospoviruses on ornamental crops, the coexistence of tospovirus with other viruses in combinations has to be studied in detail. When the vector is considered, most of the research findings of other crops are being considered for ornamental crops. However, it is essential to explore the involvement and distribution of thrips species in different ornamental crops and tospovirus interactions. Further, understanding of genome diversity, biology, geographical distribution and molecular and serological diagnostics of different tospoviruses in ornamental crops can contribute for the development of effective management strategies to prevent further spread of tospoviruses.

References

Abe H, Tomitaka Y, Shimoda T, Seo S, Sakurai T, Kugimiya S, Tsuda S, Kobayashi M (2012) Antagonistic plant defense system regulated by phytohormones assists interactions among vector insect, thrips and a tospovirus. Plant Cell Physiol 53(1):204–212

Adkins S (2003) Tomato spotted wilt virus. In: Pernezny K, Roberts PD, Murphy JF, Goldberg NP (eds) Compendium of pepper diseases. American Phytopathological Society, St. Paul, pp 39–40

Adkins S, Choi TJ, Israel BA, Bandla MD, Richmond KE, Schultz KT, Sherwood JL, German TL (1996) Baculovirus expression and processing of tomato spotted wilt tospovirus glycoproteins. Phytopathology 86(8):849–855

Antignus Y, Lapidot M, Anaim M, Cohen J, Lachman O, Pearlsman M, Raccah B, Gera A (1997) Biological and molecular characterization of tomato spotted wilt virus in Israel. Phytoparasitica 25:319–330

Basavaraj YB, Siwach J, Kumar A, Bhattarai A, Qayoom N, Baranwal VK, Jain RK (2020) First report of natural infection by Capsicum chlorosis virus on Amaryllis (*Hippeastrum hybridum*) plants from India. Plant Dis 104(11):3086–3086

Bennison J, Maulden K, Barker I, Morris J, Boonham N, Smith P, Spence N (2002) Reducing spread of TSWV on ornamentals by biological control of western flower thrips. In: Thrips and Tospoviruses: proceedings of the 7th international symposium on Thysanoptera, pp 215–219

Boevink P, Oparka KJ (2005) Virus-host interactions during movement processes. Plant Physiol 138(4):1815–1821

Brown S, Csinos A, Díaz-Pérez JC, Gitaitis R, LaHue S, Levin-Nicholson J, Martinez N, McPherson R, Mullis S, Nischwitz C, Riley D, Sherwood J, Sisson V, Stephenson M, Wells L (2005) Tospoviruses. In: Solanaceae and other crops in the coastal plain of Georgia. University Georgia. (Research Report Number 704), p 40

Carbonell A, Lisón P, Daròs JA (2019) Multi-targeting of viral RNAs with synthetic trans-acting small interfering RNAs enhances plant antiviral resistance. Plant J 100(4):720–737

Charoenvilaisiri S, Seepiban C, Bhunchoth A, Warin N, Luxananil P, Gajanandana O (2014) Development of a multiplex RT-PCR-ELISA to identify four distinct species of tospovirus. J Virol Methods 202:54–63

Cho JJ, Mau RF, German TL, Hartmann RW, Yudin LS, Gonsalves D, Provvidenti R (1989) A multidisciplinary approach to management of *tomato spotted wilt virus* in Hawaii. Plant Dis 73:375–383

Chu FH, Yeh SD (1998) Comparison of ambisense M RNA of watermelon silver mottle virus with other tospoviruses. Phytopathology 88(4):351–358

Chu FH, Chao CH, Chung MH, Chen CC, Yeh SD (2001) Completion of the genome sequence of watermelon silver mottle virus and utilization of degenerate primers for detecting tospoviruses in five serogroups. Phytopathology 91(4):361–368

Coutts BA, Thomas-Carroll ML, Jones RAC (2004) Patterns of spread of tomato spotted wilt virus in field crops of lettuce and pepper: spatial dynamics and validation of control measures. Ann Appl Biol 145(2):231–245

Daughtrey ML, Jones RK, Moyer JW, Daub ME, Baker JR (1997) Tospoviruses strike the greenhouse industry: INSV has become a major pathogen on flower crops. Plant Dis 81 (11):1220–1230

de Haan P, de Avila AC, Kormelink R, Westerbroek A, Gielen JJ, Peters D, Goldbach R (1992) The nucleotide sequence of the S RNA of Impatiens necrotic spot virus, a novel tospovirus. FEBS Lett 306(1):27–32

de Haan P, Wagemakers L, Peters D, Goldbach R (1990) The S RNA segment of tomato spotted wilt virus has an ambisense character. J Gen Virol 71:1001–1007

Duarte LM, Rivas EB, Alexandre MAV, DeÁvila AC, Nagata T, Chagas CM (1995) Chrysanthemum stem necrosis caused by a possible novel tospovirus. J Phytopathol 143:569–571

EPPO (2020) PM 7/139 (1) Tospoviruses (genus Orthotospovirus). Bulletin OEPP/EPPO Bulletin 50(2):217–240. https://doi.org/10.1111/epp.12676

Gao Y, Lei Z, Reitz SR (2012) Western flower thrips resistance to insecticides: detection, mechanisms and management strategies. Pest Manag Sci 68(8):1111–1121

Gent DH, du Toit LJ, Fichtner SF, Mohan SK, Pappu HR, Schwartz HF (2005) Iris yellow spot virus: an emerging threat to onion bulb and seed production. Plant Dis 90(12):1468–1480

Gera A, Kritzman A, Cohen J, Raccah B (1999) First report of Impatiens necrotic spot tospovirus (INSV) in Israel. Plant Dis 83(6):587

German TL, Ullman DE, Moyer JW (1992) Tospoviruses: diagnosis, molecular biology, phylogeny, and vector relationships. Annu Rev Phytopathol 30:315–346

Goldbach R, Peters D (1994) Possible causes of the emergence of tospovirus diseases. In: Seminars in virology, vol 5(2). Academic Press, pp 113–120

He Z, Guo JF, Reitz SR, Lei ZR, Wu SY (2020) A global invasion by the thrip, Frankliniella occidentalis: current virus vector status and its management. Insect science 27(4):626–645. https://doi.org/10.1111/1744-7917.12721

Huang KS, Li SL, Sun JH, Wang YC, Jan FJ, Chen TC (2018) Development of a generic method for inspection of tospoviruses. Eur J Plant Pathol 150(2):457–469

Jones RA (2004) Using epidemiological information to develop effective integrated virus disease management strategies. Virus Res 100(1):5–30

Kausche GA, Pfankuch E, Ruska H (1939) Die Sichtbarmachung von pflanzlichem virus im Übermikroskop. Naturwissenschaften 27(18):292–299

Kazinczi G, Horváth J, Takács A (2007) Tospoviruses on ornamentals Plant Viruses 1(2):142–162

Kormelink R (2011) The molecular biology of tospoviruses and resistance strategies. In: Elliott RM, Plyusin A (eds) Bunyaviridae: molecular and cellular biology. Plenum Press, New York, pp 163–191

Kormelink R, De Haan P, Meurs C, Peters D, Goldbach R (1992) The nucleotide sequence of the M RNA segment of tomato spotted wilt virus, a bunyavirus with two ambisense RNA segments. J Gen Virol 73(11):2795–2804

Law MD, Speck J, Moyer JW (1992) The M RNA of impatiens necrotic spot Tospovirus (Bunyaviridae) has an ambisense genomic organization. Virology 188(2):732–741

Li DG, Shang XY, Reitz SR, Nauen R, Lei ZR, Lee SH, Gao YL (2016) Field resistance to spinosad in western flower thrips *Frankliniella occidentalis* (Thysanoptera: Thripidae). J Integr Agric 15:2803–2808

Lin Y, Chen T, Chung M, Yeh S (2006) Sequence analysis of S RNA of Calla lily chlorotic spot virus, a new tospovirus. Acta Hortic 722:103–110

Maris PC, Joosten NN, Goldbach RW, Peters D (2004) Tomato spotted wilt virus infection improves host suitability for its vector Frankliniella occidentalis. Phytopathology 94 (7):706–711

Matteoni JA, Allen WR (1989) Symptomatology of tomato spotted wilt virus infection in florist's chrysanthemum. Can J Plant Pathol 11:373–380

Medeiros RB, Resende RDO, de Avila AC (2004) The plant virus tomato spotted wilt Tospovirus activates the immune system of its main insect vector, Frankliniella occidentalis. J Virol 78 (10):4976–4982

Morozov S, Makarova S, Erokhina T, Kopertekh L, Schiemann J, Owens R, Solovyev A (2014) Plant 4/1 protein: potential player in intracellular, cell-to-cell and long-distance signaling. Front Plant Sci 5:26

Mouden S, Sarmiento KF, Klinkhamer PG, Leiss KA (2017) Integrated pest management in western flower thrips: past, present and future. Pest Manag Sci 73:813–822

Mound LA (1998) So many thrips – so few tospoviruses? In: 7th international symposium on Thysanoptera, Reggio Calabria, Italy, pp 15–18

Mumford RA, Barker I, Wood KR (1996) The biology of the Tospoviruses. Ann Appl Biol 128:159–183

Mumford RA, Jarvis B, Morris J, Blockley A (2003) First record of Chrysanthemum stem necrosis virus (CSNV) in the UK. Plant Pathol 52:779

Nagata T, De Avila AC (2000) Transmission of chrysanthemum stem necrosis virus, a recently discovered tospovirus, by two thrips species. J Phytopathol 148(2):123–125

Naidu RA, Alabi OJ, Karthikeyan G, Nelson M, Morrell A (2007) First report of *Monarda didyma* infection with *Impatiens necrotic spot virus (Tospovirus)* in Washington State. Plant Health Progress. https://doi.org/10.1094/PHP-2007-1119-02-BR

Okuda M, Hanada K, Uematsu S, Fukumoto F, Mihira T, Ebihara Y, Iwanami T (2005) Necrotic streaks of alstroemeria (*Alstroemeria* spp.) caused by Iris yellow spot virus (IYSV). Japanese J Phytopathol 71(2):119–122

Paape M, Solovyev AG, Erokhina TN, Minina EA, Schepetilnikov MV, Lesemann DE, Schiemann J, Morozov SY, Kellmann JW (2006) At-4/1, an interactor of the tomato spotted wilt virus movement protein, belongs to a new family of plant proteins capable of directed intra- and intercellular trafficking. Mol Plant-Microbe Interact 19(8):874–883

Pappu HR, Jones RAC, Jain RK (2009) Global status of tospovirus epidemics in diverse cropping systems: successes achieved and challenges ahead. Virus Res 141(2):219–236

Persley DM, Thomas JE, Sharman M (2006) Tospoviruses – an Australian perspective. Australas Plant Pathol 35:161–180

Plyusnin A, Beaty BJ, Elliott RM, Goldbach R, Kormelink R, Lundkvist Å, Schmaljohn CS, Tesh RB (2011) Bunyaviridae. In: Virus taxonomy: ninth report of the international committee on taxonomy of viruses. Elsevier Scientific Publ. Co.

Priyanka R, Nagendran K, Aravintharaj R, Balaji CG, Mohankumar S, Renukadevi P, Karthikeyan G (2019) Characterization and management of watermelon bud necrosis virus infecting watermelon in India. Eur J Plant Pathol 153(3):759–770

Renukadevi P, Nagendran K, Nakkeeran S, Karthikeyan G, Jawaharlal M, Alice D, Malathi VG, Pappu HR (2015) First report of tomato spotted wilt virus infection of chrysanthemum in India. Plant Dis 99(8):1190

Roberts CA, Dietzgen RG, Heelan LA, Maclean DJ (2000) Real-time RT-PCR fluorescent detection of tomato spotted wilt virus. J Virol Methods 88(1):1–8

Sakimura K (1963) Frankliniella fusca, an additional vector for the tomato spotted wilt virus, with notes on Thrips tabaci, another vector. Phytopathology 53:412–415

Sastry KS, Mandal B, Hammond J, Scott SW, Briddon RW (2019) Encyclopedia of plant viruses and Viroids. Springer India, New Delhi, p 2936

Satyanarayana T, Mitchell SE, Reddy DVR, Kresovich S, Jarret R, Naidu RA, Gowda S, Demski JW (1996) The complete nucleotide sequence and genome organization of the M RNA segment of peanut bud necrosis tospovirus and comparison with other tospoviruses. J Gen Virol 77 (9):2347–2352

Shahraeen N, Ghotbi T, Mehraban AH (2002) Occurrence of Impatiens necrotic spot virus in ornamentals in Mahallat and Tehran provinces in Iran. Plant Dis 86(6):694–694

Sharman M, Thomas JE, Tree D, Persley DM (2020) Natural host range and thrips transmission of capsicum chlorosis virus in Australia. Australas Plant Pathol 49(1):45–51

Soellick TR, Uhrig JF, Bucher GL, Kellmann JW, Schreier PH (2000) The movement protein NSm of tomato spotted wilt tospovirus (TSWV): RNA binding, interaction with the TSWV N protein, and identification of interacting plant proteins. Proc Natl Acad Sci 97(5):2373–2378

Tabein S, Jansen M, Noris E, Vaira AM, Marian D, Behjatnia SAA, Accotto GP, Miozzi L (2020) The induction of an effective dsRNA-mediated resistance against tomato spotted wilt virus by exogenous application of double-stranded RNA largely depends on the selection of the viral RNA target region. Front Plant Sci 11. https://doi.org/10.3389/fpls.2020.533338

Turina M, Kormelink R, Resende RO (2016) Resistance to tospoviruses in vegetable crops: epidemiological and molecular aspects. Annu Rev Phytopathol 54:347–371

Uga H, Tsuda S (2005) A one-step reverse transcription-polymerase chain reaction system for the simultaneous detection and identification of multiple tospovirus infections. Phytopathology 95:166–171

Ullman DE (1998) Managing tospoviruses: developing a monitoring system for grower decision making. In: 14th conference on insect and disease management on ornamentals, Del Marc, CA. Society of American Florists, Alexandria, VA, pp 9–12

Ullman DE, Cho JJ, Mau RF, Hunter WB, Westcot DM, Custer DM (1992) Thrips-tomato spotted wilt virus interactions: morphological, behavioral and cellular components influencing thrips transmission. In: Advances in disease vector research. Springer, New York, pp 195–240

van Poelwijk F, Prins M, Goldbach R (1997) Completion of the impatiens necrotic spot virus genome sequence and genetic comparison of the L proteins within the family Bunyaviridae. J Gen Virol 78(3):543–546

Verhoeven JT, Roenhorst JW, Cotes I, Peters D (1996) Detection of a novel tospovirus in chrysanthemum. Acta Hortic 432:44–51

von Bargen S, Salchert K, Paape M, Piechulla B, Kellmann JW (2001) Interactions between the tomato spotted wilt virus movement protein and plant proteins showing homologies to myosin, kinesin and DnaJ-like chaperones. Plant Physiol Biochem 39(12):1083–1093

Whitfield AE, Campbell RL, Sherwood JL, Ullman DE (2003) Tissue blot immunoassay for detection of tomato spotted wilt virus in *Ranunculus asiaticus* and other ornamentals. Plant Dis 87:618–622

Whitfield AE, Ullman DE, German TL (2005) Tospovirus-thrips interactions. Annu Rev Phytopathol 43:459–489

Wijkamp I, Peters D (1993) Determination of the median latent period of two tospoviruses in Frankliniella occidentalis, using a novel leaf disk assay. Phytopathology 83:986–986

Wijkamp I, Almarza N, Goldbach R, Peters D (1995) Distinct levels of specificity in thrips transmission of tospoviruses. Phytopathology 85(10):1069–1074

Wilson CR (2001) Resistance to infection and translocation of tomato spotted wilt virus in potatoes. Plant Pathol 50(3):402–410

Yoon JB, Choi SK, Cho IS, Kwon TR, Yang CY, Seo MH, Yoon JY (2020) Epidemiology of tomato spotted wilt virus in *Chrysanthemum morifolium* in South Korea and its management using a soil-dwelling predatory mite (*Stratiolaelaps scimitus*) and essential oils. Virus Res 289:198128

Potyvirus Infecting Ornamental Plants Grown in the Neotropical Region 2

M. Amelia V. Alexandre, Ligia M. L. Duarte, Alexandre L. R. Chaves, and Elliot W. Kitajima

Abstract

This chapter presents an overview of some *Potyvirus* species infecting ornamental plants in Central America, Southern Mexico, South Florida, and South America, with an emphasis on reports from Brazil. The following topics will be covered: etiology, detection, symptomatology and management of alstroemeria mosaic virus (AlMV), bean yellow mosaic virus (BYMV), Bidens mosaic virus (BiMV), *Brugmansia suaveolens* mottle virus (BsMoV), Canna yellow streak virus (CaYSV), Catharanthus mosaic virus (CatMV), Colombian datura virus (CDV), Costus stripe mosaic virus (CoSMV), dasheen mosaic virus (DsMV), euphorbia ringspot virus (EuRSV), Gloriosa stripe mosaic virus (GSMV), Hippeastrum mosaic virus (HiMV), hyacinth mosaic virus (HyMV), konjac mosaic virus (KoMV), lily mottle virus (LMoV), Narcissus yellow stripe virus (NYSV), soybean mosaic virus (SMV), sunflower chlorotic mottle virus (SCMoV), and turnip mosaic virus (TuMV).

Keywords

Bean yellow mosaic virus · Canna yellow streak virus · Catharanthus mosaic virus · Costus stripe mosaic virus · Gloriosa stripe mosaic virus · Hippeastrum mosaic virus · Hyacinth mosaic virus · Konjac mosaic virus · Narcissus yellow stripe virus

M. A. V. Alexandre (✉) · L. M. L. Duarte · A. L. R. Chaves
Instituto Biológico, Laboratório de Fitovirologia Fisiopatológica, São Paulo, SP, Brazil
e-mail: maria.alexandre@sp.gov.br

E. W. Kitajima
Escola Superior de Agricultura Luiz de Queiroz, USP, Piracicaba, SP, Brazil

S. K. Raj et al. (eds.), *Virus Diseases of Ornamental Plants*,
https://doi.org/10.1007/978-981-16-3919-7_2

2.1 General Introduction to the Genus *Potyvirus* and Its Importance in Ornamental Horticulture

The family *Potyviridae* (potyvirids) currently comprises 12 definitive genera: *Arepavirus*, *Bevemovirus*, *Brambyvirus*, *Bymovirus*, *Celavirus*, *Ipomovirus*, *Macluravirus*, *Poacevirus*, *Potyvirus*, *Roymovirus*, *Rymovirus*, and *Tritimovirus*, including Spartina mottle virus, common reed chlorotic virus, and longan witches broom-associated virus, species not yet included in any genus (International Committee on Taxonomy of Viruses 2019).

The family *Potyviridae* was recently included in the order *Patatavirales*, class *Stelpaviricetes*, phylum *Pisuviricota*, kingdom *Orthornavirae*, realm *Riboviria*, and viruses (International Committee on Taxonomy of Viruses 2019). The genera are differentiated by biological criteria, including transmission by specific vectors, as well as by molecular and phylogenetic data (Wylie et al. 2017). The phylogeny of RdRp genes shows that potyvirids consist of at least 11 genera, the largest by far being the *Potyvirus* (Gibbs et al. 2020).

The first potyvirus probably originated 15,000–30,000 years ago in a Eurasian grass host, by acquiring crucial changes to its coat protein and HC-Pro protein, enabling it to be transmitted by migrating host-seeking aphids (Gibbs et al. 2020). Phylogeny shows that the *Potyvirus* likely diverged approximately 7250 years ago in Southwest Eurasia or North Africa (Gibbs and Ohshima 2010) and has become the largest plant-infecting genus, with more than 150 distinct virus species, described in up to 30 host families in monocots and dicots, in both cultivated and wild plants (Wylie et al. 2017). Potyviruses are non-persistently spread in nature by aphids (Wylie et al. 2017) and less frequently by seeds. Experimentally, most are mechanically transmissible. Some are either inefficiently transmitted or not transmitted at all by aphids. This is apparently due to mutations within the helper component and/or CP cistrons. Electron microscopy provided a clear taxonomic criterion to allocate viruses into the *Potyviridae* family by showing flexuous rod-shaped, 680–900 nm-long and 11–13 nm-wide particles and the consistent presence of cylindrical inclusions in the cytoplasm of infected plant cells (Fig. 2.1). These inclusions are formed by tightly apposed lamellae and appear commonly as scroll-like curved structures, which produce different configurations depending on the plane of the section. If longitudinally sectioned, they appear as a set of parallel lines and, in cross sections, as pinwheel or closely apposed lamellae. Edwardson (1974) carried out extensive and seminal studies on these inclusions, observing at least four different patterns. These cylindrical inclusions are formed by the aggregation of virus-coded helicase (cylindrical inclusion protein CI), involved in the replication process and cell-to-cell and long-distance movements (Sorel et al. 2014). Infection by some potyviruses induces crystalline nuclear inclusion (Edwardson 1974). Virions contain a single molecule of linear, positive-sense ssRNA of about 9.7 kb. The genomic RNA replicates in cytoplasm encode a polyprotein which is cleaved to produce ten proteins: P1 (translation, replication modulator), HC-Pro (suppression of gene silencing and aphid transmission), P3 (virus replication, host range, and symptom development), 6K1 (formation of replication vesicles), CI (helicase involved in viral

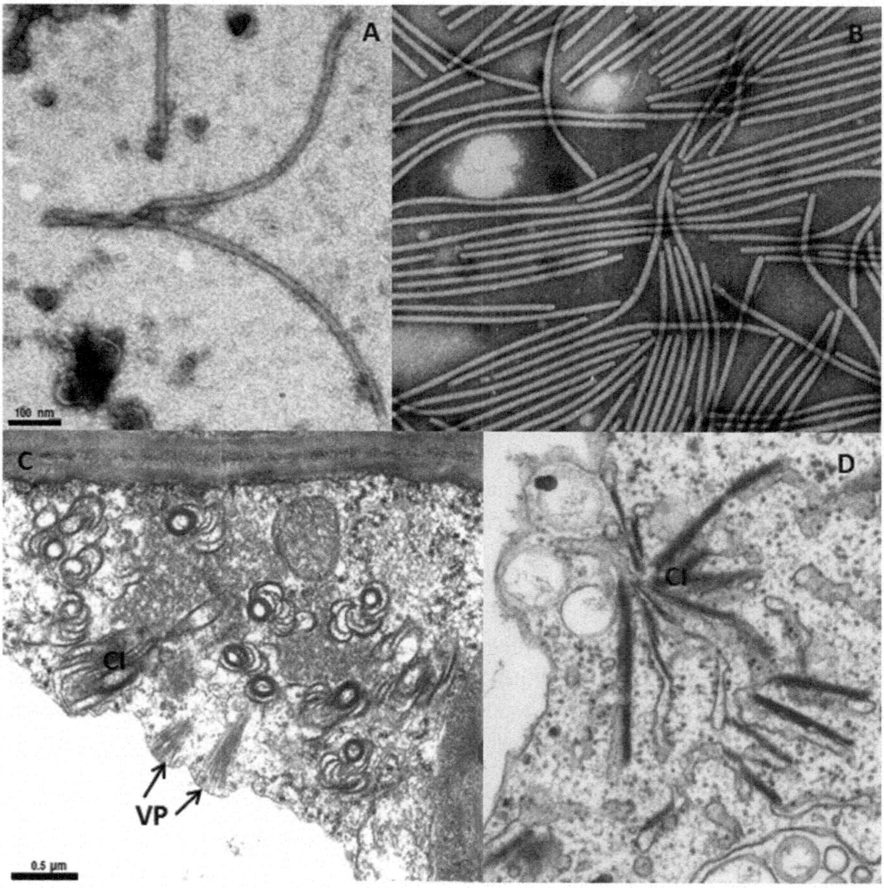

Fig. 2.1 (**a**, **b**) Transmission electron micrograph of potyvirus virions. (**a**) Negatively stained leaf extract ("leaf dip") of periwinkle (*Catharanthus roseus*) infected by Catharanthus mosaic virus (CatMV) (Maciel et al. 2011). (**b**) Purified preparation of turnip mosaic virus (TuMV) which may infect *Tropaeolum majus*. (**c**, **d**) Thin section of potyvirus-infected leaf tissues showing the presence of characteristic cylindrical inclusions (CI). (**c**) From jade vine (*Strongylodon machrobothrys*) infected by an isolate of soybean mosaic virus (SMV) (Camelo-García et al. 2021). Cylindrical inclusions, sectioned transversally, exhibit the characteristic pinwheel configuration. Viral particles can be seen in bundles (VP). (**d**) From Amazon lily (*Eucharis grandiflora*) infected by Hippeastrum mosaic virus (HiMV) (Alexandre et al. 2011). In (**c**), cylindrical inclusions are of type 1 and, in (**d**), of type 2, in the classification of Edwardson and Christie (1979)

movement and replication), 6K2 (small transmembrane protein probably anchoring the replication complex to the ER), VPg (translation, viral movement, and replication), NIa-Pro (polyprotein processing), NIb (RNA-dependent RNA polymerase), and CP (virion formation, viral movement, and vector transmission) (Wylie et al.

2017). Additionally, the P3N-PIPO protein is expressed as a result of a polymerase slippage mechanism (Chung et al. 2008; Olspert et al. 2015; Rodamilans et al. 2015).

It is widely accepted that the world is urbanizing at an increasing pace and that humans have an innate tendency to connect with nature. Appreciating the beauty of plants and flowers is part of this connection (AIPH-IVP-Report2.pdf n.d.).

In recent decades, the cultivation of ornamental plants has prompted a significant increase in the production of cut flowers, pots, and plants for domestic and urban landscaping and has become an important branch of modern agriculture, encompassing growers, service providers, and retailers alike. With the annual production of new varieties and hybrids, there is a large flow of material between countries, which can cause greater virus dispersion. In this context, vegetative propagation contributes to the spread of viruses and other diseases, and perennial ornamental crops are natural reservoirs for plant viruses, promoting their circulation and transmission to other economically important crops (Mitrofanova et al. 2018). In addition, human activities, especially trade and farming, have fostered and spread both potyviruses and their aphid vectors throughout the world, especially over the past five centuries. The global distribution of potyviruses, particularly those found on islands, indicates that they may be more frequently or effectively transmitted by seeds than experimental tests suggest (Sastry 2013).

Historical evidence of symptoms caused by potyvirids in ornamental plants includes the color break of infected tulip flowers that were reproduced in old Dutch paintings. Tulips of over 100 species were highly prized in the sixteenth century, and today there are several thousand cultivars resulting from hybridization (Mowat 1995). Viruses do not usually kill ornamental plants but often reduce their value.

In the world scenario of flower and ornamental plant producers, Colombia, Israel, the Netherlands, the USA, Ecuador, and Japan stood out during the 1970s and 1980s (Malter 1995). Latin America is also home to two global players in the field of cut flower production with an emphasis on Colombia and Ecuador which export to North America and to a lesser extent Europe and Russia. However, these two countries serve the international market, and their flowers have ended up in every corner of the world, either directly or re-exported through hubs such as the Netherlands (AIPH-IVP-Report2.pdf n.d.). Over the past 10 years, other countries have seen a sharp increase in demand for ornamental crops, but production is fragmented. This category contains countries such as China, Kenya, and Thailand, among others, each country seeking its own way to break into the market. Producers in Central America have captured their own export markets for green foliage. Mexico, and Mexico City in particular, stands out as a potential hotspot for demand. Ornamentals are culturally embedded and play an important role in religious festivities, national holidays, and major life events, and local production grew to serve the urban demand.

Brazil, a typical domestic producer, has protected its market with strict import regulations and trade tariffs and as such is not a global player in ornamentals. With 15,000 hectares of ornamental production, the country is a major global player, in terms of size, but most output is destined to domestic consumption. In 2017, revenues were ~BRL 6.9 billion (~USD 1.4 million) with 6% growth over the previous year, a favorable performance, considering the economic and financial crisis that was affecting the country (Junqueira and Peetz 2018). Brazilian flower and ornamental plants production covers an area of 13,468 ha and employs 7800 growers. Despite the considerable diversity of native species with ornamental potential, most of the species grown, commercialized and consumed in Brazil, are exotic and come from improved genetic material adapted and developed abroad, mainly by companies from the Netherlands, Germany, Japan, Thailand, and others (Junqueira and Peetz 2017). It is important to highlight that in the international market, Brazilian floriculture is specialized in the multiplication of planting material, which is re-exported to its sources in significant quantities. In this case, it is particularly noteworthy that 60% of amaryllis (*Hippeastrum* sp.) and 40% of gladiolus (*Gladiolus* × *grandiflorus*) production is exported (Tombolato et al. 2013).

Next, we will discuss on some potyviruses recently described in Brazil.

2.2 Alstroemeria Mosaic Virus (AlMV), syn. Alstroemeria Streak Virus

Alstroemeria mosaic virus (AlMV) is a potyvirus first identified in the UK from alstroemeria cultivars with symptoms of leaf chlorosis and dark red streaks on the flowers (Phillips and Brunt 1986). To date, infections caused by AlMV have been reported in Europe (Spence et al. 2000), Asia (Yasuda et al. 1998; Fuji et al. 2004, 2007; Wang and Chang 2006), Oceania (Pearson et al. 2009), North America (Gutiérrez-Estrada et al. 2000; Mosquera-Yuqui et al. 2020), and South America (Rivas et al. 2013).

AlMV has filamentous particles around 750 nm long, containing about 5% of ssRNA 9974 nt (Reference Sequence: MK440140). It is non-persistently transmitted by the aphids *Myzus persicae* and *Neotoxoptera formosana* (Yasuda et al. 1998).

Although *Alstroemeria* (Alstroemeriaceae) is a genus endemic to South America, occurring mainly in Chile, Argentina, Brazil, Ecuador, Paraguay, and Peru (van Zaayen 1995), most breeding has been conducted since the 1980s in the Netherlands, as well as in North and South America, especially Ecuador. Due to the wide variety of flower colors, alstroemeria has become one of the main cut flower crops in the world. It is important to underscore that its hybrids were produced in the Netherlands mainly from Andean species (Mosquera-Yuqui et al. 2020; Rivas et al. 2013).

In Neotropical regions, the occurrence of AlMV has been described in Brazil, Ecuador, and Mexico (Gutiérrez-Estrada et al. 2000; Mosquera-Yuqui et al. 2020; Rivas et al. 2013). In Brazil, AlMV was identified in commercial crops in São Paulo and Minas Gerais states (Rivas et al. 2013).

Symptoms associated with AlMV in alstroemeria vary considerably according to the cultivars, since they are symptomless for mosaic, chlorosis, and chlorotic spots on leaves, and color break in the flowers, as well as growing conditions and time of the year (Kitajima 2020). One of the most frequent symptoms is leaf streak mosaic

Fig. 2.2 *Alstroemeria* sp. naturally infected with Alstroemeria mosaic virus (AlMV)

(Fig. 2.2) and color break of flowers. Alstroemeria streak virus (AlSV) has been reported to be a strain of AlMV (Van der Vlugt and Bouwen 2002).

2.3 Bean Yellow Mosaic Virus (BYMV), syn. Bean Virus 2, Canna Mosaic Virus

BYMV, described by Pierce (1934), is found in most bean-producing regions of the world (Bos 1970). This economically important virus infects many leguminous (Fabaceae) crops and some members of the different families such as Araceae (Mokrá and Götzová 1994), Iridaceae (Loebenstein et al. 1995), Cannaceae (Castillo et al. 1956; Chauhan et al. 2015), Gentianaceae (Lisa and Dellavalle 1987), Amaryllidaceae (Schulze et al. 2017), Fabaceae (Provvidenti and Hunter 1975), Verbenaceae (Guaragna et al. 2004a), and Violaceae (Gorter 1977).

BYMV has filamentous particles around 750 nm long, containing about 5% of ssRNA 9532 nt (Reference Sequence: NC003492). It is non-persistently transmitted by several aphid species (Berlandier et al. 1997) and by seeds, albeit at low levels (McKirdy et al. 2000).

In ornamental plants, BYMV has been described in Mexico (Cárdenas-Alonso 1994), the USA (Provvidenti and Hunter 1975; Guaragna et al. 2004a), Argentina, Chile, and Peru (Arneodo et al. 2005). In Brazil, the virus was reported in 1972 associated with a mosaic symptom in gladiolus (Costa 1972) and subsequently in two commercial regions of São Paulo state (Alexandre et al. 2010). The virus was detected in several samples showing symptoms that varied from weak to severe mosaic, depending on the cultivar and season (Duarte and Alexandre 2017). Absence of symptoms, color break in the flowers, or mosaic with whitish spots have also been reported (Fig. 2.3).

Fig. 2.3 *Gladiolus* sp. naturally infected with bean yellow mosaic virus (BYMV)

2.4 Bidens Mosaic Virus (BiMV)

Bidens mosaic virus (BiMV) was originally described in *Bidens pilosa* weed (Kitajima et al. 1961). Later, it was also reported in other weeds such as the gallant soldier (*Galinsoga parviflora*), lilac tassel flower (*Emilia sonchifolia*), and bristly starbur (*Acanthospermum hispidum*) (Costa and Kitajima 1966; Sanches et al. 2010). However, its occurrence has also been reported in sunflower (*Helianthus annuus*), pea (*Pisum sativum*), and lettuce (*Lactuca sativa*) (Suzuki et al. 2009). In ornamental plants, BiMV has been described in tickseed (*Coreopsis lanceolata*) (Rodrigues et al. 1991) and black-eyed Susan (*Rudbeckia hirta*) (Inoue-Nagata et al. 2006), causing leaf mosaic and color break on the petals.

BiMV virions are flexuous filaments, and the complete genome consists of a single molecule of linear, positive-sense ssRNA with 9557 nt (Reference Sequence: NC023014). Except for peas, BiMV is restricted to weeds, vegetables, and ornamental species belonging to the Asteraceae family and, to date, has only been described in Brazil.

2.5 *Brugmansia suaveolens* Mottle Virus (BsMoV)

Brugmansia suaveolens mottle virus (BsMoV) was first described in São Paulo state, Brazil, in 2008, showing mottling on *Brugmansia suaveolens* leaves (Fig. 2.4) (Lucinda et al. 2008). The following year, BsMoV was accepted as a definitive species of the genus *Potyvirus* (International Committee on Taxonomy of Viruses 2019). Its occurrence has yet to be described in other parts of the world.

The virions are 750 nm-long and 12 nm-wide flexuous filaments, and the complete genome consists of a single molecule of linear, positive-sense ssRNA with 9870 nt (Reference Sequence: NC014536). BsMoV belongs to the potato virus

Fig. 2.4 *Brugmansia suaveolens* naturally infected with *Brugmansia suaveolens* mottle virus (BsMoV)

Y group and shares a common ancestor with the pepper mottle virus (Lucinda et al. 2008).

A South American genus, *Brugmansia*, includes seven species distributed from Venezuela and Colombia to northern Chile and southeastern Brazil (Hay et al. 2012). *B. suaveolens*, also known as "angel trumpet," is a shrub or small tree native to South America and widely introduced as an ornamental plant for its attractive trumpet-shaped flowers. It is present pan-tropically, common in temperate countries, and to a lesser degree all over the world (CABI 2010).

2.6 Canna Yellow Streak Virus (CaYSV)

CaYSV was first described in the UK from samples of *Canna* (var. Pink Sunburst and var. Princess Di), part of a consignment imported from Israel (Monger et al. 2007). BLAST searches of sequence databases confirmed the identity of the new virus as a member of the genus *Potyvirus*, most closely related to an isolate of Johnsongrass mosaic virus (JGMV), known as Canna yellow streak virus (CaYSV). The sequence obtained in four canna samples from different sources has been shown to be highly conserved at the nucleotide level, indicating that CaYSV likely originated from a single source and has spread worldwide via the movement of

Fig. 2.5 *Canna denudata* naturally infected with canna yellow streak virus (CaYSV)

infected material (Monger et al. 2007). CaYSV infection in *Canna* spp. plants was reported in Belgium, France, the Netherlands, the USA, Russia, and Israel (Rajakaruna et al. 2014; Zakubanskiy et al. 2017). However, in Neotropical region, to date the virus has only been detected in Brazil (Alexandre et al. 2017a).

CaYSV are 760 nm-long and 12 nm-wide flexuous filaments. The complete genome consists of a single molecule of linear, positive-sense ssRNA with 9502 nt (Reference Sequence: NC013261). The virus is poorly transmitted by mechanical inoculation using sap from canna leaves (Rajakaruna et al. 2014; Alexandre et al. 2017a) and non-persistently transmitted by aphid vectors. In addition to leaves, the virus was also isolated directly from the rhizome of the infected canna var. Panama sample (Monger et al. 2007).

Although the canna lily is commonly found in nature as well as public and private gardens in Brazil, to date there has been no reported occurrence of viral diseases in these plants. In mid-2017, *Canna paniculata* samples from a public garden in the municipality of Piracicaba, São Paulo state, Brazil, showing mosaic symptoms along leaf veins (Fig. 2.5) were analyzed for the presence of viruses. Elongated flexuous particles were observed by electron microscopy in extracts from negatively stained symptomatic canna plants (Canna Br01). Ultrathin sections showed the presence of type II lamellar inclusions, suggesting the sample was infected with a potyvirus. Analysis of the sequence of nucleotides and amino acids, corresponding to the partial sequence of the cylindrical inclusion region, showed identity percentages greater than 85% and 94%, respectively, with CaYMV isolates from the UK and the USA (Alexandre et al. 2017a).

Cannaceae is a monotypic family containing single genus *Canna*, cultivated as an ornamental plant in nearly every part of the world. *Canna* originated in tropical and subtropical America, with native species from the Southern United States to Northern Argentina. Currently, only 12 scientific species names are accepted by the Plant List (2013). In Brazil, the occurrence of four native *Canna* species (*C. glauca*, *C. indica*, *C. paniculata*, and *C. pedunculata*) has been reported.

Fig. 2.6 *Catharanthus roseus* naturally infected with Catharanthus mosaic virus (CatMV). (Courtesy of J.A.M. Rezende)

2.7 Catharanthus Mosaic Virus (CatMV)

Mosaic symptoms in periwinkle (*Catharanthus roseus*) have frequently been observed in Brazil, associated with the presence of either cucumber mosaic virus (CMV) (Espinha and Gaspar 1997) or an unidentified virus, possibly a potyvirus, based on the cytopathology of symptomatic leaves (Seabra et al. 1999). Years later, in Brazil, a potyvirus showing similar symptoms, such as leaf mosaic and malformation in addition to flower variegation in *C. roseus* (Fig. 2.6), was identified as a new species based on sequence analysis and denominated *Catharanthus* mosaic virus (CatMV) (Maciel et al. 2011). In 2015, CatMV was described in *Mandevilla* sp. (Apocynaceae) from nurseries in Minnesota and Florida (USA) (Mollov et al. 2015) and was accepted as a definitive *Potyvirus* species the following year (International Committee on Taxonomy of Viruses 2019).

The virions are flexuous filaments, and the complete genome consists of a single molecule of linear, positive-sense ssRNA with 9575 nt (Reference Sequence: NC027210). In addition to Brazil and the USA, CatMV has also been reported in Australia, in the gymnosperm *Welwitschia mirabilis* (ShuHui et al. 2015).

The Brazilian CatMV isolate was experimentally transmitted by mechanical means to periwinkle and *Nicotiana benthamiana*, resulting in systemic infection, and *Chenopodium quinoa* and *C. amaranticolor*, causing only local lesions. The aphids *Aphis gossypii* and *Myzus nicotianae* were able to non-persistently transmit CatMV. Thin sections of plants, naturally or experimentally infected by CatMV, exhibit characteristic type II lamellar inclusions in the cytoplasm, according to Edwardson (1974), and negatively stained leaf extracts show typical elongated and flexible particles. Sequence analysis of RT-PCR-amplified cDNA fragments, especially of CP genes, indicated that the sequence identity of CatMV is sufficiently different to be considered a new species (Maciel et al. 2011). In addition to

Fig. 2.7 *Mandevilla* sp. naturally infected with CatMV. (Courtesy of J.A.M. Rezende)

C. roseus, another species of Apocynaceae *Mandevilla* plant (rock trumpet) showing conspicuous mosaic symptoms (Fig. 2.7) has been reported in florists in Brazil (Favara et al. 2020). Preliminary electron microscopic examination of leaf extracts and tissue sections indicated the presence of a presumed potyvirus. High-throughput sequencing (HTS) produced a near-complete genome of two distinct CatMV isolates: M1 and M2. It is interesting to note that the former was not transmitted by aphids, whereas M2 was aphid and mechanically transmitted.

2.8 Colombian Datura Virus (CDV)

Colombian datura virus (CDV) was initially described in *Datura candida* and *D. sanguinea* (now *Brugmansia* × *candida* and *Brugmansia sanguinea*, respectively) samples collected in Putumayo, Colombia, as part of solanaceous plant taxonomic studies in the late 1960s (Kahn and Bartels 1968). The occurrence of CDV has been reported worldwide, including Neotropical countries such as Bolivia, Ecuador, and South Florida (USA), in addition to Colombia (CABI 2010; Adkins et al. 2005; Chellemi et al. 2011).

In *Brugmansia* sp. (angel's trumpet), CDV induces faint chlorotic spot symptoms and leaf mosaic and rugosity (Chellemi et al. 2011). In addition to *Brugmansia* species, the occurrence of CDV has also been reported in the Solanaceae *Mandragora autumnalis*, showing flower and leaf mottle symptoms (Pacifico et al. 2016).

The virions are 750 nm-long and 12 nm-wide flexuous filaments, and the complete genome consists of a single molecule of linear, positive-sense ssRNA with 9621 nt (Reference Sequence: NC020072). CDV is transmitted mechanically to Solanaceae hosts and non-persistently by aphids (*Myzus persicae*).

Brugmansia (*Brugmansia* spp.) is a solanaceous ornamental that produces large, fragrant, and attractive flowers. *Brugmansia* species are native to either the tropical lowlands or Andean regions of South America. They are semi-woody perennial trees or shrubs distinguishable from members of the closely related genus *Datura* by their growth pattern and fruit and seed morphology (Chellemi et al. 2011).

2.9 Costus Stripe Mosaic Virus (CoSMV), Tentative Species

Costus stripe mosaic virus (CoSMV) is a new member of the genus *Potyvirus*, family *Potyviridae*, and was first described in Brazil in *Costus spiralis* plants showing dark green stripes on their leaves (Fig. 2.8), in a public garden in the city of São Paulo (Alexandre et al. 2020).

Plants of the genus *Costus* (Costaceae) are widespread in the tropical rain forests of Asia, Africa, and the Americas. Three genera are found in Brazil, with 20 species concentrated mainly in the Amazon region. Two species (*Costus arabicus* and *C. spiralis*) can also be found throughout the Brazilian Atlantic forest, which is considered a hotspot of diversity and endemism (Myers et al. 2000; Souza and Lorenzi 2012).

CoSMV has been identified as a new species of *Potyvirus* based on morphological, biological, serological, and molecular tests. The virus has an extremely restricted host range, and when mechanically inoculated, it failed to be transmitted to indicator hosts, with the exception of *Costus spiralis* and *C. comosus*, but with a low transmission percentage (10–14%). CoSMV is a potyvirus with a number of different biological and molecular characteristics. It was not transmitted by any of the aphids tested (*Aphis solanella*, *Myzus persicae*, and *Uroleucon sonchi*), as is common for most *Potyvirus* species. It is noteworthy that LTTC rather than KITC mutations in the terminus and PIK instead of PTK in the central part of HC-Pro are likely responsible for the negative results in aphid transmission tests. In addition, a divergence in the P1 protein cleavage site was found when compared to other members of the family *Potyviridae* (Alexandre et al. 2020).

To date, the occurrence of CoSMV has only been described in Brazil.

Fig. 2.8 *Costus spiralis* naturally infected with Costus stripe mosaic virus (CoSMV)

2.10 Dasheen Mosaic Virus (DsMV)

Dasheen mosaic virus (DsMV) was described in 1978, by Zettler et al. (1978) in *Colocasia esculenta*. DsMV is the most frequently encountered virus in several cultivated Araceae species, widely used as foliage ornamentals (*Dieffenbachia*, *Caladium*, *Philodendron*, *Syngonium*), cut flowers (*Zantedeschia*, *Anthurium*), and important food staples (*Colocasia*, *Xanthosoma*, *Monstera*) in some countries (Zettler et al. 1970; Zettler and Hartman 1987; Chase and Zettler 1982; Rivas et al. 1998). Later, DsMV was described in two orchid species: *Spiranthes cernua* and *Vanilla tahitensis* (Jordan et al. 2002).

The Araceae family comprises more than 3700 species, with distribution in tropical and North America, tropical Continental and South Africa, temperate Eurasia, the Malay Archipelago, Madagascar, and the Seychelles. In Brazil, it occurs throughout the country with 35 genera and approximately 460 species and in São Paulo state with 13 genera and 77 species (Coelho 2012). A noteworthy fact is the report of one of the oldest fossil records among angiosperms, with fossil forms first appearing during the Early Cretaceous epoch, including *Spixiarum kipea*, collected in northeastern Brazil from open cast pits close to the town of Nova Olinda, Ceará State (Nauheimer et al. 2002; Coiffard et al. 2013).

DsMV has a mean particle length of about 750 nm and induces cylindrical inclusions, and symptom expression is seasonal, varying from less evident to severe. Several aphids, such as *Aphis craccivora*, *A. gossypii*, and *Myzus persicae*, are natural vectors.

Very high DsMV incidences have been reported in Australia; Egypt; Europe; the Caribbean; the Far East, North, and South America; Oceania; and Africa. The virus occurs mainly where Araceae are commercially grown, such as China, the USA, India, Italy, Nicaragua, New Zealand, and Vietnam.

In Neotropical countries, DsMV has been reported in Brazil, the USA (Florida), Costa Rica, Cuba, Nicaragua, Puerto Rico, and Venezuela associated with leaf mosaic and distortion and feathery mosaic (Fig. 2.9) (Ramírez 1985; El-Nil et al. 1977; Reyes et al. 2005; Cabrera et al. 2010).

In Brazil, DsMV was detected infecting *Anthurium* sp. in Ceará (Lima et al. 2004) and São Paulo states. DsMV was related in *Anthurium andraeanum*, *A. scherzerianum*, *Amorphophallus* sp., *Dieffenbachia* spp., *Syngonium wendlandii*, *Xanthosoma sagittifolium*, *X. atrovirens*, *and Zantedeschia aethiopica* (Rodrigues et al. 1984; Rivas et al. 1997), in *Dieffenbachia* sp. (Rivas et al. 1998).

2.11 Euphorbia Ringspot Virus (EuRSV)

Euphorbia ringspot virus (EuRSV), showing ring-shaped chlorotic spots on leaves, as well as leaf and flower deformations and reduced growth, was described in Germany, in 1976 (Bode and Lesemann 1976). In 2004, a commercially grown *Euphorbia milii* var. splendens plant in the USA exhibiting the same symptoms was shown to be infected with EuRSV (Guaragna et al. 2004b). Seven years later, the

Fig. 2.9 *Xanthosoma* sp. naturally infected with dasheen mosaic virus (DsMV)

virus was described in *E. milii* cv. splendens plants growing in Venezuela, displaying mosaic, leaf distortion, and flower color break (Marys and Romano 2011). This is the first occurrence of EuRSV in South America and Neotropicals.

The virions are 750 nm-long and 12 nm-wide flexuous filaments, and the complete genome consists of a single molecule of linear, positive-sense ssRNA with 10,154 nt (Reference Sequence: NC 031339).

Euphorbia milii (Euphorbiaceae), native to Madagascar, has long been an important ornamental crop in subtropical and tropical regions. *E. milii* cv. splendens is one of the most popular potted plants in Venezuela, valued for its brightly colored foliage and flowers (Marys and Romano 2011).

2.12 Gloriosa Stripe Mosaic Virus (GSMV), syn. Gloriosa Lily Mosaic Virus

Gloriosa stripe mosaic virus (GSMV) infection was first reported in plants of *Gloriosa* spp. from Germany, in 1974. GSMV was mechanically transmitted exclusively to gloriosa lily plants. Electron microscopy showed evidence of potyvirus infection, which was referred to as Gloriosa stripe mosaic virus (GSMV) (Koenig and Lesemann 1974). Later, GSMV was reported in Japan (1985), the Netherlands (2011), and the USA (2017) (Pham et al. 2011; Mollov et al. 2017). In Brazil, GSMV was described for the first time in 2019 (Duarte et al. 2019).

Virus-infected gloriosa lily plants exhibit stripe mosaic or mottle symptoms and color break in flowers. GSMV is transmissible to healthy *Gloriosa* by mechanical inoculation and non-persistently transmissible by aphid vectors. The virions are 760 nm-long and 13 nm-wide flexuous filaments, and the genome consists of a single molecule of linear, positive-sense ssRNA of 9455 nt (Reference Sequence: NC 038562).

Fig. 2.10 *Gloriosa superba* naturally infected with Gloriosa stripe mosaic virus (GSMV)

G. superba, also known as glory lily, flame lily, or gloriosa lily, is a species of flowering plant in the family Colchicaceae, which includes about 20 genera and 200 widely distributed species, but it is not found in South America. This ornamental species is native to tropical Africa and grows naturally in several temperate and tropical Asian countries. It is naturalized and cultivated elsewhere (Europe, Australia) and listed as a weed in Australia and the USA (Mollov et al. 2017).

In Brazil, GSMV was isolated from *G. superba* plants with leaf mosaic symptoms and color break (Fig. 2.10) purchased at four different flower markets in the city of São Paulo (São Paulo state). The virus was not mechanically transmitted to indicator plants of the Amaranthaceae, Cucurbitaceae, and Solanaceae families. A survey of GSMV occurrence in six producing regions in São Paulo state revealed 100% infection in all of the crops assessed (Duarte et al. 2019). This made it difficult to obtain healthy gloriosa lily plants to inoculate the virus isolate and reproduce the original symptoms. A few growers reported importing gloriosa lily rhizomes from Japan many years ago. However, since there are no sequences of Japanese samples available in GenBank, the origin of the Brazilian isolate cannot be accurately determined.

2.13 Hippeastrum Mosaic Virus (HiMV) syn. Amaryllis Mosaic Virus

The disease caused by Hippeastrum mosaic virus (HiMV) was reported in 1922 by Kunkel (1922), who described mosaic on the leaves of *Hippeastrum equestre* Herb. (Amaryllidaceae), which is a synonym of *H. puniceum* (The Plant List 2013). Later, HiMV was partially characterized by Brants and Van den Heuvel (1965), who

Fig. 2.11 *Hippeastrum* sp. naturally infected with Hippeastrum mosaic virus (HiMV)

described the virus in *H. hybridum* (Brants and Van den Heuvel 1965), and by Brunt (1973).

The virions are non-enveloped, usually flexuous filaments, with a clear modal length of 782 and 12 nm wide. The genome consists of a single molecule of linear, positive-sense ssRNA of 9660 nt (Reference Sequence: JQ395040 = NC_017967) (Wylie and Jones 2012).

Amaryllidaceae, former Liliaceae, comprises several genera including *Amaryllis*, *Eucharis*, and *Hippeastrum*. These last two are native to Neotropical regions. *Hippeastrum* (Leopoldii hybrid or Dutch amaryllis) is a genus of about 90 species and 600 hybrids and cultivars, originating from a much wider geographical area that extends from Argentina to Mexico and into the Caribbean. It is currently grown extensively in tropical countries. *Eucharis* (Amazon lily) is found in tropical rain forests and is adapted to low-light conditions. The bulbous plants, with broad, basal leaves narrowed into petioles, are distributed from Guatemala to Bolivia, mainly in the western Amazon basin and adjacent lower slopes of the eastern Andes. *Eucharis amazonica* is erroneously known as *E.* × *grandiflora* by horticulturists. Although they are taxonomically distinct species, both occur and are widely cultivated in Brazil (Alexandre et al. 2011).

HiMV infection in *H. hybridum* plants was first reported in the Netherlands. The virus spread to Australia, the former Czechoslovakia, Fiji, Greece (Malandrake et al. 2016), India, Japan, South Africa, the UK, China, Thailand, the USA, and Neotropical countries, including Brazil (Alexandre et al. 2011) (Fig. 2.11).

HiMV infection was also reported in *Eucharis* spp. plants from the Netherlands, India, Taiwan, and Brazil. Virus-infected Amazon lily leaves show severe rugose mosaic or light green striping, some deformation, with petals also exhibiting light green striping (Fig. 2.12). HiMV is non-persistently transmitted by aphid vectors *Aphis fabae*, *A. gossypii*, and *Myzus persicae* and mechanical sap-inoculation. The virus is not transmissible by seeds, pollen, or contact between plants (Brunt 1973).

Fig. 2.12 *Eucharis grandiflora* naturally infected with Hippeastrum mosaic virus (HiMV)

The virus-infected amaryllis leaves and flower stalks show an irregular light and dark green mosaic pattern (Duarte et al. 2009; Raj et al. 2009).

2.14 Hyacinth Mosaic Virus (HyMV)

Hyacinth mosaic virus (HyMV) was first reported in *Hyacinthus orientalis* L. from Bulgaria and the USA and detected in Australia, New Zealand, Lithuania, the Netherlands, and the UK (Pearson et al. 2009; Brunt et al. 1997). HyMV, the most important virus in hyacinth crops, is associated with leaf symptoms, flower variegation, and, depending on the cultivar, decreased bulb production, with considerable economic losses (Derks 1995). In serological tests for HyMV in hyacinths, the most reliable results from leaves were obtained from the beginning of flowering to about 1 month later (Derks et al. 1980).

The virions are 740 nm-long and 12 nm-wide flexuous filaments, and the complete genome consists of a single molecule of linear, positive-sense ssRNA with 9500 nt (Reference Sequence: NC037051). HyMV is non-persistently transmitted by aphids (*Myzus persicae* and *Macrosiphum solanifolii*) (Brunt et al. 1997).

H. orientalis (Asparagaceae, previously classified as Hyacinthaceae) is a widely cultivated ornamental species in the Northern Hemisphere (Souza and Lorenzi 2012), and bulbs imported from Europe have recently been introduced into the Brazilian market (Alexandre et al. 2017b). Although there are reports of hyacinth cultivation in central Mexico, Cuba, and Haiti (CABI 2020a), to date, the occurrence of HyMV has only been described in Brazil (Alexandre et al. 2017b), in plants displaying leaf mottling obtained at a flower market in the city of São Paulo (Fig. 2.13).

In addition to *H. orientalis*, the occurrence of HyMV has also been reported for *Veltheimia bracteata*, *Hymenocallis* sp., and *Muscari comosum*, which is a synonym

Fig. 2.13 *Hyacinthus* sp. naturally infected with hyacinth mosaic virus (HyMV)

for *Leopoldia comosa*. Thus, this virus appears to have a host range restricted to taxonomically close species (Asparagales).

2.15 Konjac Mosaic Virus (KoMV), syn. Japanese Hornwort Mosaic Virus (JHMV), Zantedeschia Mosaic Virus (ZaMV)

Konjac mosaic virus (KoMV) was the first to be described in *Amorphophallus konjac* in Japan (Shimoyama et al. 1992). Nishiguchi et al. (2006) suggested that the viruses described as Japanese hornwort mosaic virus (JHMV) and Zantedeschia mosaic virus (ZaMV) were isolates of KoMV.

KoMV has been identified in the following species belonging to the family Araceae, *Amorphophallus paeoniifolius*, *A. konjac*, *Colocasia*, *Caladium*, *Dieffenbachia* spp., *Zamioculcas zamiifolia*, and *Zantedeschia*, and reported in Japan, Taiwan, India (Padmavathi et al. 2011, 2013), Germany (Lesemann and Winter 2002), the Netherlands (Pham et al. 2002), and Hungary (Ágoston et al. 2019). JHMV infection was described in *Cryptotaenia japonica* plants from Japan (Okuno et al. 2003), and there are also reports of ZaMV infecting *Zantedeschia* (calla lily) in Korea (Kwon et al. 2002), Taiwan (Liao et al. 2020), and New Zealand (Wei et al. 2008).

It is important underscore that only KoMV is recognized as a definitive species (International Committee on Taxonomy of Viruses 2019). Isolated from konjac

Fig. 2.14 *Zamioculcas zamiifolia* naturally infected with konjac mosaic virus (KoMV)

plants, the complete nucleotide sequence of KoMV F, which can infect a relatively wide range of plants, was determined (Nishiguchi et al. 2006).

In Brazil, KoMV was described in 2013 (Alexandre et al. 2013) in *Zamioculcas* plants purchased at an ornamental plant shop in São Paulo state, with mosaic and foliar distortion symptoms (Fig. 2.14). Electron microscopy revealed the presence of elongated, flexuous viral particles in foliar extracts and cytoplasmic lamellar aggregates of type II lamellar inclusions (Edwardson's classification) in thin sections. No symptoms were induced following mechanical inoculation on *Chenopodium amaranticolor*, *C. murale*, *Gomphrena globosa*, *Nicotiana megalosiphon*, and *N. debneyii* or on the aroids *Philodendron scandens*, *P. selloum*, *Dieffenbachia amoena*, and *Colocasia esculenta* (Alexandre et al. 2013).

2.16 Lily Mottle Virus (LMoV), syn. Lily Mild Mottle Virus, Lily Strong Virus, Tulip Band Breaking Virus, Tulip Virus 1

The host range of lily mottle virus (LMoV) is restricted to Liliaceae, especially lily and tulip, inducing symptoms ranging from simple mottle to leaf mosaic and deformation, in addition to color break in tulip flowers (Dekker et al. 1993; Derks et al. 1994; Engelmann 2010; Rivas et al. 2011). LMoV was previously considered a tulip breaking virus (TBV) strain based on the host range, symptom expression in tulips, and physical and serological properties (Derks et al. 1994). The natural occurrence of LMoV, now included as a definitive *Potyvirus* species (International Committee on Taxonomy of Viruses 2019), has also been reported for Alstroemeriaceae species (*Alstroemeria* sp.) associated with leaf chlorosis and color break on petals, and in Asteraceae (*Cichorium endivia*), showing mosaic and necrotic spots on the outer leaves (Bouwen and Van Der Vlugt 2000; Lisa et al. 2002).

Fig. 2.15 *Lilium* sp. naturally infected with lily mottle virus (LMoV)

LMoV has been described in different parts of the world that produce lilies and tulips, such as the Netherlands, China, Korea, Italy, Israel, and the USA, among others. However, in Brazil, although lily bulbs are imported from the Netherlands to be commercialized as cut flowers and potted plants, the virus has been detected in single and mixed infections with lily symptomless virus (LSV) and cucumber mosaic virus (CMV). It should be noted that, in Brazil, LMoV is the most prevalent virus in lily crops. On the other hand, in Argentina, a country bordering Brazil, the most common virus in hybrid lilies is LSV in single and mixed infections.

Symptoms associated with the virus are vein clearing, leaf mottle, leaf mosaic, chlorotic and yellow streaking, and leaf curling. Some cultivars may show color break, in addition to flower malformations and asymmetry. The buds and flowers may drop prematurely. The vase life of flowers from virus-infected plants is reduced. Some plants may also have the virus without showing symptoms (Fig. 2.15). However, general symptoms are more severe in plants with mixed infections.

2.17 Narcissus Yellow Stripe Virus (NYSV), syn. Narcissus Degeneration Potyvirus, Narcissus Yellow Streak Virus, Narcissus Mosaic Virus

Narcissus yellow stripe virus (NYSV) was first described in 1908 by Darlington in *Narcissus* spp. and has been reported in Australia, China, India, Japan, New Zealand, the Netherlands, the UK, and the USA. Infected plants show conspicuous chlorotic stripes on leaves and flower stalks, flower color break, reduced bulb size, and sometimes severe stunting. Its distribution is therefore worldwide, especially in temperate regions, and, through trading, probably occurs wherever daffodils are grown.

Fig. 2.16 *Narcissus* sp. naturally infected with Narcissus yellow stripe virus (NYSV)

The virions are 755 nm-long and 12 nm-wide flexuous filaments, and the complete genome consists of a single molecule of linear, positive-sense ssRNA with 9650 nt (Reference Sequence: NC011541). The host range of NYSV is restricted to Amaryllidaceae and is non-persistently transmitted by aphids (*Aphis fabae*, *Acyrthosiphon pisum*, and *Macrosiphum euphorbiae*).

Narcissus species or daffodils (Amaryllidaceae) are native to southern Europe and northern Africa, with their center of dispersion in the Western Mediterranean, particularly on the Iberian Peninsula. They are very common in the Northern Hemisphere, particularly in Europe, where numerous hybrids are produced. Horticultural varieties slightly more resistant to hot weather have recently been introduced in Brazilian floriculture and have been well accepted in the market. However, faint stripe mosaic symptoms in *N. cyclamineus* plants obtained at a flower market in the city of São Paulo, Brazil (Fig. 2.16), have also been associated with the presence the NYSV.

2.18 Soybean Mosaic Virus (SMV)

Soybean mosaic virus (SMV), first described in the USA in 1915, is one of the viruses that most damages soybean crops *Glycine max* (CABI 2020b). SMV is widespread throughout the world, and in South America, it occurs in Argentina, Chile, Colombia, Ecuador, and Venezuela, as well as Brazil.

The virions are flexuous filaments, and the complete genome consists of a single molecule of linear, positive-sense ssRNA with 9588 nt (Reference Sequence: NC002634).

Fig. 2.17 *Strongylodon macrobotrys* naturally infected with soybean mosaic virus (SBV)

Mosaic-bearing jade vine (*Strongylodon macrobotrys*—Fabaceae) plants (Fig. 2.17) were found in São Paulo state, Brazil. Experimental transmission (mechanical and aphid) and morphological and molecular assays demonstrated that the symptoms were due to infection by an isolate of SMV. This virus has a narrow natural host range, and the isolate infected *Canavalia ensiformis* (jack bean) and jade vine, showing mild mosaic. It should be noted that jade vine, also known as turquoise jade vine, is native to the Philippines and valued as an ornamental for its blue-green inflorescence.

2.19 Sunflower Chlorotic Mottle Virus (SCMoV)

Sunflower chlorotic mottle virus (SCMoV) was first observed in Buenos Aires Province, Argentina, and is the most widely distributed potyvirus in cultivated and wild sunflowers, causing important yield losses in *Helianthus annuus*. In addition to this species, SCMoV was identified in *Helianthus petiolaris*, *Eryngium* sp., *Dipsacus fullonum*, *Ibicella lutea*, and *Zinnia elegans*. It is important to highlight that *D. fullonum* is a biennial weed that plays an important role in SCMoV disease epidemiology, which serves as a potential inoculum source of virus infection in sunflower fields. This species, known as common teasel, has been introduced to the

Fig. 2.18 *Zinnia elegans* naturally infected with sunflower chlorotic mottle virus (SCMoV)

USA, Canada, Argentina, Bolivia, Ecuador, Uruguay, Australia, and New Zealand and may be an important source to spread the virus in these countries, where it has yet to occur.

In addition to Argentina, the occurrence of SCMoV in Neotropical regions has been described in Brazil, in the northwest region of São Paulo state. The virus was isolated from *Zinnia elegans*, showing mottle (Fig. 2.18), mosaic, leaf distortion, and color break. The Brazilian SCMoV isolate was transmitted by mechanical inoculation with a host range restricted mainly to members of the family Asteraceae. In sunflower, where the infection was most severe, very young plants died.

The virions are flexuous filaments, and the complete genome consists of a single molecule of linear, positive-sense ssRNA with 9665 nt (Reference Sequence: NC014038). SCMoV is transmitted by mechanical inoculation and non-persistently by aphids, but not seeds.

Common sunflower (*H. annuus*) is an annual self-incompatible insect-pollinated plant native to North America (Mexico and the USA) where it was domesticated at least 4000 years ago. The sunflower is cultivated commercially, and, as an ornamental plant, flower color and type of inflorescence are important characteristics for its cultivation. Ray floret color can range from various shades of red to lemon yellow.

2.20 Turnip Mosaic Virus (TuMV)

The first evidence of turnip mosaic virus (TuMV) infection occurred in France in 1862, causing color break in flower petals of *Matthiola incana* (Brassicaceae). However, TuMV was first described in 1921 in the USA. Since then, there have been reports of TuMV in several species of plants in various parts of the world (Jenner et al. 2002). In the Neotropical region, TuMV was first reported in kale (Brassica oleracea) crops in Brazil in the early 1970s. In the Neotropical region, TuMV was first reported in kale (*Brassica oleracea*) crops in Brazil in the early 1970s.

Fig. 2.19 *Tropaeolum majus* naturally infected with turnip mosaic virus (TuMV)

Among the potyviruses that infect ornamental plants, TuMV is the most cosmopolitan, infecting species of different botanic families (Ohshima et al. 2002). Despite being considered the most important brassica-infecting virus worldwide, there is little information about it in the Neotropical region. It is important to underscore that recent phylogenetic studies with the complete genome of TuMV isolates suggest that this virus originally infected wild orchids in Europe and then spread in wild and cultivated brassicas in the Mediterranean, and late to other regions of the planet.

The occurrence of TuMV in ornamental plants in Neotropical regions has been described in Brazil, infecting *Tropaeolum majus* (Tropaeolaceae), showing mosaic, bubbles, and leaf distortion (Fig. 2.19). Phylogenetic studies of TuMV isolated from *T. majus* in São Paulo state, Brazil, made it possible to observe its grouping with isolates from Italy belonging to the basal group B. Epidemiologically, *T. majus* can be considered a potential reservoir of TuMV, a problem for the production of commercial brassicas, since about 85 species of aphids are reported as efficient vectors of this potyvirus.

TuMV virions are flexuous filaments, and the complete genome consists of a single molecule of linear, positive-sense ssRNA with 9835 nt (Reference Sequence: NC002509).

T. majus (garden nasturtium) originated in South America in the Andes (Peru, Bolivia, and Colombia) and was introduced into Europe in the sixteenth century. With showy, orange, red, yellow, or creamy white slightly perfumed flowers and a nectar spur at the base, garden nasturtium is widely appreciated and used as a popular ornamental garden plant. Many hybrids and horticultural cultivars have also been developed.

2.21 Management and Control of Potyviruses

Potyviruses can cause serious damage and losses in ornamental plants and other important agricultural crops. In addition to leaf mosaic and necrosis, infected ornamental plants show deformation, color break, and reduced flower size and production.

Few studies have investigated introducing resistance genes to control potyviruses in ornamental plant species. Plant transformation technology (or genetic modification) is being used to develop a few varieties of ornamental plants in order to modify (a) flower color, (b) fragrance, (c) abiotic stress resistance, (d) disease resistance, (e) pest resistance, and (f) vase life and (g) maintain quality. Thus, when growing ornamental plants, it is recommended that infected plants be eliminated in the field during the vegetative growth stage. This is important for management decisions made before the next crop because potyviruses are difficult to control during productive cycle.

In the field, the transmission and spread of potyviruses are non-persistent (stylet-borne) and mediated by several aphid species, arguably the most successful vectors of plant viruses, due to two specific characteristics: the precise introduction of virus particles through a stylet into a host cell and diverse feeding modes (phytophagous "sap-suckers": polyphagic, oligophagic, or monophagic), allowing access to the host species of several botanical families.

According to a meta-analysis, 45% of the 137 species of aphid vectors are polyphagous (feeding on plants from multiple families), 38% oligophagous (mostly feeding on plants from multiple genera within the same plant family), and a mere 17% monophagous (mostly feeding on a plant or plants from the same genus and family).

However, 21 species of aphids (38% polyphagic, 38% oligophagic, and 24% monophagic), reported as potential vectors of potyviruses in ornamental plants in Argentina, Bolivia, Brazil, Chile, Colombia, Ecuador, Mexico, Peru, Puerto Rico, Venezuela, and the USA (South Florida), are widespread throughout the Neotropical region (Table 2.1, Fig. 2.20).

The predominant aphid species in the Neotropical region are exotics and originate from temperate climates with low plant diversity. Host plants diversity can determine aphid diversity. However, a highly diverse flora negatively affects aphid diversity due to their low efficiency in locating their host plants. Consequently, the high plant diversity reported in tropical and subtropical regions resulted in the establishment of polyphagous aphid species, which favored the dissemination of potyviruses in different vegetables, fruit trees, and ornamental plants in the Neotropical region.

Non-persistent transmission means that aphids require from seconds to a few minutes to acquire viruses and minutes for transmission. Spraying with insecticides is rarely an effective control because the aphids transmit the virus before being killed. Under field conditions, potyvirus dispersion can be controlled using mineral oil to minimize aphid vector transmission efficiency. Mineral oil application, a

Table 2.1 *Potyvirus* species reported in ornamental plants in the Neotropical region and their potential vector aphid species

Potyvirus species	Ornamental host	Occurrence	Aphids vectors	Total neotropical countries with reports of aphid species	Feeding habit
Alstroemeria mosaic virus (AlMV)	*Alstroemeria* sp.	Brazil	*Myzus persicae*	10	p
		Ecuador	*Neotoxoptera formosana*	6	m
		Mexico			
Bean yellow mosaic virus (BYMV)	*Gladiolus* sp.	Argentina	*Acyrthosiphon kondoi*	5	o
	Zinnia elegans	Brazil	*Acyrthosiphon pisum*	9	o
		Chile	*Aphis craccivora*	6	p
		Mexico	*Aphis fabae*	8	p
		Peru	*Aphis gossypii*	32	p
			Aulacorthum solani	2	p
			Chaetosiphon fragaefolii	2	m
			Hyperomyzus lactucae	10	o
			Lipaphis erysimi	11	o
			Macrosiphum euphorbiae	10	p
Bidens mosaic virus (BiMV)	*Zinnia elegans*	Brazil	*Aphis coreopsidis*	8	m
	Coreopsis lanceolata				
	Rudbeckia hirta				
Brugmansia suaveolens mottle virus (BsMoV)	*Brugmansia suaveolens*	Brazil	*M. persicae*	23	p

Canna yellow streak virus (CaYSV)	*Canna paniculata*	Brazil	Unknown	–	–
Catharanthus mosaic virus (CatMV)	*Catharanthus roseus*	Brazil	*A. gossypii*	32	p
	Mandevilla sp.		*Myzus nicotianae*	1	p
Colombian datura virus (CDV)	*Brugmansia* sp.	Bolivia	*M. persicae*	23	p
	Datura arborea	Colombia			
		Ecuador			
		USA (SF)[a]			
Costus stripe mosaic virus (CoSMV)	*Costus spiralis*	Brazil	Unknown	–	–
Dasheen mosaic virus (DsMV)	*Colocasia* sp.	Puerto Rico	*A. craccivora*	6	p
	Xanthosoma sp.	Venezuela	*M. persicae*	23	p
Euphorbia ringspot virus (EuRSV)	*Euphorbia* sp.	Venezuela	*M. persicae*	23	p
Gloriosa stripe mosaic virus (GSMV)	*Gloriosa superba*	Brazil	Unknown	–	–
Hippeastrum mosaic virus (HiMV)	*Eucharis* sp.	Brazil	*A. gossypii*	32	p
	Hippeastrum × *hybridum*		*M. persicae*	23	p
Hyacinth mosaic virus (HyMV)	*Hyacinthus* sp.	Brazil	Unknown	–	–
Konjac mosaic virus (KoMV)	*Zamioculcas zamiifolia*	Brazil	*A. gossypii*	32	p
Lily mottle virus (LMoV)	*Lily* sp.	Argentina	*A. gossypii*	32	p
		Brazil	*M. euphorbiae*	10	p
			M. persicae	23	p
Narcissus yellow stripe virus (NYSV)	*Narcissus cyclamineus*	Brazil	*A. fabae*	8	p

(continued)

Table 2.1 (continued)

Potyvirus species	Ornamental host	Occurrence	Aphids vectors	Total neotropical countries with reports of aphid species	Feeding habit
			A pisum	9	o
			A. solani	2	o
			Dysaphis plantaginea	1	m
			M. euphorbiae	10	p
			Macrosiphum rosae	6	o
Soybean mosaic virus (SMV)	*Strongylodon macrobotrys*	Brazil	*Aphis spiraecola*	34	p
			A. craccivora	6	p
			A. gossypii	32	p
			Uroleucon ambrosiae	7	o
			M. euphorbiae	10	p
			M. persicae	23	p
			Rhopalosiphum maidis	25	o
			Rhopalosiphum padi	11	o
Turnip mosaic virus (TuMV)	*Tropaeolum majus*	Brazil	*A. gossypii*[b]	32	p
			Brevicoryne brassicae	18	o
			L. erysimi	11	o
			M. persicae	23	p
Sunflower chlorotic mottle virus (SuCMoV)	*Zinnia elegans*	Brazil	Unknown	–	–

Note: Table based on surveys carried out on the CAB International website (www.cabi.org) and on the review/supplementary published by Gadhave et al. (2020) and Boquel et al. (2015)

[a]South Florida (USA)

[b]TuMV can be transmitted by about 85 species of aphids; the table shows only the main species of aphid; (m) monophagous; (o) oligophagous; (p) polyphagous

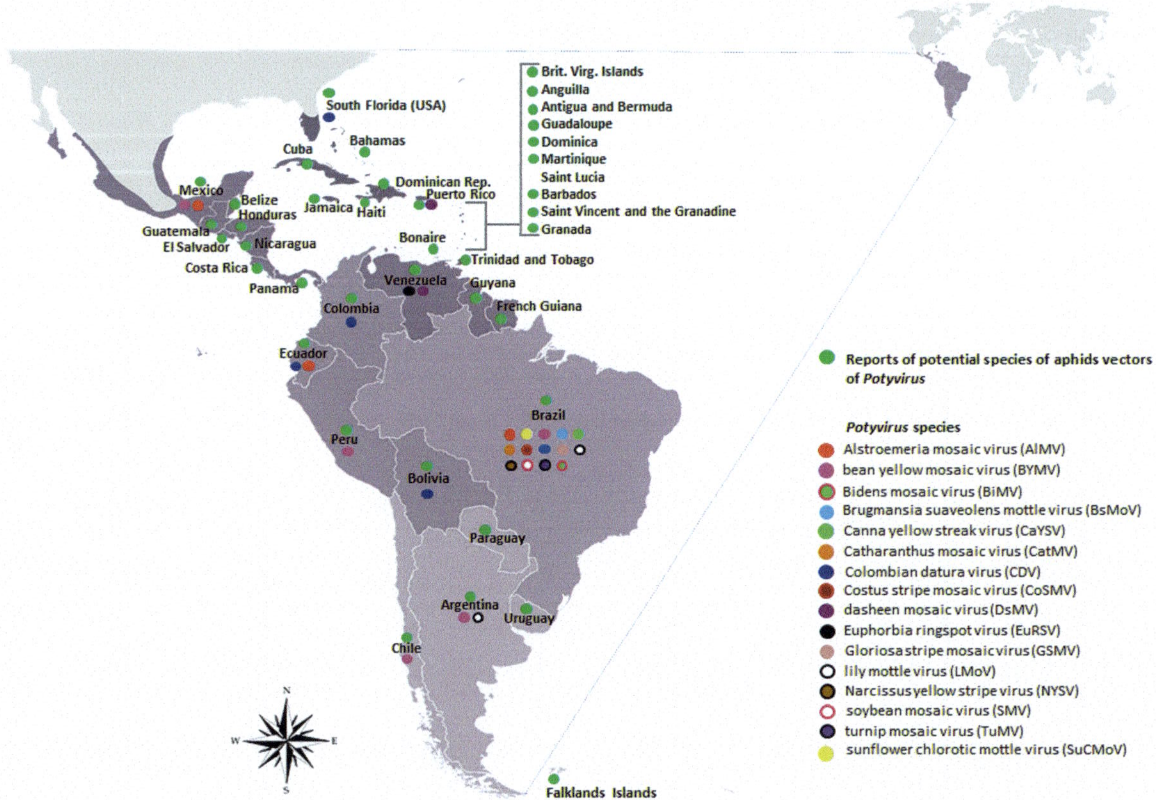

Fig. 2.20 Correlation of *Potyvirus* species reports found in ornamental plants and their potential vector aphids in the countries of the Neotropical Region

potyvirus management strategy, modifies the feeding behavior of aphids and interferes with the binding of potyvirus particles to the stylets.

Under laboratory conditions, the tissue culture techniques of different propagation organs can be associated with chemotherapy and thermotherapy to produce virus-free plants for improve crop quality and yield.

Silver reflective plastic mulches have been shown to be effective in repelling aphids in the field, thereby reducing or delaying infection. However, at high temperatures, such as those prevailing in most countries of the Neotropical region, reflective mulches may need to be removed since they can compromise plant development.

The use of barrier crops has been effective in controlling multiple potyviruses in a wide range of crops. However, studies using barriers in ornamental plant crops have yet to be carried out.

Since most species infected with potyviruses discussed in this chapter are vegetatively propagated, the most efficient management measure is the use of virus-free mother plants and seedlings.

2.22 Conclusion

Efficient control of potyviruses in ornamental plants is a challenge in the Neotropical region because it encompasses a vast area, including the USA (South Florida), Southern Mexico, Central America, the Caribbean, and South America. Since it harbors the largest number of plants in the world, comprising 7 of the 35 biodiversity hotspots, it is an attractive research setting. However, Neotropical ecological aspects are still scarcely studied, unknown, or poorly understood. This can be confirmed with respect to the control of plant viruses, because only 11 of the 37 countries that make up the Neotropical region have reported the occurrence of potyvirus species in ornamental plants (Fig. 2.20). Fourteen species of potyviruses have been reported in different ornamental plants in Brazil, due to its ability to cultivate a wide range of species in a complex production network.

However, understanding the relationships between a heterogeneous environment and the occurrence, distribution, and epidemiology of phytopathogens, such as potyviruses, is essential in identifying optimal crop management and control strategies.

References

Adkins S, Chellemi D, Annamalai M, Baker C (2005) Colombian datura virus diagnosed in *Brugmansia* spp. in Florida. Phytopathology 95(Supp):S2

Ágoston J, Almási A, Salánki K, Palkovics L (2019) First report of konjac mosaic virus in *Zantedeschia* from Hungary. J Plant Pathol 101:1217. https://doi.org/10.1007/s42161-019-00279-6

AIPH-IVP-Report2.pdf (n.d.). http://aiph.org/wp-content/uploads/2019/08/AIPH-IVP-Report2-2019.pdf. Accessed 26 Dec 2020

Alexandre MAV, Duarte LML, Rivas EB, Galleti SR (2010) Vírus detectados em plantas ornamentais no período 2004 a 2008. Rev Bras Hort Orn 16:95–100. https://doi.org/10.14295/rbho.v16i1.516

Alexandre MAV, Duarte LML, Rivas EB, Cilli A, Harakava R, Galleti SR, Kitajima EW (2011) Hippeastrum mosaic virus diagnosed in *Hippeastrum* and *Eucharis* in Brazil. J Plant Pathol 93:643–649

Alexandre MAV, Duarte LML, Rivas EB, Kitajima EW, Harakava R (2013) First report of Konjac mosaic virus in *Zamioculcas zamiifolia*. Plant Dis 97(11):1517. https://doi.org/10.1094/PDIS-05-13-0537-PDN. PMID: 30708485

Alexandre MAV, Duarte LML, Chaves ALR, Ramos AF, Harakava R, Kitajima EW (2017a) Canna paniculata as natural host of canna yellow streak virus in Brazil. Aust Plant Dis 12:38–41. https://doi.org/10.1007/s13314-017-0264-3

Alexandre MAV, Duarte LML, Rodrigues LK, Ramos AF, Harakava R (2017b) Hyacinth mosaic virus infecting *Hyacinthus* sp. plants in Brazil. Trop Plant Pathol 42:51–54. https://doi.org/10.1007/s40858-016-0121-5

Alexandre MAV, Duarte LML, Ramos-González PL, Chaves ALR, Harakava R, Rodrigues LK, Chabi-Jesus C, Kitajima EW (2020) Costus stripe mosaic virus, a tentative new member of the *Potyvirus* genus. Arch Virol 165(11):2541–2548. https://doi.org/10.1007/s00705-020-04788-z

Arneodo JD, Breuil S, Conci VC (2005) Detection of Bean yellow mosaic virus and Cucumber mosaic virus infecting gladiolus in Argentina. AgriScientia 22:87–89

Berlandier FA, Thackray DJ, Jones RAC, Latham LJ, Cartwright L (1997) Determining the relative roles of different aphid species as vectors of cucumber mosaic and bean yellow mosaic viruses in lupins. Ann Appl Biol 131:297–314. https://doi.org/10.1111/j.1744-7348.1997.tb05158.x

Bode O, Lesemann DE (1976) Euphorbia ringspot virus, a new virus in *Euphorbia milii* × *lophogona*. Acta Hortic 9:161–165

Boquel S, Zhang J, Goyer C, Giguère MA, Clark C, Pelletier Y (2015) Effect of insecticide-treated potato plants on aphid behavior and potato virus Y acquisition. Pest Manag Sci 71:1106–1112

Bos L (1970) Bean yellow mosaic virus. C.M.I./A.A.B. description of plant viruses 40. Association of Applied Biologists, Warwickshire

Bouwen I, Van Der Vlugt RAA (2000) Natural infection of *Alstroemeria brasiliensis* with Lily mottle virus. Plant Dis 84:103

Brants DH, Van den Heuvel J (1965) Investigation of Hippeastrum mosaic virus in *Hippeastrum hybridum*. Neth J Plant Pathol 71:145–151

Brunt AA (1973) Hippeastrum mosaic virus. C.M.I./A.A.B. Description of plant viruses 117. Association of Applied Biologists, Warwickshire

Brunt AA, Crabtree K, Dallwitz MJ, Gibbs AJ, Watson L (1997) Viruses of plants—descriptions and lists from the VIDE database. CAB International, Wallingford, pp 675–676

CABI (2020a) https://en.wikipedia.org/wiki/Hyacinth_(plant)#Distribution. Accessed 5 May 2020

CABI (2020b) Invasive species compendium. CAB International, Wallingford. www.cabi.org/isc. Accessed 5 May 2020

CABI (2010) Colombian datura virus. Distribution maps of plant diseases 2010 No.October Map 1086 (Edition 1). https://doi.org/10.1079/DMPD/20103281431

Cabrera D, González JE, Portal O, Hernández R (2010) Influencia del virus del mosaico de la malanga sobre el contenido de clorofilas en *Xanthosoma nigrum* (Vell.) genotipo inivit m 95-1. Rev Protección Veg 25(3):194–196

Camelo-Garcia VM, Esquivel-Fariña A, Ferro CG, Kitajima EW, Rezende JAM (2021) Strongylodon macrobotrys: new host of soybean mosaic virus in Brazil. Plant Dis 105:first look. https://doi.org/10.1094/PDIS-07-20-1607-PDN

Cárdenas-Alonso MR (1994) Virus diseases of ornamentals in Mexico, and management alternatives. Revista Chapingo, serie Horticultura 1:124–130

Castillo BS, Yarwood CE, Gold AH (1956) Canna mosaic virus. Plant Dis Rep 40:169–172

Chase AR, Zettler FW (1982) Dasheen mosaic virus infection of dieffenbachia cultivars. Plant Dis 66:891–893

Chauhan RP, Hamon HF, Rajakaruna P, Webb MA, Payton M, Verchot J (2015) Reliable detection for *Bean yellow mosaic virus*, *Canna yellow streak virus*, and *Canna yellow mottle virus* in canna varieties with red foliage. Plant Dis 99:188–194. https://doi.org/10.1094/PDIS-05-14-0538-RE

Chellemi DO, Webster CG, Baker CA, Annamalai M, Achor D, Adkins S (2011) Widespread occurrence and low genetic diversity of Colombian datura virus in *Brugmansia* suggest an anthropogenic role in virus selection and spread. Plant Dis 95:755–761

Chung BY, Miller WA, Atkins JF, Firth AE (2008) An overlapping essential gene in the *Potyviridae*. Proc Natl Acad Sci U S A 105:5897–5902. https://doi.org/10.1073/pnas.0800468105

Coelho MAN (2012) Araceae. In: Wanderley MGL, Martins SE, Romanini RP, Melhem TS, Shepherd GJ, Giulietii AM, Pirani JR, Kirizawa M, Melo MMRF, Cordeiro I, Kinoshita LS (eds) Flora Fanerogâmica do Estado de São Paulo, vol 7. Instituto de Botânica, São Paulo, pp 27–72

Coiffard C, Mohr BAR, Bernardes-de-Oliveira MEC (2013) The Early Cretaceous Aroid, Spixiarum *kipea* gen. et sp. nov., and implications on early dispersal and ecology of basal monocots. Taxon 62(5):997–1008. 12p. https://doi.org/10.12705/625.21

Costa AS (1972) Investigações sobre moléstias do feijoeiro no Brasil. In: Simpósio Brasileiro de Feijão, 1., 1971, Campinas. Anais. Viçosa: UFV, vol 2, pp 303–384

Costa AS, Kitajima EW (1966) Vírus do mosaico do picão ataca girassol. Suplemento Agrícola: O Estado de São Paulo 605:12–13

Dekker EL, Derks AFLM, Asjes CJ, Lemmers MC, Bol JF, Langeveld SA (1993) Characterization of potyviruses from tulip and lily which cause flower-breaking. J Gen Virol 74:881–887. https://doi.org/10.1099/0022-1317-74-5-881

Derks AFLM (1995) Hyacinth. In: Loebenstein G, Lawson RH, Brunt AA (eds) Virus and virus-like diseases of bulb and flower crops. Wiley, Chichester, pp 298–302

Derks AFLM, Vink-van den A, Jenny L (1980) Hyacinth mosaic virus: symptoms in hyacinths, serological detection, and relationships with other potyviruses. Acta Hortic 109:495–502. https://doi.org/10.17660/ActaHortic.1980.109.74

Derks AFLM, Lemmers MEC, Van Gemen BA (1994) Lily mottle virus in lilies: characterization, strains and its differentiation from Tulip breaking virus in tulips. Acta Hortic 377:81–88. https://doi.org/10.17660/actahortic.1994.377.30

Duarte LML, Alexandre MAV (2017) Principais vírus em plantas ornamentais. In: Alexandre MAV, Duarte LML, Campos AEC (eds) Plantas Ornamentais: Doenças e Pragas, 2nd edn. Editora Devir, São Paulo, pp 83–169. ISBN 978-85-7532-642-8

Duarte LML, Alexandre MAV, Harakava R (2009) Identificação molecular do *Hippeastrum mosaic virus* em Amarílis (*Hippeastrum* sp.). Trop Plant Pathol 34(Suppl):S274

Duarte LML, Alexandre MAV, Chaves ALR, Harakava R, Alvim GB, Kitajima EW (2019) Widespread incidence of Gloriosa stripe mosaic virus infecting *Gloriosa superba* in São Paulo state, Brazil. Trop Plant Pathol 44:297–301. https://doi.org/10.1007/s40858-018-0265-6

Edwardson JR (1974) Some properties of the potato virus-Y group, Florida agricultural experiment station monograph series 4. Institute of Food and Agricultural Sciences/University of Florida, Gainesville

Edwardson JR, Christie RG (1979) Light microscopy of inclusions induced by viruses infecting peppers (*Capsicum annuum*). Fitopatol Bras 4:341–373

El-Nil MMA, Zettler FW, Hiebert E (1977) Purification, serology and some physical properties of dasheen mosaic virus. Phytopathology 67:1445–1450

Engelmann J (2010) Hamacher J. In: Mahy BWJ, VanRegenmortel MHV (eds) Desk encyclopedia of plant and fungal virology. Elsevier, Oxford, pp 436–458

Espinha LM, Gaspar JO (1997) Caracterização parcial do vírus do mosaico do pepino ("cucumber mosaic virus"- CMV) isolado de *Catharanthus roseus*. Fitopatol Bras 22:209–212

Favara GM, Camelo-García VM, Silva JMF, Silva TNZ, Mituti T, Nagata T, Kitajima EW, Rezende JAM (2020) Biological and molecular characterization of isolates of catharanthus mosaic virus infecting *Mandevilla* sp. Trop Plant Pathol 45:461–465. https://doi.org/10.1007/s40858-020-00346-9

Fuji S, Terami F, Furuya H, Naito H, Fukumoto F (2004) Nucleotide sequence of the coat protein genes of *Alstroemeria mosaic virus* and Amazon lily mosaic virus, a tentative species of genus *Potyvirus*. Arch Virol 149:1843–1849

Fuji S-I, Mochizuki N, Fujinaga M, Ikeda M, Shinoda K, Uematsu S, Furuya H, Naito H, Fukumoto F (2007) Incidence of viruses in *Alstroemeria* plants cultivated in Japan and characterization of *Broad bean wilt virus-2, Cucumber mosaic virus and Youcai mosaic virus*. J Gen Plant Pathol 73:216–221

Gadhave KR, Gautam S, Rasmussen DA, Srinivasan R (2020) Aphid transmission of *Potyvirus*: the largest plant-infecting RNA virus genus (supplementary). Viruses 12(773):1–57

Gibbs A, Ohshima K (2010) Potyviruses and the digital revolution. Annu Rev Phytopathol 48 (1):205–225. https://doi.org/10.1146/annurev-phyto-073009-114404

Gibbs AJ, Hajizadeh M, Ohshima K, Jones RAC (2020) The potyviruses: an evolutionary synthesis is emerging. Viruses 2(2):132–161. https://doi.org/10.3390/v12020132

Gorter GJMA (1977) Index of plant pathogens and diseases they cause in cultivated plants in South Africa. Sci Bull 392:177

Guaragna MA, Jordan RL, Putnam MLF (2004a) First report of bean yellow mosaic virus (pea mosaic virus strain) in *Verbena* x *hybrida*. Plant Dis 88(5):574. https://doi.org/10.1094/PDIS.2004.88.5.574C

Guaragna MA, Ambrose J, Jordan RL (2004b) Detection and characterization of euphorbia ringspot potyvirus. Phytopathology 94:S36

Gutiérrez-Estrada A, Zavaleta-Mejía E, Gaytán-Acuña EA, Herrera-Guadarrama AJ (2000) Virus associated with *Alstroemeria* in México. Rev Mex Fitopatol 17:97–103

Hay A, Gottschalk M, Holguín A (2012) Huanduj: Brugmansia. Florilegium, Glebe. Royal Botanic Gardens, Kew; 424 p

Inoue-Nagata AK, Oliveira PA, Dutra LS, Nagata T (2006) *Bidens mosaic virus* is a member of the *Potato virus Y* species. Virus Genes 33:45–49

International Committee on Taxonomy of Viruses (2019). https://talk.ictvonline.org/ictv-reports/ictv_online_report/positive-sense-rna-viruses/w/potyviridae. Accessed 20 Dec 2020

Jenner CE, Wang X, Ponz F, Walsh JA (2002) A fitness cost for *Turnip mosaic virus* to overcome host resistance. Virus Res 86:1–6

Jordan R, Guaragna MA, Kinard G, Lynn S (2002) Detection and first report of *Dasheen mosaic virus* and a second potyvirus infecting the terrestrial orchid *Spiranthes cernua*. Acta Hortic 568:253–260. https://doi.org/10.17660/ActaHortic.2002.568.37

Junqueira AH, Peetz MS (2017) Brazilian consumption of flowers and ornamental plants: habits, practices, and trends. Ornam Hortic 23(2):178–184. https://doi.org/10.14295/oh.v23i2.1070

Junqueira AH, Peetz MS (2018) Sustainability in Brazilian floriculture: introductory notes to a systemic approach. Ornam Hortic 24(2):155–162. https://doi.org/10.14295/oh.v24i2.1253

Kahn RP, Bartels R (1968) The Colombian datura virus—a new virus in the Potato virus Y group. Phytopathology 58:587–592

Kitajima EW (2020) An annotated list of plant viruses and viroids described in Brazil (1926-2018). Biota Neotropica 20(2):e20190932. https://doi.org/10.1590/1676-0611-BN-2019-0932

Kitajima EW, Carvalho AMB, Costa AS (1961) Morfologia do vírus do mosaico do picão. Bragantia 20(13):503–512

Koenig R, Lesemann D (1974) A potyvirus from *Gloriosa rothschildiana*. Phytopathol Z 80:136–142

Kunkel LO (1922) Ameboid bodies associated with Hippeastrum mosaic. Science 55:73

Kwon SB, Ha JH, Yoon JY, Ryu KH (2002) Zantedeschia mosaic virus causing leaf mosaic symptom in calla lily is a new potyvirus. Arch Virol 147:2281–2289

Lesemann DE, Winter S (2002) Konjac mosaic virus, Dasheen mosaic virus and unknown potyviruses infecting Zantedeschia spp. Acta Hortic 568:135–141. https://doi.org/10.17660/ActaHortic.2002.568.19

Liao M-T, Mou D-F, Chang Y-C, Tsai C-W (2020) Vector transmission of konjac mosaic virus to calla lily (*Zantedeschia* spp.) by aphids. Ann Appl Biol 177(3):367–373. https://doi.org/10.1111/aab.12628

Lima RCA, Lima JAA, Aguiar JR (2004) Serological identification of *Dasheen mosaic virus* in *Anthurium* sp. in the State of Ceará. Fitopatol Bras 29(1):105. https://doi.org/10.1590/S0100-41582004000100020

Lisa V, Dellavalle G (1987) Bean yellow mosaic virus in *Lisianthus russelianus*. Plant Pathol 36:214–215

Lisa V, Vetten HJ, Lesemann D-E, Gotta P (2002) Occurrence of *Lily mottle virus* in escarole. Plant Dis 86:329. https://doi.org/10.1094/PDIS.2002.86.3.329A

Loebenstein G, Lawson RH, Brunt AA (eds) (1995) Virus and virus-like diseases of bulb and flower crops. Wiley, Chichester

Lucinda N, Nagata T, Inoue-Nagata AK, Salaroli AB, Kitajima EW (2008) Brugmansia suaveolens mottle virus, a novel potyvirus causing leaf mottling of *Brugmansia suaveolens* in Brazil. Arch Virol 153:1971–1976. https://doi.org/10.1007/s00705-008-0211-x

Maciel SC, Silva RF, Reis MS, Jadão AS, Rosa DD, Giampan JS, Kitajima EW, Rezende JAM, Camargo LEA (2011) Characterization of a new potyvirus causing mosaic and flower variegation in *Catharanthus roseus* in Brazil. Sci Agric 68:687–690

Malandrake I, Driessen A, Varveri C, Vassilakos N (2016) First report of Hippeastrum mosaic virus in *Hippeastrum* sp in Greece. Plant Dis 100(4):869

Malter AJ (1995) The economic importance of ornamentals. In: Loebenstein G, Lawson RH, Brunt AA (eds) Virus and virus-like diseases of bulb and flower crops. Chichester, Wiley, pp 1–3

Marys E, Romano M (2011) Occurrence of Euphorbia ringspot virus in *Euphorbia milii* cv. splendens in Venezuela. J Phytopathol 159:66–68. https://doi.org/10.1111/j.1439-0434.2010.01709.x

McKirdy SJ, Jones RAC, Latham LJ, Coutts BA (2000) Bean yellow mosaic potyvirus infection of alternative annual pasture, forage, and cool season crop legumes: susceptibility, sensitivity, and seed transmission. Aust J Agric Res 51:325–346. https://doi.org/10.1071/AR99110

Mitrofanova IV, Zakubansky AV, Mitrofanova OV (2018) Viruses infecting main ornamental plants: an overview. Ornam Hortic 24(2):95–102. https://doi.org/10.14295/oh.v24i2.1199

Mokrá V, Götzová B (1994) Identification of virus infections in *Dieffenbachia* and *Zantedeschia* in Czechoslovakia. Acta Hortic 377:361–362

Mollov D, Guaragna MA, Lockhart B, Rezende JAM, Jordan R (2015) First report of Catharanthus mosaic virus in *Mandevilla* in the United States. Plant Dis 99(1):165

Mollov D, Grinstead S, Tahir MN, Reinsel M, Hammond J (2017) First report of multiple isolates of gloriosa stripe mosaic virus in *Gloriosa superba* in the United States. Plant Dis 101:1070

Monger WA, Harju V, Skelton A, Seal SE, Mumford RA (2007) Canna yellow streak virus: a new potyvirus associated with severe symptoms in canna. Arch Virol 152:1527–1530. https://doi.org/10.1007/s00705-007-0977-2

Mosquera-Yuqui F, Garrido P, Flores FJ (2020) Molecular characterization and complete genome of alstroemeria mosaic virus (AlMV). Virus Genes 56:87–93

Mowat WP (1995) Tulip. In: Loebenstein G, Lawson RH, Brunt AA (eds) Virus and virus-like diseases of bulb and flower crops. Wiley, Chichester, pp 352–383

Myers N, Mittermeier RA, Mittermeier CG, Fonseca GAB, Kent J (2000) Biodiversity hotspots for conservation priorities. Nature 403:853–858

Nauheimer L, Metzler D, Renner SS (2002) Global history of the ancient monocot family Araceae inferred with models accounting for past continental positions and previous ranges based on fossil. New Phytopatol 195:938–950. https://doi.org/10.1111/j.1469-8137.2012.04220.x

Nishiguchi M, Yamasaki S, Lu X-Z et al (2006) Konjak mosaic virus: the complete nucleotide sequence of the genomic RNA and its comparison with other potyviruses. Arch Virol 151(8):643–650. https://doi.org/10.1007/s00705-006-0735-x

Ohshima K, Yamaguchi Y, Hirota R, Hamamoto T, Tomimura K, Zhongyang T, Sano T, Azuhata F, Walsh JA, Fletcher J, Chen J, Gera A, Gibbs A (2002) The molecular evolution of *Turnip mosaic virus*; evidence of host adaptation, genetic recombination and geographical spread. J Gen Virol 83:1511–1521

Okuno K, Hama T, Takeshita M, Furuya N, Takanami Y (2003) New Potyvirus isolated from *Cryptotaenia japonica*. J Gen Plant Pathol 69:138–142

Olspert A, Chung BY, Atkins JF, Carr JP, Firth AE (2015) Transcriptional slippage in the positive-sense RNA virus family *Potyviridae*. EMBO Rep 16:995–1004. https://doi.org/10.15252/embr.201540509

Pacifico D, Crucitti D, Stigliano E, Ciuffo M, Vallino M, Carimi F (2016) First report of Colombian datura virus in *Mandragora autumnalis* in Sicily, Italy. Plant Dis 100:2338. https://doi.org/10.1094/PDIS-04-16-0445-PDN

Padmavathi M, Srinivas KP, Subba Reddy CV, Ramesh B, Navodayam K, Krishnaprasadji J, Babu Ratan P, Sreenivasulu P (2011) Konjac mosaic virus naturally infecting three aroid plant species in Andhra Pradesh, India. J Phytopathol 159:133–135

Padmavathi M, Srinivas KP, Hema M, Sreenivasulu P (2013) First report of Konjac mosaic virus in elephant foot yam (*Amorphophallus paeoniifolius*) from India. Australasian Plant Dis Notes 8:27–29. https://doi.org/10.1007/s13314-012-0088-0

Pearson MN, Cohen D, Cowell SJ, Jones D, Blouin A, Lebasc BSM, Shillerc JB, Cloverc GRG (2009) A survey of viruses of flower bulbs in New Zealand. Australas Plant Pathol 38:305–309

Pham K, Langeveld SA, Lemmers MEC, Derks AFLM (2002) Detection and identification of potyviruses in *Zantedeschia*. Acta Hortic 568:143–148. https://doi.org/10.17660/ActaHortic.2002.568.20

Pham KTK, Kock MJD, Lemmers MEC, Derks AFLM (2011) Molecular identification of potyviruses infecting ornamentals bulbs by the analysis of coat protein (CP) sequences. Acta Hortic 901:167–172

Phillips S, Brunt AA (1986) Four viruses of alstroemeria in Britain. Acta Hortic 177:227–234

Pierce WB (1934) Viruses of the bean. Phytopathology 24(2):87–115

Provvidenti R, Hunter JE (1975) *Bean yellow mosaic virus* infection in *Cladrastis lutea*, an ornamental leguminous tree. Plant Dis Reptr 59:86–87

Raj SK, Snehi SK, Kumar S, Khan MS (2009) First molecular detection and identification of a potyvirus associated with severe mosaic disease of amaryllis (*Hippeastrum hybridum* Hort.) in India. Australas Plant Dis Notes 4:50–53

Rajakaruna P, Shafiekhani M, Kim T, Payton M, Chauhan R, Verchot J (2014) Production of discernable disease phenotypes in canna by five plant viruses to the genera *Potyvirus*, *Cucumovirus* and *Badnavirus*. Plant Pathol J 63:821–830

Ramírez P (1985) Aislamiento y caracterización del virus del mosaico del dasheen (DMV) en Costa Rica. Turrialba 35:279–283

Reyes G, Nyman M, Rönnberg-Wästljung AC (2005) Agronomic performance of three cocoyam (*Xanthosoma violaceum* Schott) genotypes grown in Nicaragua. Euphytica 142:265–272

Rivas EB, Duarte LML, Alexandre MAV, Galleti SR (1997) Dasheen mosaic virus in *Anthurium* species. Virus Rev Res 2:192–193

Rivas EB, Alexandre MAV, Duarte LML (1998) Infecção natural de 'Dasheen mosaic virus' em *Dieffenbachia* sp. Summa Phytopathol 24:71

Rivas EB, Bôdi ECA, Vaz Lobo RS (2011) *Lily mottle virus* no Brasil, vol 10. Instituto Biológico, São Paulo, pp 1–9. http://www.biologico.sp.gov.br/docs/dt/lily_mottle_virus.pdf

Rivas EB, Bôdi ECA, Harakava R (2013) Primeira ocorrência de *Alstroemeria mosaic virus* em cultivos comerciais de alstroemérias. Summa Phytopathol 39(Suppl):S254

Rodamilans B, Valli A, Mingot A, San León D, Baulcombe D, López-Moya JJ, García JA (2015) RNA polymerase slippage as a mechanism for the production of frameshift gene products in plant viruses of the *Potyviridae* family. J Virol 89(13):6965–6967. https://doi.org/10.1128/JVI00337-15

Rodrigues MGR, Kitajima EW, Lin MT (1984) Mosaico em aráceas comestíveis e ornamentais causado pelo vírus do mosaico do inhame ("dasheen mosaic virus"). Fitopatol Bras 9:291–299

Rodrigues MGR, Marinho VLA, Ribeiro SG, Kitajima EW (1991) A mosaic in tickseed (*Coreopsis lanceolata*) caused by *Bidens mosaic virus*. Fitopatologia Bras 16(1):114–117

Sanches MM, Spadotti DMA, De Marchi BR, Pavan MA, Krause-Sakate R (2010) *Bidens mosaic virus*: detection by RT-PCR and identification of *Galinsoga parviflora* as a new natural host of the virus. Summa Phytopathol 36:304–307

Sastry KS (2013) Seed-borne plant virus diseases. Springer, New Delhi, pp 10–30

Schulze A, Roberts R, Pietersen G (2017) First report of detection of Bean yellow mosaic virus (BYMV) on *Tropaeolum majus*, *Hippeastrum* spp. and *Liatris* spp. in South Africa. Plant Dis 101(5):846–847

Seabra PV, MAV A, Galleti SR, Rivas EB, Duarte LML (1999) Ocorrência de um Potyvirus em *Catharanthus roseus*. Arq Inst Biol 66:116. (Abst)

Shimoyama J, Kamuya-Iwaki M, Hanada K, Gunji T (1992) Konjak mosaic virus, a new potyvirus infecting konjak, *Amorphophallus konjac*. Ann Phytopathol Soc Jpn 58(2):706–712

ShuHui K, Hua L, Admiraal R, Jones MGK, Wylie SJ (2015) Catharanthus mosaic virus: a potyvirus from a gymnosperm, *Welwitschia mirabilis*. Virus Res 203:41–46

Sorel M, Garcia JA, German-Retana S (2014) The potyviridae cylindrical inclusion helicase: a key multipartner and multifunctional protein. Mol Plant Microbe Interact 3:215–226

Souza CS, Lorenzi H (2012) Botânica Sistemática. Guia ilustrado para identificação das famílias de Fanerógamas nativas e exóticas no Brasil, baseado em APG III, 3rd edn. Instituto Plantarum, Nova Odessa

Spence NJ, Mills PR, Barbara DJ (2000) A survey of viruses of *Alstroemeria* in the UK and the characterisation of carlaviruses infecting *Alstroemeria*. Eur J Plant Pathol 106:843–847

Suzuki GS, Rosa RAC, Sanches MM, Nozaki DN, Pavan MA, Krause-Sakate R (2009) Characterization of an isolate of *Bidens mosaic virus* (BiMV) from lettuce. Summa Phytopathol 35 (3):231–233

The Plant List (2013) Version 1. Published on the Internet; http://www.theplantlist.org/. Accessed 30 Nov 2020

Tombolato AFC, Uzzo RP, Junqueira AH, Peetz MS, Stancato GC (2013) Geophyte research and production in Brazil. In: Kamenetsky R, Okubo H (eds) Ornamental geophytes: from basic science to sustainable production. CRC Press, Boca Raton, pp 435–448

Van der Vlugt RAA, Bouwen I (2002) Alstroemeria streak virus is an isolate of *Alstroemeria* mosaic potyvirus. Phytomed Dtsch Phytomed Ges Sonderh 1:31

Wang CY, Chang YC (2006) First identification of *Alstroemeria mosaic virus* in Taiwan. Plant Pathol 55:566

Wei T, Pearson MN, Cohen D, Tang JZ, Clover GRG (2008) First report of Zantedeschia mosaic virus infecting a *Zantedeschia* sp. N Z Plant Dis 92:1253

Wylie SJ, Jones MG (2012) Complete genome sequences of seven carlavirus and potyvirus isolates from *Narcissus* and *Hippeastrum* plants in Australia, and proposals to clarify their naming. Arch Virol 157(8):1471–1480

Wylie SJ, Adam M, Chalam C, Kreuze J, López-Moya JJ, Ohshima K, Praveen S, Rabenstein F, Stenger D, Wang A, Zerbini FM (2017) ICTV report consortium, *ICTV virus taxonomy profile:* Potyviridae. J Gen Virol 98:352–354

Yasuda S, Saka K, Natsuaki KT (1998) Characterization and serodiagnosis of alstroemeria mosaic potyvirus. Jpn J Trop Agric 42:85–93

van Zaayen A (1995) Alstroemeria. In: Loebenstein G, Lawson RH, Brunt AA (eds) Virus and virus-like diseases of bulb and flowers. Chichester, Wiley, pp 237–249

Zakubanskiy AV, Mitrofanova IV, Chirkov SN (2017) Molecular characterization of viruses infecting canna in Russia. Eur J Plant Pathol 149:923–931. https://doi.org/10.1007/s10658-017-1241-6

Zettler FW, Hartman RD (1987) Dasheen mosaic virus as a pathogen of cultivated aroids and control of the virus by tissue culture. Plant Dis 71:958–963

Zettler FW, Foxe MJ, Hartman RD, Edwardson JR, Christie RG (1970) Filamentous viruses infecting dasheen and other araceous plants. Phytopathology 60:983–987

Zettler FW, Abo El-Nil MM, Hartman RD (1978) Dasheen mosaic virus. CMI/AAB Description of plant viruses 191. Association of Applied Biologists, Warwickshire

Viruses of Some Bulbous Ornamental Flowers from *Liliaceae* Family

3

Nikolay Manchev Petrov, Mariya Ivanova Stoyanova, and Rajarshi Kumar Gaur

Abstract

Ornamental bulbous plants from family *Liliaceae* are an increasingly important floricultural crop all over the world. Virus diseases are of great economic importance in bulbous flowers because most of them are propagated vegetatively by tissue culture, scions, bulbs or other tissues. Once infected with a virus, the plants of the *Liliaceae* family react to the infection with specific symptoms such as reduced growth, chlorotic spots and deformed flowers, which makes them unrealisable on the market. This chapter deals with the most important viruses on bulbous ornamental plants mainly from genera *Cucumovirus*, *Potyvirus*, *Potexvirus*, *Tospovirus* and *Nepovirus*.

Keywords

Ornamental plants · Viruses · *Liliaceae* · Tulip · *Lilium*

3.1 Introduction

Ornamental bulbous plants like these from genera *Allium*, *Anemone*, *Begonia*, *Daffodil*, *Gladiolus*, *Narcissus*, *Tulipa*, *Lilium*, *Crocus*, *Hyacinth*, *Iris*, *Scilla*, etc. are an increasingly important floricultural crop all over the world. Bulbous

N. M. Petrov (✉)
New Bulgarian University, Sofia, Bulgaria

M. I. Stoyanova
Institute of Soil Science, Agrotechnologies and Plant Protection (ISSAPP) "Nikola Pushkarov", Sofia, Bulgaria

R. K. Gaur
Deen Dayal Upadhyay Gorakhpur University, Gorakhpur, Uttar Pradesh, India

S. K. Raj et al. (eds.), *Virus Diseases of Ornamental Plants*,
https://doi.org/10.1007/978-981-16-3919-7_3

ornamental flowers suffer from diseases caused by viruses, bacteria, fungi, insect pests, thrips, bulb mites and foliar nematodes as well as from physical disorders like black sprouts, upper leaf necrosis and mineral deficiencies. Important fungi for flower growers are the soil-borne *Fusarium* (Shahin et al. 2009), *Phytophthora* (Yang et al. 2010), *Pythium* and *Rhizoctonia solani* (van Os 2003; Dijst and Schneider 1996).

Virus diseases are of great economic importance in bulbous flowers because most of them are propagated vegetatively by tissue culture, scions, bulbs or other tissues. Flowering bulbous plants are susceptible to various viral infections that cause significant economic losses. Once infected with a virus, the plants of the *Liliaceae* family react to the infection with specific symptoms such as reduced growth, chlorotic spots and deformed flowers, which makes them unrealisable on the market. Symptoms on the leaves and flowers of plants reduce their value and demand, even if the viruses cause minor damage. Symptoms are more and more severe if the plants are co-infected with other virus like Lily symptomless virus (LSV) (Loebenstein et al. 1995). LSV can reduce the yield of tulip bulbs (Derks and Asjes 1975).

Due to their vegetative propagation, flowering bulbs have serious problems with the spread and long-distance transport of viral infections in the globalised world, as they survive in dormant buds. Some of the important viruses that infect bulbous ornamental plants are Cucumber mosaic virus (CMV), Tomato spotted wilt virus (TSWV), Lily mottle virus (LMoV), Lily virus X (LVH) and Tulip virus X (TVX), Tulip mild mottle mosaic virus (TMMMV), Tobacco necrosis virus (TNV), Lily symptomless virus (LSV), Tulip breaking virus (TBV), Tulip band-breaking virus (TBBV), Plantago asiatica mosaic virus (PlAMV), Lily yellow mosaic virus (LYMV), Lily latent virus (LLV) and Arabis mosaic virus (AMV) (Aravintharaj et al. 2017; Asjes and Segers 1983; Brierley and Smith 1944; Chen et al. 2013; De Best et al. 2000; Derks 1995a; McWhorter 1940; Memelink et al. 1990; Parrella et al. 2003; Sochacki and Komorowska 2012; Li et al. 2018).

3.2 Characteristics of Virus Infections of *Liliaceae* Species

Viruses represent the widespread and very important group of pathogens among other groups of lilies (Slate 1986). At least 13 viruses have been reported to infect *Lilium* spp., but only 3, CMV, LSV and TBV, are common for North America (Allen 1980).

3.2.1 Cucumber Mosaic Cucumovirus (CMV)

CMV is a pathogen, with an extremely wide variety of hospitable species, including ornamental plants. It is identified in 13 families of ornamental plant species of the genera *Lilium*, *Tulipa*, *Asclepias*, *Canna*, *Crocus*, *Dahlia*, *Delphinium*, *Gaillardia*, *Gladiolus*, *Iris*, *Lupinus*, *Crocosmia*, *Muscari*, *Narcissus*, *Phlox*, *Pelargonium*, *Primula*, *Rudbeckia* and *Viola* (Samuitiene and Navalinskiene 2008). It is aphid-transmitted virus (by *Myzus persicae* and *Aphis gossypii*) (Francki et al. 1979) and is reported on lilies in several countries—Belgium (Verhoyen and Horvat 1972), Great Britain (Mowat and Stefanac 1974), the Netherlands (Asjes et al. 1973), Italy (Benetti and Tomassoli 1988), Korea (Chang and Chung 1987), Japan (Hagita 1989; Hagita et al. 1989) and Taiwan (Chang and Pang 1991).

CMV is an RNA virus that belongs to the genus *Cucumovirus*, family *Bromoviridae*, and has spherical icosahedral capsids about 30 nm in diameter. This virus is transmitted worldwide in a non-persistent manner by many aphid species. CMV causes a variety of symptoms in ornamental species that can reduce their aesthetic commercial qualities, such as deformation and flower discoloration in gladioli (Derks 1995a; Lisa and Gera 1995; Stein 1995). The viral symptoms in lilies are vein-clearing and leaf curl, and in mixed infection with LSV, which is also transmitted by aphid vectors, it can cause severe necrosis (Derks 1995b).

Several diseases of lilies caused by different CMV strains have been described as follows:

3.2.1.1 Celery Mosaic

Initially in 1934 the causative agent of the disease was identified as Celery virus 1 (Wellman 1934). The following year, it was found that many cultivated lilies were infected with this virus by inoculation with infected vectors (aphids) (Wellman 1935). The virus was later identified as a CMV strain (Price 1935).

3.2.1.2 Lily Mosaic

The first description of a virus-like disease on lilies was in 1896 (Stewart 1896). The first report of the mechanical transmission of the pathogen from infected plants was in 1928 (Guterman 1928). In 1937, it was reported that this lily disease was caused by CMV (Price 1937).

3.2.1.3 Necrotic Flecking

In 1939, conclusive evidence of the involvement of more than one virus in the mosaic complex was provided (Wellman 1935). After 10 years, the participants in this viral complex—CMV and LSV—were identified (Brierley and Smith 1944). Due to the severity of symptoms of infected lilies, such as necrotic spots caused by mixed CMV and LSV infection, this has become an important factor leading to the demise of the lily industry in the United States (Bickey 1963; Shippy 1937). The CMV reservoir for lilies was discovered in Florida on a plant from genus *Commelina* (Wellman 1935; Boolittle and Wellman 1934). Changes in cultural practices since the 1970s and shifts in crop production have controlled the spread of CMV (Zitter and Simons 1980). However, *Commelina* spp. remain major potential reservoirs of CMV for crops such as gladioli and lilies (Baker and Zettler 1988).

3.2.1.4 Lily Ringspot

In hybrid lilies imported from the United States to England, disease symptoms such as lily ringspots have been found (Smith 1950). The causative agent of the disease is defined as CMV. Later, in 1962, CMV infection was confirmed in Georgia lilies with symptoms similar to those of ring lilies (Brierley 1962).

In the cultivation of Italian lilies, in which a serious reduction in yield and quality of flowers has been observed, CMV infection is found. In such cases, CMV is always present with LSV in mixed infection. The CMV virus isolate in lilies is significantly different from other strains of CMV because it cannot infect different cucumber cultivars. CMV virus isolate from lilies is difficult to isolate but is purified and obtained antiserum for diagnosis (Benetti and Tomassoli 1988).

Four virus isolates (called Cas, CB, P26 and Simp2) have been identified in naturally infected CMV lilies, which are distinguished based on their serological properties and differences in RNA3 containing coat and movement protein genes. These lily isolates were compared to eight other CMV isolates from dahlia, delphinium, impatiens, honeysuckle, cucumber and redcurrant plants. Based on their specific recognition and binding to monoclonal antibodies, the Cas and CB isolates were classified in group I of CMV. The other two isolates, P26 and Simp2, showed more similarity to group II isolates. ELISA results with group II-specific monoclonal antibodies confirmed that Simp2 isolate belonged to group II, while P26 did not react with any of the monoclonal antibodies. To explain this lack of response, a mutation was found at position 138 of the threonine responsible for the serological properties of isolate P26 (Berniak et al. 2010).

A new CMV strain was identified from *Lilium longiflorum*-Ly2-CMV, which was compared with well-characterised viral strains from both groups by host reaction in indicator plants, serological properties, RT-PCR and restriction analysis of the viral capsid genes. Differences were found between the symptoms of Ly2-CMV and other CMV isolates in indicator plants—tobacco and *Datura stramonium*. Ly2-CMV causes local necrotic ringspots on the leaves of *Nicotiana tabacum* cv. Xanthi, Burley 21 and *D. stramonium* but cannot cause systemic infection. *Nicotiana benthamiana* responds only to systemic infection (Jung et al. 2000).

3.2.2 Lily Latent Virus (LLV)

The virus isolated from a healthy-looking lily that subsequently caused symptoms on inoculated tulips is called the Lily latent virus. It was originally thought to be identical to Tulip virus I but was later distinguished based on the lack of viruplasts in LLV-infected tulips (McWhorter 1940; McWhorter 1937).

3.2.3 Lily Symptomless Virus (LSV)

In 1944, a new asymptomatic virus, called Lily symptomless virus (LSV), was described (Brierley and Smith 1944). Like CMV, this virus has spread widely due to the lack of symptoms. In addition to the necrotic spots caused by the LSV in mixed CMV infection, it also increases the host's susceptibility to root rot caused by fungi of the genus *Pythium* (Raabe 1975).

LSV is a member of the genus *Carlavirus* of the family *Betaflexiviridae*. The viral genome is a single molecule of linear single-stranded RNA with a length of approximately 8.4 kb. The virus has hosts that are restricted to the *Liliaceae* family and are transmitted by aphids in a non-persistent way (Fisher et al. 2013).

LSV alone can cause leaf curl-stripe symptoms when the plants are grown at lower ambient temperatures, while at higher temperatures, the symptoms are masked and become dormant. LSV in mixed TBV infection can cause the formation of brown rings in the bulbs of lilies (Berks 1976). LSV is found everywhere where commercially lilies are grown, but unlike them, it does not infect tulips as often, although it is isolated from their leaves (Verhoyen and Horvat 1972). LSV is transmitted by four species of aphids in a non-persistent manner, with *Myzus persicae* being the most effective of these (Mowat and Stefanac 1974). No differentiated strains of the virus have been reported. The main hosts of LSV are limited to *Liliaceae* (Allen 1972).

The acceptable frequency of viruses such as LSV, which are usually asymptomatic in field-grown plants, can create problems in unfavourable light and growing conditions for flower production in greenhouses. The reduction in yield and quality of bulbs caused by lily viruses necessitates a further reduction of viruses in propagating bulbs. The spread of LSV transmitted by aphids usually occurs very rapidly. By reducing the access of viral vectors and through tissue cultures, very low levels of virus presence in bulbs are achieved. The viruses spread mainly in June and July, less in May and at least in August and September. Therefore, routine spraying of mixtures of mineral oil and pyrethroid insecticide should be done weekly in May, June and July and in August and September to be sprayed in 2 weeks. There are differences in the efficiency of different brands of mineral oils. The propagation of lilies is done by the scaling of bulbs and by tissue culture procedures which enables the rapid bulking of virus-tested and other stocks of high quality which have the health status required (Asjes 2000).

The commercial significance of LSV secondary infections for commercial producers of *Lilium* cut flowers was assessed by comparing plant growth and flowering quality produced from healthy and LSV-infected bulbs. The characteristics that are most economically important for producers of cut flowers (the number of inflorescences per stem, the quality of the flower and the life of the cut plant) are relatively weakly affected by the presence of LSV infection in any variety and are unlikely to arouse serious commercial interest (Blake and Wilson 1996).

LSV was purified from infected plants by clarification with chloroform, precipitation with polyethylene glycol and NaCl and differential centrifugation. Antisera against intact and degraded pyrrolidine LSV particles were prepared. Intact and degraded LSV have very little or no common antigenic determinants. The sensitivity of the microprecipitin test and the single immunodiffusion drop test are approximately the same but lower than those of electron microscopy. When testing lilies for LSV, the most reliable results were obtained from samples taken in the period of 2 weeks after flowering and from leaves growing at a level about a quarter of the distance from the top of the stem. Unlike secondary infections, primary infections are more successfully detected in stored bulbs than in leaves taken from plants during the previous growing season. In tulips, LSV is better found in flowers than in leaves. Experimentally infected tulips remain asymptomatic (Derks and Vink-Van den Abeele 1980).

In 2012, the plant *Lilium martagon* 'Pink Taurade' was discovered with symptoms of stunting and an interveinal chlorosis, resembling a lack of nutrients (Fisher 2013).

3.2.4 Lily Mottle Virus (LMoV)

Lily mottle virus (LMoV), which belongs to the genus *Potyvirus*, family *Potyviridae*, is one of the main viruses infecting lilies. Antiviral genomes are a single-stranded RNA molecule approximately 9.7 kb long, encoding a single, large open reading frame that transforms into a polyprotein that undergoes proteolytic cleavage into ten separate products. The virus is transmitted by aphid vector. Viral symptoms vary depending on the susceptibility of different varieties and hybrids, ranging from leaf spot or mosaic, vein clearing, chlorotic and yellow streaking, leaf curling and necrotic spots to milder forms of leaf symptoms. Plants may even be asymptomatic at some stages of growth. In 2010, symptoms such as vein clearing, leaf mottle, leaf mosaic and reddish brownish necrotic spots were observed in a collection of lilies (Fisher et al. 2013; Rizzo et al. 2012).

There are three isolates of LMV, which are distinguished by herbaceous test plants. They cannot be distinguished by immunoblotting, immunospecific electron microscopy and nucleic acid hybridisation assays. LMV isolates are distinguished from TBV by antisera prepared against antivirals that cause colour degradation in tulips and cDNA probes (Derks et al. 1994).

In samples of lilies from a field for growing cut flowers in India, LMoV has been identified by serological and molecular methods (Aravintharaj et al. 2017).

3.2.5 Plantago Asiatica Mosaic Virus (PLAMV)

In early 2011, *Potexvirus* was detected by RT-PCR in imported lily bulbs, the causative agent was identified as Plantago asiatica mosaic virus (PlAMV) family *Alphaflexiviridae*, and this was confirmed by genome analysis. PlAMV is much more common in roots than in bulbs (Chen et al. 2013). The genomic sequences of PlAMV were sequenced. The genomic size of all isolates is about 6102 nucleotides. Pairwise analyses confirm the close relationship between PlAMV and TVX. Very low identities are observed between the individual PlAMV isolates. These values are the lowest among different isolates of the same species of all potexviruses (Komatsu et al. 2008).

Asian and Oriental hybrid lilies are bulbous ornamental plants valued for their beautiful colour. After their flowering, but before the aging of the leaves, necrotic streaks were observed in the middle stem leaves of several plants. Mechanical inoculation of the collected leaf samples of these lilies on *N. benthamiana*, *Nicotiana glutinosa* and *Chenopodium quinoa* resulted in chlorotic or necrotic local lesions on

C. quinoa and systemic mosaic with necrotic spots, stripes or apical necrosis of *N. benthamiana*. Electron microscopy reveals potexvirus-like flexible particles (Hammond et al. 2015).

In the winter of 2013, in a greenhouse complex in Hungary, symptoms of necrosis of the veins of the leaves of *Lilium* sp., drying of the lower leaves and a gradual decrease and death of flower production were observed. Symptoms become more pronounced in darker conditions. Similar symptoms have been reported with *Lilium* sp. in Dutch greenhouses in 2010, and the pathogen was identified as PlAMV. The virus was first isolated from imported lily bulbs in 2012 (Pajtli et al. 2015). In the winter of 2013 and 2014, severe necrotic streaks were observed in the leaves of the middle stems on plants of lily hybrids in several greenhouses in Italy (Parrella et al. 2015). PlAMV was first discovered in Chile by asymptomatic lilium plants in 2013. Symptoms range from brown necrotic streaks associated with improper leaf chlorosis to brown streaks in the stem and necrosis in tepal tips (Vidal et al. 2016).

When lily plants become infected with PlAMV, necrotic streaks appear on the veins of the leaves, flower production decreases and the plants gradually die (Pajtli et al. 2015). The virus was found in lilies in Hungary and southern Italy (Pajtli et al. 2015; Parrella et al. 2015). Severe necrotic leaf stripes were observed on Asian hybrid lily plants (Asiatic 'Tiny Padhye') in the greenhouse of the Chinese Academy of Agricultural Sciences (Xu et al. 2017; Kimura et al. 1990).

3.2.6 Lily Virus X and Tulip Virus X

In 1980, Lily virus X (LVX) was described, a *Potexvirus* isolated from lilies that infects some dicotyledonous herbaceous species. LVX is considered potentially hazardous to lilies and has been introduced into Dutch certification schemes (Stone 1980; European and Mediterranean Plant Protection Organization 1993). LVX was partially characterised by nucleotide analysis (Memelink et al. 1990). LVX has not been reported in native or cultivated lilies, with the exception of those found in several varieties of lilies grown in Japan (Kimura et al. 1990).

Tulip virus X (TVX) was first isolated from tulips in Scotland and is now found in all tulip-growing areas in the Netherlands and other countries, including Japan and New Zealand. In the spring of 2011, tulips were found in Poland with symptoms typical of TVX infection. The leaves show chlorotic and white, elongated stripes and spots, which are typical symptoms of TVX infection. Lily virus X and Tulip virus X were confirmed as *Potexviruses* serologically with antiserum (Sochacki and Komorowska 2012).

3.2.7 Tulip Breaking Virus and Other Potyviruses of Lilies and Tulips

Tulips were first described as ornamental plants in 1575 by Carclus Clusius. They are bulbous plants that belong to the genus *Tulipa* (Jelitto et al. 1995). This genus has over 102 species with over 6700 varieties, most of which are hybrids of *Tulipa gesneriana* (WCSP 2020; Bryan 2005). Tulips are propagated vegetatively, while only plant lovers grow tulips from seeds (Hop 2018; Hertogh and le Nard 1993). Vegetative propagation aims to obtain a relatively uniform crop with a low degree of vegetative propagation, which makes this crop susceptible to various diseases, especially viruses. Tulip breaking virus is economically the most important viral disease in modern tulip breeding, while in the seventeenth century, such infected tulips were highly valued and became a symbol of status (Baardse and Krabbendam 1967). In the spring, when infected plants emerge, the leaves express symptoms like crooked or cigar shaped and are often smaller than normal. Some varieties show a purple-burgundy or light and dark green stripe on the abaxial side and near the edges of the leaves. In the case of white, cream and yellow colours, Tulip breaking virus infection causes shiny spots on the petals that are difficult to spot in the field. In the case of varieties with a green stigma, the infected plants become lighter, and initially the light green flowers become white. The change in colour intensifies with the progress of flowering (De Best et al. 2000).

Various tulip potiviruses have been identified, including Tulip breaking virus (TBV), Tulip band-breaking virus (TBBV), Tulip top-breaking virus (TTBV) that is a strain of Turnip mosaic virus (TuMV-TTB), Rembrandt tulip-breaking virus (ReTBV) and Lily mottle virus (LMoV) (De Best et al. 2000). It is thought that *Potyvirus* infection is the leading cause of tulip breaking (Dekker et al. 1993; Gleason et al. 2009).

Five major viruses that cause tulip breaking have been isolated from tulips and lilies. TBV, TTBV and ReTBV have been shown to be potyviruses by serological and molecular methods. The transmission of aphid viruses causes enormous economic damages annually in the flower bulb sector. Mineral oils and pyrethroids can be applied weekly during the growing season to reduce the transmission of the virus through the aphid vector in the flower bulbs. For this reason, the dynamics of the aphid population in the fields of tulips is observed during three growing seasons. The time of TBV spread is related to the dynamics of aphid population and meteorological conditions. TBV transmission usually begins in the spring, in April, while the first aphids are not discovered until May (de Kock et al. 2011).

Because tulips with flower break symptoms are similar to those induced by TBV, illustrated by Dutch Renaissance artists, the virus was believed to cause 'the oldest plant viral disease' (McKay and Warner 1933). This viral agent has also been shown to be transmitted by at least three species of aphids (McKay et al. 1929).

Primary TBV and LSV infections can be easily detected in the infected lily bulbs using ELISA. The test is reliable especially during the period from November to March, when a large number of bulb samples are tested for viruses (Van Schadewijk 1986).

TBV-li isolate from *Lilium asiatica* hybrid cv. Enchantment was isolated in Japan, which is closely related to TBV. It is believed that these two strains should be classified into a separate species different from the TBV strains in tulips (Yamaji et al. 2001).

3.2.8 Arabis Mosaic Virus (AMV)

Arabis mosaic virus (AMV) is a *Nepovirus* pathogenic to many plant species, including hosts from eight families of ornamental species of the genera *Arum, Camassia, Crocus, Dahlia, Dicentra, Dieffenbachia, Eryngium, Liatris, Lychnis, Muscari, Iris* and *Phlox*. The virus can be detected by indicator plant assay, electron microscopy, DAS-ELISA and RT PCR. It has also been identified in mixed infections with other viruses in naturally infected host plants (Samuitiene et al. 2008).

AMV causes necrotic mosaic on the leaves and ringspots on the bulbs of *Lilium tigrinum*. The initial symptoms of infected bulbs may be spongy roots and large necrotic areas on the cream-colored bulb scales. AMV manages to infect at a very high frequency about 70%, successive series of transplanted plants in soil contaminated with the virus. Under experimental conditions with thinned soil, it was found that an almost undetectable number of nematodes of the species *Xiphinema diversicaudatum* can very effectively infect lilies with AMV up to 70%. Comprehensive virus control measures applicable under the conditions of the manufacturers shall be used. In general, disinfection of soil with dichloropropene, dazomet, methyl bromide and other disinfectants is relatively effective. However, if untested contaminated material from previous sowings is introduced, this may increase the number of AMV-infected nematodes and lead to soil disinfection failure (Asjes and Segers 1983).

3.2.9 Tomato Spotted Wilt Virus (TSWV)

TSWV is the most widespread plant virus and has the most species of plant hosts. He is responsible for the numerous epidemics of vegetable and flower crops in different regions of the world, which cause great economic losses. The virus is a polyphagous and is one of the most dangerous viruses for farmers due to its high efficiency of infection transmission, strong mobility and activity of its vectors—thrips, its rapid variability and difficulties in controlling vectors. The virus is mainly controlled by following preventive measures and good plant protection integrated practices to eliminate the weed reservoirs of the virus in combination with measures to manage vector populations. For this reason, updating TSWV's plant species and host varieties is also very valuable information for producers to take adequate and timely measures. TSWV lists more than 1090 species of plant hosts belonging to 15 families of monocotyledonous plants and 70 families of dicotyledonous plants (Parrella et al. 2003).

3.2.10 Lily Yellow Mosaic Virus (LYMV)

Next-generation sequencing of small RNAs was performed on samples of lilies showing virus-like symptoms of leaf yellowing, torsion and brownish necrotic spots in order to identify the causative viral agents. Thus, CMV LSV and a hitherto unrecorded *Potyvirus*, tentatively called the Lily yellow mosaic virus (LYMV), were discovered. LYMV genomic RNA is composed of 9811 nt encoding a large polyprotein with a high degree of amino acid sequence identity to Thunberg fritillary mosaic virus (55%), Bean yellow mosaic virus (52%) and Clover yellow vein virus (51%) (Li et al. 2018).

3.3 Virus Incidence and Disease Control

The reliability and accuracy of epidemiological information are of great importance for the development of effective strategies for the control of plant viruses. Viruses of *Liliaceae* plants cause significant damage to plants by reducing marketable yields and reducing their life as cut flowers for a vase. It is extremely important to start with the identification of the causal agent of the disease and its spread. *Liliaceae* viruses have been causing diseases for at least two centuries, with LSV being considered the most common in commercial establishments, and the plants themselves tend to periodically express viral symptoms, some of which are masked at high ambient temperatures (McWhorter and Allen 1967). On the other hand, TBV causes visible symptoms in some varieties, thus making them unfit for sale. LSV mainly causes latent infections of lilies, which significantly complicates its monitoring and control (Asjes et al. 1973).

One of the strategies and mass programs for elimination of viruses on lilies is through the use of cultures of top meristem (Walkey and Webb 1968). About 0.1–0.3 mm meristematic tissue from the shoot tip of CMV-infected plants is used, which is virus-free (Mori and Hosokawa 1977).

To improve the production of more virus-free plants, a combination of chemotherapy with subsequent use of meristem cultures has been proposed (van Aartrijk et al. 1990). Thermotherapy has not been widely used to produce virus-free lily seedlings, although it dramatically reduces LSV and TBV titres when grown at temperatures above 30 °C (Cohen et al. 1985).

3.3.1 Production of Virus-Free Lilies

Most bulbous plants are propagated by bulbous scales. For this reason, bulb growers should use virus-free bulbs from meristem crops in their production by growing these young bulbs in sheltered conditions in greenhouses. Each bulb determined for propagation must be tested for the presence of viruses by PCR and ELISA. Virus-free lilies can be obtained by a combination of three techniques—meristem tissue, shoot tip tissue and thermotherapy (Nesi et al. 2009).

For the use of chemotherapeutics, the effect of ribavirin on the elimination of CMV, LSV and LMoV in lily plants in vitro cultivation was studied. The effect of ribavirin depends on the concentration used, the type of virus, the viral concentration in the plant tissue and the specific genotype of the host plant (Sochacki et al. 2012).

Bulb growers produce virus-free seedlings under protected conditions in greenhouses. After leaving the protected conditions, the bulbs find themselves in the field and are unprotected by various viral vectors such as soil nematodes and aphids. On the other hand, to reduce the risk of infection with mechanically transmitted viruses, contact with juice on surfaces of various agricultural objects such as knives, boxes, machines, clothing, etc. is prevented. Transmission of viruses can also occur during various pre-processes in agriculture such as washing and root treatment. For the proper production of healthy lilies for planting, growers must use plant material that has been cultivated in protected conditions and tested for viruses at the beginning of growing the plants. Testing of plants for the presence of viruses is necessary to know the phytosanitary condition of the plant stock, taking samples for testing from the correct tissue and at the exact stage of development of the bulb. ELISA and PCR are mainly used to detect the viruses (de Kock 2013).

One of the new alternative methods to block the replication of different potyviruses in plant is by introducing small interfering RNAs and inducing posttranscriptional gene silencing of essential viral genes like HC-Pro (Petrov et al. 2015).

3.3.2 Virus Vector Control

The spread of most viruses on lilies is carried out by above-ground aphids. Mixtures of mineral and pyrethroid insecticides are used to limit their spread, with the aim of limiting the population and vector density of infected aphids with CMV, TBV, LMoV and LSV. Mineral oil is the most effective component in mixtures, with 1-week spraying being more effective than 2-week spraying (Asjes 1991). Preliminary soil testing is required before sowing to reduce the risk of the presence of soil nematodes and fungi as vectors of various lily viruses. For this reason, the soil is disinfected and is sometimes effective for AMV (Asjes and Segers 1983). Virus-infected sap passes from bulb to bulb mechanically in damaged or injured parts of plants. This is a common phenomenon in the spread of PlAMV and TVX during bulb processing (de Kock 2013). Contaminated surfaces and contaminated water can lead to a high rate of PlAMV contamination. The water is disinfected by heating at 65 °C for 10 min. The surfaces of the boxes and machines can also be cleaned with hot water above 65 °C or with disinfectants.

Acknowledgements The work on this chapter was supported by the Bulgarian Science Fund under contract DNTS India 01/1.

References

Allen TC (1972) Lily symptomless virus. No. 96. In: Descriptions of plant viruses. Commonwealth Mycological Institute/Association of Applied Biologists, Kew

Allen TC (1980) Diagnosis of virus diseases of lilies. North Am Lily Soc Yearbook 33:41–43

Aravintharaj R, Balaji CG, Nagendran K, Priyanka R, Karthikeyan G (2017) First report of lily mottle virus on lily (*Lilium* spp.) in southern India. Virus Dis 28 222–223

Asjes CJ (1991) Control of air-borne field spread of tulip breaking virus, lily symptomless virus and lily virus X in lilies by mineral oils, synthetic pyrethroids, and a nematicide in the Netherlands. Neth J Plant Pathol 97(3):129–138
Asjes CJ (2000) Control of aphid-borne lily symptomless virus and lily mottle virus in Lilium in the Netherlands. Virus Res 71:23–32
Asjes CJ, Segers LC (1983) Incidence and control of necrotic leaf mosaic caused by Arabis mosaic virus in *Lilium tigrinum* splendens in the Netherlands. Phytopathol Z 196:115–126
Asjes CJ, de Vos NP, van Slogteren DHM (1973) Brown ring formation and streak mottle, two distinct syndromes in lilies associated with complex infections of lily symptomless virus and tulip breaking virus. Neth J Pi Pathol 79:23–35
Baardse AA, Krabbendam P (1967) Eenvoudig Leerboek Voor de Bloembollenteelt: Algemeen Gedeelte, Tulp, Hyacint, Narcis, Gladiool en Boliris. In: Leidraad Voor Het Land- en Tuinbouwonderwijs, 2nd edn. Tjeenk Willink, Zwolle
Baker CA, Zettler FW (1988) Viruses infecting wild and cultivated species in the Commelinaceae. Plant Dis 72:513–518
Benetti MP, Tomassoli L (1988) Cucumber mosaic virus in lilies in Italy. Acta Hortic 234:465–468
Berks AFLM (1976) Tulip breaking virus, the only cause of brown ring formation in Lilium Mid Century hybrid Enchantment. Acta Hortic 59:25–26
Berniak H, Kaminska M, Malinowski T (2010) Cucumber mosaic virus groups IA and II are represented among isolates from naturally infected lilies. Eur J Plant Pathol 127:305–309
Bickey RB (1963) Easter lilies. Sunshine State Agric Res Rptr 8:101
Blake MR, Wilson CR (1996) Significance of secondary infections with lily symptomless carlavirus to cut-flower production of *Lilium*. Ann Appl Biol 129:39–45
Boolittle SP, Wellman FL (1934) Commelina nudiflora, a monocotyledonous host of a celery mosaic in Florida. Phytopathology 24:48–61
Brierley P (1962) A lily ringspot virus from Georgia. Plant Bis Rptr 46:625–626
Brierley P, Smith FF (1944) Studies on lily virus diseases: the necrotic fleck complex in *Lilium longiflorum*. Phytopathology 34:529–555
Bryan JE (2005) Timber Press pocket guide to bulbs. Timber Press, Portland. ISBN 978-0-88192-725-2
Chang MU, Chung JD (1987) Studies on viruses isolated from Lilium spp. in Korea. Korean J. Plant Pathol 3:223–235
Chang CA, Pang J-H (1991) Isolation of cucumber mosaic virus from lily plants showing necrotic stripe symptoms. Plant Prot Bull Taiwan, ROC (Abstr.) 33:425
Chen CC, Jhang YL, Lin BY, Chiang FL, Cheng YH, Deng TC (2013) Serological reagent preparation and improvement of serological method to detect the Plantago asiatica mosaic virus in lily. J Taiwan Agric Res 62:268–279
Cohen D, Milne KS, Hyland MJ (1985) In vitro manipulation of virus concentrations in hybrid lilies. Acta Hortic 164:319–324
De Best ALIC, Zwart MJ, van Aartrijk J, van den Ende JE, Peeters JMM (2000) Ziekten en afwijkingen bij bolgewassen: *Liliaceae*, vol 1. Laboratorium voor Bloembollenonderzoek, Lisse
de Kock M (2013) Right time sheet sampling crucial for accurate test results (in Dutch). BloembollenVisie 274:24–25
de Kock M, Stijger CCMM, Pham KTK, Lemmers MEC, van Dam M (2011) Non-persistent TBV transmission in correlation to aphid population dynamics in tulip flower bulbs. Acta Hortic 901:191–197
Dekker EL, Derks AFLM, Asjes CJ, Lemmers MEC, Bol JF, Langeveld SA (1993) Characterization of potyviruses from tulip and lily which cause flower-breaking. J Gen Virol 74:881–887
Derks AFLM (1995a) Chapter 25. *Hippeastrum* (*Amaryllis*). In: Loebenstein G, Lawson RH, Brunt AA (eds) Virus and virus-like diseases of bulb and flower crops. Wiley/Balaban, Chichester, pp 293–297
Derks AFLM (1995b) Chapter 28. Lily. In: Loebenstein G, Lawson RH, Brunt AA (eds) Virus and virus-like diseases of bulb and flower crops. Wiley/Balaban, Chichester, pp 313–321

Derks AFLM, Asjes CJ (1975) Lily symptomless virus in tulip. Neth J Plant Pathol 81:14–21. https://doi.org/10.1007/BF02650324

Derks AFLM, Vink-Van den Abeele JL (1980) Purification of lily symptomless virus. Use and value of antisera against intact and pyrrolidone-degraded virus for testing lilies and tulips. Neth J Plant Pathol 86:239–250

Derks AFLM, Lemmers MEC, van Gemen BA (1994) Lily mottle virus in lilies: characterization, strains and its differentiation from Tulip breaking virus in tulips. Acta Hortic 377:281–288

Dijst G, Schneider JHM (1996) Flower bulbs diseases incited by *Rhizoctonia* species. In: Sneh B, Jabaji-Hare S, Neate S, Dijst G (eds) Rhizoctonia species: taxonomy, molecular biology, ecology, pathology and disease control. Kluwer Academic, Dordrecht, pp 279–288

European and Mediterranean Plant Protection Organization (1993) Certification scheme: pathogen-tested material of lily. Bull OEPP/EPPO 23:215–224

Fisher JR (2013) Identification of lily symptomless virus infecting *Lilium martagon* 'Pink Taurade' in Ohio. Plant health progress no. August, p PHP-2013-0814-01-BR

Fisher JR, Rutherford SC, Fleming NK (2013) Identification of lily mottle virus and lily symptomless virus co-infecting Asiatic lily in Ohio. Online Plant Health Prog 14(1). https://doi.org/10.1094/PHP-2013-1017-01-BR

Francki RIB, Mossop BW, Hatta T (1979) Cucumber mosaic virus. In: Descriptions of plant viruses. Commonwealth Mycological Institute/Association of Applied Biologists, Kew

Gleason ML, Daughtrey ML, Chase AR, Moorman GW, Mueller DS (2009) Diseases of herbaceous perennials. APS Press, St. Paul. ISBN 978-0-89054-374-0

Guterman CEF (1928) A preliminary report on mechanical transmission of the mosaic of *Lilium auratum*. Phytopathology 18:1025–1026

Hagita T (1989) Detection of cucumber mosaic virus and lily symptomless virus from bulb scales of Maximowicz's lily by enzyme-linked immunosorbent assay. Ann Phytopathol Soc Japan 55:344–348

Hagita T, Kodama F, Akai J (1989) The virus diseases of lily in Hokkaido. Ann Phytopathol Soc Japan 55:1–8

Hammond J, Bampi D, Reinsel MD (2015) First report of Plantago asiatica mosaic virus in imported Asiatic and oriental lilies (*Lilium* hybrids) in the United States. Plant Dis 99:292

Hertogh AA, le Nard M (1993) Tulipa. In: de Hertogh AA, le Nard M (eds) The physiology of flower bulbs: a comprehensive treatise on the physiology and utilization of ornamental flowering bulbous and tuberous plants. Elsevier, Amsterdam, pp 517–682. ISBN 978-0-444-87498-6

Hop M (2018) De Veredeling van bloembollen. Stichting Bollenacademie, Hillegom. ISBN 8717247013442

Jelitto L, Schacht W, Epp ME, Baumgardt JP, Fessler A (1995) Hardy herbaceous perennials. Timber Press, Portland. ISBN 978-0-88192-159-5

Jung HJ, Ueda S, Ryu KH, Lee SY, Choi JK (2000) A novel strain of Cucumber mosaic virus isolated from *Lilium longiflorum*. Plant Pathol J 16:306–311

Kimura S, Goto M, Saito N, Fujiwara Y (1990) Detection of lily virus X (LVX) from several lily cultivarieties grown in Japan. Res Bull Plant Prot Serv Japan 26:79–81

Komatsu K, Yamaji Y, Ozeki J, Hashimoto M, Kagiwada S, Takahashi S, Namba S (2008) Nucleotide sequence analysis of seven Japanese isolates of Plantago asiatica mosaic virus (PlAMV): a unique potexvirus with significantly high genomic and biological variability within the species. Arch Virol 153:193–198

Li Y, Jia A, Qiao Y, Xiang J, Zhang Y, Wang W (2018) Virome analysis of lily plants reveals a new potyvirus. Arch Virol 163(4):1079–1082

Lisa V, Gera A (1995) Chapter 40. Lisianthus. In: Loebenstein G, Lawson RH, Brunt AA (eds) Virus and virus-like diseases of bulb and flower crops. Wiley/Balaban, Chichester, pp 443–448

Loebenstein G, Lawson RH, Brunt AA (1995) Virus and virus-like diseases of bulb and flower crops. Wiley, Chichester

McKay MB, Warner MF (1933) Historical sketch of tulip mosaic or breaking, the oldest known plant virus disease. Natl Hortic Mag 12:179–216
McKay MB, Brierley P, Dykstra TP (1929) Tulip "breaking" is proved to be caused by mosaic infection. Yearbook Agric 1928:596–597
McWhorter FP (1937) A latent virus of lily. Science 86:179
McWhorter FP (1940) Separation of tulip 1 virus from lily-latent by cytological methods. Phytopathology (Abstr) 30:788
McWhorter FP, Allen TC (1967) Viruses. In: Kiplinger DC, Langhans RW (eds) Easter lilies. The New York and Ohio Lily Schools, Wooster, pp 111–118
Memelink J, van der Vlugt CIM, Linthorst HJM, Berks AFLM, Asjes CJ, Bol JF (1990) Homologies between the genomes of a carlavirus (lily symptomless virus) and a potexvirus (lily virus X) from lily plants. J Gen Virol 71:917–924
Mori K, Hosokawa D (1977) Localization of viruses in apical meristem and production of virus-free plants by means of meristem and tissue culture. Acta Hortic 78:389–396
Mowat WP, Stefanac Z (1974) Aphid-transmitted viruses from lilies in Britain. Ann Appl Biol 76:281–287
Nesi B, Trinchello D, Lazzereschi S, Grassotti A (2009) Production of lily symptomless virus-free plants by shoot meristem tip culture and in vitro thermotherapy. Hortic Sci 44(1):217–219
Pajtli E, Eke S, Palkovics L (2015) First report of the Plantago asiatica mosaic virus (PlAMV) incidence on *Lilium* sp. in Hungary. Plant Dis 99:1288
Parrella G, Gognalons P, Gebre-Selassi K, Vovlas C, Marchoux G (2003) An update of the host range of tomato spotted wilt virus. J Plant Pathol 85(4):227–264
Parrella G, Greco B, Pasqualini A, Nappo AG (2015) Plantago asiatica mosaic virus found in protected crops of lily hybrids in Southern Italy. Plant Dis 99:1289
Petrov N, Stoyanova M, Andonova R, Teneva A (2015) Induction of resistance to potato virus Y strain NTN in potato plants through RNAi. Biotechnol Biotechnol Equip 29(1):21–26
Price WC (1935) Classification of southern celery mosaic virus. Phytopathology 25:947–954
Price WC (1937) Classification of lily-mosaic virus. Phytopathology 27:561–569
Raabe RD (1975) Increased susceptibility of Easter lilies to Pythium root rot as a result of infection by necrotic fleck virus complex. Acta Hortic 47:91–97
Rizzo D, Stefani L, Paoli M, Lazzereschi S, Nesi B, Pecchioli S et al (2012) Occurrence of Lily mottle virus on Lilium in Italy. Plant Dis 96:771
Samuitiene M, Navalinskiene M (2008) Occurrence of Cucumber mosaic cucumovirus on ornamental plants in Lithuania. Zemdirbyste-Agric 95:135–143
Samuitiene M, Navalinskiene M, Jackeviciene E (2008) Arabis mosaic virus on ornamental plants. Biologia 54:264–268
Shahin A, Arens P, van Heusden S, Van Tuyl JM (2009) Conversion of molecular markers linked to *Fusarium* and virus resistance in Asiatic lily hybrids. Acta Hortic 836:131–136
Shippy W (1937) Factors affecting Easter lily flower production in Florida. Fla Agric Exp Sta Bull 312:1–19
Slate GL (1986) Diseases of garden lilies and their control. North Am Lily Soc Yearbook 39:74–87
Smith KM (1950) Some new virus diseases of ornamental plants. J Royal Hortic Soc 75:350–353
Sochacki D, Komorowska B (2012) First report of *Tulip virus X* on Tulip in Poland. Plant Dis 96:594
Sochacki D, Gabryszewska E, Kamińska M (2012) The application of ribavirin for elimination of viruses in lily plants growing in vitro. In: Bioforsk FOKUS 7(9). The 13th international symposium on virus diseases of ornamental plants, Norway, 24-29 June 2012
Stein A (1995) Chapter 24. Gladiolus. In: Loebenstein G, Lawson RH, Brunt AA (eds) Virus and virus-like diseases of bulb and flower crops. Wiley/Balaban, Chichester, pp 281–292
Stewart FC (1896) Two destructive lily diseases. In: New York state agricultural experiment station annual report, pp 520–524
Stone OM (1980) Two new potexviruses from monocotyledons. Acta Hortic 110:59–63

van Aartrijk J, Blom-Barnhoorn GJ, van der Linde PCG (1990) Lilies. In: Ammirato PV, Evans DA, Sharp WR, Bajaj YPS (eds) Handbook of plant cell culture, Ornamental species, vol 5. McGraw-Hill, New York, pp 535–576

van Os GJ (2003) Ecology and control of *Pythium* root rot in cultivation of bulbs (in Dutch). Gewasbescherming Jaargang 34(6):190–193

Van Schadewijk AR (1986) Detection of tulip breaking virus and lily symptomless virus in lily bulbs by means of ELISA. Acta Hortic 177:121–128

Verhoyen M, Horvat F (1972) The deformation of lily flowers: a serious disease provoked by a complex of viruses; methods for identification. Ann Phytopathol 4:311–323

Vidal AK, Camps R, Besoian XA (2016) First report of necrotic streaking of Asiatic lilies caused by Plantago asiatica mosaic virus in Chile. Plant Dis 100:1799

Walkey DGA, Webb MJW (1968) Virus in plant apical meristem. J Gen Virol 3:311–313

WCSP (2020) World checklist of selected plant families. Royal Botanic Gardens, Kew

Wellman FL (1934) Identification of celery virus, the cause of southern celery mosaic. Phytopathology 24:695–725

Wellman FL (1935) The host range of the southern celery-mosaic virus. Phytopathology 25:377–404

Xu LF, Ming J, Yuan SX (2017) First report of Plantago asiatica mosaic virus in lily hybrids in China. Plant Dis 101:263

Yamaji Y, Lu X, Kagiwada S, Ohshima K, Nama S (2001) Molecular evidence that a lily-infecting strain of Tulip breaking virus from Japan is a strain of Lily Mottle virus. Eur J Plant Pathol 107:833–837

Yang XM, Wang JH, Qu SP, Wang LH, Peng LC (2010) First report of Phytophthora blight of lily caused by *Phytophthora nicotianae* in China. Plant Dis 94(6):782

Zitter TA, Simons JN (1980) Management of viruses by alternation of vector efficiency and by cultural practices. Annu Rev Phytopathol 18:289–310

Current Status of Major Viruses, Viroids, and Phytoplasma Infecting Petunia: Diagnosis, Characterization, and Commercial Impact

4

Karmveer Kumar Gautam, Susheel Kumar, and Shri Krishna Raj

Abstract

Petunia (*Petunia hybrida*) is a widely cultivated and popular ornamental plant used in flower beds, hanging baskets, and window boxes and in gardens for its beautiful blooms of various colors throughout the world. Besides its floriculture importance, it is a model plant for genetic transformation research and is also reported as an important source of an anti-microbial agent. Its cultivation has been affected due to several viral, viroid, and phytoplasma pathogens of genera *Alfamovirus*, *Ilarvirus*, *Cucumovirus*, *Nepovirus*, *Fabavirus*, *Alphacarmovirus*, *Begomovirus*, *Pospiviroid*, *Macluravirus*, *Orthotospovirus*, *Tombusvirus*, *Tymovirus*, *Petuvirus*, and *Tobamovirus* (Sastry et al., Encylopedia of plant viruses and viroid. Springer, Berlin: 1731–1793, 2019; Gautam et al., Phytoparasitica 40:425–431, 2012) which cause considerable economic losses to the petunia industry. In this chapter, we described the current status of major viruses, viroids, and phytoplasma infecting petunia and their diagnosis, characterization, and commercial impact.

Keywords

Petunia hybrida · Viruses · Viroid · Phytoplasma

4.1 Introduction

Petunia (*Petunia hybrida* Hort.) belongs to the family Solanaceae. It is originated from America. Most of the varieties of *P. hybrida* are grown in gardens worldwide for their attractive beautiful blooms of various colors. Besides its floriculture

K. K. Gautam · S. Kumar · S. K. Raj (✉)
Plant Molecular Virology Laboratory, CSIR-National Botanical Research Institute, Lucknow, Uttar Pradesh, India

S. K. Raj et al. (eds.), *Virus Diseases of Ornamental Plants*,
https://doi.org/10.1007/978-981-16-3919-7_4

importance, it is an important source of an anti-microbial agent and also showed the mildest anti-oxidation activity (Gautam et al. 2012). Its leaves yield an important insecticide widely used against a broad range of insects. Petunias are also used as a genetic model. Several pathogens are commonly found in petunias, including bacteria, fungi, oomycetes, and viruses. The cultivation of petunia has been affected due to several viral/vioroid pathogens belonging family Bromoviridae: *Alfalfa mosaic virus* (AMV) *Tobacco streak virus* (TSV) *Cucumber mosaic virus* (CMV) *Tomato aspermy virus* (TAV), Secoviridae: *Arabis mosaic virus* (ArMV) *Tomato black ring virus* (TBRV) *Cherry leaf roll virus* (CLRV) *Tobacco ringspot virus* (TRSV), Secoviridae: *Petunia ringspot virus* (BBWV-1) *Broad bean wilt virus 2* (BBWV-2), Tombusviridae: *Calibrachoa mottle virus* (CMoV), Geminiviridae: *Chilli leaf curl virus* (ChiLCV) *Tomato yellow leaf curl virus* (TYLCV), Potyviridae: *Colombian datura virus* (CDV) or *Petunia flower mottle virus* (PFMoV) Potato virus Y (PVY) *Tobacco etch virus* (TEV) *Turnip mosaic virus* (TuMV) *Artichoke latent virus* (ArLV), Tospoviridae: *Impatiens necrotic spot orthotospovirus* (INSV) *Tomato spotted wilt orthotospovirus* (TSWV), Family: Tombusviridae: *Petunia asteroid mosaic virus* (PetAMV), Tymoviridae: *Petunia vein banding virus* (PetVBV), Caulimoviridae: *Petunia vein clearing virus* (PVCV), Virgaviridae: *Tobacco mild green mosaic virus* (TMGMV) *Tobacco mosaic virus* (TMV) *Tomato mosaic virus* (ToMV) *Turnip vein-clearing virus* (TVCV), Pospiviroidae: *Chrysanthemum stunt viroid* (CSVd) *Citrus exocortis viroid* (CEVd) *Potato spindle tuber viroid* (PSTVd) *Tomato chlorotic dwarf viroid* (TCDVd) and Phytoplasma: *Petunia flat stem phytoplasma* which cause considerable economic losses to the petunia. (Lesemann 1996; Mavric et al. 1996; Cohen and Sikron 1999; Alexandre et al. 2000a, b; Liu et al. 2003; Sabanadzovic et al. 2008, Gautam et al. 2012, Sastry et al. 2019). These viruses have been found as both single and mixed infections causing symptoms of mosaic, chlorosis, necrosis, leaf deformation, stunting, flower color breaking, and vein clearing (Lesemann 1996; Sanchez-Cuevas and Nameth 2002). The mode of transmission, characterized symptoms on petunia of respective pathogens and reported country are summarised in Table 4.1.

4.2 Major Viruses Infecting Petunia: Diagnosis and Characterization

4.2.1 Detection of Tobacco Mosaic Virus in Petunia and Tobacco by Light Microscopy Using a Simplified Inclusion Body Staining Technique

Tobacco mosaic virus (TMV) causes significant losses in many economically important crops. The virus caused flower break, mild mosaic, and brown, brittle stems in petunia and mosaic and mottling symptoms in tobacco leaves in northern Thailand. This study evaluated virus inclusion visualization by light microscopy (VIVLM) with virus confirmation by reverse transcription polymerase chain reaction

Table 4.1 Some important virus, viroid, and phytoplasma pathogens infecting petunia

Viruses infected on petunia	Symptoms	Reported from (country)	References
Genus, *Alfamovirus*; family, *Bromoviridae* (Transmitted by: aphid and mechanical sap-inoculation)			
Alfalfa mosaic virus (AMV)	Foliar ringspot and flower break on varieties with colored flowers	India, Germany, the UK, Lithuania, and New Zealand	Mali (1980), Lesemann (1996), Spence et al. (1996), Navalinskiene and Samuitiene (2008), Fletcher 2001
P Genus, *Ilarvirus*; family, *Bromoviridae* (The virus is transmitted by thrips vectors			
Tobacco streak virus (TSV)	Mosaic, leaf deformation, and stunting	Iran	Ghotbi (2006)
Genus, *Cucumovirus*; family, *Bromoviridae* (Transmitted by: aphid species and also by mechanical sap-inoculation)			
Cucumber mosaic virus (CMV)	Mild to severe mosaic and yellowing on leaves and color breaking on the flowers. Some petunia varieties exhibit necrosis of leaves and petioles; and apical growth was arrested resulting in a stunted appearance	Worldwide	Di Franco and Gallitelli (1985), Fletcher (1987), Srivastava et al. (1991), Lesemann (1996), Mavric et al. (1996), Lu (2005), Horvath et al. (2006), Gautam et al. (2012)
Tomato aspermy virus (TAV) or Chrysanthemum mild mottle virus; Chrysanthemum aspermy virus	–	Japan, Germany, Iran	Natsuaki et al. (1994), Lesemann (1996), Maddahian et al. (2017)
Genus, *Nepovirus*; family, *Secoviridae* (Transmitted by: nematode vectors and also by mechanical sap-inoculation)			
Arabis mosaic virus (ArMV)	Mottle or systemic chlorosis symptoms in leaves	UK and Italy	Brown (1986), Gerola et al. (1966)
Tomato black ring virus (TBRV)	Systemic chlorosis or veinal necrosis symptoms	England	Brown et al. (1989)
Cherry leaf roll virus (CLRV)	Mosaic and/or necrosis symptoms	–	Werner et al. (1997), von Bargen et al. (2009)
Tobacco ringspot virus (TRSV)	Ringspot symptoms on the foliage	Iran	Ghotbi and Shahraeen (2009)
Genus, *Fabavirus*; family, *Secoviridae* (Transmitted by: aphid species and also by mechanical sap-inoculation)			

(continued)

Table 4.1 (continued)

Viruses infected on petunia	Symptoms	Reported from (country)	References
Petunia ringspot virus (BBWV-1)	Ringspots and oak leaf patterns	Spain, Germany	Rubio (1959), Lesemann (1996)
Broad bean wilt virus 2 (BBWV-2)	Ringspots and oak leaf patterns	Germany	Lesemann (1996)
Genus, *Alphacarmovirus*; family, *Tombusviridae* (Transmitted by: aphid species and also by mechanical sap-inoculation)			
Calibrachoa mottle virus (CMoV)	Light mottling and chlorotic blotching	USA	Liu et al. (2003)
Genus, *Begomovirus*; family, *Geminiviridae* (Transmitted by: grafting)			
Chilli leaf curl virus (ChiLCV)	–	Oman and India	Al-Shehi et al. (2014), Nehra and Gaur (2015)
Tomato yellow leaf curl virus (TYLCV)	Leaf chlorosis and distortion, apical distortion, and swellings of the veins on the underside of the leaf; plants infected when young may not develop flowers	Israel	Sikron et al. (1995), Lesemann (1996)
Genus, *Potyvirus*; family, *Potyviridae* (Transmitted by: aphid species and also by mechanical sap-inoculation)			
Colombian datura virus (CDV) or *Petunia flower mottle virus* (PFMoV)	Severe flower mottle and chlorotic mottled leaves	Germany	Feldhoff et al. (1998)
Potato virus Y (PVY)	Leaf mosaic, yellow mottling, vein clearing, and distortion of leaves and stems. Color breaking in flowers was often observed, and the overall growth of the whole plant may be stunted	Worldwide	Sharp (1994), Lesemann (1996), Bellardi et al. (1996), Mavric et al. (1996), Boonham et al. (1999), Baracsi et al. (2006), Horvath et al. (2006), Spence et al. (1996), Bellardi et al. (1996)
Tobacco etch virus (TEV)	Streaks of pigment in white tissue of star-type flowers, with different degrees of pigment in the flowers developing at different times	USA	Hammond et al. (2011)
Turnip mosaic virus (TuMV)	Flower break chlorotic spots, mosaic, vein yellowing, and stunting	Germany	Lesemann (1996), Farzadfar et al. (2005)
Genus, *Macluravirus*; family, *Potyviridae* (Transmitted by: aphid species and also by mechanical sap-inoculation)			

(continued)

Table 4.1 (continued)

Viruses infected on petunia	Symptoms	Reported from (country)	References
Artichoke latent virus (ArLV)	Systemic mosaic and stunting	Italy	Di Franco and Gallitelli (1985)
Genus, *Orthotospovirus*; family, *Tospoviridae* (Transmitted by: thrips vector *Frankliniella occidentalis*)			
Impatiens necrotic spot virus (INSV)	Necrotic ringspots and/or necrotic flecking of the leaves	Worldwide	Daughtrey (1996), Daughtrey et al. (1997), Sanchez-Cuevas and Nameth (2002)
Tomato spotted wilt virus (TSWV)	Necrotic ringspots and/or necrotic flecking of the leaves	USA	Hausbeck et al. (1992), Daughtrey et al. (1997), Sanchez-Cuevas and Nameth (2002)
Genus, *Tombusvirus*; family, *Tombusviridae* (The virus is also transmissible by grafting)			
Petunia asteroid mosaic virus (PetAMV)	Yellow stellate spots on leaves, leaf distortion, and flower color breaking	Italy, Canada, Germany, Switzerland, the UK	Ambrosino et al. (1967), Lovisolo (1957)
Genus, *Tymovirus*; family, *Tymoviridae* (The virus is transmitted by beetle vectors)			
Petunia vein banding virus (PetVBV)	Chlorotic and necrotic spots and systemic vein-banding symptoms	Brazil	Alexandre et al. (2000a)
Genus, *Petuvirus*; family, *Caulimoviridae* (Transmitted by: grafting and mechanical)			
Petunia vein clearing virus (PVCV)	Chlorotic vein-clearing, with sunken leaf veins	Worldwide	Lesemann and Casper (1973) Lesemann (1996), Lockhart and Lesemann (1998), Gera et al. (2000), Zeidan et al. (2001), Harper et al. (2003), Richert-Poggeler et al. (2003), Richert-Poggeler and Lesemann (2007)
Genus, *Tobamovirus*; family, *Virgaviridae* (Transmitted by: mechanical sap-inoculation)			
Tobacco mild green mosaic virus (TMGMV)	Yellow mosaic of leaves, with some leaf distortion and vein mosaic, deformed flowers, and light color break of the petals	France and Israel	Parrella et al. (2006), Gera et al. (2007), Zeidan et al. (2008)
Tobacco mosaic virus (TMV)	Severe to mild mosaic in some petunia cultivars; symptoms	Worldwide in distribution	Lesemann (1996), Alexandre et al. (2000b), Spence et al.

(continued)

Table 4.1 (continued)

Viruses infected on petunia	Symptoms	Reported from (country)	References
	like veinal chlorosis, necrosis, mottling, and blistering with dark green areas were also noticed		(2001), Sanchez-Cuevas and Nameth (2002), Horvath et al. (2006), Chung et al. (2007), Kim et al. (2014)
Tomato mosaic virus (ToMV)	Leaf mottling, leaf distortion, vein-clearing, stunting, and flower break	Italy	Bellardi et al. (1996), Lesemann (1996)
Turnip vein-clearing virus (TVCV)	Leaf mottling and distortion, as well as discoloration or deformation of the petals	Northeastern Mississippi, the USA	Sabanadzovic et al. (2008)
Genus, *Pospiviroid*; family, *Pospiviroidae* (The viroid is mechanically sap transmissible)			
Chrysanthemum stunt viroid (CSVd)	Show mosaic on the malformed leaves and a generalized reduction in growth	The Netherlands	Verhoeven et al. (1998), Cervena et al. (2011), Matsushita (2013)
Citrus exocortis viroid (CEVd)		Australia	Van Brunschot et al. (2014)
Potato spindle tuber viroid (PSTVd)	–	Czech Republic	Mertelik et al.(2010), Cervena et al. (2011), Matousek et al. (2014), Matsushita and Tsuda (2014)
Tomato chlorotic dwarf viroid (TCDVd)	Symptomless	The Netherlands, Slovenia, Japan, Belgium, the USA, and the UK	Verhoeven et al. (2007), James et al. (2008), Virscek Marn and Mavric Plesko (2010), Cervena et al. (2011), Shiraishi et al. (2013), Van Bogaert et al. (2017)
Phytoplasma			
Petunia flat stem phytoplasma	Flattened stem with flower malformation or phyllody	Korea	Chung and Huh (2008)

(RT-PCR). VIVLM was accomplished by staining leaf epidermal tissue with toluidine blue O and basic fuchsine and detected *Tobamovirus* in all symptomatic samples. TMV infection was confirmed by RT-PCR; there was a 91% agreement between VIVLM and RT-PCR in *Tobamovirus* detection. Detection of other plant virus groups by this simplified virus inclusion visualization technique may be possible (Khamphirapaeng et al. 2017).

4.2.2 Characterization of a Petunia Strain of Turnip Vein-Clearing Virus

Virus-like symptoms were observed in February 2005 on Double Wave™ petunias (*Petunia* x *hybrida* Hort. Vilm.-Andr.) grown in a greenhouse in northeastern Mississippi. Electron microscope observations of leaf dips from symptomatic plants showed the presence of rod-shaped virus-like particles. Infected petunia samples reacted weakly with an immunostrip test to Tobacco mosaic virus (TMV), but not in ELISA using a polyclonal anti-TMV-c commercial kit. The dsRNA patterns from four infected petunias were identical and resembled those of tobamoviruses although they differed slightly in size from the replicative forms of TMV extracted from infected tobacco plants. Purified dsRNAs were used for random-primer cloning of the viral genome. A partial sequence of the viral 3′ end showed that the virus is a strain of Turnip vein-clearing virus (TVCV), reported also as Tobacco mosaic virus-crucifer strain, sharing 96% common nucleotides and 98% identical amino acids. To the best of our knowledge, TVCV has not been reported previously either from petunias or from Mississippi in general (Sabanadzovic et al. 2008).

4.2.3 Characterization of a New Nepovirus Causing a Leaf Mottling Disease in Petunia Hybrid

In 2012, *Petunia hybrida* cuttings that originated from South America and displayed virus-like symptoms were obtained from a commercial grower. The infected plants exhibited foliar interveinal chlorosis and mottling. The symptoms were distinct from those previously associated with viruses infecting petunia; therefore, it prompted further investigation. Here, we identified and molecularly characterized a new virus infecting petunia causing a leaf mottling disease and determined its taxonomic relationship to other viruses. This new virus was provisionally named Petunia chlorotic mottle virus (PCMoV) and showed close similarity to members of the genus *Nepovirus*, subgroup A. Icosahedral virus-like particles were isolated from *Petunia hybrida* cuttings with interveinal chlorotic mottling. The virus was transmitted by mechanical inoculation from infected to healthy *P. hybrida* and was found to contain a bipartite RNA genome of 7.6 and 3.8 kilobases. Full genomic sequence was obtained by high-throughput sequencing combined with RACE amplification of the 5′-termini of RNAs 1 and 2 and reverse transcription PCR amplification of the 3′-termini with oligo-dT and sequence specific primers. Based on particle morphology, genome organization, and phylogenetic analyses, it was concluded that the new virus is a member of the genus *Nepovirus*, subgroup A. This new virus causing a leaf mottling disease of petunia was provisionally named Petunia chlorotic mottle virus (PCMoV) (Bratsch et al. 2017).

4.2.4 Characterization of Petunia Flower Mottle Virus (PetFMV): A New *Potyvirus* Infecting *Petunia x hybrida*

With the introduction of cutting-grown *Petunia x hybrida* plants on the European market, a new potyvirus which showed no serological reaction with antisera against any other potyviruses infecting petunias was discovered. Infected leaves contained flexuous rod-shaped virus particles of 750–800 nm in length and inclusion bodies (pinwheel structures) typical for potyviruses in ultrathin leaf sections. The purified coat protein with a Mr. of approximately 36 kDa could be detected in Western immunoblots with a specific antibody to the coat protein of the petunia-infecting virus. The 3′ end of the viral genome encompassing the 3′ non-coding region, the coat protein gene, and part of the NIb gene was amplified from infected leaf material by IC/PCR using degenerate and specific primers. Sequences of PCR-generated cDNA clones were compared to other known sequences of potyviruses. Maximum homology of 56% was found in the 3′ non-coding region between the petunia isolate and other potyviruses. A maximum homology of 69% was found between the amino acid sequence of the coat protein of the petunia isolate and corresponding sequences of other potyviruses. These data indicate that the petunia-infecting virus is a previously undescribed potyvirus and the name petunia flower mottle virus (PetFMV) is suggested (Feldhoff et al. 1998).

4.2.5 First Report of *Calibrachoa Mottle Virus* Infecting Petunia

During the spring of 2003, petunia samples from Florida and California tested positive for CbMV by enzyme-linked immunosorbent assay (ELISA) and lateral flow immunoassay (ImmunoStrips). These samples also tested positive by carmovirus group-specific polymerase chain reaction (PCR) primers and by immunocapture PCR. All plant samples that tested positive for CbMV were symptomless except one symptomatic sample that also tested positive for *Tobacco mosaic virus*. From samples that tested positive for CbMV only, mechanical inoculations were made to *Chenopodium quinoa* at a USDA-ARS greenhouse in Salinas, CA. Representative single, local lesions were used to inoculate additional *C. quinoa* plants. The resulting local lesions from these inoculations were freeze-dried and further used as virus inoculum (CbMV petunia). Similar inoculum was made with CbMV isolated from *Calibrachoa* plants (CbMV calibrachoa). Virus-free *Petunia hybrida* cultivars Surfinia 'Baby Pink' and Surfinia 'Violet' (Jackson and Perkins Inc., Somis, CA) were mechanically inoculated with CbMV petunia and CbMV calibrachoa. Four weeks post-inoculation, all plants were tested using ELISA for the presence of CbMV. In greenhouse conditions, 14.3% of 'Baby Pink' plants were positive for CbMV petunia, whereas none were positive for CbMV calibrachoa. 'Violet' plants were 64.3 and 33.3% positive for CbMV petunia and CbMV calibrachoa, respectively. None of the positive plants expressed virus-like symptoms. Virus particles resembling those of CbMV were observed from infected petunia plants with transmission electron microscopy in leaf-dip preparations. This is the first report of CbMV infecting petunia (Liu et al. 2007).

4.2.6 *Petunia* Species as Virus Hosts and Characterization of *Potato Virus Y* (PVY) Strains Isolated from Petunias in Hungary

Symptomatic petunias collected from two different parts of Hungary were identified *Potato virus Y*, based on symptoms, serology, electron microscopy, and CP gene study. The analyzed region of the CP gene of the isolate from the 'Surfinia Revolution Purple' variety showed 98% homology with a PVY^{H}: PVY^{NTN} isolate (accession number m95491), while the isolate from the 'Bravo Rose' variety showed 97% homology with a PVY^{FR}: PVY^{N} isolate (accession number d00441). This is the first report on the natural occurrence of the NTN strain of *Potato virus Y* (PVY^{NTN}) on petunia in Hungary (Baracsi et al. 2006).

4.2.7 Biological and Molecular Characterization of *Cucumber Mosaic Virus* Infecting Petunia and Its Management

In 2012, Gautam and co-workers characterized *Cucumber mosaic virus* in petunia having severe mosaic, yellowing, and stunting symptoms growing at NBRI and in various gardens of Lucknow, India. They were detected based on host response assay, positive sap and aphid transmissions, successful establishment of Koch's postulate, the presence of 28 nm isometric cored virus particles during electron microscopic observations, positive RT-PCR amplification of the CP gene and RNA3 genome, highest sequence identities, and closest phylogenetic relationships with CMV strains of subgroup IB; the virus isolate causing mosaic disease in petunia has been identified as a member of CMV which belongs to subgroup IB (Gautam et al. 2012).

4.3 Viroids Infecting Petunia: Diagnosis and Characterization

4.3.1 First Report of Tomato Chlorotic Dwarf Viroid Isolated from Symptomless Petunia Plants (*Petunia* spp.) in Japan

A viroid was detected for the first time in symptomless petunia plants (*Petunia* spp.) and identified as *Tomato chlorotic dwarf viroid* (TCDVd) based on an analysis of the complete genomic sequence. These petunia plants are a likely source of inoculum for tomato or potato plants because TCDVd induces severe symptoms on these plants. The genomic sequence of this petunia isolate from Japan shared 100% identity with petunia isolates from the Netherlands and the United Kingdom and a tomato isolate from Japan. Phylogenetic analysis showed that all petunia isolates and the tomato isolate from Japan formed a monophyletic clade (Shiraishi et al. 2013).

4.3.2 Natural Infection of Petunia by Chrysanthemum Stunt Viroid

In 1995, the Dutch Inspection Service for Floriculture and Arboriculture (NAKB) detected a viroid by return-polyacrylamide gel electrophoresis (R-PAGE) analysis of nucleic acids extracted from a plant of *Petunia hybrida* Surfinia 'Purple' obtained from a commercial garden center. This infected plant was later sent to the Plant Protection Service in Wageningen for definitive identification of the viroid as the particular petunia plant showed mosaic on the malformed leaves and a generalized reduction in growth. A viroid that behaved similar to chrysanthemum stunt viroid (CSVd) during return-polyacrylamide gel electrophoresis has been detected in petunia. Subsequent transmission studies as well as reverse transcription polymerase chain reaction and sequence analysis showed that the viroid is indeed a strain of CSVd. As long as the viroid is absent from plants used for vegetative propagation, it appears not to pose a serious threat to petunia cultivation (Verhoeven et al. 1998).

4.4 Phytoplasma Infecting Petunia: Diagnosis and Characterization

4.4.1 Occurrence of Petunia Flattened Stem Caused by Phytoplasma

This study describes a phytoplasmal disease occurring in petunia leaves grown in the glasshouse of the National Horticultural Research Institute, Suwon, Korea. Abnormal growth like flattened stem with flower malformation or phyllody was observed from the plant. The DNA extracted from the diseased leaves was amplified using a universal primer pair of P1/P6 derived from the conserved 16S rRNA gene of *Mollicutes* giving the expected polymerase chain reaction (PCR) product of 1.5 kb. In the nested PCR assays, the expected DNA fragment of 1.1 kb was amplified with the specific primer pair R16F1/R16R1 that was designed on the basis of aster yellows (AY) phytoplasma 16S rDNA sequences. The 1.1 kb PCR products were cloned, and nucleotide sequences were determined, and the sequences of the cloned 168 rRNA gene were deposited in the GenBank database under the accession no. of EU267779. Analysis of the homology percent of the 168 rDNA of PFS-K showed the closest relationship with Hydrangea phyllody phytoplasma (AY265215), Brassica napus phytoplasma (EU123466), and AY phytoplasma CHRY (AY180956). Phytoplasma isolated from the diseased petunia was designated as Petunia flat stem phytoplasma Korean isolate (PFS-K) in this study. Flattened stem occurring in petunia was confirmed as infection of AY group of phytoplasma by determination of 16S rRNA gene sequences of phytoplasma and microscopic observation of phytoplasma bodies. This is the first report on the phytoplasmal disease in petunia in Korea (Chung BN and Huh KY, 2008).

4.5 Management of Viruses, Viroid, and Phytoplasma Infecting Petunia

Management of viral diseases is much more difficult than that of diseases caused by other pathogens (Verma et al. 2002) because the viral diseases have a complex disease cycle, efficient vector transmission, and no effective viricides. Integration of various approaches like the avoidance of sources of infection, control of vectors, cultural practices (conventional), and use of resistant host plants (non-conventional) has been used for the management of viral diseases of plants (Fig. 4.1).

4.5.1 Management of CMV in Petunia

Raj and their group's attempts were made to standardize transformation protocol of petunia and development of transgenic petunia plants using the coat protein gene of CMV for management of CMV infection in petunia. The results obtained out of the management study demonstrated successful standardization of transformation protocol and development of T0 generation transgenic petunia plants expressing and translating the coat protein gene of CMV. During the study, total 26 putative transgenic petunia plants were developed, among them 19 plants showed positive integration of coat protein gene of CMV when tested by PCR using CMV-CP gene-specific primers followed by hybridization with the radiolabeled probe of CMV CP. The transgenic petunia plants expressing the CMV-CP gene also showed

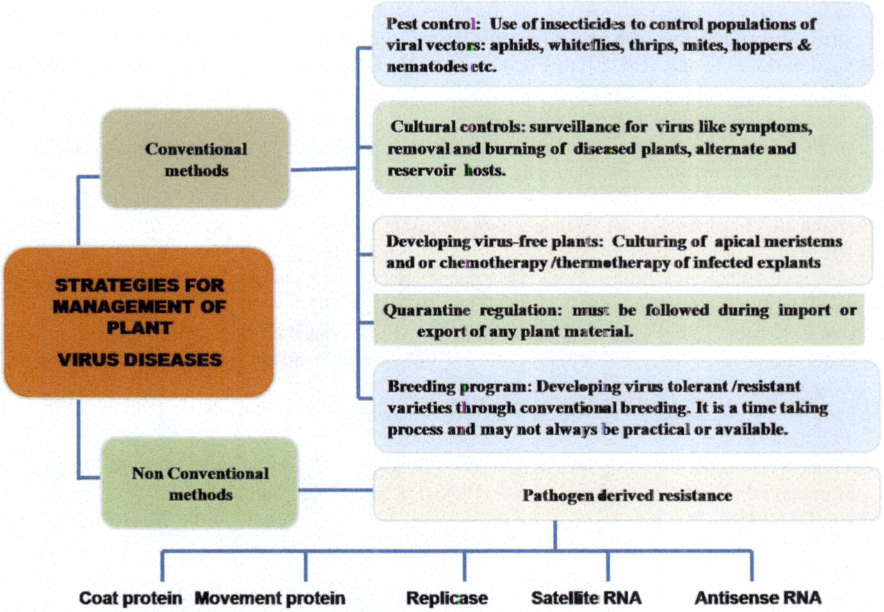

Fig. 4.1 Strategies of management of plant viruses

translation of the expected size (26 KDa) protein when evaluated by Western blot immunoassay using the antiserum of CMV. The expression of coat protein did not evoke any abnormal phenotype. The seed from these T0 generation transgenic petunia plants were collected to raise the T1 progeny and evaluation for resistance/ tolerance against challenge inoculations of CMV.

4.6 Conclusion

Petunia is a very popular ornamental plant in India as well as worldwide. Its popularity increases day by day due to their spectrum of colors. So it is a very economic ornamental plant for floriculture industries. But it is infected by so many viruses (Table 4.1). It becomes a day-by-day lovely host for viruses. Aphid, thrips, and vegetative propagation are important probable mode of virus transmissions in petunia. *Bromoviridae* and *Secoviridae* are the most common family of viruses infecting petunia. Conventional and non-conventional methods are used for controlling viruses. Coat protein-mediated pathogen-derived resistance method is very useful in controlling viruses infecting petunia. Composing data in this book chapter would be helpful for management of petunia for petunia grower as well as researchers.

References

Alexandre MAV, Duarte LML, Rivas EB, Barradas MM, Koenig R (2000a) Petunia vein banding virus: characterization of a new tymovirus from Petunia x hybrida. Plant Dis 84:739–742

Alexandre MAV, Soares RM, Rivas EB, Duarte LML, Chagas CM, Saunal H, Van Regenmortel MHV, Richtzehain LJ (2000b) Characterization of a strain of Tobacco mosaic virus from Petunia. J Phytopathol 148:601–607

Al-Shehi AA, Akhtar S, Khan AJ (2014) Identification of Chili leaf curl virus causing leaf curl disease of Petunia in Oman. Plant Dis 98:572

Ambrosino C, Appiano A, Rialdi G, Papa G, Redolfi R, Carrara M (1967) Caractterizzazione chimica e chimicofisica del Petunia asteroid mosaic virus (PAMV). Atti Acad Sci Fis Torino 101:301

Bratsch S, Lockhart B, Mollov D (2017) Characterization of a new Nepovirus causing a leaf mottling disease in petunia hybrida. Plant Dis 101:1017–1021

Baracsi E, Kriston E, Palkovics L, Toth EK, Takacs AP, Horvath J (2006) Petunia species as virus hosts and characterization of Potato virus Y (PVY) strains isolated from petunias in Hungary. Acta Hortic 722:271–276

Bellardi MG, Rubiesautonell C, Vicchi V (1996) Virus infection of Surfinia in Italy. Acta Hortic 432:306–310

Boonham N, Hims M, Barker I, Spence N (1999) Potato virus Y from petunia can cause symptoms of potato tuber necrotic ringspot disease (PTNRD). Eur J Plant Pathol 105:617–621

Brown DJ (1986) The transmission of two strains of arabis mosaic virus from England by populations of Xiphinema diversicaudatum (Nematoda: Dorylaimoidea) from ten countries. Rev Nemato l(9):83–87

Brown DJF, Murant AF, Trudgill DL (1989) Differences between isolates of the English serotype of tomato black ring virus in their transmissibility by an English population of Longidorus attenuatus (Nematoda: Dorylaimoidea). Rev Nématol 12:51–56

Cervena G, Necekalova J, Mikulkova H, Levkanicova Z, Mertelik J (2011) Viroids on petunia and other solanaceous crops in the Czech Republic. Acta Hortic 35(901):35–40
Chung BN, Kim JS, Cho JD, Cheong SR, Jeong MI (2007) Tobacco mosaic virus detected in vegetatively propagate Petunia hybrids 'Surfinia'. Plant Pathol J 23:34–36
Daughtrey ML (1996) Detection and identification of tospoviruses in greenhouses. Acta Hortic 431:90–98
Daughtrey ML, Jones RK, Moyer JW, Daub ME, Baker JR (1997) Tospoviruses strike the greenhouse industry: INSV has become a major pathogen on flower crops. Plant Dis 81 (11):1220–1230
Di Franco A, Gallitelli D (1985) Petunia hybrida Vilm., a natural host of cucumber mosaic virus CMV and artichoke latent virus ALV in Apulia. Inf Fitopatol 35:45–46
Farzadfar S, Ohshima K, Pourrahim R, Golnaraghi AR, Jalali S, Ahoonmanesh A (2005) Occurrence of Turnip mosaic virus on ornamental crops in Iran. Plant Pathol 54:261
Feldhoff A, Wetzel T, Peters D, Kellner R, Krczal G (1998) Characterization of Petunia flower mottle virus (PetFMV), a new potyvirus infecting Petunia x hybrida. Arch Virol 143:475–488
Fletcher JD (1987) New plant disease records in New Zealand: additional hosts of alfalfa mosaic virus and cucumber mosaic virus. N Z J Agric Res 30:505–506
Fletcher JD (2001) New hosts of Alfalfa mosaic virus, Cucumber mosaic virus, Potato virus Y, Soybean dwarf virus, and Tomato spotted wilt virus in New Zealand. N Z J Crop Hortic Sci 29:213–217
Gautam KK, Kumar S, Raj SK (2012) Molecular characterization of Cucumber mosaic virus strain causing severe mosaic disease on petunia in India. Phytoparasitica 40:425–431
Gera A, Sikron N, Cohen J, Zeidan M (2000) First report of Petunia vein clearing virus in Israel. Plant Dis 84:201
Gera A, Beckelman H, Lipsker Z, Maslenin L, Spiegel S, Ziedan M (2007) Identification of tobacco mild green mosaic virus in Torenia, Petunia and Calibrachoa. Olam Haperach Sept-Octo:58–60
Gerola FM, Bassi M, Belli GG (1966) Some observations on the shape and localization of different viruses in experimentally infected plants, and on the fine structure of the host cells IV. — Arabis mosaic virus in Petunia hybrida Hort. Hortic Caryologia Int J Cytol Cytosyst Cytogen 19:481–491
Ghotbi T (2006) First report on incidence of Tobacco streak virus (TSV) on ornamental plants in Iran. Iranian J Plant Pathol 42:159–160
Ghotbi T, Shahraeen N (2009) Natural incidence and infectivity level of three nepoviruses in ornamental crops in Iran. J Plant Breed Crop Sci 1(3):39–44
Hammond J, Jordan R, Vaira AM (2011) Virus diseases of ornamentals, in soils, plant growth and crop production. In: UNESCO-EOLSS Joint Committee (ed) Encyclopedia of Life Support Systems (EOLSS), Developed under the auspices of the UNESCO. Eolss Publishers, Oxford. http://www.eolss.net
Harper G, Richert-Poggeler KR, Hohn T, Hull R (2003) Detection of Petunia vein clearing virus: modal for the detection of episomal viruses with endogenous Pararetrovirus sequence counter parts. J Virol Methods 107:177–184
Hausbeck MK, Welliver RA, Derr MA, Gildow FE (1992) Tomato spotted wilt virus survey among green house ornamentals in Pennsylvania. Plant Dis 76:795–800
Horvath J, Baracsi E, Takacs A, Kazinczi G, Gaborjanyi R, Krajczinger R (2006) Virus infection of ornamental plants in Hungary. Cereal Res Commun 34:485–488
James T, Mulholland V, Jeffries C, Chard J (2008) First report of tomato chlorotic dwarf viroid infecting commercial petunia stocks in the United Kingdom. Plant Pathol 57:400
Khamphirapaeng P, Cheewangkoon R, McGovern RJ, Wong SM, To-Anun, C (2017) Detection of tobacco mosaic virus in petunia and tobacco cells using inclusion body staining. Int J Agric Technol 13(2):163–168
Kim BS, Ruhl G, Creswell T, Loesch-Fries LS (2014) Molecular identification of a Tobacco mosaic virus isolate from imported petunias. Plant Health Prog. https://doi.org/10.1094/PHP-BR-14-0018

Lesemann D-E (1996) Viruses recently detected in vegetatively propagated Petunia. Acta Hortic 432:88–95

Lesemann D-E, Casper R (1973) Electron microscopy of Petunia vein clearing virus, an isometric plant virus associated with specific inclusions in petunia cells. Phytopathology 63:1118–1124

Liu H, Sears JL, Bandla M, Harness AM, Kulemeka B (2003) First report of Calibrachoa mottle virus (CbMV) infecting petunia. Plant Dis 87:1538

Liu HY, Sears JL, Bandla M, Harness AM, Kulemeka B (2007) First Report of *Calibrachoa mottle virus* Infecting Petunia Disease Notes. https://doi.org/10.1094/PDIS.2003.87.12.1538A

Lockhart BEL, Lesemann D-E (1998) Occurrence of Petunia vein-clearing virus in the USA. Plant Dis 82:262

Lovisolo O (1957) Petunia: nuovo ospite naturale del virus del rachitismo cespuglioao del pomodoro. Boll Stn Patol Veg Roma Ser III 14:103–109

Lu YC (2005) Characterization of Cucumber mosaic virus isolated from petunia hybrid. Master Thesis, Graduate Institute of Biological Science National Kaohsiung Normal University. Kaohsiung. 69pp

Maddahian M, Massumi H, Heydarnejad J, Pour AH, Varsani A (2017) Characterization of Iranian Tomato aspermy virus isolates with a variant 2b gene sequence. Trop Plant Pathol 42:475–484

Mali VR (1980) A ringspot disease of petunia caused by alfalfa mosaic virus. Indian Phytopathol 33:212–216

Matousek J, Piernikarczyk RJJ, Dedic P, Mertelík J, Uhlířová K, Duraisamy G, Orctová L, Kloudová K, Ptáček J, Steger G (2014) Characterization of Potato spindle tuber viroid (PSTVd) incidence and new variants from ornamentals. Eur J Plant Pathol 138:93–101

Matsushita Y (2013) Chrysanthemum stunt viroid. JARQ 47:237–242

Matsushita Y, Tsuda S (2014) Distribution of Potato spindle tuber viroid in reproductive organs of petunia during its developmental stages. Phytopathology 104:964–969

Mavric I, Ravnikar M, Blatnik A (1996) Viruses infecting trailing petunias in Slovenia. Phytoparasitica 24:328–329

Mertelik J, Kloudova K, Cervena G, Necekalova J, Mikulkova H, Levkanicova Z, Dedic P, Ptacek J (2010) First report of Potato spindle tuber viroid (PSTVd) in Brugmansia spp., Solanum jasminoides, Solanum muricatum and Petunia spp. in the Czech Republic. Plant Pathol 59:392

Natsuaki T, Wang WQ, Okuda S (1994) Characterization of Chrysanthemum mild mottle virus isolates from sweet pepper (Capsicum annuum) and petunia (Petunia x hybrida). Bull Coll Agric Utsunomiya Univ 15:11–19

Navalinskiene M, Samuitiene M (2008) Natural infection of Alfalfa mosaic alfamovirus in Petunia hybrida. Bot Lithuanica 14:57–63

Nehra C, Gaur RK (2015) Molecular characterization of Chilli leaf curl viruses infecting new host plant Petunia hybrid in India. Virus Genes 50:58–62

Parrella G, Verdin E, Gognalons P, Marchoux G (2006) Detection and characterization of tobacco mild green mosaic virus (TMGMV) large type isolate from trailing petunia in France. Commun Agric Appl Biol Sci 71:1237–1244

Richert-Poggeler KR, Lesemann DE (2007) Petunia vein clearing virus. AAB Descriptions of Plant viruses. No. 417

Richert-Poggeler KR, Noreen F, Schwarzacher T, Harper G, Hohn T (2003) Induction of infectious petunia vein clearing (pararetro) virus from endogenous provirus in petunia. EMBO J 22:4836–4845

Rubio M (1959) Nota previa sobre un virus de tipo ‘ringspot’ encontrado en Petunia hybrida Vilm. Microbiol Esp 12:325–330

Sabanadzovic S, Abou Ghanem-Sabanadzovic N, Henn A, Lawrence A (2008) Characterization of a Petunia strain of Turnip vein-clearing virus. J Plant Pathol 90(3):505–509

Sastry KS, Mandal B, Hammond J, Scott SW, Briddon RW (2019) Encylopedia of plant viruses and viroid. Springer, Berlin, pp 1731–1793

Sanchez-Cuevas M-C, Nameth SGP (2002) Virus-associated diseases of double petunia: frequency and distribution in Ohio greenhouses. Hortic Science 37:543–546

Sharp C (1994) Potato virus YN (PVYN) in petunia Surfinia, Plant Disease Notice No 79. Ministry of Agriculture, Fisheries and Food, UK

Shiraishi T, Maejima K, Komatsu K, Hashimoto M, Okano Y, Kitazawa Y, Yamaji Y, Namba S (2013) First report of tomato chlorotic dwarf viroid isolated from symptomless petunia plants (Petunia spp.) in Japan. J Gen Plant Pathol 79:214–216

Sikron N, Cohen J, Shoval S, Gera A (1995) Virus diseases in Petunia. Phytoparasitica 23:273

Spence NJ, Barker I, O'Neill T, Bennison J, Wright D (1996) Viruses in trailing petunias, HDC Fact Sheet No. 04/96. Horticultural Development Council, UK

Spence NJ, Sealy I, Mills PR, Foster GD (2001) Characterization of a tobamovirus from trailing petunias. Eur J Plant Pathol 107:633–638

Srivastava A, Chandra G, Raizada RK, Srivastava KM, Singh BP (1991) Characterization of a virus causing necrosis in Petunia hybrida and establishment of its relationships using ELISA. Indian J Exp Biol 29:591–593

Van Bogaert N, Smagghe G, Maes M, De Backer M, De Jonghe K (2017) Phylogeny of five predominant pospiviroid species in Belgium. Eur J Plant Pathol 149:25–33

Van Brunschot SL, Persley DM, Roberts A (2014) Thomas JE. First report of pospiviroids infecting ornamental plants in Australia: Potato spindle tuber viroid in Solanum laxum (synonym S. jasminoides) and Citrus exocortis viroid in Petunia spp. New Dis Rep 29:3

Verhoeven JTJ, Arts MSJ, Owens RA, Roenhorst JW (1998) Natural infection of petunia by Chrysanthemum stunt viroid. Eur J Plant Pathol 104:383–386

Verhoeven JTJ, Jansen CCC, Werkman AW, Roenhorst JW (2007) First report of Tomato chlorotic dwarf viroid in Petunia hybrida from the United States of America. Plant Dis 91:324

Virscek Marn M, Mavric Plesko I (2010) First report of Tomato chlorotic dwarf viroid in Petunia spp in Slovenia. Plant Dis 94:1171

von Bargen S, Grubits E, Jalkanen R, Büttner C (2009) Cherry leaf roll virus – an emerging virus in Finland? Silva Fenn 43:727–738

Werner R, Mühlbach H-P, Büttner C (1997) Detection of cherry leaf roll nepovirus (CLRV) in birch, beech and petunia by immunocapture RT-PCR using a conserved primer pair. Eur J For Pathol 27:309–318

Zeidan M, Sikron N, Cohen J, Gera A (2001) Improved detection of Petunia vein clearing caulimovirus. Hortic Science 35:1279–1282

Zeidan M, Beckelman H, Holdengreber V, Maslenin L, Sobolov I, Katzir P, Tam Y, Spiegel S, Gera A (2008) Detection of Tobacco mild green mosaic virus in ornamental crops. In: Proceedings of the 12th international symposium and virus diseases of ornamental plants. International Society for Horticultural Sciences, Haarlem, 20–24, April 2008, p 52

5 Detection, Characterization, and Management of Potyviruses Infecting Gladiolus, an Important Plant for Floriculture Trade

Charanjeet Kaur and Shri Krishna Raj

Abstract

The chapter describes about the brief history and background, detection, characterization, and management of viruses infecting gladiolus. It also describes about the methods of elimination of potyviruses from infected gladiolus corm explants through in vitro chemotherapy, thermotherapy, and electrotherapy. In this chapter, the work published by researchers all over the world and by virology group at NBRI, Lucknow, has been summarized.

Keywords

Potyviruses · Gladiolus · Host range · Genome · Diagnosis · Management

5.1 Brief History and Background of Gladiolus

Gladiolus (*Gladiolus* sp., of family Iridaceae) is an integral component of the global floriculture industry, and it is one of the top six export flowers (Anonymous 1997). Gladiolus was first found in the late 1820s near the end of its range in KwaZulu-Natal. Species cultivated in Holland were given the name *G. natalensis*. Professor C. G. C. Reinwardt of Leyden distributed plants to growers under this name

C. Kaur
Genetics and Environmental Biology Laboratory, Department of Plant Science, M.J.P. Rohilkhand University, Bareilly, India

Plant Molecular Virology Laboratory, CSIR-National Botanical Research Institute, Lucknow, Uttar Pradesh, India

S. K. Raj (✉)
Plant Molecular Virology Laboratory, CSIR-National Botanical Research Institute, Lucknow, Uttar Pradesh, India

S. K. Raj et al. (eds.), *Virus Diseases of Ornamental Plants*,
https://doi.org/10.1007/978-981-16-3919-7_5

(blackherbals.com n.d.). The genus is found in two Mediterranean climate biomes: the Cape of Southern Africa and the Mediterranean Basin (Kew Science n.d.). Its origins are in Tropical and Southern Africa. Gladioli were introduced to Europe for the first time in the late seventeenth century. It is now widely used as a key component of the floriculture industry all over the world. *Gladiolus* species were first discovered in Asia Minor about 2000 years ago (Cantor and Tolety 2011). Gladiolus can be found almost anywhere throughout sub-Saharan Africa's grasslands, savannas, and woodlands, as well as in Arabia and Madagascar. There are several subspecies, the most prominent of which are two that are only found in tropical Africa. *Gladiolus* has a surprising range of distribution across Africa, which may be due to human activity.

5.2 Characteristics of Gladiolus

Gladiolus is an herbaceous plant with rounded, symmetrical corms encased in many layers of brownish, fibrous tunics. Their stems are usually unbranched, and they produce one to nine small, sword-shaped, longitudinally grooved leaves that are enclosed in a sheath. The lowest leaf is called a cataphyll. The cross section of the leaf blades may be plane or cruciform. Flowers in unmodified wild species range in size from very small to 40 mm across, with inflorescences bearing one too many flowers. In trade, the magnificent giant flower spikes are items that have evolved through hybridization and selection (Gladiolus n.d.). The flower spikes are wide present on one-sided bisexual flowers with two leathery, green bracts on each side.

Tepals are made up of sepals and petals that are almost similar in appearance. They form a tube-shaped structure when they are joined at the root. The largest tepal is the dorsal tepal, which arches over the three stamens. The tepals on the outside are smaller. The stamens that are funnel-shaped are attached to the base of the perianth (Gladiolus n.d.). The style has three filiform, spoon-shaped branches, each extending towards the apex.

The ovary is three-locular, with oblong or globose capsules containing numerous winged brown seeds that dehisce longitudinally. The unique pellet-like structure in the middle, which is the real seed without the fine coat, must be visible. This attribute is wrinkled and black in some seeds. These seeds will not germinate. Pink to reddish or light purple with white, contrasting markings, or white to cream or orange to red (Gladiolus n.d.), these flowers come in a variety of color. The reproductive structure of the flower varies greatly between varieties.

The variability exists concerning quantitative characters (length of androecium, filament, anther, style, stigma lobe, angle of filament to anthor, pollen size) and qualitative characters (pollen shape, anther color, androecium, and gynoecium color), observed after a thorough study on morphological characteristics of 35 varieties of *Gladiolus* in CSIR National Botanical Research Institute, Lucknow, India (Akansha et al. 2014).

5.3 Uses and Importance of Gladiolus

Gladiolus is a high-value, vegetatively propagated ornamental plant with a prominent position in the global flower industry (Benschop et al. 2010). The growth of large-flowered *Gladiolus* cultivars, which are now among the world's most popular cut flower crops, was aided by hybridization and breeding programs. It is great for both the garden and floral arrangements. Cut spikes have a good holding consistency since the florets open in series for a long time. During 2006, a total of 19,900 gladiolus stems were imported into the European market (excluding the Netherlands) for 0.52 USD per stem (Chandel and Deepika 2010) per period. Japan grew 82,760 cut gladiolus stems for 0.45 USD per stem and imported 28,800 stems for 0.27 USD per stem from the Netherlands and Taiwan. Singapore paid 0.44 USD and 0.61 USD for gladiolus stems imported from China and Malaysia, respectively (Anonymous 2006).

Gladiolus is the second most common cut flower in India, both in terms of area and yield (Ahmad et al. 2008). *Gladiolus* originated in tropical Africa and South Africa and now its widespread distribution across Asia may be due to human activity consisting winter blooming, summer blooming, Nanus hybrids, and modern summer-blooming gladiolus (The University of Arizona n.d.). It is grown on 11,660 area of land in the region, with an approximate cut flower production of 106 crores. Uttar Pradesh, Chhattisgarh, Haryana, West Bengal, Odisha, and Maharashtra are the major gladiolus-producing states in India, though they are also grown in Karnataka, Andhra Pradesh, Uttarakhand, and Sikkim. Uttar Pradesh is India's top gladiolus (National Horticultural Board n.d.), with nearly 200 acres of land under commercial gladiolus (Roy 2000).

5.4 Viruses Infecting Gladiolus

Gladiolus is a virus reservoir that survives in infected corms (small daughter corms), spreads vegetatively, and is transmitted by mobile vectors (Duraisamy et al. 2011). *Gladiolus* plants can be infected by a variety of viruses in natural climatic conditions (Table 5.1) (Fig. 5.1) (Stein 1995).

The symptoms on various gladiolus plants—leaves, inflorescence, florets, and corms of naturally infected gladiolus plants—are shown in Fig. 5.1.

Viruses are a significant issue for gladiolus since the plants are propagated each year from corms that contain a virus. Gladiolus is commonly infected by a variety of viruses, the most common of which are as follows: (a) aphid-transmitted *Potyvirus*, *Bean yellow mosaic virus* (BYMV), which causes overall stunting, color breaking, flower distortion, and decreased flower and cormel development (Dubey et al. 2009; Dorrigiv et al. 2013; Kaur et al. 2011); (b) *Cucumber mosaic virus* (CMV), aphid-transmitted *Cucumovirus*, which causes extreme mosaic in leaves, stunted plants, and color breaking in flowers (Raj et al. 2002; Torres et al. 2013; Dubey et al. 2008, 2010; Katoch et al. 2003); and (c) *Tobacco rattle virus* (TRV), a nematode-transmitted *Tobravirus*, which causes notched leaf blade margin symptoms (Bozarth

Table 5.1 Summary of major viruses and their symptoms on gladiolus (Shodhganga n.d.)

Virus	Family	Symptom on gladiolus
Bean yellow mosaic virus	*Potyviridae*	Plants exhibit overall stunting, color breaking, flower distortion, reduced flower, and cormel production
Ornithogalum mosaic virus	*Potyviridae*	Symptoms on leaves were mosaic, chlorotic spots on leaves, and floral deformations
Pea mosaic virus	*Potyviridae*	Plants exhibit overall stunting, color breaking, flower distortion, reduced flower, and cormel production
Clover yellow vein virus	*Potyviridae*	Mosaic symptoms
Cucumber mosaic virus	*Bromoviridae*	Major disease symptoms are mosaic, color breaking, stunting of spikes, and reduction in flower size
Tomato aspermy virus	*Bromoviridae*	Severe mosaic, necrotic dots along the mid-rib of leaves, and the drastic reduction in quality and quantity of the flowers
Tobacco streak virus	*Bromoviridae*	Stunting of flower spikes and occasionally mosaic, ringspots, and necrotic streaks on leaves
Tobacco rattle virus	*Virgaviridae*	Notched leaf blade margin symptoms
Tobacco mosaic virus	*Virgaviridae*	Mosaic symptoms on leaves
Tobacco ringspot virus	*Secoviridae*	Leaves on affected plants develop necrotic or chlorotic ringspot patterns. The flowers are not affected. Mild strains of the virus may cause mild chlorotic mottling
Arabis mosaic virus	*Secoviridae*	Mild mosaic or lack of symptoms
Tomato ringspot virus	*Secoviridae*	Stunting, flower break, ring patterns, and streaking of leaves
Artichoke Italian latent nepovirus	*Secoviridae*	Yellow rings and line pattern
Tomato black ring virus	*Secoviridae*	Mild or lack of visible symptoms
Broad bean wilt virus	*Secoviridae*	Flower color breaking, leaf mosaics, notched leaf, and dwarfing or lack of visible symptoms
Tomato spotted wilt virus	*Bunyaviridae*	Symptoms include leaf distortion; mottling of leaves; vein clearing; as well as wavy lines and/or concentric rings on foliage, flowers, and petioles
Impatiens necrotic spot virus	*Bunyaviridae*	Similar to *Tomato spotted wilt virus*
Cycas necrotic stunt virus	*Comoviridae*	Symptomless
Strawberry latent ringspot virus	Unassigned family	

and Corbett 1957); BYMV and CMV are very common plant viruses that have been found infecting gladiolus in a variety of locations around the world.

Other viruses reported in gladiolus from different countries are *Tomato ringspot virus* (TRSV) from the USA (Bozarth and Corbett 1957); *Tomato black ring virus*

Fig. 5.1 Symptoms on leaves, inflorescence, florets, and corms of naturally infected gladiolus plants (**a**, **c**, **e**, **g**) as compared with the healthy plants (**b**, **d**, **f**, **h**) (Kaur et al. 2015)

from Poland (Kamińska 1978); *Tomato spotted wilt virus* from Australia (Lee et al. 1979); *Tobacco mosaic virus* and *Tobacco ringspot virus* from Japan (Fukumoto et al. 1982); *Arabis mosaic virus*, *Strawberry latent ringspot virus*, *Tobacco ringspot virus*, and *Tobacco streak virus* from Italy (Bellardi and Marani 1985; Bellardi et al. 1986, 1987); *Tobacco rattle virus* (TRV) from Holland, Israel, Egypt, and Poland (Stein 1995); and *Clover yellow vein virus* (CYVV) from Korea (Park et al. 1998).

Gladiolus is a reservoir for a variety of viruses, but *Bean yellow mosaic virus* (BYMV) has the highest prevalence in gladioli worldwide, which is associated with mosaic symptoms on leaves, color breaking in flowers, and decreased vigor (Fry 1953). BYMV is a member of the *Potyvirus* genus, which causes serious diseases in many leguminous and ornamental plants (Shukla et al. 1994).

5.4.1 Potyviridae Family

Based on vector transmission, the family *Potyviridae* was initially divided into six genera: *Potyvirus*, *Rymovirus*, *Bymovirus*, *Macluravirus*, *Tritimovirus*, and *Ipomovirus* (Berger et al. 2005). There are currently eight genera recognized in the

Table 5.2 Current classification of the family *Potyviridae* (Shodhganga n.d.)

Genus	Type species	Genome	Vector	Species
Potyvirus	*Potato virus Y*	Monopartite	Aphids	146
Rymovirus	*Ryegrass mosaic virus*	Monopartite	Mites	3
Bymovirus	*Barley yellow mosaic virus*	Bipartite	Fungus	6
Macluravirus	*Maclura mosaic virus*	Monopartite	Aphids	6
Tritimovirus	*Wheat streak mosaic virus*	Monopartite	Mites	5
Ipomovirus	*Sweet potato mild mottle virus*	Monopartite	Whitefly	6
Brambyvirus	*Blackberry virus Y*	Monopartite	Unknown	1
Poacevirus	*Triticum mosaic virus*	Monopartite	Mites	2
Unassigned	*Spartina mottle virus, Rose yellow mosaic virus*	Monopartite	Unknown	2
				177

family (Adams et al. 2005) including four former and two new genera, *Brambyvirus* and *Poacevirus* (Table 5.2).

The genome structure, vector transmission, and genome sequence are used to differentiate these genera. Aphids spread potyviruses which are macluraviruses, while mites spread rymoviruses, tritimoviruses, and poaceviruses. Ipomoviruses and begomoviruses are transmitted by whiteflies and fungi, respectively. However, the vector for *Brambyvirus* transmission is unknown.

Species demarcation in the family *Potyviridae* has always been problematic (Van Regenmortel 2000). Traditional criteria to discriminate between species and isolates are predominantly based on serology and biological criteria such as host range, cross protection, and symptomatology (Shukla et al. 1994). Current approaches have adopted the use of the genomic composition of the virus, mainly the coat protein (CP) and the 3′ non-translated region (NTR) sequences (Van Regenmortel 2000). According to the ICTV ninth Report (Adams et al. 2005) throughout the family, species are distinguished by the criteria as follows:

1. Genome sequence relatedness: different species have CP aa sequence identity less than about 80% and nt sequence identity less than 76% either in the CP or over the whole genome; there are also differences in polyprotein cleavage sites.
2. Host range and key host reactions and lack of cross protection.
3. Different inclusion body morphology and antigenic properties: serological relatedness may help in distinguishing species.

Potyviruses are the largest and economically most important group of plant-infecting viruses (Shukla et al. 1994); currently, the genus *Potyvirus* contains about 146 recognized members (Adams et al. 2005) which represent more than 30% of all known plant viruses. Potyviruses are transmitted by aphids in a non-persistent manner (Shukla et al. 1994); some of them can also be transmitted through the seed of some of their hosts (Johansen et al. 1994). Furthermore, they are transmitted in infected plant material such as cuttings and tubers (Shukla et al. 1994).

The significance of potyviruses can be seen from the rising number of the potyvirus publications from the "Biological Abstracts" database which reflect the relevant rising number of potyviruses. During the decade 1969–1978, there were only 30 publications related to potyviruses concerning a total of 46 potyviruses of which 22 were well characterized, while the other 19 were tentative potyvirus species. The number of potyvirus papers noticeably increased during the following decade, 1979–1988, during which time 704 papers related to potyvirus were published. In 1987, 77% of the 630 known sense ssRNA plant viruses were recognized as potyviruses (Zaitlin and Hull 1987). During the next 18 years, a total of 1743 potyvirus-relevant papers were published, and the total number of potyviruses increased to 111 definitive species and 86 tentative species. Also, new viruses (Ciuffo et al. 2006; Turina et al. 2006), new strains (James et al. 2003; Tan et al. 2005), and new recombinants (Zhong et al. 2005; Martín et al. 2006) are appearing continuously.

It is not just the number of records of *Potyvirus* species that is increasing but also the hosts they are known to infect. Individual potyvirus may have restricted host ranges, but as a group, they flourish in a wide range of crops (Hollings and Brunt 1981a). Some potyviruses do have a relatively wide host range; Watermelon mosaic virus, for example, infects plants in 23 families of dicotyledonous species and occurs naturally in cucurbit and legume crops (Purcifull et al. 1984). The hosts of potyvirus listed include 503 plant species belonging to 59 plant families, most of which are economically important plants (Brunt et al. 1996).

5.4.2 Biological Properties of Potyviruses

5.4.2.1 Inclusion Body Formation

During infection, all members of the *Potyviridae* family produce cytoplasmic cylindrical inclusion (CI) bodies (Adams et al. 2005). The CI consists of a 70 kDa viral protein with ATPase and helicase activities. Nuclear inclusion bodies are co-crystals of two virally encoded proteins: NIa and NIb, which are found in equimolar quantities in certain potyviruses. The small nuclear inclusion (NIa) protein (49 kDa) is a polyprotein that contains VPg and proteinase. During certain potyvirus infections, amorphous inclusion bodies can be seen in the cytoplasm, and possibly other non-structural proteins are aggregations of the protein HC-Pro.

5.4.2.2 Host Range

Members of the *Potyvirus* family have a limited host range that is restricted to closely related genera. *Potato virus Y* (PVY) can infect species in the Chenopodiaceae, Commelinaceae, and Solanaceae families [cryptogram of PVY: R/1: 3.1/6: E/E: S/Ap]. There are members with even broader host ranges, such as the *Lettuce mosaic virus* (LMV) and *Bean yellow mosaic virus*, which can infect plants from 8 and 13 different families, respectively. Mechanical inoculation is a simple way to spread the virus to most hosts. Many viruses can be found all over the world. Seed transmission can help with distribution in some cases. Various species vectored members of the *Potyviridae* family. Aphid vectors of the genera *Potyvirus* and *Macluravirus* transmit in a non-persistent and non-circulative manner. For aphid transmission, a helper component and a specific CP aa triplet (i.e., DAG for some potyviruses) are needed. Eriophyid mites spread rymoviruses and tritimoviruses in a semi-persistent manner. Bymoviruses are spread by root-infecting vectors of the Plasmodiophorales order, which was formerly known as fungi but is now classified as Cercozoa. Whiteflies tend to be the carriers of ipomoviruses. *Brambyvirus* and *Poacevirus* transmission vectors are uncertain.

5.4.3 Genome Organization of Potyviruses

The genus *Potyvirus* comprises the largest group of plant RNA viruses, positive (Fig. 5.2) (Shukla et al. 1994). Total 168 potyviruses are being recognized by the International Committee on Taxonomy of Viruses (TAXONOMY n.d.); the species demarcation requirements for potyviruses are based on the complete sequence of the broad open reading frames (ORFs) with a nucleotide sequence identity of 76% and an amino acid sequence identity of 82% (Wylie et al. 2010).

Positive-sense RNA is linear and single-stranded in members of the *Potyviridae* family. They have a monopartite or bipartite genome with a total genome length of 8500–12,000 nucleotides. A poly(A) tract can be found at the 3′ terminus. A genome-linked protein is found at the 5′ terminus. Polyproteins are formed from RNA molecules, which are then transformed into a variety of structural proteins (Potyviridae n.d.). Seven genera (*Potyvirus*, *Macluravirus*, *Ipomovirus*, *Rymovirus*, *Tritimovirus*, *Brambyvirus*, and *Poacevirus*) have a monopartite genome containing only one RNA molecule, whereas only one genus (*Bymovirus*) has a bipartite genome containing two RNA molecules, RNA-1 and RNA-2 (Shukla et al. 1998). *Bymovirus* RNA-1 is similar to the C-terminal two-thirds of monopartite genomes and encodes proteins similar to P3, 6K1, CI, 6K2, NIa, NIb, and CP, while *Bymovirus* RNA-2 encodes a polyprotein that is processed into two proteins, P1 and P2. P1 is related to monopartite viruses' HC-Pro, while P2 is related to furoviruses' capsid read-through protein and is necessary for virus transmission by fungi. Bymovirus RNA-1 and RNA-2 both have a VPg connected to the 5′ terminal nucleotide, as well as a 5′ UTR, a 3′ UTR, and a poly-A tail. Rymoviruses and potyviruses share a high degree of sequence identity, so they belong in the genus *Potyvirus*, according to sequence analysis (Adams et al. 2005; Shukla et al. 1998).

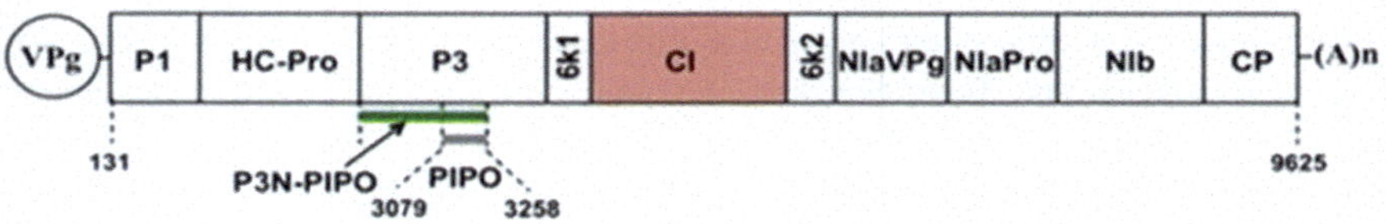

Fig. 5.2 Schematic representation of a potyvirus genome (Kaur et al. 2015)

The monopartite genera genome organization is very similar to one another. They have a 10 kb genome with a 5′ untranslated region (5′ UTR), a major single ORF, and a 3′ UTR region with a poly-A tail at the end. The main ORF encodes a large polyprotein that is co-translationally processed into ten functional proteins. The first protein (P1), helper component protein (HC-Pro), third protein (P3), 6K1, cylindrical inclusion protein (CI), 6K2, VPg (viral protein genome-linked), NIa-Pro (major protease of small nuclear inclusion protein—NIa), NIb (large nuclear inclusion protein), and CP (coat protein) are listed in descending order (5′–3′).

Potyvirus virions (Shukla et al. 1998) are flexuous filamentous particles that lack an envelope and measure 11–13 nm in diameter and 680–900 nm in length. A potyvirus genome consists up of a 9.7 kb positive-sense single-stranded RNA with a viral genome-linked protein (VPg) covalently bound to the 5′ end and a polyadenylated tail attached to the 3′ end. A potyvirus genome has a single ORF that encodes a polypeptide that is expected to be translated into 11 different proteins. The RNA polymerase slippage mechanism encodes a P3N-PIPO protein through the small ORF labeled "pretty interesting potyvirus ORF" (PIPO). PIPO is needed for the *Potyviridae* virus intercellular movement (Chung et al. 2008). The large ORF codes for a polyprotein, which is then self-cleaved into ten polypeptides by proteolysis at specific cleavage sites (Urcuqui-Inchima et al. 2001). A P1 protein, a helper-component proteinase (HC-Pro), a P3 protein, a 6 kDa peptide (6K1), a cylindrical inclusion protein, a second 6 kDa peptide (6K2), a VPg, a nuclear inclusion "a" (NIa) protease, a nuclear inclusion "b" protein (NIb), and a coat protein (CP) are among the proteins involved. The majority of potyvirus proteins has multiple roles and often collaborates during viral infection.

5.4.4 Bean Yellow Mosaic Virus (BYMV)

Bean yellow mosaic virus (BYMV) is the most common in gladiolus and is linked to mosaic symptoms on leaves, color breaking in flowers, and decreased plant vigor; BYMV is a member of the *Potyvirus* genus, which belongs to the *Potyviridae* family (Ward et al. 1995). BYMV particles, which are flexuous in form and range in size from 750 to 900 nm in length and 11 to 12 nm in width, cause characteristic cylindrical inclusions in host cells, and aphids spread the disease in a non-persistent manner (Milne 1988). It has a broad natural host range that includes both monocots and dicots (Bos 1970).

The BYMV genome contains 9.5 kb of single-stranded RNA (Kehoe et al. 2014). The genome is a set of genetic information that contains a 5′-UTR, a big open reading frame (ORF), a putatively small overlapping ORF, and an overlapping 3′-UTR zone (Wylie and Jones 2009). The ORFs are thought to code for ten mature gene products: P1 proteinase; helper part proteinase (HC-Pro); P3 protein; 6K1 protein; cylindrical inclusion (CI) protein; 6K2 protein; nuclear inclusion a (NIa) protein, a polyprotein that is further processed to the viral protein genome-linked (VPg) and the NIa proteinase (NIa-Pro); and nuclear inclusion b (NIb) protein, a polyprotein that is further processed to the viral protein genome-linked (VPg)

(Urcuqui-Inchima et al. 2001). A translation frame change in a slippery region of the P3 cistron in BYMV results in an additional gene product called P3N-PIPO (Chung et al. 2008). Based on coat protein (CP), gene sequences of the isolates of their original hosts suggested the presence of seven phylogenetic groupings of BYMV. Further, 40 complete BYMV genomes were analyzed which covered all of the virus's original hosts (except canna, due to the lack of complete genome) and nine phylogenetic classes (I–IX) proposed. BYMV general group (Wylie et al. 2010) came from a variety of places and was divided into three groups: I, II, and IV (Kehoe et al. 2014). Extensive recombinations were discovered among diverse a BYMV genome sequence (Kehoe et al. 2014) which is likely to have important evolutionary implications for the virus. Also, recombination plays a major role in BYMV host specialization; *Gladiolus* corms are used to spread the plant vegetatively. When a plant is infected with a virus at any point, it passes the virus on to the next generation, and the cycle continues for many generations through vegetative propagation through the mother stocks' dissemination. As a result, a variety's entire population may be infected with the same virus (APS Publications n.d.). Since *Gladiolus* is a vegetatively propagated crop, diagnosing virus (BYMV) infection in *Gladiolus* is critical for viral disease management. Virus diagnosis or identification is needed at several levels, including in the planting material, the current crop population, insect vectors, and the planting material created for the next crop, among others. Hence, proper diagnosis of the virus appears to be essential for successful virus disease control. Plant viruses have been detected using a variety of cytological, serological, and molecular techniques. Symptomatology, host range, and mode of transmission are the first signs of viral infection. With the emergence of viruses, techniques for purification, and visualization of morphology, as well as plant identification, viruses became more available (Wylie et al. 2008; Francki et al. 1985).

5.5 Diagnosis of Potyviruses

In plants, the diagnosis of viral infection in various parts of the plant, especially in propagating material, has always been emphasized because it indirectly aids in the control of the infection. Immunodiagnostics, molecular biology has provided methods for detecting viruses, such as the use of DNA/RNA probes, the polymerase chain reaction, etc.

5.5.1 Serological Detection of Potyviruses in Gladiolus

5.5.1.1 By Western Blot Immunoassay

To detect the potyvirus in fection in naturally infected gladiolus and experimentally inoculated gladiolus, *V. faba* and *P. peruviana* by sap, the Western blot immunoassay was performed using antiserum of three potyviruses: Narcissus potyvirus, *Sugarcane mosaic virus*, and *Bean yellow mosaic virus*. The proteins extracted from crude saps and blotted on membrane from SDS-PAGE were treated with

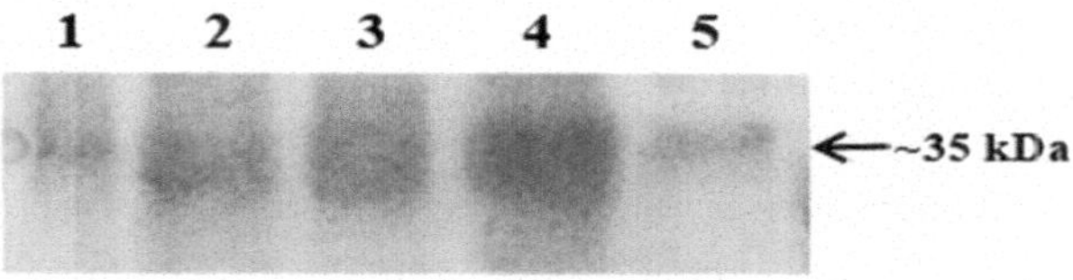

Fig. 5.3 Western blot immunoassay using BYMV antiserum showing positive bands of ~35 kDa in naturally infected gladiolus and experimentally inoculated gladiolus, *V. faba* and *P. peruviana*, similar to BYMV-infected *V. faba*, taken as a positive control (Charanjeet 2016)

antiserum of narcissus potyvirus, ScMV and BYMV. Western blot immunoassay using antiserum of BYMV resulted in the expected ~35 kDa bands in naturally infected gladiolus and experimentally inoculated gladiolus, *V. faba* and *P. peruviana* plants, which were similar to BYMV-infected *V. faba* taken as a positive control (Fig. 5.3) (Charanjeet 2016).

These results of Western blot immunoassay provided proper evidences for the presence of BYMV in naturally infected gladiolus as well as experimentally inoculated gladiolus and *V. faba* and *P. peruviana* plants. Presence of BYMV in *V. faba* and *P. peruviana* plants also suggested that both the plants may serve as the alternate hosts of the BYMV. *V. faba* and *P. peruviana* have already been reported as the susceptible host of BYMV from India (Charanjeet 2016).

5.5.1.2 By DAS-ELISA Tests

The viruses are tested by using a double-antibody sandwich (DAS-ELISA) including negative and positive controls. For many reasons, serological relationships between different potyviruses using polyclonal antibodies are complicated. The difficulties with the serology of members of this family are due to intrinsic complications associated with potyvirus coat proteins and particles, rather than issues with serological techniques (Shukla and Ward 1989). The majority of definitive members are serologically linked to at least one other member of the group and, in certain cases, several others. Antisera of the same virus prepared under different conditions (laboratories, dissociated CP vs. intact virions, immunization procedures) may have very different specificities. Strains of the same species may have very different serological affinities (Shukla and Ward 1989; Shukla et al. 1998).

Potyvirus serology has a well-established basis. The CP gene's N- and C-termini are surface-located, with the N-terminus being the most variable and immunodominant site. As a result of the epitopes found in this area, workers produce virus-specific antibodies. CP's N- and C-termini degrade quickly during purification and storage. Because the CP core region of different potyviruses is conserved, antibodies that can detect a wide range of potyviruses can be generated (Shukla and Ward 1989; Shukla et al. 1998). Antibodies and monoclonal antibodies (pAbs and mAbs) against the majority of economically significant potyviruses are currently available in the laboratory and commercially. Even though most genus members are serologically related to at least one other member of the group, and in several cases to several others, the predicted serological relationships between several associated pairs have yet to be observed (Hollings and Brunt 1981b). For instance, *Bean yellow mosaic virus* (BYMV) is serologically similar to the *Lettuce mosaic virus* (LMV) and *Bean common mosaic virus* (BCMV) (APS Publications n.d.), but LMV and BCMV have no known serological relationship. Since serology represents protein structure, the complexity of potyvirus serology can only be overcome through a detailed understanding of their coat-protein structure variation. Serological methods, especially mAbs, are commonly used in the diagnosis of potyviruses in general (Van der Vlugt et al. 1999; Koch and Salomon 1994; Kantrong and Sako 1993, b; Desbiez et al. 2002; Balamurali Krishnan et al. 2002; Crosslin et al. 2005; Hammond and Hammond 2003; Karyeija et al. 2000; Ounouna et al. 2002; Villamor et al. 2003; Mink and Silbernagel 1992; Mink et al. 1999).

5.5.2 Detection of Potyvirus by Reverse Transcription Polymerase Chain Reaction

RT-PCR-based methods for the detection and identification of potyviruses rely on degenerate primers designed to conserved genome sequences to unique primers designed from recognized sequences (Bateson and Dale 1995; Colinet and Kummert 1993). Since most published potyvirus sequences come from the 3′ region of the genome, universal primers for identifying potyviruses have mostly been based on conserved sequences in the CP gene, such as the WCIEN box or QMKAA motif (Gibbs and Mackenzie 1997; Chen and Adams 2001). The consensus motif (GNNSGQPSTVVDN) in the NIb gene has recently been discovered to be highly conserved among *Potyviridae* members (Gibbs and Mackenzie 1997). Several members of the family have forward degenerate primers corresponding to the GNNSGQP sequence of this motif.

5.5.2.1 Detection of Potyvirus in Gladiolus Plant Parts by RT-PCR

To detect potyvirus in different plant parts of gladiolus, total RNA was extracted from leaf, stem, corm, bract, and petal tissues of naturally infected gladiolus showing severe mosaic symptoms. RT-PCR was performed using potyvirus coat protein gene-specific primers (9502/CPUP), and during electrophoresis, the successful amplification of expected ~800 bp amplicons of potyvirus CP gene in leaf, stem, corm, bract, and petal samples of infected gladiolus was obtained which were similar to the amplicon obtained in positive control (BYMV-infected gladiolus leaf tissue). However, there was no amplification in the healthy control (Fig. 5.4).

The potyvirus was detected from all tested plant parts: leaf, stem, corm, bract, and petal of gladiolus. The presence of specific band indicates that potyvirus was present in whole gladiolus plant; however, the virus load in corms was less (Charanjeet 2016).

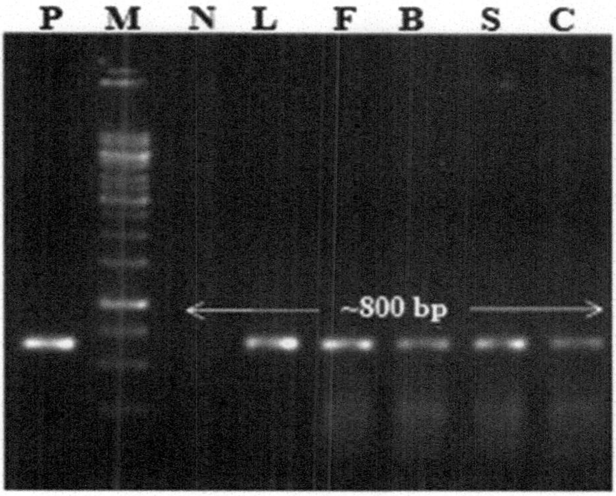

Fig. 5.4 Agarose gel electrophoresis of RT-PCR products of various gladiolus plants parts. *P* BYMV positive control (infected gladiolus), *M* 1 kb DNA ladder marker, *N* healthy gladiolus leaf, *L* leaf, *F* flower petal, *B* bracts, *S* stem, and *C* corms of infected gladiolus (Charanjeet 2016)

5.5.3 Nucleic Acid Spot Hybridization (NASH) Tests

Dot blot and tissue print are the most commonly used hybridization techniques. The viral RNA is immobilized onto a nylon membrane and then hybridized with labeled cDNA probes synthesized using RT-PCR in conventional procedures. For the rapid detection and identification of six potyviruses, a modified hybridization technique called reverse dot blot hybridization had been developed. In this process, DIG-labeled RT-PCR products amplified by potyvirus degenerate primers were hybridized with cDNA probes synthesized by RT-PCR with species-specific primers immobilized onto a nylon membrane. This technique is similar to a microarray-based approach for diagnosing potato RNA viruses developed (Boonham et al. 2003).

5.5.3.1 Detection of Potyvirus in Gladiolus by NASH Test

Though the RT-PCR is sensitive and reliable method for detection of viruses, however, it has some limitations that many samples cannot be handled at a time; therefore, DNA probe-based detection method was developed, and nucleic acid spot hybridization (NASH) assay was standardized, which is cost effective and can detect hundreds of samples at a time. Since the coat protein (CP) region is highly conserved among potyviruses, therefore, a clone of the CP gene region of BYMV gladiolus isolate (JF682236) was chosen, and the radiolabeled nucleic acid probe was generated to detect the BYMV in various naturally infected samples of gladiolus and *V. faba*, experimentally inoculated hosts of BYMV infecting gladiolus, and viruliferous aphids.

For detection of BYMV, approximately 1.0μg of total nucleic acids isolated from 12 samples each of naturally infected gladiolus and *V. faba* were blotted onto nylon membrane and allowed to hybridize with the probe prepared from CP of BYMV. The BYMV was detected in naturally infected gladiolus and *V. faba* samples; however, the intensity of signal from samples varied in each sample as compared

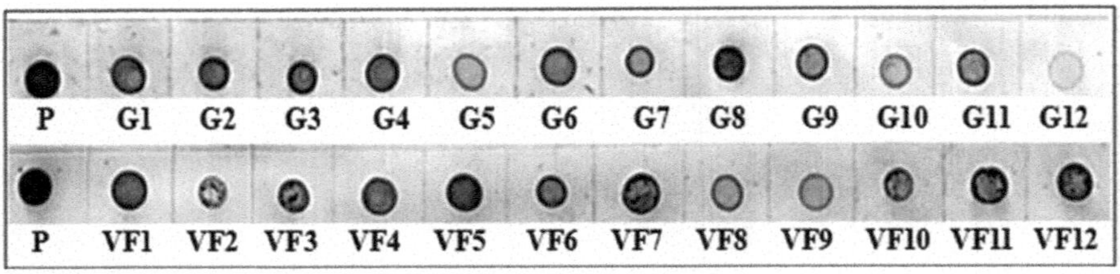

Fig. 5.5 Detection of BYMV by NASH method using probe developed from coat protein region of BYMV gladiolus (JF682236) showing positive signals of hybridization in RNA samples of all symptomatic gladiolus (G1–G12) and *V. faba* (VF1–VF12) and as in positive controls (P) (Charanjeet 2016)

to the positive control (clone of BYMV-CP) indicating variable viral load in tested samples. The strong signals in positive control (clone of BYMV-CP) indicated the authenticity of NASH (Fig. 5.5) (Charanjeet 2016).

5.6 Characterization of Potyviruses

Viruses causing infections lead to the damage of crops; to address the issues, molecular characterization of potyvirus (BYMV) is done. Molecular characterization can be done by sequencing the genome of the viruses. The infection-causing (target) region has to be amplified with primers using RT-PCR (Singh 2008). The sequencing of clones is done by BLAST analysis, and the highest sequence homology from the database will confirm the relation of plant viral infection of gladiolus with the virus.

5.6.1 Biological Characterization

Symptomatic samples with mosaic-like symptoms on leaves and sepals, as well as color breaking in flowers, are to be collected for the analysis. Gladiolus plants are then tested using DAC-ELISA using *Bean yellow mosaic virus* polyclonal antibodies (BYMV) (Reddy et al. 2019). The virus-positive gladiolus plant samples are mechanically transmitted for further virus inoculation. Maceration of symptomatic gladiolus leaf tissue in potassium phosphate buffer with a sterile pestle and mortar on ice was done. The sap is further crushed, filtered, and inoculated. For symptom speech, the inoculated plants should be held under insect-proof glass. For 30 days after infection, the production of local and systemic symptoms on host plants has to be monitored (Singh 2008).

During mechanical inoculations, using crude sap obtained from leaf tissue of diseased gladiolus (showing severe mosaic symptoms) induced necrotic local lesions on leaves of *C. amaranticolor*, *Vicia faba*, and *V. radiata* at 7–10 dpi. It also induced

Fig. 5.6 Sap inoculation of gladiolus potyvirus inducing necrotic local lesion at 10 dpi on *C. amaranticolor* (**a**), *V. faba* (**b**), and *V. radiata* (**c**) and systemic mosaic symptoms on gladiolus (**g**), *V. faba* (**d**), *P. hybrida* (**e**), and *C. annum* (**f**) at 30 dpi (Charanjeet 2016)

systemic mosaic symptoms on *V. faba*, *Petunia hybrida*, and *Capsicum annum* at 25–30 dpi (Fig. 5.6).

These results indicated that the disease was mechanically transmissible to *C. amaranticolor*, *V. faba*, *V. radiata*, *P. hybrida*, and *C. annum* species. These results are in accordance to the transmission and host range studies of a BYMV strain infecting gladiolus described earlier from India (Charanjeet 2016).

5.6.2 Electron Microscopy for Virus Morphology

The virus has to be partly purified from an infected gladiolus sample with mosaic symptoms that has been obtained and processed so that it can be seen under an electron microscope for the virus morphology. The partially purified potyvirus preparations of infected gladiolus leaves were observed under transmission electron microscope (TEM). During TEM study, the flexuous rod-shaped virus particles of size 720 × 11 nm were observed in the partially purified virus preparation. The size and shape of the virus particles were similar to those reported earlier for potyviruses, indicating the virus infecting gladiolus to be a potyvirus (Charanjeet 2016).

5.6.3 Molecular Characterization of Potyviruses in Gladiolus

For molecular characterization of potyvirus in *Gladiolus*, total genomic RNAs from infected gladiolus samples were extracted, and the first strand of viral cDNA has been synthesized using degenerate downstream primer (Pot I) of family *Potyviridae*.

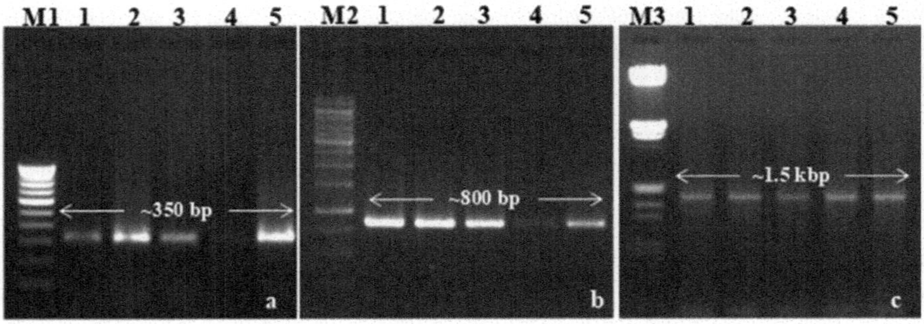

Fig. 5.7 1.5% Agarose gel electrophoresis of RT-PCR products of potyvirus-infected gladiolus samples (1–5) using three different primers. M1 = 100 bp DNA ladder marker, M2 = 1 kb DNA ladder marker; M3 = λ DNA *Eco*RI/*Hind*III marker, (**a**) RT-PCR product with primers pot MJ I/II, (**b**) RT-PCR product with primers pot CPUP/9502, and (**c**)RT-PCR product with primers pot I/II

Then RT-PCRs were performed with cDNA templates using three sets of primer pairs of conserved core CP region of potyviruses (Pot MJI/II), complete CP (Pot 9502/CPUP), and partial NIb to 3′-UTR region (Pot I/II).

The RT-PCR products were electrophoresed in 1.5% agarose gel, which resulted in positive amplicons of ~350 bp, ~800 bp, and ~1.5 kb band probably of the core CP, complete CP, and partial NIb to 3′-UTR regions, respectively, in all five infected gladiolus samples (Fig. 5.7). The obtained amplicons of ~350 bp, ~800 bp, and ~1.5 kb were eluted, cloned, and sequenced. The sequence data obtained, of all three amplified regions understudy, were then analyzed through BLASTn program for comparing the complete genome sequence of BYMV to all available sequences in GenBank (Altschul et al. 1990). BLASTn analysis resulted in 85–97% nucleotide identities with BYMV isolates reported worldwide; hence the sequences were identified as isolates of BYMV, and the sequence data were submitted in NCBI GenBank under the accessions EU249374 (330 bp), JF682236 (681 bp), and JQ686721 (1549 bp), respectively, as BYMV isolates from gladiolus. Multiple sequence alignment was used to confirm the virus identity as *Bean yellow mosaic virus* (BYMV) (Sharma et al. 2015).

Further, the entire genome of BYMV infecting gladiolus was amplified by RT-PCR using primer pairs that overlapped the viral genome. The PCR-amplified fragments are then cloned, sequenced, and assembled using various bioinformatics programs, and the full genome sequence data was submitted to GenBank under accession KM114059 (*G. dalenii* cv. Sylvia: CK-GL2, 9532 nt).

The complete genome sequence of BYMV infecting *G. dalenii* cv. Sylvia was 9532 nt (CK-GL2, Acc: KM114059) including poly(A) tail. The CK-GL2 isolate contained 5′-UTR (1–186 nt), P1 (187–1035 nt), HC-Pro (1036–2412 nt), P3 (2413–3456 nt), PIPO (2589–2699 nt), 6K1 (3457–3615 nt), CI (3616–5520 nt), 6K2 (5521–5679 nt), NIa-VPg (5680–6252 nt), NIa-Pro (6253–6981 nt), NIb (6982–8538 nt), CP (8539–9357 nt), and 3′-UTR (9358–9532 nt) region. These genes (excluding 5′- and 3′-UTR regions) encode P1 protein (283aa, 32.66 kDa), HC-Pro protein (458aa, 52.1 kDa), P3 protein (348aa, 40.78 kDa), P3N-PIPO protein (34aa, 3.97 kDa), 6K1 protein (53aa, 5.98 kDa), CI protein (635aa, 71.08 kDa), 6K2 protein (53aa, 5.98 kDa), VPg protein (191aa, 22.23 kDa), NIa protease (243aa, 27.22 kDa), NIb protein (519aa, 59.06 kDa), and coat protein (273aa, 31.21 kDa), respectively (Charanjeet 2016).

5.7 Management of Potyvirus Infecting Gladiolus

Gladiolus is propagated through corms and virus accumulates and survives in the mother corms of gladiolus year after year. Under natural conditions, virus is transmitted by aphids and by vegetative propagations (Dubey et al. 2009), and once it infects plants, they cannot be controlled by any chemical treatment in the fields. However, they can be managed by the use of some strategies that prevent their transmission and infection in other plants of the field. Therefore, the conventional methods for control of insect vectors, the development of quick detection techniques, and the elimination of virus (BYMV) from gladiolus were essential for the production of healthy and vigor plants and better quality blooms of gladiolus (Charanjeet 2016).

5.7.1 Conventional Method for Management of Virus Disease

5.7.1.1 Vector Management by a Chemical Insecticide

Aphids (*A. craccivora* and *A. gossypii*) are known vectors of BYMV and many other viruses that affect many crops including gladiolus by sucking the plant sap and by transmitting the virus; therefore, control of aphids is essential. In an earlier study, the populations of aphids and whiteflies were greatly reduced by three sprayings of confidor insecticide which ultimately enhanced the chili production in field conditions. Therefore, similar practice has been attempted for minimizing the populations of aphids from gladiolus and *V. faba* fields. Three sprayings of confidor insecticide (0.2% v/v in water) at 21 days interval were attempted in field which minimized the aphid population and improved the biomass and ultimately the yield of gladiolus and *V. faba* (Charanjeet 2016).

During management studies of BYMV, an improvement in plant growth of gladiolus and *V. faba* was observed as a result of three sprayings of confidor insecticide. It was clear from the above results that the application of 0.2% confidor effectively reduced the infection of mosaic disease and good plant growth was observed. Therefore, three sprayings of confidor insecticide at 21 days interval were found best for minimizing the aphid populations and its transmitted potyviruses, and this practice is recommended for its possible management of mosaic diseases caused by BYMV in gladiolus and *V. faba* (Charanjeet 2016).

5.7.1.2 Biological Control of Aphid Vectors BYMV

Although the confidor insecticide at 21 days interval was found best for minimizing the aphid populations and possible management of mosaic diseases caused by

BYMV in gladiolus and *V. faba*, use of insecticides causes residual effect on human/ cattle health; therefore, it may not be considered as an eco-friendly or healthy approach. Hence, the eco-friendly approaches of virus-disease management are needed to be developed which neither has adverse effect on human health nor possesses hazards to the environment.

During biological control study, the control of aphid vectors population has been attempted by the ladybird (*Coccinella transversalis*), a predator of aphid. For this, the adult and larvae of *C. transversalis* had been allowed to feed the aphid (*A. craccivora*) population on *V. faba*, and their feeding behavior was recorded. *C. transversalis* was found to be most efficient predator of aphid population and fed approximately 20 aphids per minute. It was also found that the larvae feed more aphids than that of adults. It was also observed that aphids quickly migrate away from the ladybirds. These observations may be utilized for minimizing the aphid population, indirectly minimizing the load of the virus in nature (Charanjeet 2016).

5.7.2 Elimination of Potyvirus from Gladiolus and Development of Virus-Free Plants

For the elimination of virus infection from *Gladiolus*, in vitro cultivation of meristematic tissue coupled with therapies has been used to grow virus-free plants (Kaur et al. 2019). The use of such therapies to produce virus-free plants in vitro has been described as an effective method. In gladiolus, elimination of BYMV can be done by using various therapies alone or in combination with one another for better result.

5.7.2.1 Thermotherapy

Thermotherapy is the most commonly used technique for the elimination of viruses from infected explants. Under this treatment, the explants have to be surface sterilized and then transferred in to the MS media for incubation at high temperature for 30 days. The infected gladiolus when subjected to thermotherapy at a higher temperature then maximum of the explant could not survive. The regeneration efficiency is found to be low in all gladiolus cultivars. Some plant species have poor growth after being exposed to high temperatures, as examined by several potyvirus elimination studies (Kaur et al. 2019).

5.7.2.2 In Vitro Chemotherapy

For in vitro chemotherapy of infected gladiolus cormels, an antiviral agent, ribavirin, was amended in the MSg3 medium, and explants were kept in dark for 30 days for germination. Various concentrations of ribavirin (30, 40, 50, and 60 mg/L) were amended in MSg3 medium, and the response of germination of explants in ten gladiolus cultivars, viz., Shagun, Tiger Flame, Snow Princess, Regency, Aldebaran, Vink's Glory, Decisso, Sylvia, True Love, and Promise, was observed after 30 days.

The maximum explants of all cultivars were germinated in 30 mg/L concentration of ribavirin. At 40 mg/L ribavirin concentration, Snow Princess and Promise cultivars failed to grow, and other cultivars also responded poorly; in 50 mg/L ribavirin, only Tiger Flame responded with a very poor regeneration rate, while 60 mg/L ribavirin was lethal for explants of all gladiolus cultivars. At 30 mg/L ribavirin concentration, explants of all of the 10 gladiolus cultivars germinated, but Shagun, Vink's Glory, Tiger Flame, Decisso, and Aldebaran were most regenerative. The 30 mg/L ribavirin was found optimum for these 5 gladiolus cultivars and chosen for the elimination of BYMV from their infected explants (Charanjeet 2016).

5.7.2.3 Electrotherapy

For electrotherapy, explants have to be immersed in buffer (1× TAE buffer, most commonly used) in an electrophoresis tank to expose with electric current of 10, 20, and 30 ampere for 20 min approximately (Kaur et al. 2019; Kumar et al. 2009). Explants then have to be sterilized and incubated in Ms. media. After electrotherapy, the explants should be put on ribavirin-added Ms medium for a combination of chemotherapy and electrotherapy. Both treated explants, as well as untreated control explants, will be cultured in a culture room at 23–25 °C with 16 h of light. Plants when treated with electrotherapy have shown successful regeneration rate mostly at low ampere, but regeneration has been decreased with increasing electric current (Kumar et al. 2009). It has also been observed that electrotherapy can induce physiological changes in the *Gladiolus* (Lozoya et al. 1996; Meybodi et al. 2011).

In all cultivars, the combination of chemotherapy and electrotherapy resulted in the highest rate of regeneration and percentage of BYMV-free plants (Mahmoud et al. 2009; Charanjeet 2016).

RT-PCR is used to monitor regenerated plantlets obtained from various treatments for the presence or absence of BYMV in three phases. In preliminary testing, total RNA is extracted from gladiolus plantlets and tested using BYMV-specific primers via RT-PCR (as described above). The plantlets that do not have BYMV would be multiply in Ms. medium and tested for the existence of BYMV using RT-PCR as a second stage virus test. After that, the plantlets that are virus-free were allowed to induce in Ms medium. After 60 days, the plantlets are tested for the third stage using leaf tissue and RT-PCR.

The plantlets collected from thermotherapy, chemotherapy, electrotherapy, and combinations of therapies, as well as their respective controls, have been separated from the culture medium, and the excess medium has been washed away. The plants were transferred to glasshouse conditions after 2 weeks and observed for regeneration frequency (Kaur et al. 2019; Charanjeet 2016).

5.7.2.4 Combination of Electrotherapy and In Vitro Chemotherapy

The combination of these therapies were also attempted to eliminate BYMV from infected gladiolus cormels; combination of electrotherapy and in vitro chemotherapy resulted in the increased regeneration efficiency of the aforementioned gladiolus cultivars; however, the combination of thermotherapy with electrotherapy and in vitro chemotherapy did not work well, and the percentage of surviving explants was very less; hence, the combinations of thermotherapy with chemo- and electrotherapies were not applied to eliminate BYMV from gladiolus cultivars (Charanjeet 2016).

5.8 Conclusion

Viruses are a major issue in gladiolus since the plants are propagated by corms that may contain the virus, causing substantial plant mortality. Several RNA viruses, including AMV, BYMV, BBWV, CMV, INSPV, OrMV, SLRSV, TAV, TMV, TRV, TRSV, and TSWV, have been found to naturally infect *Gladiolus*. The cultivation of ornamentals, especially *Gladiolus* cultivars, necessitates a significant improvement. The screening methods used in breeding programs have contributed to the development of cultivars resistant to BYMV and other viruses. The study of *Gladiolus* cultivar's resistance to BYMV was also difficult due to a lack of appropriate virus detection techniques in history. The development of screening methods for stable resistance to both viruses has been a major concern. The procedure used to test for resistance is inconclusive due to the complexity of viral diseases and host interactions. ELISA and RT-PCR diagnostic techniques were developed and standardized for detecting BYMV infection in *Gladiolus* cultivars to screen them. The utility of diagnostic techniques for the specific identification of BYMV affecting *Gladiolus* cultivars and the generation of virus-free plants was clearly explained. Because of the cross-reactivity of antibodies in potyviruses, cDNA probes are more appropriate and sensitive. Plant viruses, on the other hand, can be successfully detected using the PCR process. BYMV was eradicated from contaminated explants using three separate therapies: thermo-, chemo-, and electrotherapies alone, as well as in combination of chemo- and electrotherapies. Electrotherapy is a form of treatment that uses electricity. In three commercially relevant cases, chemotherapy has been successfully used. The existing chemotherapy and electrotherapy to remove BYMV from gladiolus are effective and can be used at a commercial level for virus disease management in ornamental plants and would be beneficial to gladiolus growers in providing healthy planting material for better quality production in the region.

References

Adams MJ, Antoniw JF, Fauquet CM (2005) Molecular criteria for genus and species discrimination within the family *Potyviridae*. Arch Virol 150:459–479

Ahmad T, Ahmad I, Oasim M (2008) Present status and prospects of *Gladiolus* cultivation in Punjab. Pakistan J Tekirdag Agric Faculty 5(3):227–238

Akansha R, Kumar S, Singh S, Rastogi RR, Roy RK (2014) Studies on reproductive parts of flower and palynology of Gladiolus (*Gladiolus* L) varieties. Int J Curr Microbiol App Sci 3 (10):823–829

Altschul SF, Gish W, Miller W, Myers EW, Lipman DJ (1990) Basic local alignment search tool. J Mol Biol 215:403–410

Anonymous (1997) Gene banks for 150 Gladiolus varieties developed. Agric News III:139

Anonymous (2006) Cut flower and plants: European and Asian Markets Market News Service International Trade Center (UNCTAD/WTO) Issue No M2, 10th March 2006, Geneva, Switzerland, p 28

APS Publications (n.d.). https://apsjournals.apsnet.org/doi/10.1094/PDIS-10-18-1839-RE

Balamurali Krishnan M, Doraisamy S, Ganapathy T, Viswanathan R (2002) Serological specificity and titer of Sugarcane mosaic virus polyclonal antisera raised under varying immunization procedures and bleeding time. J Plant Dis Protect 109:646–654

Bateson MF, Dale JL (1995) *Banana bract mosaic virus*: characterization using potyvirus specific degenerate PCR primers. Arch Virol 140:515–527

Bellardi MG, Marani F (1985) *Nepoviruses* isolated from *Gladiolus* in Italy. Acta Hortic 164:297–308

Bellardi MG, Canova A, Gelli C (1986) Comparative studies on *Gladiolus* isolates of *Arabis mosaic virus* (ArMV). Phytopathol Mediterr 25:85–91

Bellardi MG, Vicchi V, Gelli C (1987) Stunting of gladiolus flower spike associated with *Tobacco Streak Virus* (TSV). Phytopathol Mediterr 26:73–80

Benschop M, Kamenetsky R, Le Nard M, Okubo H, De Hertogh A (2010) the global flower bulb industry: production, utilization, research. Horticult Rev 36:1–115

Berger PH, Adams MJ, Barnett OW, Brunt AA, Hammond J, Hill JH, Jordan RL, Kashiwazaki S, Rybicki E, Spence N, Stenger DC, Ohki ST, Uyeda I, van Jaayen A, Valkonen J, Vetten HJ (2005) *Potyviridae*. In: Fauquet CM, Mayo MA, Maniloff J, Desselberger U, Ball LA (eds) Virus taxonomy VIIIth report of the International Committee on Taxonomy of Viruses. Elsevier/Academic Press, London

blackherbals.com (n.d.). http://www.blackherbals.com/gladiolus.html

Boonham N, Walsh K, Smith P, Madagan K, Graham I, Barker I (2003) Detection of potato viruses using microarray technology towards a generic method for plant viral disease diagnosis. J Virol Methods 108:181–187

Bos L (1970) *Bean yellow mosaic virus*. Descriptions of plant viruses no. 40. Association of Applied Biology, Wellesbourne

Bozarth RH, Corbett MK (1957) *Tomato ringspot virus* associated with stunt or stub disease of gladiolus in Florida. Plant Dis Rep 54:125–127

Brunt A, Crabtree K, Dallwitz MJ, Gibbs A, Watson L, Zurcher E (1996) Plant viruses online descriptions and lists from the VIDE database. CAB International, Wallingford

Cantor M, Tolety J (2011) Gladiolus. In: Wild crop relatives: genomic and breeding resources, vol 10. Springer, Berlin, pp 133–159

Chandel S, Deepika R (2010) Recent advances in management and control of Fusarium yellows in *Gladiolus* species. J Fruit Ornamental Plant Res 18(2):361–380

Charanjeet K (2016) Molecular detection and identification of potyvirus/es infecting gladiolus and their disease management. PhD thesis, Department of Botany, University of Lucknow, Lucknow, India

Chen J, Adams MJ (2001) A universal PCR primer to detect members of the *Potyviridae* and its use to examine the taxonomic status of several members of the family. Arch Virol 146:757–766

Chung BY-W, Miller WA, Atkins JF, Firth AE (2008) An overlapping essential gene in the *Potyviridae*. Proc Natl Acad Sci 105:5897–5902

Ciuffo M, Masenga V, Turina M (2006) Characterization of a Potyvirus isolated from *Tradescantia fluminensis* in northern Italy. Arch Virol 151:1235–1241

Colinet D, Kummert J (1993) Identification of a *sweet potato feathery mottle virus* isolate from China (SPFMV-Ch) by the polymerase chain reaction with degenerate primers. J Virol Methods 45:149–159

Crosslin JM, Hamm PB, Shiel PJ, Hane DC, Brown CR, Berger PH (2005) Serological and molecular detection of *tobacco veinal necrosis* isolates of *Potato virus Y* (PVYN) from potatoes grown in the western United States. Am J Potato Res 82:263–269

Desbiez C, Wipf-Scheibel C, Lecoq H (2002) Biological and serological variability, evolution and molecular epidemiology of *Zucchini yellow mosaic virus* (ZYMV, *Potyvirus*) with special reference to Caribbean islands. Virus Res 85:5–16

Dorrigiv R, Jafarpour B, Rastegar MF (2013) Detection of some virus pathogens of gladiolus in Iran international. J Agric Crop Sci 5:1653–1658

Dubey VK, Aminuddin A, Singh VP (2008) First report of a subgroup IA *Cucumber mosaic virus* isolate from *Gladiolus* in India. Australas Plant Dis 3:35–37
Dubey VK, Aminuddin A, Singh VP (2009) First report of *Bean yellow mosaic virus* in *Gladiolus* in Indian. J Plant Pathol 91(4):787–791
Dubey VK, Aminuddin A, Singh VP (2010) Molecular characterization of *Cucumber mosaic virus* infecting *Gladiolus*, revealing its phylogeny distinct from the Indian isolate and alike the Fny strain of CMV. Virus Genes 41:126–134
Duraisamy GS, Pokorný R, Holková L (2011) Possibility of *Bean yellow mosaic virus* detection in *Gladiolus* plants by different methods. J Plant Dis Protect 118:2–6
Francki RIB, Milna RG, Hatta T (1985) Pot virus group. In: Atlas of plant viruses, vol II. CRC Press, Boca Raton, pp 183–217
Fry PRN (1953) Two virus diseases of gladiolus. N Z J Sci Technol Sect A 34:460–464
Fukumoto F, Ito Y, Tochihara H (1982) Viruses isolated from *Gladiolus* in Japan. Ann Phytopathol Soc Japan 48:68–71
Gibbs A, Mackenzie A (1997) A primer pair for amplifying part of the genome of all potyvirids by RTPCR. J Virol Methods 63:9–16
Gladiolus (n.d.). http://en.wikipedia.org/wiki/Gladiolus
Hammond J, Hammond RW (2003) The complete nucleotide sequence of isolate BYMV-GDD of *Bean yellow mosaic virus*, and comparison to other potyviruses. Arch Virol 148:2461–2470
Hollings M, Brunt AA (1981a) Potyvirus. In: Kurstak E (ed) Handbook of plant virus infections: comparative diagnosis. Elsevier/North-Holland, Amsterdam, pp 147–170
Hollings M, Brunt AA (1981b) Potyviruses. In: Kurstak E (ed) Handbook of plant virus infections: comparative diagnosis. Elsevier/North-Holland, Amsterdam, pp 731–807
James D, Varga A, Thompson D, Hayes S (2003) Detection of a new and unusual isolate of *Plum pox virus* in Plum (*Prunus domestica*). Plant Dis 87:1119–1124
Johansen E, Edwards MC, Hampton RO (1994) Seed transmission of viruses: current perspectives. Annu Rev Phytopathol 32:363–386
Kamińska M (1978) Some properties of isolates of *Tomato black ring virus*. Zeszyty Problemowe Postepow Nauk Rolniczych 214:109–117
Kantrong S, Sako N (1993) Characterization of epitopes recognized by monoclonal antibodies to aphid transmissible and non-transmissible strains of *turnip mosaic virus*. Arch Virol 133:11–20
Karyeija RF, Kreuze JF, Gibson RW, Valkonen JPT (2000) Two serotypes of *Sweet potato feathery mottle virus* in Uganda and their interaction with resistant sweet potato cultivars. Phytopathology 90:1250–1255
Katoch M, Ram R, Zaidi AA (2003) First report of *Tobacco ringspot virus* occurring in *Gladiolus* in India. Plant Pathol 52:789
Kaur C, Raj SK, Snehi SK, Goel AK, Roy RK (2011) Natural occurrence of *Ornithogalum mosaic virus* newly reported on gladiolus in India. New Dis Rep 24:2
Kaur C, Kumar S, Raj SK, Chauhan PS, Sharma N (2015) Characterization of a new isolate of *Bean yellow mosaic virus* group-IV associated with mosaic disease of Gladiolus in India. J Plant Pathol Microbiol 6:309
Kaur C, Raj R, Kumar S, Purshottam DK, Agrawal L, Chauhan PS, Raj SK (2019) Elimination of *Bean yellow mosaic virus* from infected cormels of three cultivars of gladiolus using thermo-, electro- and chemotherapy. 3 Biotech 9:154
Kehoe MA, Coutts BA, Buirchell BJ, Jones RAC (2014) Plant virology and next generation sequencing: experiences with a Potyvirus. PLoS One 9(8):e104580
Kew Science (n.d.). www.kew.org/kewscientist/KewScientist41.pdf
Koch M, Salomon R (1994) Serological detection of *onion yellow dwarf virus* in garlic. Plant Disease 78:785–788
Kumar S, Khan MS, Raj SK, Sharma AK (2009) Elimination of mixed infection of *Cucumber mosaic* and *Tomato aspermy virus* from *Chrysanthemum morifolium* Ramat. cv. Pooja by shoot meristem culture. Sci Hortic 119:108–112

Lee TC, Francki RIB, Hatta T (1979) A serious disease of gladiolus in Australia caused by *Tomato spotted wilt virus*. Plant Dis Rep 63:343–348

Lozoya S, Abello FJ, Grarcia De La G (1996) Electrotherapy shoot tip culture eliminate *potato virus* X in potatoes. Am Potato J 73:149–154

Mahmoud YM, Hosseny MH, Abdel-Ghaf MH (2009) Evaluation of some therapies to eliminate Potato Y *potyvirus* from potato plants. Int J Virol 5:64–76

Martín V, Perales C, Abia D, Ortíz AR, Domingo E, Briones C (2006) Microarray-based identification of antigenic variants of foot-and-mouth disease virus: a bioinformatics quality assessment. BMC Bioinform 7:1–12

Meybodi DE, Mozafari J, Babaeiyan N, Rahimian H (2011) Application of electrotherapy for the elimination of potato *potyviruses*. J Agric Sci Technol 13:921–927

Milne RG (1988) Taxonomy of the rod-shaped filamentous viruses. In: Milne RG (ed) The plant viruses, 4, the *filamentous plant viruses*. Plenum Press, New York, pp 3–50

Mink GI, Silbernagel MJ (1992) Serological and biological relationships among viruses in the *bean common mosaic-virus* subgroup. Arch Virol Suppl 5:397–406

Mink GI, Vetten HJ, Wyatt SD, Berger PH, Silbernagel MJ (1999) Three epitopes located on the coat protein amino terminus of viruses in the *bean common mosaic potyvirus* subgroup. Arch Virol 144:1173–1189

National Horticultural Board (n.d.). http://www.nhb.gov.in/area-pro/NHB Database 2015.pdf

Ounouna H, Kerlan C, Lafaye P, Loukili MJ, Gaaied A (2002) Production of monoclonal antibodies against synthetic peptides of the N-terminal region of *Potato virus Y* coat protein and their use in PVY strain differentiation. Plant Pathol 51:487–494

Park IS, Kim KW, Kyun H, Chang MU (1998) The viruses in *Gladiolus hybrids* cultivated in Korea 1 *Bean yellow mosaic virus* and *Clover yellow vein virus*. Korean J Plant Pathol 14:74–82

Potyviridae (n.d.). http://www.dpvweb.net/notes/showfamily.php?family=Potyviridae#genera

Purcifull D, Hiebert E, Edwardson J (1984) *Watermelon mosaic virus* 2CMI/AAB Descr No 293, p 7

Raj SK, Srivastava A, Chandra G, Singh BP (2002) Characterization of *Cucumber mosaic virus* isolate infecting *gladiolus* cultivars and comparative evaluation of serological and molecular methods for sensitive diagnosis. Curr Sci 83(9):1132–1137

Reddy PH, Venkataravanappa V, Swarnalatha P, Ashwathappa KV, Krishna RM (2019) Characterization of recombinant Bean yellow mosaic virus belongs to group-IV infecting Gladiolus. Int J Curr Microbiol App Sci 8:1644–1657

Roy RK (2000) NBRI: A National Institute for Conservation and Exploitation of Plants in India, vol 3(4). Resource Centre, India

Sharma PN, Sharma V, Sharma A, Rajput K, Sharma SK (2015) Identification and molecular characterization of *Bean yellow mosaic virus* infecting French bean in Himachal Pradesh. Virus Dis 26:315–318

Shodhganga (n.d.). http://hdl.handle.net/10603/123859

Shukla DD, Ward CW (1989) Identification and classification of potyviruses based on coat protein sequence data and serology. Brief review. Arch Virol 106:171–200

Shukla DD, Ward CW, Brunt AA (1994) The Potyviridae. CAB International, Wallingford

Shukla DD, Ward CW, Brunt AA, Berger PH (1998) *Potyviridae* family. In: Description of plant viruses, no 366. Association of Applied Biologists (AAB), Warwickshire

Singh BR (2008) Characterization of a potyvirus infecting gladiolus crops and development of immunological and molecular based diagnostic system. PhD thesis, Department of Agriculture Biology, Aligarh Muslim University, Aligarh, India

Stein A (1995) *Gladiolus*. In: Loebenstein G, Lawson RH, Brunt AA (eds) Virus and virus-like diseases of bulb and flower crops. Wiley, Chichester

Tan ZY, Gibbs AJ, Tomitaka Y, Sanchez F, Ponz F, Ohshima K (2005) Mutations in Turnip mosaic virus genomes that have adapted to *Raphanus sativus*. J Gen Virol 86:501–510

TAXONOMY (n.d.). https://talk.ictvonline.org/taxonomy/

The University of Arizona (n.d.). http://ag.arizona.edu/daves/types.html

Torres DP, Silva MA, Pinho DB, Pereira OL, Furtado GQ (2013) First report of Curvularia *gladioli* causing a leaf spot on *Gladiolus grandiflorus* in Brazil. Plant Dis 97:847

Turina M, Ciuffo M, Lenzi R, Rostagno L, Mela L, Derin E, Palmano S (2006) Characterization of four viral species belonging to the family *Potyviridae*. Phytopathology 96:560–566

Urcuqui-Inchima S, Haenni A, Bernardi F (2001) Potyvirus proteins: a wealth of functions. Virus Res 74:157–175

Van der Vlugt RAA, Steffens P, Cuperus C, Barg E, Lesemann DE, Bos L, Vetten HJ (1999) Further evidence that *shallot yellow stripe virus* (SYSV) is a distinct potyvirus and re-identification of *Welsh onion yellow stripe virus* as an SYSV strain. Phytopathology 89:148–155

Van Regenmortel MHV (2000) Virus taxonomy: the classification and nomenclature of viruses: the seventh report of the International Committee on the Taxonomy of Viruses San Diego, USA. Academic, Cambridge

Villamor DEV, Bajet NB, Lockhart BE (2003) Purification and production of antiserum against *abaca mosaic potyvirus* virus coat protein mutants. Philipp Agric Sci 86:368–376

Ward C, Weiller G, Shukla D, Gibbs A (1995) Molecular systematics of the Potyviridae, the largest plant virus family. In: Gibbs A, Calisher C, García-Arenal F (eds) Molecular basis of virus evolution. Cambridge University Press, Cambridge, pp 477–500. https://doi.org/10.1017/CBO9780511661686.034

Wylie SJ, Jones RAC (2009) Role of recombination in the evolution of host specialization within *Bean yellow mosaic virus*. Phytopathology 99:512–518

Wylie SJ, Coutts BA, Jones MGK, Jones RAC (2008) Phylogenetic analysis of *Bean yellow mosaic virus* isolates from four continents: the relationship between the seven groups found and their hosts and origins. Plant Disease 92:1596–1603

Wylie SJ, Coutts BA, Jones MGK, Jones RAC (2010) Phylogenetic analysis of *Bean yellow mosaic virus* isolates from four continents: the relationship between the seven groups found and their hosts and origins. Plant Dis 92(12):1596–1603

Zaitlin M, Hull R (1987) Plant virus-host interactions. Annu Rev Phytopathol 38:291–315

Zhong Y, Guo A, Li C, Zhuang B, Lai M, Wei C (2005) Identification of a naturally occurring recombinant isolate of *Sugarcane mosaic virus* causing maize dwarf mosaic disease. Virus Genes 30:75–83

Current Status of Three Virus Genera (*Badnavirus*, *Cucumovirus*, and *Potyvirus*) in *Canna* Species in India

6

Aarti Kumari, Charanjeet Kaur, Susheel Kumar, Puneet Singh Chauhan, and Shri Krishna Raj

Abstract

Canna is considered as an ornamental plant because of its attractive green striped, streaked, spotted, or variegated foliage and beautiful yellow or orange flowers with darker red to brown splotches. Various phytopathogens such as fungi, bacteria, phytoplasma, and viruses are known to infect and cause diseases in canna. Among them, viruses are the most important cause of diseases, which can drastically reduce the quality of foliage and flowers in canna species. Various viruses belonging to three genera *Badnavirus*, *Potyvirus*, and *Cucumovirus* are reported worldwide on canna cultivars. In this chapter, we describe the detection, characterization, and management of five viruses among three virus genera reported to infect canna species worldwide: *Bean yellow mosaic virus*, *Canna yellow streak virus* of family *Potyviridae* (genus *Potyvirus*); *Canna yellow mottle virus* of family *Caulimoviridae* (genus *Badnavirus*); *Cucumber mosaic virus* and *Tomato aspermy virus* of family *Bromoviridae* (genus *Cucumovirus*). *Bean yellow mosaic virus* and *Canna yellow mottle virus* are the most common in canna species grown in India.

Keywords

Canna lily, Virus characterization, Disease management, *Bean yellow mosaic virus*, *Canna yellow mottle virus*, *Cucumber mosaic virus*

A. Kumari (✉) · C. Kaur · S. Kumar · S. K. Raj (✉)
Plant Molecular Virology Laboratory, CSIR-National Botanical Research Institute, Lucknow, Uttar Pradesh, India

P. S. Chauhan
Microbial Technology Division, Council of Scientific and Industrial Research-National Botanical Research Institute, Lucknow, Uttar Pradesh, India

S. K. Raj et al. (eds.), *Virus Diseases of Ornamental Plants*,
https://doi.org/10.1007/978-981-16-3919-7_6

6.1 Introduction

Canna is the genus of Cannaceae family which belongs to the order Zingiberales. Cannas are large tropical and subtropical perennial herbs with a rhizomatous rootstock that allows them to spread slowly outward from where they are planted. Each individual stem consists of a central stalk with 10–12 leaves arranged alternately or spirally along it. Leaves of cannas are usually large banana-like, tropical-looking, and bold. Most cultivars have rich emerald green leaves, but some have purple/red or variegated leaves. The purple or red leaf color is usually quite dark and may cover the entire leaf, just the outside margin, or occasionally just the midrib. The canna flower is very exotic, and its color ranges from pale yellow to orange, blood red, and all shades in between (salmon, apricot, and pink). Canna flowers may be striped, streaked, spotted, or splotched with contrasting colors. The most common form is a yellow or orange flower with darker red to brown splotches on it. Because of beautiful flower and foliage, canna is considered as an ornamental plant.

Besides ornamental value, cannas are also used for food, bioethanol production, medicine (Mishra et al. 2011), phytoremediation (Darsini et al. 2015), etc. With the advancement in knowledge about canna and its uses for several aspects (horticulture, industrial, and medicinal), the economic importance of canna has been enhanced. To date, 20 species have been reported in genus *Canna* (Tanaka 2008) and more than 500 varieties by different methods, hybridization of wild species, radiation treatments, etc. Nowadays, a large number of hybrids have been generated, cultivated, and naturalized around the globe by the horticultural trade.

Because canna production is so lucrative, there are farmers and nurseries dedicated solely to its production. The specific issue that the canna industry faces is virus diseases. Infection of viruses causes serious diseases to crop and non-crop plants and is responsible for huge losses in crop production and quality deterioration of ornamental plants throughout the world. There are many reports of RNA and DNA virus infections on canna cultivars worldwide which belong to three genera (*Badnavirus*, *Potyvirus*, and *Cucumovirus*). The most common virus diseases which have been reported in *Canna* are caused by *Canna yellow mottle virus* (CaYMV; *Badnavirus*), *Canna yellow streak virus* (CaYSV; *Potyvirus*), *Bean yellow mosaic virus* (BYMV; *Potyvirus*), *Cucumber mosaic virus* (CMV; *Cucumovirus*), and *Tomato aspermy virus* (TAV; *Cucumovirus*) (Rajakaruna et al. 2014).

6.2 Symptoms Caused by Virus/es

The virus/es are reported as single infections, as well as mixed infections in canna (Kumari et al. 2016). Disease symptoms such as foliar necrosis, striations, mottling, and discoloration have been reported (Lockhart 1988; Fisher et al. 1997; Monger et al. 2007, 2010; Pappu et al. 2008; Kumari et al. 2016) (Fig. 6.1). However, the foliar symptoms can vary significantly, making it difficult to discriminate the causal agent for the various disease phenotypes.

Fig. 6.1 Canna leaves exhibiting streak (**a**), mottle (**b**), mosaic (**c**), and streak mosaic (**d**) symptoms as compared to a healthy leaf (**e**)

6.2.1 *Badnavirus*

Canna yellow mottle virus (CaYMV), a monopartite double-stranded DNA virus of genus *Badnavirus* (family *Caulimoviridae*), is reported to infect several canna species and commonly associated with the leaf mottling symptoms in canna (Rajakaruna et al. 2014; Kumari et al. 2015a, b) (Fig. 6.1). Badnaviruses are non-enveloped bacilliform particles and circular dsDNA genomes of 7.0–7.6 kbp (Fig. 6.2a, b). Badnavirus genomes have three open reading frames (ORFs). ORF1 encodes a protein of unknown function, ORF2 encodes the virion-associated protein (Fauquet et al. 2005), and ORF3 encodes a 208-kDa polyprotein, which contains a movement protein, coat protein, aspartic protease, reverse transcriptase, and RNase H (Gawel and Jarret 1991).

6.3 Host Range and Mode of Transmission of *Badnavirus*

Badnavirus has diverse host range because it can infect monocot as well as dicot plants. Investigations from different parts of the world indicated that there has been a continuous increase in the disease incidence caused by *Badnavirus* on monocot as well as dicot plants, viz., *Canna yellow mottle virus* from canna species (Borroto et al. 2007); *Piper betle* (Kumari et al. 2015a, b); *Aucuba bacilliform virus* from *Aucuba japonica* (http://sdb.im.ac.cn/vide/descr052.htm); *Commelina yellow mottle badnavirus* from *Commelina diffusa* (http://sdb.im.ac.cn/vide/descr243.htm); *Sugarcane bacilliform virus* from sugarcane (http://pvo.bio-mirror.cn/descr768.htm); *Piper yellow mottle virus* from black pepper (Lockhart et al. 1997); *Banana streak virus* from banana (Natsuaki and Furuya 2007); integrated badnavirus in fig germplasm (Laney et al. 2012); and *Yam badnavirus* from yam (*Dioscorea sansibarensis*) plant (Seal and Muller 2007). *Badnavirus* is also reported in red clover (*Trifolium pratense*) (Franova and Jakesova 2012). The virus could be transmitted by mechanical inoculation, by grafting, and by seed but not by contact between hosts.

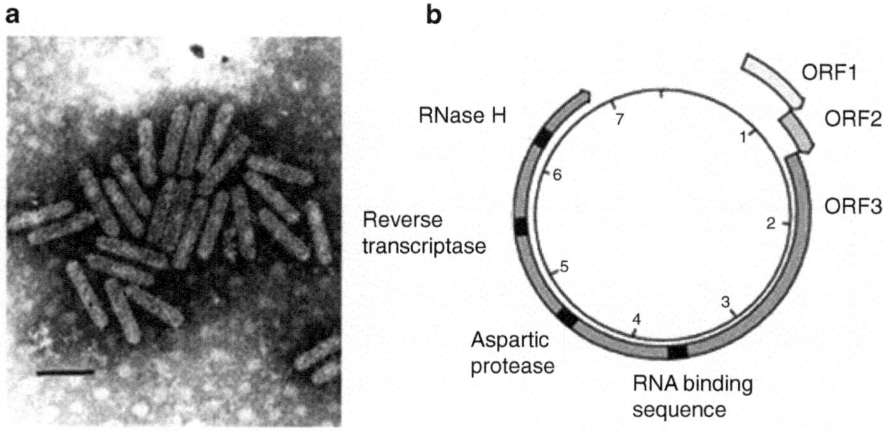

Fig. 6.2 Bacilliform morphology (**a**) and genomic organization of *Badnavirus* (**b**)

Badnavirus can also be transmitted by insect vector mainly mealybug, in a persistent or semi-persistent manner (Gambley et al. 2008).

6.4 CaYMV (*Badnavirus*) in Canna

CaYMV has also been reported to cause yellow mottle disease in canna species in Japan (Yamashita et al. 1985); Florida (Momol et al. 2004); the USA (Pappu et al. 2008); Italy and the Netherlands (Marino et al. 2007); and Austria (Borroto et al. 2007) and in India (Kumari et al. 2014, 2015a, b) and the USA (Rajakaruna et al. 2014) (Table 6.1).

6.4.1 Cucumovirus

Cucumber mosaic virus (CMV) is the type member of genus *Cucumovirus*. CMV is a tripartite plus-sense RNA virus, and genomic RNAs are packaged in separate particles. CMV contains five open reading frames (ORFs). The 1a and 2a ORFs, encoded on RNAs 1 and 2, respectively, are the viral components of the replicase. The 2b ORF, a gene overlapping the 2a ORF, is expressed from a sub-genomic RNA, RNA 4A, and encodes a suppressor of post-transcriptional gene silencing. RNA 3 encodes the 3a protein, the viral movement protein, and the CP, expressed from sub-genomic RNA 4. Phylogeny estimations with the CP ORF, as well as rearrangements in the 5′ nontranslated region (NTR) of RNA 3, divided CMV strains into three subgroups: IA, IB, and II (Roossinck et al. 1999) (Fig. 6.3).

Table 6.1 Summary of major viral diseases occurring on *Canna*

Genus/ viruses	Country	Symptomatology	Citations
Badnavirus			
CaYMV	Kenya	Yellow/green chlorotic mottle, necrotic mottle, and veinal streaking	Agnerch et al. (2015)
CaYMV	USA	Mosaic	Chauhan et al. (2015)
CaYMV	USA	–	Rajakaruna et al. (2014)
CaYMV	India	Necrotic striping, severe veinal chlorosis, yellow mosaic, and veinal streaking	Kumari et al. (2014)
CaYMV	Washington	Mottling, general yellowing, and veinal chlorosis	Pappu et al. (2008)
CaYMV	Italy	Veinal yellowing and chlorotic mottle	Marino et al. (2007)
	The Netherlands		
CaYMV	Austria	Vein streaking	Borroto et al. (2007)
CaYMV	Florida	Veinal necrosis and mottling	Momol et al. (2004)
CaYMV	North America	Yellow mottle	Lockhart (1988)
CaYMV	Japan	–	Yamashita et al. (1985)
Potyvirus			
BYMV	USA	Mosaic	Chauhan et al. (2015)
BYMV	USA	Mosaic, yellowing, and necrosis	Rajakaruna et al. (2014)
BYMV	India	Severe mosaic	Kumari et al. (2015a, b)
CaYSV	USA	Severe streaking	Chauhan et al. (2015)
CaYSV	UK	Severe streaking	Monger et al. (2007)
CaYSV	USA	–	Rajakaruna et al. (2014)
Cucumovirus			
CMV	USA	Mild mottling	Rajakaruna et al. (2014)
CMV	Iran	Mosaic symptoms	Saidi and Safaeizadeh (2012)
CMV	Columbus	–	Fisher et al. (1997)
TAV	Australia	–	Borroto et al. (2007)
TAV	USA	Mild mosaic and yellowing	Rajakaruna et al. (2014)

6.5 Host Range and Mode of Transmission of *Cucumovirus*

The host range includes all type of cultures food and feed product, ornamental, etc. Vegetable host are muskmelon, squash, peppers, turnip, pumpkin, potato, carrot, etc. Ornamental host are chrysanthemum, lily, marigold, petunia, gerbera, amaranth, aster, gladiolus, etc. Typically, it shows symptoms like diffuse chlorotic dots near the veins and severe chlorosis. CMV has been reported on *Jatropha curcas* (Raj et al. 2008), *Rauwolfia serpentina* (Raj et al. 2007), and *Cymbopogon citratus* (Raj

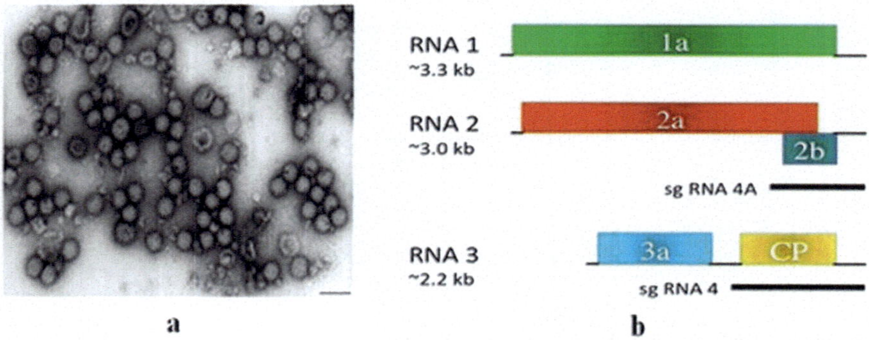

Fig. 6.3 Electron micrograph of CMV (**a**) and genome organization of CMV (**b**)

et al. 2007). The most notable characteristic of cucumoviruses for its natural spread is that it can be transmitted by numerous species of aphids (Kennedy et al. 1962; Palukaitis et al. 1992). More than 85 different species, especially *Aphis gossypii*, *Myzus persicae*, and *Macrosiphoniella sanborni* are known to be capable of transmitting cucumoviruses in more than 1200 plant species (Roossinck 2002). The virus is transmitted from infected plants to healthy ones in a non-persistent or stylet-borne manner (Kennedy et al. 1962; Palukaitis et al. 1992).

6.6 Cucumber Mosaic Virus (CMV) in Canna

Canna has been reported to be infected by CMV in Iran (Saidi and Safaeizadeh 2012), the USA (Rajakaruna et al. 2014), and India (Kumari et al. 2016). The other member of genus *Cucumovirus* infecting canna is *Tomato aspermy virus* which has also been reported (Hollings and Stone 1971) (Table 6.1).

6.6.1 *Potyvirus*

Potyvirus, another RNA virus infecting canna, is the largest genus in the family *Potyviridae*. These viruses are non-enveloped rod-shaped flexuous particles 680–900 nm long and 11–13 nm in diameter, encapsidating a genome of about 9.7 kb. *Potyvirus* is a monopartite, positive-sense, single-stranded RNA genome containing single ORF, which encodes a 350 kDa polyprotein (Fig. 6.4).

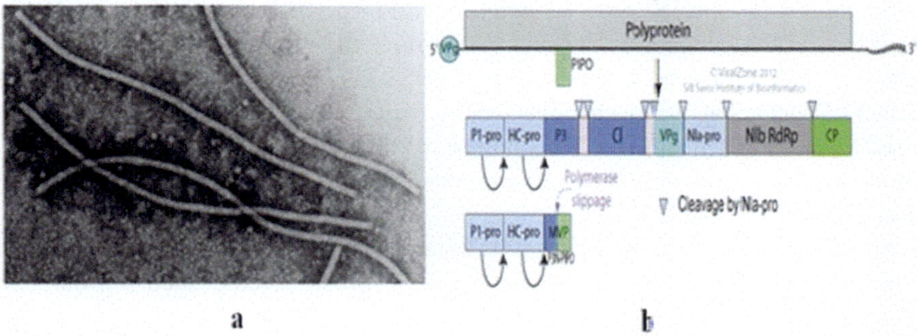

Fig. 6.4 Flexuous rod-shaped morphology (**a**) and genomic organization of *Potyvirus* (**b**)

6.7 Host Range and Mode of Transmission of *Potyvirus*

The viruses of the genus *Potyvirus* have a narrow host range. Plant species of families Chenopodiaceae, Commelinaceae, and Solanaceae are susceptible to get infection of potyviruses. Some members of *Potyvirus* such as *Lettuce mosaic virus* (LMV) and *Bean yellow mosaic virus* (BYMV) have broad host range that can infect 8 and 13 different plant families, respectively. *Potyvirus* is transmitted by aphids in a non-persistent manner.

6.8 *Potyviruses* in Canna

Two members of *Potyvirus*, i.e., Bean yellow mosaic virus (BYMV) and Canna yellow streak virus (CaYSV), are reported in canna. BYMV is reported from USA (Rajakaruna et al. 2014; Chauhan et al. 2015) and India (Kumari et al. 2015a, b), while CaYSV is reported from USA (Rajakaruna et al. 2014; Chauhan et al. 2015) and UK (Monger et al. 2007) (Table 6.1).

6.9 Management Strategies for Viruses in *Canna* Species

Viruses cause significant losses of the major crops, around the world. Therefore, infection of viruses to the crops is the bottleneck for their production. Badnaviruses have got attention globally and are currently considered as economically important plant pathogens since they can cause destructive losses to many crops: banana, rice, cocoa, sugarcane, piper betel, etc. (Huang and Hartung 2001; Harper et al. 2005; Kumari et al. 2015a, b). The genus *Potyvirus* is one of the most devastating groups of plant pathogen in the world, and yield losses due to infection of this virus generally range from 8 to 35%; however, losses as high as 94% have been reported (Ward et al. 1994). Cucumoviruses have wide host range and cause serious yield loss in banana which ranges 45–62%, 50–90% in legumes worldwide (Subrahmanyam et al. 1992), etc. Since canna is being propagated by rhizome, the mother stock, once infected, acts as a source for disease spread in successive generations. Virus-infected plants cannot be controlled by any chemical means, they can be only managed by strategies based on measures that prevent infection or spread to new areas (Verma et al. 2004). To prevent spread of viruses, several methods have been suggested for control of viral diseases.

Spread of viruses can be reduced by cleaning and eradication of virus-infected plant material from the fields. Perennial weeds should be eradicated from around gardens and fields to eliminate possible source of *Badnavirus*, *Potyvirus*, and *Cucumovirus*. Virus spread can also be controlled biologically by using ladybird (*Coccinella transversalis*) which is a natural predator of aphids (virus vector). By using ladybird, virus spread and chemical mist in the environment can also be reduced. The best way to decrease the multiplication of viruses is the use of virus-free plant material. A number of virus elimination protocols have been developed to obtain healthy or virus-free plants such as applying heat therapy (30–37 °C) (Panattoni et al. 2013; Kaur et al. 2019), electrotherapy (Meybodi et al. 2011; Kaur et al. 2019), use of meristem tip cultures, and cold treatment and chemotherapy (Panattoni et al. 2013; Kaur et al. 2019). Virus-free plants can also be developed by using genetic engineering approach. In this approach, virus gene (e.g., CP gene) is utilized for generating virus-resistant transgenic plants (Pratap et al. 2012).

6.10 Conclusion

Canna seems to be a very important plant not only because of its ornamental value, but also it is being utilized in the production of ethanol and medicine. The cultivation of canna is greatly reduced because of the diseases caused by bacteria, fungi, virus, and other pests. Cannas are susceptible to certain viruses. Five major viruses (CaYMV, CaYSV, BYMV, CMV, and TAV) of three genera *Badnavirus*, *Potyvirus*, and *Cucumovirus* have been reported to infect cannas worldwide (Table 6.1). Because virus-infected plants cannot be cured by chemical means, therefore these infected plants behave like source of virus and spread through insects (e.g., aphids, mealybug). Spread of the viruses can be managed by cleaning and eradication of infected plants or by providing virus-free plants. Plenty of virus elimination protocols have been developed in other ornamental or non-ornamental plants, but in case of canna, very less study is reported. To reduce the spread of viruses in canna globally, more study is required in future for development of virus-free canna plants.

References

Agneroh TA, Bratsch SA, Lockhart BE (2015) First report of *Canna yellow mottle virus* in Kenya. Plant Health Prog 16(1):34–35

Borroto FEG, Maghuly F, Fellner A, Laimer M (2007) Determination of viral infections in an Austrian collection of *Canna indica*. J Plant Dis Protect 115(3):102–103
Chauhan R, Hamon H, Rajakaruna P, Webb M, Payton M, Verchot J (2015) Reliable detection for *Bean yellow mosaic virus*, *Canna yellow streak virus* and *Canna yellow mottle virus* in canna varieties with red foliage. Plant Dis 99(2):188–194
Darsini IP, Shamshad AS, Paul MJ (2015) *Canna Indica* (L.): a plant with potential healing powers: a review. Int J Pharma Bio Sci 6(2):1–8
Fauquet CM, Mayo MA, Maniloff J, Desselberger U, Ball LA (2005) Virus taxonomy, classification and nomenclature of viruses. 8th report of the International Committee on the Taxonomy of Viruses, vol 83. Elsevier Academic Press, San Diego, pp 988–992
Fisher JR, Sanchez-Cuevas MC, Nameth ST, Woods VL, Ellet CW (1997) First report of *Cucumber mosaic virus* in *Eryngium amethystinum*, *Canna* spp., and *Aquilegia* hybrids in Ohio. Plant Dis 81(11):1331
Franova J, Jakesova H (2012) First report of Bacilliform Badnavirus-like virus particles in red clover. J Phytopathol 160(10):588–590
Gambley CF, Geering ADW, Steele V, Thomas JE (2008) Identification of viral and non-viral reverse transcribing elements in pineapple (*Ananas comosus*), including members of two new badnavirus species. Arch Virol 153(8):1599–1604
Gawel NJ, Jarret RL (1991) A modified CTAB DNA extraction procedure for Musa and Ipomoea. Plant Mol Biol Rep 9:262–266
Harper G, Hart D, Moult S, Hull R, Geering A, Thomas J (2005) The diversity of Banana streak virus isolates in Uganda. Arch Virol 150(12):2407–2420
Hollings M, Stone OM (1971). *Tomato aspermy virus*. In: CMI0AAB descriptions of plant viruses no. 79. Commonwealth Mycological Institute/Association of Applied Biologists, Kew.
Huang Q, Hartung JS (2001) Cloning and sequence analysis of an infectious clone of *Citrus yellow mosaic virus* that can infect sweet orange via Agrobacterium-mediated inoculation. J Gen Virol 82(10):2549–2558
Kaur C, Raj R, Kumar S, Purshottam DK, Agrawal L, Chauhan PS, Raj SK (2019) Elimination of Bean yellow mosaic virus from infected cormels of three cultivars of gladiolus using thermo-, electro- and chemotherapy. 3 Biotech 9(4):154
Kennedy JS, Day MF, Eastop VF (1962) A conspectus of aphids as vectors of plant viruses, vol 95. Commonwealth Institute of Entomology, London, p 336
Kumari A, Kumar S, Raj SK (2014) First report of *Canna yellow mottle virus* on *Canna* from India. New Dis Rep 29(1):9
Kumari A, Kaur C, Kumar S, Raj SK, Roy RK, Nautiyal CS (2015a) First report of *Bean yellow mosaic virus* causing a mosaic disease of *Canna* sp. in India. Plant Dis 99(6):897
Kumari A, Kumar S, Raj SK, Johri JK, Nautiyal CS (2015b) First report of *Canna yellow mottle virus* associated with yellow vein mosaic disease of Betel Vine (*Piper betel*) in India. Plant Dis 99(8):1189
Kumari A, Raj R, Kumar S, Chauhan PS, Raj SK (2016) Coexistence of three virus genera (Badnavirus, Potyvirus and Cucumovirus) in Canna species in India. Ann Virol Res 2:1008
Laney AG, Hassan M, Tzanetakis IE (2012) An integrated badnavirus is prevalent in fig germplasm. Phytopathology 1029(12):1182–1189
Lockhart BEL (1988) Occurrence of *Canna yellow mottle virus* in North America. Acta Horticult 234(7):69–72
Lockhart BEL, Kiratiya-Angul K, Jones P, Eng L, Silva PD, Olszewski NE, Lockhart N, Deema N, Sangalang J (1997) Identification of *Piper yellow mottle virus*, a mealybug-transmitted badnavirus infecting Piper spp. in Southeast Asia. Eur J Plant Pathol 103:303–311
Marino MT, Ragozzino E, Lockhart BEL, Miglino R, Alioto D (2007) First report of *Canna yellow mottle virus* (CaYMV) in Italy and in the Netherlands. Plant Pathol 57(2):394
Meybodi DE, Mozafari J, Babaeiyan N, Rahimian H (2011) Application of electrotherapy for the elimination of potato potyviruses. J Agric Sci Technol 13:921–927

Mishra T, Goyal AK, Middha SK, Sen A (2011) Antioxidative properties of *Canna edulis* Ker-Gawl. Indian J Nat Prod Resour 2(3):315–321

Momol MT, Lockhart BEL, Dankers H, Adkins S (2004) *Canna yellow mottle virus* detected in canna in Florida. Plant Health Prog. https://doi.org/10.1094/PHP-2004-0809-01-HN

Monger WA, Harju V, Skelton A, Seal SE, Mumford RA (2007) *Canna yellow streak virus*: a new potyvirus associated with severe streaking symptoms in canna. Arch Virol 152(8):1527–1530

Monger WA, Alicai T, Ndunguru J, Kinyua ZM, Potts M, Reeder RH, Miano DW, Adams IP, Boonham N, Glover RH, Smith J (2010) The complete genome sequence of the Tanzanian strain of *Cassava brown streak virus* and comparison with the Ugandan strain sequence. Arch Virol 155(3):429–433

Natsuaki KT, Furuya N (2007) The genera *Babuvirus* and *Badnavirus* in Asia. Plant Pathol J 23 (4):227–232

Palukaitis P, Roossinck MJ, Dietzgen RG, Francki RIB (1992) *Cucumber mosaic virus*. Adv Virus Res 41(C):248–281

Panattoni A, Luvisi A, Triolo E (2013) Review elimination of viruses in plants: twenty years of progress. Spanish J Agric Res 11(1):173–188

Pappu HR, Druffel KB, Eastwell KC (2008) *Canna yellow mottle virus* in *Canna* spp. in Washington State. Plant Dis 92(7):1136

Pratap D, Raj SK, Kumar S, Kumar S, Snehi SK, Gautam KK, Sharma AK (2012) Coat protein-mediated transgenic resistance in tomato against a IB subgroup Cucumber mosaic virus strain. Phytoparasitica 40(4):375–382

Raj SK, Kumar S, Pratap D, Vishnoi R, Snehi SK (2007) Natural occurrence of *Cucumber mosaic virus* on *Rauwolfia serpentina*, a new record. Plant Dis 91(3):322

Raj SK, Kumar S, Snehi SK (2008) First report of *Cucumber mosaic virus* on *Jatropha curcas* in India. Plant Dis 92(1):171

Rajakaruna P, Shafiekhani M, Kim T, Payton M, Chauhan R, Verchot J (2014) Production of discernible disease phenotypes in *Canna* by five plant viruses belonging to the genera *Potyvirus*, *Cucumovirus* and *Badnavirus*. Plant Pathol 63(4):821–830

Roossinck MJ (2002) Evolutionary history of *Cucumber mosaic virus* deduced by phylogenetic analyses. J Virol 76(7):3382–3387

Roossinck MJ, Zhang L, Hellward K (1999) Rearrangements in the 5′ nontranslated region and phylogenetic analyses of *Cucumber mosaic virus* RNA3 indicate radial evolution of three subgroups. J Virol 73(8):6752–6758

Saidi A, Safaeizadeh M (2012) First report of *Cucumber mosaic virus* on *Canna indica* in Iran. Austral Plant Dis Notes 7(1):119–121

Seal S, Muller E (2007) Molecular analysis of a full-length sequence of a new yam badnavirus from *Dioscorea sansibarensis*. Arch Virol 152(4):819–825

Subrahmanyam PS, Wongkaew S, Reddy DVR (1992) Field diagnosis of groundnut diseases. Information bulletin no. 36. International Crop Research Institute for the Semi-Arid Tropics, Patancheru, p 84

Tanaka N (2008) A new species of the genus Canna (*Cannaceae*) from eastern Honduras. J Jpn Bot 83:7–10

Verma N, Ram R, Hallan V, Kumar K, Zaidi AA (2004) Production of *Cucumber mosaic virus*-free chrysanthemums by meristem tip culture. Crop Protect 23(5):469–473

Ward CW, Wieller GF, Shukla DD, Gibbs AJ (1994) Molecular evolution of Potyviridae: the largest plant virus family. In: Gibbs AJ, Calisheh CH, Garcia-Arenal F (eds) Molecular basis of virus evolution. Cambridge University Press, Cambridge

Yamashita S, Natsuaki T, Doi Y, Yora K (1985) *Canna yellow mottle virus*, a non-enveloped small-bacilliform virus in *Canna* sp. Annu Phytopathol Soc Jpn 51(5):642–646

Molecular Characterization and Diversity Analysis of Viruses Infecting Orchids Plants: Global Scenario

7

Malyaj R. Prajapati, Jitender Singh, Koushlesh Ranjan, R. P. Pant, and V. K. Baranwal

Abstract

Orchid family is the largest plant family which consists of a diverse group of flowering plants, with blooms that are often colorful and fragrant. Orchids are commercially traded as ornamental plants, medicinal products, and food. Climates allows to grow most of the climatic environment like humid, subtropical or moderate climates allows to grow most of the orchids species. The worldwide commercial production of these ornamental plants is based on clonal propagation, which leads to proliferation of pathogens and disease outbreaks. The viruses infecting orchid plants have a big influence on the produce and the quality of bulbs, force flowers and are relentless intimidation to the propagation. To control the viral pathogens, specific diagnostic procedure is required which can be fulfilled by advanced molecular techniques. Viruses from orchid's crop can be diagnosed by techniques such as electron microscopy (EM), immunosorbent electron microscopy (ISEM), enzyme-linked immunosorbent assay (ELISA), nucleic acid hybridization, PCR, RT-PCR, next-generation sequencing (NGS) techniques, etc. The expansion of NGS in detection of known and emergent viruses from infected material is proving to be a major breakthrough for the detection of asymptomatic and multiple viruses. Globally, studies on orchid

M. R. Prajapati · J. Singh (✉)
College of Biotechnology, Sardar Vallabhbhai Patel University of Agriculture & Technology, Meerut, Uttar Pradesh, India

K. Ranjan
Department of Veterinary Physiology and Biochemistry, College of Veterinary and Animal Sciences, Sardar Vallabhbhai Patel University of Agriculture & Technology, Meerut, Uttar Pradesh, India

R. P. Pant · V. K. Baranwal
Division of Plant Pathology, ICAR-Indian Agricultural Research Institute, New Delhi, India

S. K. Raj et al. (eds.), *Virus Diseases of Ornamental Plants*,
https://doi.org/10.1007/978-981-16-3919-7_7

reported that currently more than 60 viruses are known to infect orchid plants. This chapter describes latest research work on genome analysis and diversity study of viruses infecting orchid plants belonging to different virus groups.

Keywords

Orchids · Virus · Genome · ORF · RT-PCR · Phylogenetic analysis · Next-generation sequencing

7.1 Introduction

Ornamental plants are economically important and cultivated worldwide. The international business of ornamental plants is constantly expanding which includes cultivation of cut foliage plants and potted flowers and proliferation of plant material. Orchid family (Orchidaceae) is the second largest family of flowering plants with approximately 28,000 species and more than 850 genera. Development and registration of more than 100,000 hybrids have also contributed substantially to the development of international trade in potted plants as well as in cut flowers. Orchids display an enormous range of diversity in shape, size, color, and prolonged vase life, which makes orchids cultivating a highly money-spinning industry globally. The plants long to genera like *Arachnis*, *Aranda*, *Ascocentrum*, *Ascocenda*, *Cattleya*, *Cymbidium*, *Dendrobium*, *Laelia*, *Mokara*, *Oncidium*, *Paphiopedilum*, *Phalaenopsis*, *Renenthera*, and *Vanda* are used for potted as well for cut flowers (Pant et al. 2020).

The orchid-based floriculture industriousness is well instituted in South Asian countries—Thailand, Singapore, Malaysia, and Vietnam including other countries like Holland, Taiwan, and Hawaii—and contributes significantly to the growth of their economies (Pant et al. 2020). The diversity increases towards the tropic, where the epiphytic species predominate and constitute almost 73% of the family. Colombia is the country with greatest number of orchid species in America (3000 spp.) followed by Ecuador and Brazil (2500 spp. each one).

For germination process, the orchid seeds are completely dependent on a symbiotic relationship with soil-borne fungi, i.e., *Rhizoctonia* spp. The fungi usually attack inside the cytoplasm of orchid cells to obtain the nutrition and in turn assist the orchid seeds to germinate. Viral diseases possess serious threat to the orchid industry because they reduce the general vigor of orchid plants yielding in the decrease decorative value of flower and aesthetic value of propagated material which in turn reduces its marketability causing serious economic losses. Plants like orchids consist of several species and cultivars which also reflect huge diversity of viruses infecting these ornamental plants. The vegetative propagation of orchid plants significantly contributes to the unrestricted spread of plant viruses. Perennial ornamental crops may act as natural reservoirs for numerous plant viruses. Such plants facilitate the transmission of viruses to other economically important crops. The study based on molecular investigation of viruses infecting ornamentals is an important step towards effective diagnosis and prevention of these viruses.

The first evidence of virus associate with orchid plant was reported in 1943 in Australia by Magee. This virus was characterized as rod-shaped particles causing mosaic disease of orchids. Later on, Nobrega (1947) reported cucumber mosaic virus (CMV) in *Dendrobium* orchids from Brazil. Cymbidium mosaic virus was identified from *Cymbidium* spp. from California described by Jenson (1950). Subsequently, *Odontoglossum ringspot virus* (ORSV) was identified in *Odontoglossum grande* plant from the USA (Jenson and Gold 1951). As of 2020, more than 60 plant viruses have been informed to affect orchid plants. Among them, 44 orchid-infecting viruses have been identified and classified (Table 7.1). The remaining orchid-associated viruses are still unclassified (Table 7.2) (Lee et al. 2017). Up to now, 147 orchid virus sequences are available in GenBank database.

7.2 Detection of Plant Viruses

Virus detection and identification are always challenging for plant virologists. During the last few decades, a number of diagnostic methods have been reported for plant virus detection (Smith 1931). Initially, methods introduced were based on biological properties of the viruses, but they were later upgraded with addition of the microscopy (light microscopy and electron microscopy) and serological techniques (Slogteren and Slogteren 1957; Ball and Brakke 1968). Later on, the diagnosis of plant viruses was revolutionized by the application of ELISA (Clark and Adams 1977a), several molecular-based techniques (Mumford et al. 2003; Webster et al. 2004; James et al. 2006), and NGS (Singh et al. 2020a, b). As the expertise of diagnosis developed, the challenges also increased enormously. As on date, more than 1200 different plant viruses infecting plants are reported (Fauquet et al. 2005), among them more than 50 viruses are associated with orchid plants (Pant et al. 2020). The increasing viral genomic diversity is a massive challenge, posing serious threats in ornamental plants including orchids. The seriousness of this plant disease increases multifold due movement of live plant materials across the international borders. Previously it is showed that mother plant infected with viruses will instantly spread in the new populations (Boonham et al. 2007).

Keeping in view of losses caused by viruses, delicate and trustworthy diagnostics are essential for identifying the source plants, germplasm, and tissue culture-grown plants. Previously reports on the presence of viruses in commercial nurseries suggest the unintentional spread of disease via infected planting material.

7.2.1 Conventional and Molecular Diagnosis

The initial indication of viral infection in plant is phenotype changes in leaf, root, stem, etc. of host plant which should be further confirmed by other suitable tests. In orchid plants, symptoms of viral infections may vary with different environmental

Table 7.1 Orchid viruses classified on the basis of decisive information

No.	Name of virus	Abbreviation	Genus	References
1	*Cymbidium mosaic virus*	CymMV	*Potexvirus*	Francki (1970)
2	*Odontoglossum ringspot virus*	ORSV	*Tobamovirus*	Seoh et al. (1998)
3	*Tobacco mosaic virus*—orchid	TMV-orchid	*Tobamovirus*	Lawson and Hsu (1995)
4	*Cucumber mosaic virus*	CMV	*Cucumovirus*	Inouye (1969)
5	*Bean yellow mosaic virus*	BYMV	*Potyvirus*	Lesemann and Koenig (1985)
6	*Calanthe mild mosaic virus*	CalMMV	*Potyvirus*	Gara et al. (1998a)
7	*Ceratobium mosaic virus*	CerMV	*Potyvirus*	Mackenzie et al. (1998)
8	*Clover yellow vein virus*	ClYVV	*Potyvirus*	Inouye et al. (1998)
9	*Cypripedium virus Y*	CypVY	*Potyvirus*	Gribbs et al. (2000)
10	Dendrobium mosaic virus	DeMV	*Potyvirus*	Inouye (1976), Hu et al. (1995)
11	Dendrobium severe mosaic virus	DeSMV	*Potyvirus*	Gara et al. (1998a)
12	*Diurus virus Y*	DiVY	*Potyvirus*	Gibbs et al. (2000)
13	Habenaria mosaic virus	HaMV	*Potyvirus*	Inouye et al. (1998)
14	Pecteilis mosaic virus	PcMV	*Potyvirus*	Yora et al. (1983)
15	*Pleione virus Y*	PlVY	*Potyvirus*	Gibbs et al. (2000)
16	Pterostylis virus Y	PtVY	*Potyvirus*	Gibbs et al. (2000)
17	*Rhopalanthe virus Y*	RhVY	*Potyvirus*	Gibbs et al. (2000)
19	*Turnip mosaic virus*	TuMV	*Potyvirus*	Lesemann and Koenig (1985)
20	Vanilla mosaic virus	VanMV	*Potyvirus*	Wisler et al. (1987)
21	Vanilla necrosis virus	VanNV (VNV)	*Potyvirus*	Pearson and Pone (1988)
22	*Watermelon mosaic virus 2*	WMV-2	*Potyvirus*	Gara et al. (1997)
23	Dendrobium vein necrosis virus	DVNV	*Closterovirus*	Lesemann (1977)
24	*Tobacco rattle virus*	TRV	*Tobravirus*	Lesemann and Vetten (1985)
25	*Cymbidium ringspot virus*	CymRSV	*Tombusvirus*	Hollings et al. (1977)
26	*Tomato ringspot virus*	TomRSV	*Nepovirus*	Goff and Corbett (1977)
27	Orchid fleck virus	OFV	*Rhabdovirus*	Chang et al. (1976)
28	*Tomato spotted wilt virus*	TSWV	*Tospovirus*	Hu et al. (1993)
29	*Impatiens necrotic spot virus*	INSV	*Tospovirus*	Lawson and Hsu (1995)
30	Calanthe mosaic virus	CalMV	*Carlavirus*	Yamamoto and Ishii (1981)
31	*Bean common mosaic virus*	BCMV	*Potyvirus*	Grisoni et al. (2004)
32	*Colombian datura virus*	CDV	*Potyvirus*	Fry et al. (2004)
33	*Cowpea aphid-borne mosaic virus*	CABMV	*Potyvirus*	Grisoni et al. 2006

(continued)

Table 7.1 (continued)

No.	Name of virus	Abbreviation	Genus	References
34	*Dasheen mosaic virus*	DsMV	*Potyvirus*	Jordan et al. (2004)
35	*Ornithogalum mosaic virus*	OrMV	*Potyvirus*	Grisoni et al. 2006
36	*Spiranthes mosaic virus 2*	SpiMV2	*Potyvirus*	Guaragna et al. (2006)
37	*Spiranthes mosaic virus 3*	SpiMV3	*Potyvirus*	Guaragna et al. (2006)
38	*Wisteria vein mosaic virus*	WVMV	*Potyvirus*	Grisoni et al. (2006)
39	Capsicum chlorosis virus	CaCV	*Tospovirus*	Zheng et al. (2008b)
40	Phalaenopsis chlorotic spot virus (=*Basella rugose mosaic virus*)	PhCSV (BaRMV)	*Potyvirus*	Zheng et al. (2008c)
41	*Carnation mottle virus*	CarMV	*Carmovirus*	Zheng et al. (2008d), Zheng et al. (2011)
42	Cymbidium chlorotic mosaic virus	CymCMV	*Sobemovirus*	Kondo et al. (2015)
43	**Donkey orchid virus A**		*Potyvirus*	Wylie et al. (2013)
44	**Donkey orchid symptomless virus**	DOSV	*Potyvirus*	Gara et al. (1998a)

Table 7.2 Unclassified viruses affecting orchid plants

No.	Virus	Abbreviation	References
1	Aranthera filamentous virus	—	Inouye (2008)
2	Colmanara mosaic virus	ColMV	Inouye (2008)
3	Cymbidium mild mosaic virus	CymMMV	Doi and Yora (1978)
4	Cymbidium potyvirus	—	Singh et al. (2007)
5	Dendrobium leaf spot virus	—	Peters (1977)
6	Dendrobium rhabdovirus	—	Lawson and Ali (1975)
7	Dendrobium virus	—	Lawson and Hsu (1995)
8	Grammatophyllum bacilliform virus	GBV	Inouye (2008)
9	Laelia red leaf spot virus	LRLV	Lawson and Hsu (1995)
10	Long orchid rhabdovirus	—	Inouye (2008)
11	Masdevallia isometric virus	—	Inouye (2008)
12	Phalaenopsis bacilliform virus	—	Inouye (2008)
13	Phalaenopsis necrosis virus	—	Lesemann and Begtrup (1971)
14	Short orchid rhabdovirus	—	Inouye (2008)
15	Trichopilia isometric virus	—	Inouye (2008)
16	Pleione flower breaking virus	PlFBV	Komínek et al. (2019)

conditions of growing and species of viruses. The electron microscopy (EM) is developed as a structural analysis and diagnostics tool for viruses (Martelli and Russo 1984; Hill 1984; Milne 1972). The majority of plant viruses may exist in

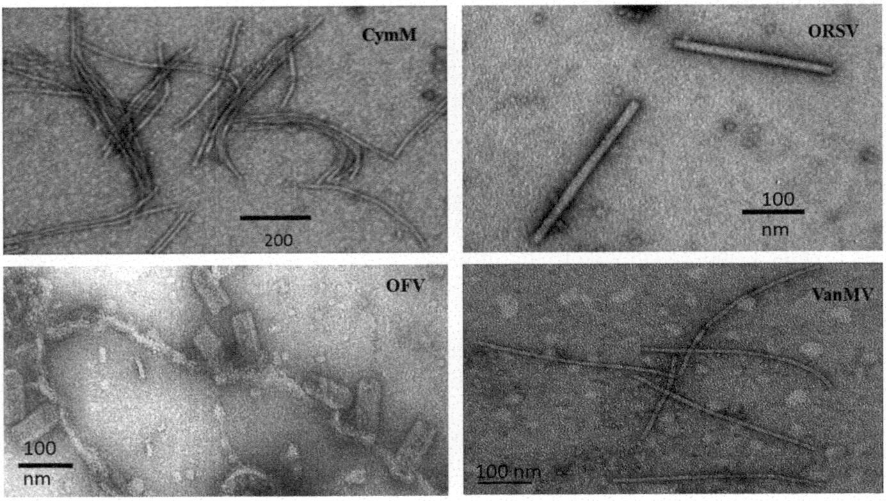

Fig. 7.1 Electron micrograph of orchid-infected viruses

adequate concentrations in infected plant tissue which may be visualized under EM. Several methods have been employed for direct visualization of virus particle using EM such as tissue extraction, leaf dip method, and partially purified and purified virus preparation methods. Principally, all the abovementioned methods assist in release of virus particles from virus-infected host cells. The EM technique has been used for diagnosis of numerous orchid viruses especially OFV, CymMV, CaCV, CalMMV, and ORSV (Fig. 7.1). Ultrathin sectioning has also been used for visualization of location of viruses and virus-induced inclusion bodies in host tissue. Limitation of EM is that it may only demonstrate the shape and size of the virus particle. The presence of virus in plant tissue samples can be confirmed by ISEM or other diagnostic assays. Previous studies demonstrated that occasionally bioassay and serological tests may fail in the identification of ORSV and CymMV. However, in such conditions, EM may still provide accurate identification of virus (Lawson and Brannigan 1986).

Molecular (DNA amplification and molecular hybridization) and specific and rapid serological techniques (ELISA) have been developed for the plant virus detection. Molecular hybridization based on sequence-specific DNA or RNA probes binds with viral nucleic acids, due to their base complementarity sequence. However, ELISA technique utilizes selective binding of viral proteins with specific antibodies (Clark and Adams 1977b). The genome of these viruses can be further studied by first-generation Sanger sequencing followed by NCBI database search which facilitate the more precise identification and comparison with previously reported similar viruses. The sequence analysis-based approaches also facilitate the genotyping of newly isolated virus populations, study of genetic diversity, and their evolution.

In one of the studies, a multiplex RT-PCR assay was validated for simultaneous recognition of three orchid viruses, viz., Odontoglossum ringspot virus (ORSV),

Cymbidium mosaic virus (CymMV), and Orchid fleck virus (OFV). The assay was performed against 31 collected plant samples from 6 different orchid genera, and the result was also compared with results of transmission electron microscopy (TEM). The study revealed that for detection of OFV, RT-PCR assay is more sensitive in comparison to TEM (Ali et al. 2014). Similarly, CP and RdRp gene and their respective TaqMan probes were used in real-time quantitative PCR (qPCR) for the recognition of ORSV and CymMV (Eun and Wong 2000). In another study, Liu et al. (2009) used SYBR green-based real-time RT-PCR assay for quantitative detection of sequence containing CP gene of Cymbidium mosaic virus (CymMV) in China. The assay was found sensitive for CymMV detection with detection limit of two copies of viral genome per tube. Thus, SYBR green-based real-time RT-PCR assay can be used for characterization of CymMV with high specificity, sensitivity, and precision (Liu et al. 2009).

7.2.2 Next-Generation Sequencing

Next-generation sequencing (NGS) technologies have led to a revolution in plant virus diagnosis (Maree et al. 2018). The specialty of NGS does not require prior knowledge of viral genome sequences. Moreover, it can be used for sequencing of millions or billions of nucleic acid fragments in parallel leading to detection of several viruses at a time in an infected host plant (Villamor et al. 2016). In 2009, it was first time used for plant virus detection (Al Rwahnih et al. 2009; Adams et al. 2009). For identification of plant viruses using NGS techniques, several sample preparation methods are used such as extraction of double-stranded RNA (dsRNA); virus-derived small interfering RNA (vsiRNA); total RNA (totRNA); ribosomal RNA-depleted total RNA (rRNA-depleted tRNA); RNA from purified or partially purified viral particles; RNA after subtractive hybridization with healthy plant RNA; and polyadenylated RNA (poly(A) RNA) (Wu et al. 2015; Roossinck et al. 2015; Adams and Fox 2016). Moreover, scientific development in different NGS platforms also led to revolution in discovery of new plant viruses. In one of the study, Illumina sequencing platform was used for comparison of two RNA sequence inputs (ribosomal RNA-depleted total RNA vs. small RNAs) for the detection of plant viruses (Pecman et al. 2017). The sequence comparison from different samples provides information about several viruses having differences in genome organization. The viral genome sequences obtained from the samples usually depends on genome organization of virus and number of viral sequence reads in the sequence metadata. Apart from identification of new plant viruses, NGS can also be used for the study of epidemiology, interactions between viruses, and genetic diversity along with the study of evolutionary mechanisms of viruses (Singh et al. 2020a, b). The complete genome sequence-based study using Sanger and high throughput sequencing assisted in identification of a novel potyvirus i.e. *Pleione flower breaking virus* (PlFBV) in orchids of the genus Pleione in Netherlands and in the Czech Republic (Komínek et al. 2019) (Fig. 7.2).

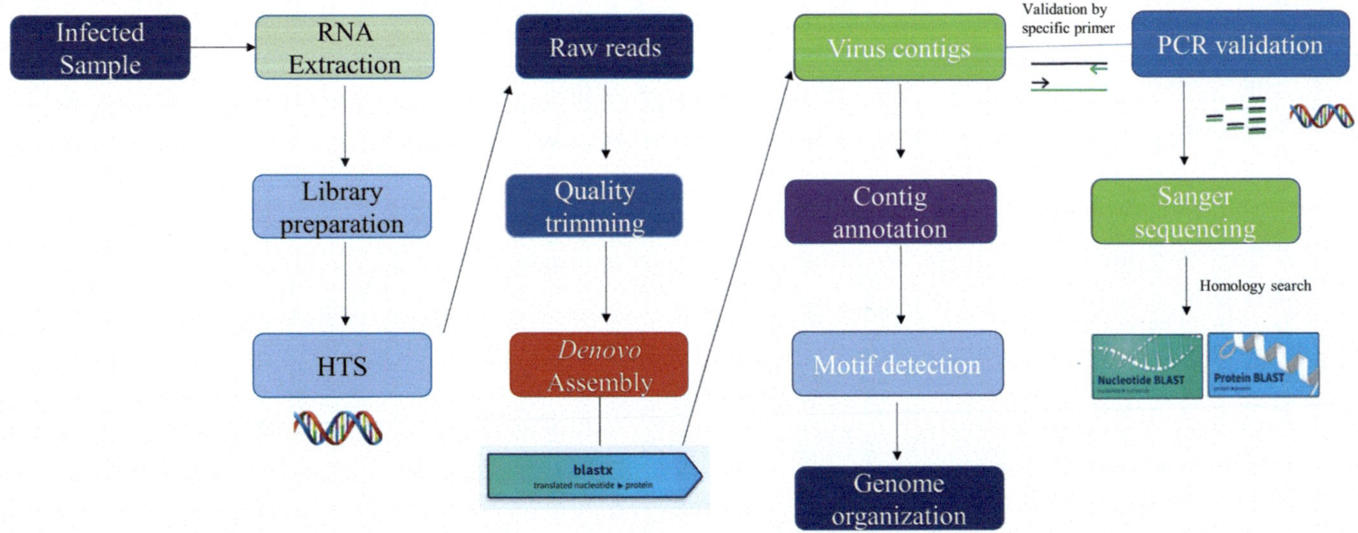

Fig. 7.2 Methodology for identification and characterization of orchid-infecting viruses using NGS approaches

Pai et al. (2020) described the genome-wide analysis of small RNAs from Cymbidium mosaic virus (CymMV) and Odontoglossum ringspot virus (ORSV) (Pai et al. 2020). These are most prevalent and economically important viruses which infect orchid plants worldwide. The high-throughput RNA-seq-based study also revealed the presence of an unusual plant virus, donkey orchid symptomless virus (DOSV), from asymptomatic plant of common donkey orchid (Wylie et al. 2013). Further genetic study revealed that DOSV has significant level of genetic deviation from known plant viruses. The CP and replicase gene of DOSV showed homology with those from alphaflexiviruses, while the MP gene showed resemblances with tombus-like viruses. Finally, DOSV was classified into the genus *Platypuvirus* under the family *Alphaflexiviridae*. It indicates an evolutionary link between tombus-like viruses and alphaflexiviruses (Wylie et al. 2013).

7.3 Characterization of Orchid Viruses

7.3.1 *Cymbidium Mosaic Virus* (CymMV)

Cymbidium mosaic virus (CymMV) is the most important virus infecting orchids and causing economic losses globally. The orchid plants infected with CymMV are characterized by specific symptoms such as chlorotic streaks with pitted dark spots on leaves. It is a member of the genus *Potexvirus* under the family *Flexiviridae* (Adams et al. 2004). The length of the complete genome is approximately 6227 bp ssRNA with 5 open reading frames (ORFs) (Wong et al. 1997). ORF1 encodes a replicase (1620 residues), ORF2 a triple gene block 1 (TGB 1) (260 residues), ORF3 a triple gene block 2 (TGB 2) (130 residues), ORF4 a triple gene block 3 (TGB 3) (280 residues), and ORF5 a viral coat protein (CP) (240 residues). ORF1 contains three conserved domains: a methyl transferase domain, RNA helicase domain, and RNA-dependent RNA polymerase (RdRp). TGB1, TGB2, and TGB3 proteins are involved in suppression of RNA silencing and cell-to-cell movement (Verchot-Lubicz 2005). CP is also involved in cell-to-cell movement and encapsidation of viral RNA (Adams et al. 2004). The GenBank database presently contains the 14 complete sequences and 157 CP sequences of CymMV. The BLASTn (Altschul et al. 1990) searches to the NCBI databases revealed that *Cymbidium mosaic virus* (CymMV) shared 87.6–97.6% nucleotide sequence identities with other CymMV isolates reported from the USA, China, Japan, India, Korea, Malaysia, Taiwan, Singapore, and Canada. Similar results were obtained with neighbor joining (NJ)-based phylogenetic tree of complete sequence of CymMV from different regions of the world using MEGA X program (Kumar et al. 2018) (Fig. 7.3).

In India, CymMV-like symptoms such as leaf necrosis, flower color break, and necrotic spotting were detected in several species of orchids including *Cymbidium* hybrid (Arabian Sea), *Cymbidium aloifolium*, *Epidendrum* sp., *C. iridioides*, *Liparis bootanensis*, *Pholidota imbricata*, and *Phaius tankervilleae*. Later on, RT-PCR followed by nucleic acid sequencing confirmed the presence of RdRp gene of CymMV with 94% sequence homology in Korea (Seoh et al. 1998). The variability

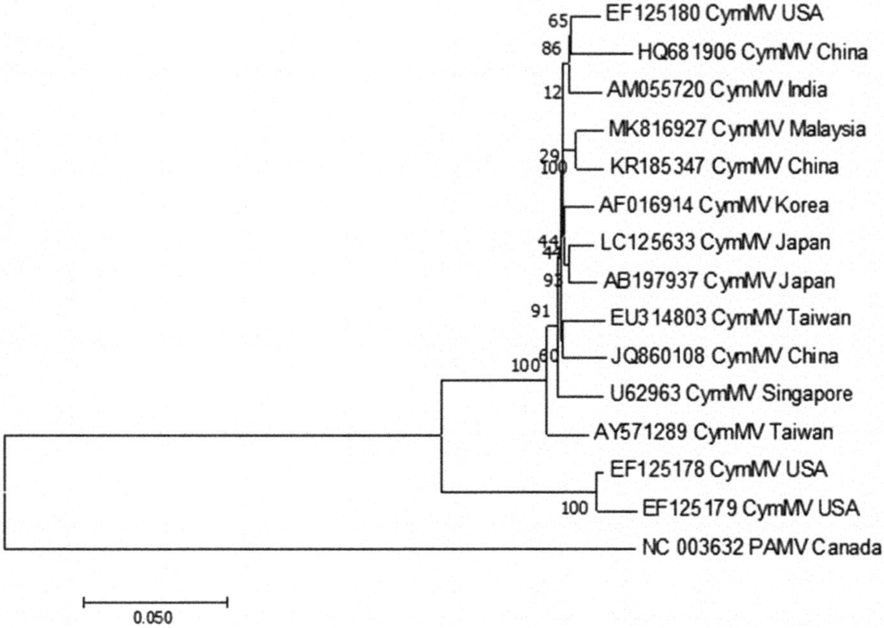

Fig. 7.3 Phylogenetic of CymMV based on complete genome sequence and PAMV (NC 003632) used as outgroup. CymMV (USA-EF125180; EF125178; EF12517; China-HQ681906; KR18347; EU3414803; JQ860108; India-AM055720; Malaysia-MK816927; Korea-AF016914; Japan-LC125633; AB197937; Taiwan-AB197937; AY571289; Singapore-U62963; USA-PAMV (NC_003632))

study of nucleotide (nt) and amino acid (aa) sequences of coat protein (CP) of 157 Cymbidium mosaic virus (CymMV) isolates was done in Korea which showed 74.9–98.3% and 52.7–100% homology at nt and aa levels, respectively, with previously reported CymMV isolates from different regions of the world (Yoon et al. 2012). The analysis of CymMV sequence data from Korea, Taiwan, and Singapore showed the homology of 89.1%, 99.7%, 93.2%, and 100% at the nt and aa levels, respectively (Ajjikuttira et al. 2002). However, during sequence analysis, variability was not detected in any specific regions of the viruses, and the C-terminal was found to be more divergent compared to N-terminal region.

7.3.2 *Odontoglossum Ringspot Virus* (ORSV)

ORSV is the most important viral infection of orchids worldwide (Sherpa et al. 2004) and belongs to the genus *Tobamovirus*. The infection of ORSV is characterized by appearance of specific symptoms such as ringspot, stripe breaking, mosaic on leaves, and color burst and color breaking on *Cattleya* and *Cymbidium* flowers (Lawson and Hsu 1995; Kondo et al. 2003; Blanchfield et al. 2001). The ORSV-infected *Phalaenopsis* and *Oncidium* plants showing normal growth

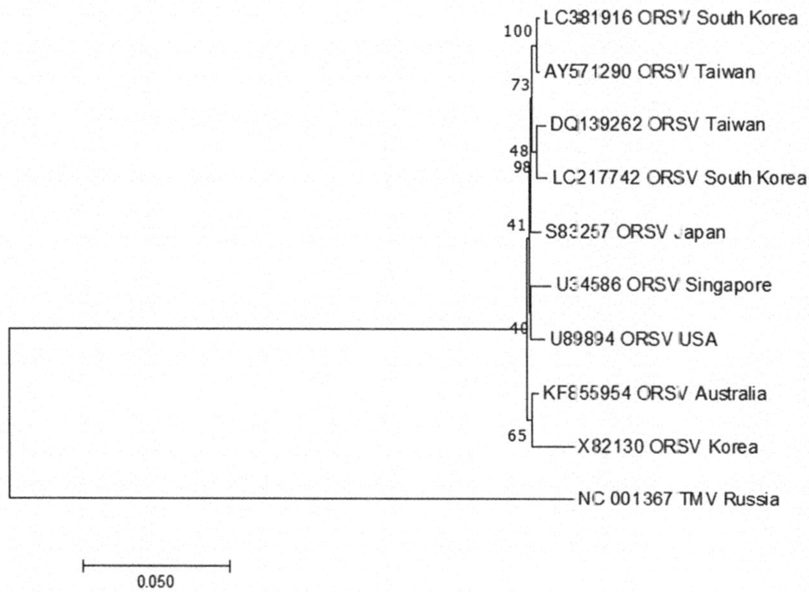

Fig. 7.4 Phylogenetic of ORSV based on complete genome sequences and TMV (NC 001367) used as outgroup. ORSV (South Korea-LC381916; LC217742; Taiwan-AY571290; DQ139262; Japan-S83257; Singapore-U34586; USA-U89894; Australia-KF8559454; Korea-X82130); TMV (Russia-NC_001367)

and flower structure with no obvious disease symptoms were reported, whereas ORSV and CymMV co-infected *Phalaenopsis* and *Oncidium* showed enhanced symptoms (Navalinskiene et al. 2005) and approximately 31 orchid genera (Jensen and Gold 1951; Zaitlin 1976) with vanilla plant (Grisoni et al. 2006). ORSV is considered as the most common and economically important virus infecting orchids (Jensen and Gold 1951; Zettler et al. 1990). The length of the complete genome of ORSV is approximately 6600 bp ssRNA with 4 open reading frames (ORFs) (Ikegami et al. 1995). ORF1 encodes a 181 kDa read-through protein with 1596 aa residues, ORF2 a 181 kDa replicase (1112 aa residues), ORF3 a movement protein (303 aa residues), and ORF4 a coat protein (CP) (158 aa residues). The GenBank database presently contains 8 complete sequences and 122 CP sequences of ORSV. BLASTn (Altschul et al. 1990) analysis among ORSV strains in NCBI databases revealed that this virus shares 87.6–97.6% nucleotide sequence identities with other ORSV isolates reported from the USA, China, Japan, India, Korea, Malaysia, Taiwan, Singapore, and Canada. Similar results were obtained with neighbor joining (NJ)-based phylogenetic tree of complete sequence of ORSV from different regions of the world using MEGA X program (Kumar et al. 2018) (Fig. 7.4). However, very little information is available on incidence of this virus in orchid plant in India. Sherpa et al. (2004) described the development of a slot-blot hybridization assay for ORSV.

The CP gene nucleotide sequence was determined for Indian ORSV isolate (AJ564563) which shared 96–100% homology with other ORSV isolates at nucleotide as well as amino acid. RT-PCR revealed that the orchid samples found to be symptomless may also remain infected with the virus. Due to high sensitivity and efficacy, RT-PCR was reported to be useful for the certification and marketing of virus-free orchid plant material to commercial growers. Among the ORSV isolates, collected from Korea, Singapore, and Taiwan, 95.5–100% and 93–100% sequence homology at nucleotide and amino acid level, respectively, were reported. Therefore, the higher level of sequence conservation suggested that ORSV coat protein genes might be suitable candidate to study the resistance against virus in orchid plants cultivated in different regions of the world (Ajjikuttira et al. 2002).

7.3.3 *Orchid Fleck Virus* (OFV)

Orchid fleck virus (OFV) belongs to genus *Dichorhavirus*. The infection with OFV led to chlorotic ringspots and fleck-like symptoms in orchid plants. The virion particle of OFV is morphologically characterized by non-enveloped, bacilliform architect of about 40 nm × 100–150 nm (Fig. 7.6). The complete genome of OFV consists of 2 RNA molecules, RNA1 (6431 bp) and RNA2 (6001 bp), with complementary terminal sequences (Ajjikuttira et al. 2002). Moreover, it has five open reading frames (ORFs) in RNA1 and a single ORF in RNA2 molecule. The molecular study revealed that some of the proteins of OFV have protein sequence homology with rhabdoviruses. However, OFV differs from viruses of *Rhabdoviridae* family in having a bipartite genome. The GenBank database presently contains the 06 complete sequences of OFV. BLASTn (Altschul et al. 1990) searches to the NCBI database showed that OFV shared 82.28–100% nucleotide sequence identities with other OFV isolates reported from China, Japan, South Korea, and Mexico. Similar results were obtained with neighbor joining (NJ)-based phylogenetic tree of complete sequence of RNA 1 (Fig. 7.5) and RNA 2

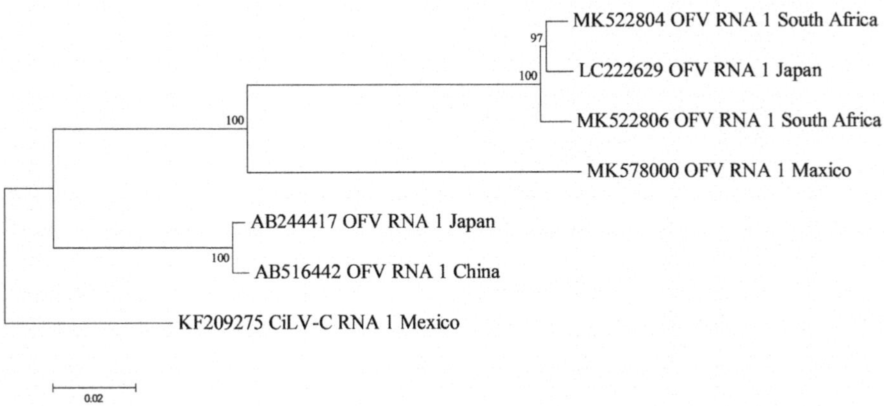

Fig. 7.5 Phylogenetic of OFV based on complete RNA 1 sequence and CiLV-C RNA 1 (KF209275) used as outgroup. OFV RNA-1 (South Africa-MK522804; MK522806; Japan-LC222629; AB244417; Maxico-MK578000; China-AB516442); CiLV-C (Mexico-KF209275)

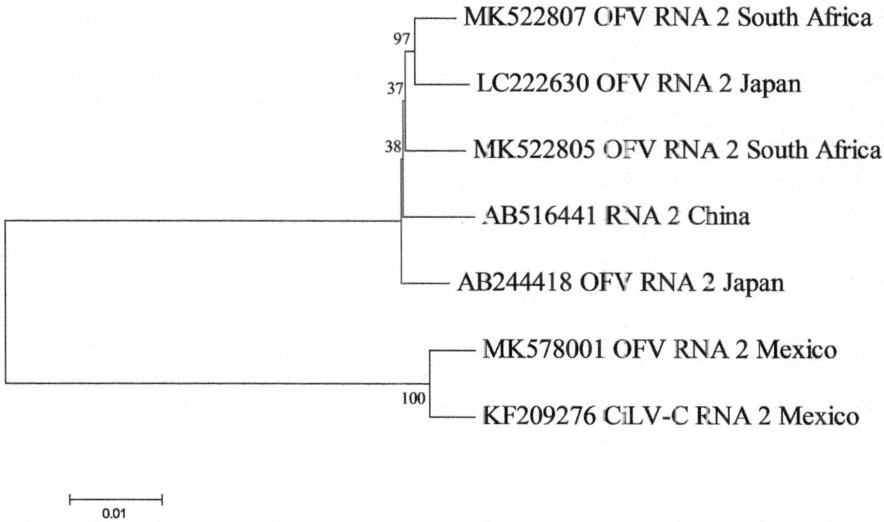

Fig. 7.6 Phylogenetic of OFV based on complete RNA 2 sequence and CiLV-C RNA 2 (KF209276) used as outgroup. OFV RNA-2 (South Africa-MK522807; MK522805; Japan-LC222630; AB244418; China-AB516441; Mexico-MK578001); CiLV-C (Mexico-KF209276)

(Fig. 7.6) of OFV from different regions of the world using MEGA X program (Kumar et al. 2018).

7.3.4 Potyviruses

Potyvirus genus is the largest group of single-strand positive-sense RNA viruses infecting plants (Revers and García 2015). According to the International Committee on Taxonomy of Viruses (ICTV) report, there are at least 168 potyvirus species currently being recognized (ICTV, 2018). Sixteen potyviruses, BYMV, *Colombian datura virus* (CDV), *Calanthe mild mosaic virus* (CalMMV), *Ceratobium mosaic virus* (CerMV), *Dasheen mosaic virus* (DsMV), *Clover yellow vein virus* (ClYVV), *Spiranthes mosaic virus* 2 (SpiMV 2), *Dendrobium mosaic virus* (DeMV), *Habenaria mosaic virus* (HaMV), *Pecteilis mosaic virus* (PcMV), *Diurus virus Y* (DVY), *Turnip mosaic virus* (TuMV), *Vanilla mosaic virus* (VanMV), SpiMV 3, *Vanilla necrosis virus* (VNV), and *Sarcochilus virus Y* (SVY), were reported from orchid plants (Table 7.1) (Pant et al. 2020). However, potyviruses have narrow host range in comparison to several other known plant viruses (Wylie et al. 2017). They are primarily transmitted via aphids in a non-persistent manner (Ng and Perry 2004).

The virion particle of potyvirus is flexuous filamentous in structure with no envelope. It often measures 680–900 nm in length and 11–13 nm in diameter. The genome of potyvirus consists of a single-stranded, positive-sense RNA of about

9.7 kb in size where a viral genome-linked protein (VPg) is covalently connected with 5′ end and a polyadenylated tail is bound to 3′ end (Adams et al. 2012). The virus genome encodes 11 different functional proteins. The small ORF is termed as "pretty interesting potyvirus ORF" (PIPO) which encodes a protein termed as P3N-PIPO via the RNA polymerase slippage mechanism. The PIPO plays a major role in intercellular movement of the viruses (Rodamilans et al. 2015). On the other hand, the large ORF encodes a polyprotein which is self-cleaved into ten different polypeptides at specific cleavage sites (Wylie et al. 2017). These polypeptides include a P1 protein, a P3 protein, a 6 kDa peptide (6K1), a helper component-proteinase (HC-Pro), a cylindrical inclusion protein, a nuclear inclusion "b" protein (NIb), a second 6 kDa peptide (6K2), VPg, a nuclear inclusion "a" (NIa) protease, and a coat protein (CP) (Wylie et al. 2017; Adams et al. 2012). Most of the potyvirus proteins perform multiple functions and cooperate with each other in viral infection cycle.

7.3.4.1 *Habenaria Mosaic Virus* (HaMV)

Habenaria radiata (Thurnberg) K. Spreng is a wild orchid plant growing in wet grasslands in Japan. It is cultivated as an ornamental plant due to its beautiful bird-shaped flower petals (Mitsukuri et al. 2009). The complete genome sequence-based analyses have revealed the presence of *Habenaria mosaic virus* (HaMV) in terrestrial orchids (*Habenaria radiata*). The complete genome sequence of HaMV is approximately 9.5 kb with single ORF encoding a large polyprotein of 3054 amino acids. Sequence comparison with potyvirus showed that polypeptide chain of HaMV has typical genomic features of a potyvirus. Moreover, putative proteolytic cleavage sites were also identified in amino acid sequence of HaMV by sequence comparison with potyviruses. The HaMV polyprotein showed 58% amino acid sequence identity with tobacco vein banding mosaic virus (Kondo et al. 2014). The GenBank database presently contains three complete sequences of HaMV. BLASTn (Altschul et al. 1990) analysis of HaMV strains revealed that the virus shares 81.16% nucleotide sequence identities with other HaMV isolates reported from South Korea. In neighbor joining (NJ)-based phylogenetic tree, the HaMV clustered within same clade with other isolate of *Habenaria mosaic virus* (HaMV) reported from South Korea and other potyviruses like *Tobacco vein banding mosaic virus* (TVBMV) from China, and *Chilli veinal mottle virus* (ChiVMV) from China using MEGA X program (Kumar et al. 2018) (Fig. 7.7).

7.3.4.2 *Calanthe Mild Mosaic Virus*

Calanthe mild mosaic virus (CalMMV) was reported from Japan. It causes symptoms like flower color breaking and mild leaf mosaic in *Calanthe* plants. The transmission of CalMMV was found mechanically through the aphid *Myzus persicae* in a non-persistent manner to *Phalaenopsis* sp., *Calanthe* sp., and *Tetragonia expansa*. The virus consists of 764 nm long flexuous virion particles and forms cylindrical intracellular cytoplasmic inclusions. The CalMMV genome consists of a single RNA of mol. weight 3.1×10^6 which encodes single polypeptide of 32.0 kDa.

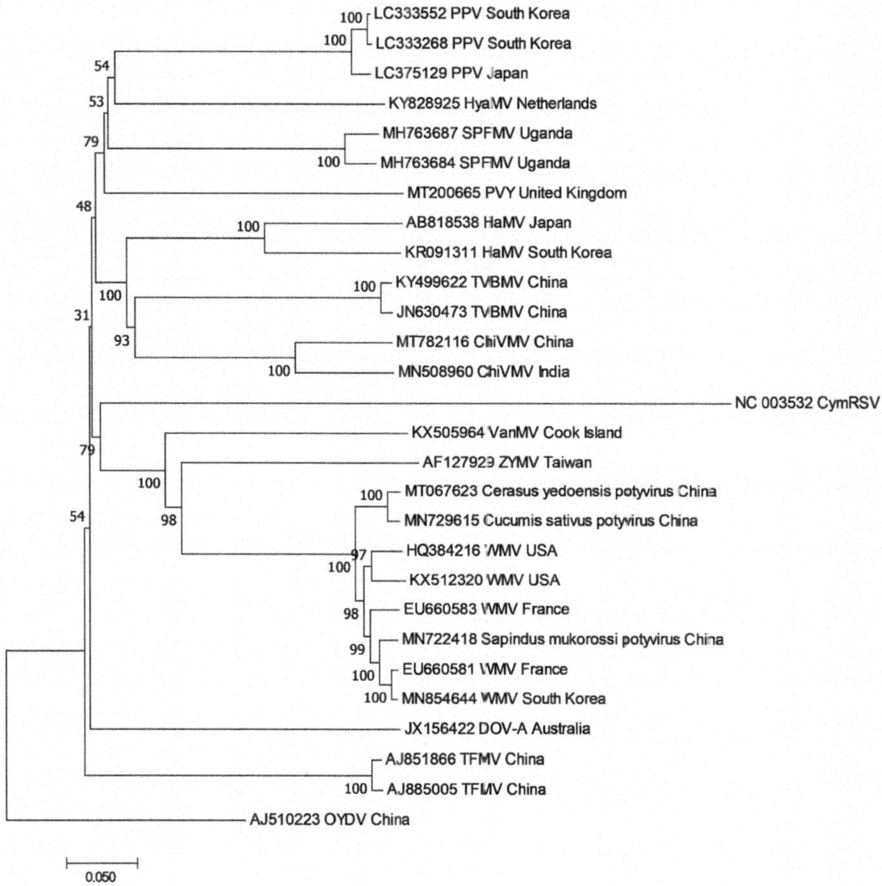

Fig. 7.7 Phylogenetic analysis of orchid-infecting *Potyvirus*: PPV (South Korea-LC333552; LC333268; Japan-LC375129); HyaMV (Netherlands-KY828925); SPFMV (Uganda-MH73687; MH763684); PVY (UK-MT200665); HaMV (Japan-AB818538; South Korea-KR091311); TVBMV (China-KY499622; JN630473); ChiVMV (China-MT782116; India-MN508960); CymRSV (NC_003532); VanMV (Cook Island-KX505964); ZYMV (Taiwan-AF127929); CYP (China-MT067623); CSP (China-MN729615); WMV (USA-HQ384216; KX512320; France-EU660583; EU660581; South Korea-MN854664); DOV-A (Australia-JX156422); TFMV (China-AJ851866; AJ225005); OYDV (China-AJ510223)

Immunoelectron microscopy revealed that CalMMV is distantly associated with Japanese isolate of dendrobium mosaic potyvirus (DeMV). However, it did not show any serological relationship with other potyviruses. The amino acid sequence of the CP gene of CalMMV was found 73% homologous with DeMV. However, it was found less than 66% homologous with other potyviruses (Gara et al. 1998a, b).

Phaius tankervilleae is a ground orchid species widely distributed in the Indian subcontinent, Malaysia, China, the USA, Japan, the Philippines, and Australia.

During surveillance program in December 2016 in Sikkim and Darjeeling hills, several members from *P. tankervilleae* were observed with streak symptoms and mild mosaic symptoms in germplasm collection center of the National Research Centre for Orchids, Pakyong, Sikkim. The negative staining preparations of symptomatic leaf samples visualized under the electron microscope revealed the presence of flexuous filamentous particles of about 800 nm long and 12 nm wide. Further study by RT-PCR (Gibbs et al. 2000) showed the absence of Odontoglossum ringspot virus and Cymbidium mosaic virus in the sample. Pair-wise sequence alignment of both the amplicons revealed 94 and 87% nt as well as 95 and 98% amino acid sequence identity with the corresponding CP gene sequence of CalMMV Japan isolate (804 bp) and partial CP gene sequence of Cymbidium potyvirus isolate (453 bp) (AJ871476), respectively. The amino acid sequences (DAG and WCIEN) were found conserved among the majority of potyviruses (Pant et al. 2020).

7.3.4.3 *Vanilla Mosaic Virus* (VanMV)

Vanilla mosaic virus (VanMV) belongs to *Potyvirus* genus that was first identified from French Polynesia (Zettler et al. 1987). There have been no reports of VanMV as yet outside the South Pacific region. Serological and sequence-based analysis revealed that VanMV is a specific strain of Dasheen mosaic virus (DsMV) that infects *Vanilla* spp. only. However, DsMV infects aroid species in tropical regions with a global distribution (Farreyrol et al. 2006). VanMV is transmissible to *V. pompona* by *Myzus persicae* and mechanically transmissible to *V. tahitensis*. However, VanMV is not mechanically transmissible to a wide range of other plant species, including *V. pompona* and Araceae species, while DsMV is not transmissible to *Vanilla* spp. (Puli'uvea et al. 2017). The length of complete genome sequence of VanMV is approx. 9.8 kb containing only 1 open reading frame (ORF) encoding a polyprotein of 3139 amino acids. The GenBank database presently contains only 1 complete sequence and 05 Nib/CP sequences of VanMV.

In India, *Zinnia bicolor* plant leaf samples showing symptoms of chlorotic rings and mosaic-like pattern were collected from Navsari, Gujarat. The RT-PCR assay from the symptomatic leaf extracts using potyvirus degenerative primer against the 3′ end of the NIb gene and the 5′ end of the coat protein (CP) gene revealed the presence of *Vanilla distortion mosaic virus* (VDMV) (Hsu et al. 2005). Later on, a VDMV-specific primer pair against the entire CP gene (804 nts) and partial NIb (143 nts) and 5′ UTR (153 nts) was designed to amplify the 1100 nt. The sequence analysis of this product revealed a maximum of 88% (nts) and 94% (aa) identity with VDMV from *Vanilla planifolia* in India which was the first report of VDMV in ornamental *Zinnia bicolor* in India. The BLASTn (Altschul et al. 1990) analysis revealed that *Vanilla mosaic virus* shares 75.18 to 84.37% nucleotide sequence identities with other potyvirus reported worldwide. The neighbor joining (NJ)-based phylogenetic tree analysis of complete sequence of *Vanilla mosaic virus* infecting orchid clustered in the same clade with other potyviruses: *Zucchini yellow mosaic virus* (ZYMV), *Cerasus yedoensis potyvirus*, and *Watermelon mosaic virus* (WMV) reported from different regions of the world (Fig. 7.7).

7.3.5 *Donkey Orchid Symptomless Virus* (DOSV)

Donkey orchid symptomless virus (DOSV) was identified from terrestrial orchid *Diuris longifolia* (common donkey orchid) and *Caladenia latifolia* (pink fairy orchid) (Ong et al. 2016). The genome analysis revealed ORF encoding seven putative proteins of apparently unrelated origins. The DOSV genome, ORF1 encodes a 69 kDa protein that overlapped with replicase and provided low identity with movement proteins (MPs) of tymoviruses. ORF2 encoded a 157 kDa replicase protein and 22 kDa coat protein (ORF4) which shared significant similarity with other viruses of family Alphaflexiviridae. ORF3 encoded a 44 kDa protein and shared low identity with autophagy protein and myosin from squirrelpox virus. ORF5 codes for a 27 kDa protein and reported to share no identity with any known proteins. A 14 kDa protein (ORF6) revealed limited sequence identity of 26% over a limited region of the envelope glycoprotein of Crimean-Congo hemorrhagic fever virus (*Bunyaviridae*). ORF7 (putative 25 kDa movement protein) shared limited identity with 3A-like movement proteins of viruses from *Tombusviridae* and *Virgaviridae*. Further, sequence analysis of putative replicase protein of DOSV suggested a common evolutionary origin of this virus with potexvirus-like virus in *Tymoviridae* and *Alphaflexiviridae*. The movement protein of DOSV shared an evolutionary history with that of dianthoviruses. Virus transmissibility of DOSV was studied in *Nicotiana benthamiana* plants via systemic infection. Presently, DOSV is not readily classified in current lower order virus taxa (Wylie et al. 2013). The GenBank database presently contains only 04 complete sequences of DOVS.

7.3.6 Groundnut Bud Necrosis Virus

Phalaenopsis orchid is an attractive ornamental plant with a long vase life. Some of the tospoviruses including *Impatiens necrotic spot virus* (INSV), *Capsicum chlorosis virus*, and *Tomato spotted wilt virus* (TSWV) are reported to infect *Phalaenopsis* plant (Zheng et al. 2008a; Chen et al. 2010). In 2016–2017, some of the *Phalaenopsis* plants at the National Research Centre for Orchids, Sikkim, with symptoms of mild mosaic, mild chlorotic spots, and dark green patches on leaves were collected and subjected to transmission electron microscopy which revealed the presence of 80–110 nm tospovirus-like particles. Further, study on these samples using DAC-ELISA with polyclonal antibodies against GBNV detects all the tospoviruses of serogroup IV (Holkar et al. 2017); INSV and TSWV (Agdia Inc., USA) showed that all the symptomatic samples reacted with polyclonal antibodies against GBNV. These ELISA-positive samples were allowed for sap inoculation in cowpea (*Vigna unguiculata*) and tobacco (*Nicotiana benthamiana*) plants where chlorosis, systemic mottling, and downward curling of leaves on tobacco and chlorotic local lesions on cowpea were observed post 4–5 days inoculation. Further, molecular study using RT-PCR using N gene-specific primers was performed to identify the GBNV and WBNV (Basavaraj et al. 2017). The N gene sequence analysis revealed that it shared

99% nt and 100% aa identity with the similar regions of other GBNV isolates which confirmed the association of GBNV with *Phalaenopsis* plants in India.

7.4 Conclusion

Orchids are best known for their eye-catching flowers and domination in the international cut flower trade; it ranks sixth in the international market. They are perhaps the luxurious attractive plants known today. This study highlights the need to study the evolution roles of indigenous and exotic viruses as well as assess phyto-health risks to conservation of ex situ and wild orchid populations. Virus diseases on orchids have increased over the years owing to increasing trade and exchange of germplasm. The main reason for cross-country spread is the import of infected plant materials by bypassing quarantine procedures. Today, more than 60 viruses are known to infect orchids all over the world, and the list is growing. CymMV, ORSV, and OFV are the most prevalent and economically important viruses of orchids worldwide. Recently, DOV-A and DOSV are reported in orchid. These viruses are highly contagious, stable, and found in very high concentrations in host tissue and transmitted only by mechanical means. Once the parent stock is infected, it becomes extremely difficult to control these viruses. Therefore, an integrated approach using virus-free propagation stock, strict sanitary measure, use of protein and nucleic acid-based diagnostics, and genetically engineered orchid resistance to orchid viruses may provide management of orchid viruses. These studies provide orchid-infecting virus population and insight into the intermediate stages of virus evolution. Accurate diagnosis and detection of orchid viruses is very important for controlling virus disease. If a virus is transmitted by an insect vector, application of insecticides will restrict viral spread. Orchids are propagated mainly via vegetative tissues, providing easy and rapid paths for viral transmission. Despite the fact that many orchid-infecting viruses have been identified, many field symptoms in orchids could be caused by uncharacterized viruses. As with many field virologists, a complete survey of viral pathogens in orchids requires continued research to overcome various limitations. A plant virus often causes different symptoms in various plants, and different viruses might cause similar symptoms in a single plant. Furthermore, a plant might be co-infected by multiple viruses and exhibit complex symptoms. Hence, it is difficult to identify a novel virus based on symptoms alone. To accurately diagnose viral diseases, ELISA using antisera against plant viruses, PCR, and NGS is often employed for quick screening. However, antisera against orchid viruses are not widely distributed. In addition, some viruses could be difficult to be isolated from affected plants, posing a great challenge. Without fulfilling Koch's postulates, one could not conclusively demonstrate that a viral agent is indeed responsible for the observed symptoms. Accurate identification and characterization of orchid infecting viruses requires tremendous work. These data suggest a more recent evolutionary history of orchid-infecting viruses.

References

Adams I, Fox A (2016) Diagnosis of plant viruses using next-generation sequencing and metagenomic analysis, Current research topics in plant virology. Springer, Cham, pp 323–335

Adams MJ, Antoniw JF, Bar-Joseph M, Brunt AA, Candresse T, Foster GD, Martelli GP, Milne RG, Zavriev SK, Fauquet CM (2004) The new plant virus family *Flexiviridae* and assessment of molecular criteria for species demarcation. Arch Virol 149:1045–1460

Adams IP, Glover RH, Monger WA, Mumford R, Jackeviciene E, Navalinskiene M, Boonham N (2009) Next-generation sequencing and metagenomic analysis: a universal diagnostic tool in plant virology. Mol Plant Pathol 10(4):537–545

Adams MJ, Zerbini FM, French R, Rabenstein F, Stenger DC, Valkonen JPT (2012) Family Potyviridae. In: King AMQ, Adams MJ, Carstens EB, Lefkowitz EJ (eds) Virus taxonomy, ninth report of the International Committee on Taxonomy of Viruses. Elsevier/Academic Press, London, pp 1069–1089

Ajjikuttira P, Lim-Ho C, Woon M et al (2002) Genetic variability in the coat protein genes of two orchid viruses: Cymbidium mosaic virus and Odontoglossum ringspot virus. Arch Virol 147:1943–1954

Al Rwahnih M, Daubert S, Golino D, Rowhani A (2009) Deep sequencing analysis of RNAs from a grapevine showing Syrah decline symptoms reveals a multiple virus infection that includes a novel virus. Virology 387(2):395–401

Ali RN, Dann AL, Cross PA, Wilson CR (2014) Multiplex RT-PCR detection of three common viruses infecting orchids. Arch Virol 159(11):3095–3099

Altschul SF, Gish W, Miller W, Myers EW, Lipman DJ (1990) Basic local alignment search tool. J Mol Biol 215(3):403–410

Ball EM, Brakke MK (1968) Leaves dip serology for electron microscopic identification of plant viruses. Virology 36:152–155

Basavaraj Y, Kumar A, Holkar SK, Jain RK, Mandal B (2017) First report of groundnut bud necrosis virus infecting periwinkle (Catharanthus roseus) in India. Plant Dis 101(8):1559–1559

Blanchfield AL, Mackenzie AM, Gibbs A, Kondo H, Tamada T, Wilson CR (2001) Identification of *Orchid fleck virus* by reverse transcriptase-polymerase chain reaction and analysis of isolate relationships. J Phytopatol 149:713–718

Boonham N, Tomlinson J, Mumford R (2007) Microarrays for rapid identification of plant viruses. Annu Rev Phytopathol 45:307–328

Chang MU, Arai K, Doi Y, Yora K (1976) Morphology and intracellular appearance of orchid fleck virus. Ann Phytopathol Soc Jpn 42:156–157

Chen TC, Lu YY, Cheng YH, Li JT, Yeh YC, Kang YC et al (2010) Serological relationship between melon yellow spot virus and watermelon silver mottle virus and differential detection of the two viruses in cucurbits. Arch Virol 155(7):1085–1095

Clark MF, Adams AM (1977a) Characteristics of micro plate method of ELISA for the detection of plant viruses. J Gen Virol 34:475–483

Clark MF, Adams AN (1977b) Characteristics of the microplate method of enzyme-linked immunosorbent assay for the detection of plant viruses. J Gen Virol 34(3):475–483

Doi Y, Yora K (1978) Cymbidium mild mosaic virus, isolation of Cymbidium mild mosaic virus. Korean J Plant Prot 17:131–138

Eun AJC, Wong SM (2000) Molecular beacons: a new approach to plant virus detection. Phytopathology 90(3):269–275

Farreyrol K, Pearson MN, Grisoni M, Cohen D, Beck D (2006) Vanilla mosaic virus isolates from French Polynesia and the Cook Islands are Dasheen mosaic virus strains that exclusively infect vanilla. Arch Virol 151:905–919

Fauquet CM, Mayo MA, Maniloff J, Desselberger U, Ball LA (eds) (2005) Virus taxonomy: VIIIth report of the International Committee on Taxonomy of Viruses. Academic Press, New York

Francki RIB (1970) Cymbidium mosaic virus. CMI/AAB descriptions of plant viruses no 27. Association of Applied Biologists, Warwickshire

Fry CR, Zimmerman MT, Scott SW (2004) Occurrence of *Colombian datura virus* in the terrestrial orchid, *Spiranthes cernua*. J Phytopathol 152:200–203

Gara IW, Kondo H, Maeda T et al (1997) Stunt disease of *Habenaria radiata* caused by a strain of watermelon mosaic virus 2. Ann Phytopathol Soc Jpn 63:113–117

Gara IW, Kondo H, Maeda T et al (1998a) Dendrobium severe mosaic virus, a distinct member of potyvirus genus. Ann Phytopathol Soc Jpn 64:422

Gara IW, Kondo H, Maeda T et al (1998b) Calanthe mild mosaic virus, a new potyvirus causing a mild mosaic disease of *Calanthe* orchid in Japan. J Phytopathol 146:357–363

Gibbs A, Mackenzie A, Blanchfield A et al (2000) Viruses of orchids in Australia: their identification, biology and control. Aust Ochid Rev 65:10–21

Goff LM, Corbett MK (1977) Association of tomato ringspot virus with a chlorotic leaf streak of *Cymbidium* orchids. Phytopathology 67:1096–1100

Gribbs A, Mackenzie A, Blanchfield A et al (2000) Viruses of orchids in Australia: their identification, biology and control. Aust Ochid Rev 65:10–21

Grisoni M, Davidson F, Hyrondelle C et al (2004) Nature, incidence and symptomatology of viruses infecting *Vanilla tahitensis* in French Polynesia. Plant Dis 88:119–124

Grisoni M, Moles M, Farreyrol K, Rassaby L, Davis R, Pearson M (2006) Identification of potyviruses infecting vanilla by direct sequencing of a short RT-PCR amplicon. Plant Pathol 55(4):523–529

Guaragna MA, Ndum O, Jordan R (2006) Detection and characterization of two previously undescribed potyviruses in the terrestrial orchid *Spiranthes cernua*. Acta Hortic 722:209–218

Hill SA (1984) Methods in plant virology (no. 581.234 H5)

Holkar SK, Kumar R, Yogita M, Katiyar A, Jain RK, Mandal B (2017) Diagnostic assays for two closely related tospovirus species, watermelon bud necrosis virus and groundnut bud necrosis virus and identification of new natural hosts. J Plant Biochem Biotechnol 26(1):43–51

Hollings M, Stone OM, Barton RJ (1977) Pathology, soil transmission and characterization of Cymbidium ringspot, a virus from *Cymbidium* orchids and white clover (*Trifolium repens*). Ann Appl Biol 85:233–248

Hsu YC, Yeh TJ, Chang YC (2005) A new combination of RT-PCR and reverse dot blot hybridization for rapid detection and identification of potyviruses. J Virol Methods 128 (1–2):54–60

Hu JS, Ferreira S, Wang M, Xu MQ (1993) Detection of *Cymbidium mosaic virus*, *Odontoglossum ringspot virus*, tomato spotted wilt virus, and potyviruses infecting orchids in Hawaii. Plant Dis 77:464–468

Hu JS, Ferreria S, Wang M et al (1995) Purification, host range, serology and partial sequencing of Dendrobium mosaic potyvirus, a new member of the bean common mosaic virus subgroup. Phytopathology 85:542–546

Ikegami M, Isomura Y, Matumoto Y, Chatani M, Inouye N (1995) The complete nucleotide sequence of odontoglossum ringspot virus (Cy-1 strain) genomic RNA. Microbiol Immunol 39(12):995–1001

Inouye N (1969) Cucumber mosaic virus isolated from *Dendrobium*. Agric Res 53:49–60

Inouye N (1976) Dendrobium mosaic virus. Ber Ohara Inst Landw Biol Okayama Univ 16:165–174

Inouye N (2008) Viruses of orchids: symptoms, diagnosis, spread and control (Rural Culture Association, Tokyo, through Japan Uni Agency, Inc., Tokyo), p 176

Inouye N, Maeda T, Mitsuhata K, Gara IW (1998) Habenaria mosaic virus, a new member of potyvirus from *Habenaria radiata* orchid. Bull Res Inst Bioresour Okayama Univ 5:155–168

James D, Varga A, Pallas V, Candresse T (2006) Strategies for simultaneous detection of multiple plant viruses. Can J Plant Pathol 28:16–29

Jensen DD, Gold HA (1951) A virus ring spot of *Odontoglossum* orchid: symptoms, transmission and electron microscopy. Phytopathology 41:648–653

Jenson DD (1950) Mosaic of Cymbidium orchids. Phytopathology 40:966–967

Jenson DD, Gold HA (1951) A virus ringspot of Odontoglossum orchid: symptoms, transmission and electron microscopy. Phytopathology 41:648

Jordan R, Guaragna MA, Ndum O (2004) Detection and characterization of two previously undescribed potyviruses in the terrestrial orchid Spiranthes cernua. In: XI international symposium on virus diseases of ornamental plants 722, pp 209–218

Komínek P, Massart S, Pham K, van Leeuwen P, Komínková M (2019) Characterisation of a novel virus infecting orchid of the genus Pleione. Virus Res 261:56–59

Kondo H, Matsumoto J, Maeda N, Tamada T (2003) Orchid fleck virus: Brevipalpus californicus transmission, biological properties, and genome structure. Exp Appl Acarol 30:215–223

Kondo H, Maeda T, Gara IW, Chiba S, Maruyama K, Tamada T, Suzuki N (2014) Complete genome sequence of Habenaria mosaic virus, a new potyvirus infecting a terrestrial orchid (Habenaria radiata) in Japan. Arch Virol 159(1):163–166

Kondo H, Takemoto S, Maruyama K et al (2015) Cymbidium chlorotic mosaic virus, a new sobemovirus isolated from a spring orchid (*Cymbidium goeringii*) in Japan. Arch Virol 160:2099–2104

Kumar S, Stecher G, Li M, Knyaz C, Tamura K (2018) MEGA X: molecular evolutionary genetics analysis across computing platforms. Mol Biol Evol 35(6):1547–1549

Lawson RH, Ali S (1975) Ultrastructure of white-streaked *Dendrobium phalaenopsis* orchid flowers infected with a rhabdovirus. J Ultrastruct Res 53:345–353

Lawson RH, Brannigan M (1986) Virus disease of orchids. In: Handbook on orchid pests and disease. American Orchid Society, Florida, pp 2–49

Lawson RH, Hsu HT (1995) Orchid. In: Loebenstein G, Lawson RH, Brunt AA (eds) Virus and virus-like disease of bulb and flower crops. Wiley, Chichester

Lee CH, Zheng YX, Jan FJ (2017) The orchid-infecting viruses found in the 21st century. In: Orchid biotechnology III, pp 145–164

Lesemann DE (1977) Long, filamentous virus-like particles associated with vein necrosis of *Dendrobium phalaenopsis*. J Phytopathol 89:330–339

Lesemann D, Begtrup J (1971) Electron microscopic demonstration of a bacilliform virus in *Phalaenopsis*. Phytopathol Z 71:257–269

Lesemann DE, Koenig R (1985) Identification of bean yellow mosaic virus in *Masdevallia*. Acta Hortic 164:347–354

Lesemann DE, Vetten HJ (1985) The occurrence of tobacco rattle and turnip mosaic viruses in *Orchis* sp., and of an unidentified potyvirus in *Cypripedium calceolus*. Acta Hortic 164:45–54

Liu A, Zhao Y, Ruan S, Shen G (2009) Quantitative detection of Cymbidium mosaic virus by real time PCR. Front Biol China 4(3):314–320

Mackenzie AM, Nolan M, Wei KJ, Clements MA, Gowanlock D, Wallace BJ, Gibbs AJ (1998) Ceratobium mosaic potyvirus: another virus from orchids. Arch Virol 143(5):903–914

Maree HJ, Fox A, Al Rwahnih M, Boonham N, Candresse T (2018) Application of HTS for routine plant virus diagnostics: state of the art and challenges. Front Plant Sci 9:1082

Martelli GP, Russo M (1984) Use of thin sectioning for visualization and identification of plant viruses. In: Methods in virology, vol 8. Elsevier, Amsterdam, pp 143–224

Milne RG (1972) Electron microscopy of viruses. In: Kado CI, Agarwal O (eds) Principals and techniques in plant virology. Van Nostrand Reinhold Co, New York, pp 76–126

Mitsukuri K, Arita T, Johkan M, Yamasaki S, Mishiba K, Oda M (2009) Effects of type of explant and dark preconditioning on bud formation in Habenaria radiata (Thunb.) in vitro. HortScience 44:523–525

Mumford RA, Jarvis B, Morris J, Blockley A (2003) The first record of Chrysanthemum stem necrosis virus (CSNV) from the UK. Plant Pathol 52:779

Navalinskiene M, Raugalas J, Samuitiene M (2005) Viral diseases of flower plants. 16. Identification of viruses affecting orchids (*Cymbidium* Sw.). Biologia 2:29–34

Ng JCK, Perry KL (2004) Transmission of plant viruses by aphid vectors. Mol Plant Pathol 5:505–511

Nobrega NR (1947) Uma doenca de virus em orquidea. O Biologico 13:62

Ong JWL, Phillips RD, Dixon KW, Jones MG, Wylie SJ (2016) Characterization of the first two viruses described from wild populations of hammer orchids (Drakaea spp.) in Australia. Plant Pathol 65(1):163–172

Pai H, Jean WH, Lee YS, Chang YCA, Lin NS (2020) Genome-wide analysis of small RNAs from Odontoglossum ringspot virus and Cymbidium mosaic virus synergistically infecting Phalaenopsis. Mol Plant Pathol 21(2):188–205

Pant RP, Rashmi ER, Manjunath N, Baranwal VK (2020) Status of orchid viruses in India and management strategies for them. In: Applied plant virology. Academic Press, New York, pp 725–745

Pearson MN, Pone SP (1988) Viruses of Vanilla in the Kingdom of Tonga. Austral Plant Pathol 17:59–60

Pecman A, Kutnjak D, Gutiérrez-Aguirre I, Adams I, Fox A, Boonham N, Ravnikar M (2017) Next generation sequencing for detection and discovery of plant viruses and viroids: comparison of two approaches. Front Microbiol 8:1998

Peters KR (1977) Orchid viruses: a new rhabdovirus in Laelia red leafspots. J Ultrastruct Res 58 (2):166–171

Puli'uvea C, Khan S, Chang WL, Valmonte G, Pearson MN, Higgins CM (2017) First complete genome sequence of vanilla mosaic strain of Dasheen mosaic virus isolated from the Cook Islands. Arch Virol 162(2):591–595

Revers F, García JA (2015) Molecular biology of potyviruses. Adv Virus Res 92:101–199

Rodamilans B, Valli A, Mingot A, San León D, Baulcombe D, Lopez-Moya JJ, Garcia JA (2015) RNA polymerase slippage as a mechanism for the production of frameshift gene products in plant viruses of the Potyviridae family. J Virol 89:6965–6967

Roossinck MJ, Martin DP, Roumagnac P (2015) Plant virus metagenomics: advances in virus discovery. Phytopathology 105(6):716–727

Seoh ML, Wong SM, Zhang L (1998) Simultaneous TD/RT-PCR detection of cymbidium mosaic potexvirus and odontoglossum ringspot tobamovirus with a single pair of primers. J Virol Methods 72(2):197–204

Sherpa AR, Hallan V, Zaidi AA (2004) Cloning and sequencing of coat protein gene of an Indian Odontoglossum ringspot virus isolate. Acta Virol 48(4):267–269

Singh MK, Sherpa AR, Hallan V, Zaidi AA (2007) A potyvirus in Cymbidium spp. in northern India. Aust Plant Dis Notes 2(1):11–13

Singh J, Singh MK, Ranjan K, Kumar A, Kumar P, Sirohi A, Baranwal VK (2020a) First complete genome sequence of garlic virus X infecting Allium sativum-G282 from India. Genomics 112 (2):1861–1865

Singh J, Truong TN, An D, Prajapati MR, Manav A, Quoc NB et al (2020b) Complete genome sequence and genetic organization of a garlic virus D infecting garlic (Allium sativum) from northern India. Acta Virol 64(4):427–432

Slogteren EV, Slogteren DHMV (1957) Serological identification of plant viruses and serological diagnosis of virus diseases of plants. Annu Rev Microbiol 11:149–164

Smith KM (1931) On the composite nature of certain potato virus diseases of the mosaic group as revealed by the use of plant indicators and selective methods of transmission. Proc R Soc Lond Ser B 109(762):251–267

Verchot-Lubicz J (2005) A new cell to cell transport model for potexviruses. Mol Plant-Microbe Interact 18:283–290

Villamor DEV, Mekuria TA, Pillai SS, Eastwell KC (2016) High-throughput sequencing identifies novel viruses in nectarine: insights to the etiology of stem-pitting disease. Phytopathology 106 (5):519–527

Webster CG, Wylie SJ, Jones GK (2004) Diagnosis of plant viral pathogens. Curr Sci 86:1604–1607

Wisler GC, Zettler FW, Mu L (1987) Virus infections of *Vanilla* and other orchids in French Polynesia. Plant Dis 71:1125–1129

Wong SM, Mahtani PH, Lee KC, Yu HH, Tan Y, Neo KK et al (1997) Cymbidium mosaic potexvirus RNA: complete nucleotide sequence and phylogenetic analysis. Arch Virol 142 (2):383–391

Wu Q, Ding SW, Zhang Y, Zhu S (2015) Identification of viruses and viroids by next-generation sequencing and homology-dependent and homology-independent algorithms. Annu Rev Phytopathol 53:425–444

Wylie SJ, Li H, Jones MG (2013) Donkey orchid symptomless virus: a viral 'platypus' from Australian terrestrial orchids. PLoS One 8(11):e79587

Wylie SJ, Adams M, Chalam C, Kreuze J, López-Moya JJ, Ohshima K, Praveen S, Rabenstein F, Stenger D, Wang A, Zerbini FM (2017) ICTV virus taxonomy profile: Potyviridae. J Gen Virol 98:352–354

Yamamoto T, Ishii M (1981) Mosaic disease of Calanthe discolor Lind [in Japan]. In: Proceedings of the association for plant protection of Shikoku

Yoon JY, Chung BN, Choi GS, Choi SK (2012) Genetic variability in the coat protein genes of Cymbidium mosaic virus isolates from orchids. Virus Genes 44(2):323–328

Yora K, Saito Y, Doi Y et al (1983) Handbook of plant viruses. Asakura Shoten, Tokyo, p 584

Zaitlin M (1976) Viral cross protection: more understanding is needed. Phytopathology 66:382–383

Zettler FW, Tsai JH, Faan HC, Ke L (1987) Dasheen mosaic virus infecting taro in People's Republic of China. Plant Dis 71:837

Zettler FW, Ko NJ, Wisler GC, Elliott MS, Wong SM (1990) Viruses of orchids and their control. Plant Dis 74(9):621–626

Zheng YX, Chen CC, Jan FJ (2008a). Identification and characterization of three new Phalaenopsis orchid-infecting viruses. In: XII international symposium on virus diseases of ornamental plants 901, pp 127–132

Zheng YX, Chen CC, Yang CJ, Yeh SD, Jan FJ (2008b) Identification and characterization of a tospovirus causing chlorotic ringspots on Phalaenopsis orchids. Eur J Plant Pathol 120 (2):199–209

Zheng YX, Chen C, Chen YK, Jan FJ (2008c) Identification and characterization of a potyvirus causing chlorotic spots on *Phalaenopsis* orchids. Eur J Plant Pathol 121:87–95

Zheng YX, Chen CC, Jan FJ (2008d) *Phalaenopsis* orchids showing chlorotic rings, a new disease caused by *carnation mottle virus*. J Plant Pathol 90:192(S2)

Zheng YX, Chen CC, Jan FJ (2011) First report of *Carnation mottle virus* in *Phalaenopsis* orchids. Plant Dis 95:354

Detection and Characterization of Plant Viruses Infecting *Hibiscus rosa-sinensis* L. 8

Smriti Mall, Vineeta Pandey, Aarshi Srivastava, and Rajarshi Kumar Gaur

Abstract

An ornamental plant *Hibiscus rosa-sinensis* is cultivated as a topical perennial shrub. Viral pathogenesis upon this flowering plant shows variable symptoms like mosaic, vein banding chlorotic spot, ringspots, etc., and its host range, transmissibility, and morphology are the backbone for viral identification. Vector-virus relationship manipulates the vector behavior and activates proteins which on one half facilitate effective transmission adding in viral diseases and on other half promote genetic behavior, recombination, and evolution of viruses. Disease-causing microorganisms require diagnostic techniques which is necessary for efficient and specific for viral detection which may also contribute to early disease management. Further different approaches such as use of RT-PCR, various microscopic technique, chemical analysis, high-throughput sequencing, and serological and molecular methods are also described for virus identification. Failure in eco-agriculture practice led to considerable financial loss affecting directly dependent developing countries and indirectly developed nations. Therefore, their management is essential to avoid risk. The vast view and understanding of viral disease precisely on *Hibiscus rosa-sinensis* and also the peak of approaches engaged for management are summarized in this chapter.

Keywords

Ornamental plants · Hibiscus · Plant virus · Vector-virus relationship · Detection

S. Mall (✉)
Department of Botany, DDU Gorakhpur University, Gorakhpur, UP, India

V. Pandey · A. Srivastava · R. K. Gaur
Department of Biotechnology, DDU Gorakhpur University, Gorakhpur, UP, India

S. K. Raj et al. (eds.), *Virus Diseases of Ornamental Plants*,
https://doi.org/10.1007/978-981-16-3919-7_8

8.1 Introduction

Plant viruses are omnipresent to all the habitats that support primitive life along with most important emerging infectious viral diseases of plants. Amid the diseases that affect plants, viral diseases are significantly important as it can be rapidly transmitted by different mode without any effective approach to treat the virus-infected plants (Hull 2014). *Tobacco mosaic virus* (TMV) was the earliest virus that has been studied and was discovered by Martinus Beijerinck in 1898. These pathogenic viruses of plant are nucleoproteins parasitic with variations in size, physical structure, chemical composition, and number of genomic units. They are vectored by insects, mites, fungi, nematode, and parasitic plants and also transmitted by vegetative propagation, pollen, and seed and also by contact. Plant viruses migrate within a plant via plasmodesmata for between-cell movement and also via phloem for transport. Plant viruses could show divergent symptoms like mosaic patterns, streaking, and hypo- or hyperplasia and stunting vein clearing. Moreover, plants can also maintain viruses in them without any accessible symptoms. Furthermore, as obligate parasites of host plants, these viruses could cause utmost injury to plants. Ribonucleic acid (RNA) or deoxyribonucleic acid (DNA) and a protein capsid are established as a nucleic acid composition in these viruses. Classification of plant viruses is based on their chemical compounds and morphologic, biologic, pathologic, and molecular characteristics. Simultaneously, these plant viruses with their host cell create a close biologic unit. Concurrently, some artificial and natural substances are supposed to inhibit virus replication and their cell-to-cell movement, and although these substances cannot act as viricides in vivo, still viral infection can cause biologic decline which contributes to the reduction in the competitive ability of weeds in agroecosystems. Therefore, viruses can indirectly contribute to weed control also.

Hibiscus rosa-sinensis L. (family Malvaceae) is a tropical perennial shrub often grown as an ornamental plant in tropical as well as subtropical region and is admired for its aesthetic value; in Indian landscape gardens from ancient times, this plant is popular. Among the medicinally important flowering plants, hibiscus is widely available and engaged by practitioner of natural health for contraception, menorrhagia, venereal diseases, controlling uterine bleeding, fever, and cough, refrigerant and vitalize in palpitation. It is reported to contain anthocyanins, flavonoids, cyclopeptide alkaloid, and vitamins. The pharmacological and clinical studies confirmed the therapeutic value of 1172 *Hibiscus rosa-sinensis* (Jadhav et al. 2009). However, due to viral diseases that results in yield losses in the form of reduced flowering along with immature, irregular, and deformed flowering. Leaves that are affected are severely curled, reducing the aesthetic value with stunted plants (Rajeshwari et al. 2005). In this reveiw, we are going to discuss the esteemed effect, geographical distribution, identification, and characterization of viruses and the novel emerging approaches for viral disease management of plant virus on *Hibiscus rosa-sinensis* L.

8.2 Virus Disease of *Hibiscus*

Ornamental plants are economically important which are very popular throughout the world. Demand of these ornamentals plants is expanding constantly in international market. Moreover, it includes production of cut and potted flowers, cut foliage plants, as well as propagation material. However, both decorative value and quality of propagated material obtained from ornamental plants are diminished by viruses and viroids infection. Due to this reason wide range of ornamental plant species along with cultivars and their expanded geographical distribution, increases the diversity of virus infected them. However, viral phytopathogens can also spread from vegetative propagation. Natural reservoirs for plant viruses that promote circulation of viruses and their transmission to other economically important crops are generally perennial ornamental crops, and *Hibiscus rosa-sinensis* L. is one of them. Effective investigation, detection, and prevention of diversity among these viruses and viroids associated with ornamental plants are essential to conserve the further spread of infections (Mitrofanova et al. 2018).

Hibiscus rosa-sinensis Linn (*family*: Malvaceae), an attractive horticultural plant, is a tropical perennial shrub frequently cultivated as an ornamental plant. Five plant viruses that utmost infect *H. rosa-sinensis* that have been characterized previously are *Hibiscus chlorotic ringspot virus* (HCRSV, genus *Carmovirus*), *Hibiscus yellow mosaic virus* (genus *Tobamovirus*), *Eggplant mottled dwarf virus* (EMDV, genus *Nucleorhabdovirus*), *Hibiscus latent ringspot virus* (HLRSV, genus *Nepovirus*), and *Okra mosaic virus* (OkMV, genus *Tymovirus*) (Regenmortel et al. 2000) from China. In recent year, two novel tobamoviruses infecting *H. rosa-sinensis* were characterized in Singapore, Florida and two new begmoviruses were reported in Brazil (Adkins et al. 2006; Quadros et al. 2019). Virus isolated from *H. rosa-sinensis* in China was analyzed by biological and serological means along with virion properties and coat protein (*CP*) gene sequences, concluding that it was a *Tomato mosaic virus* (ToMV) that infects *H. rosa-sinensis*. Multiple studies have been done on RNA, DNA, and especially multiple alignments of amino acid sequences that demonstrated several substitutions in CPs of hibiscus-infecting ToMV isolates which might contribute to the variations in symptoms induced by them.

8.3 Geographical Distribution

The rapid spread of virus associated with *Hibiscus* draws a lot of attention from different countries. Virology experts are trying to understand the different mechanisms as the virus disease occurred in almost all part of the countries. Table 8.1 gives the valuable information field symptomatology and geographical spread of the virus associated with *Hibiscus*.

Table 8.1 Worldwide distributions of viruses infecting *Hibiscus rosa-sinensis*

Genus	Virus	Symptoms	Geographical origin	References
Tobamovirus	Hibiscus latent ringspot virus	Leaf mosaic and mottle	Western Nigeria	Brunt et al. (1980)
	Hibiscus latent Singapore virus	Chlorotic mottling	Singapore	Srinivasan et al. (2002)
	Hibiscus latent Fort Pierce virus	Diffuse chlorotic spots and mottle	Florida	Adkins et al. (2003)
	Tomato mosaic virus	Mosaic	China	Huang et al. (2004)
	Hibiscus latent Fort Pierce virus	Diffuse chlorotic spots and mottle	New Mexico	Allen et al. (2005)
	Hibiscus latent Fort Pierce virus—J	Diffuse chlorotic spots and mottle	Japan	Yoshida et al. (2014)
	Tomato mosaic virus—Ab10	Mosaic	Iran	Parizipour and Keshavarz-Tohid (2020)
	Tomato mosaic virus—Ah14	Mosaic	Iran	Parizipour and Keshavarz-Tohid (2020)
	Tomato mosaic virus—D09	Mosaic	Iran	Parizipour and Keshavarz-Tohid (2020)
	Tomato mosaic virus—M06	Mosaic	Iran	Parizipour and Keshavarz-Tohid (2020)
Carmovirus	Hibiscus chlorotic ringspot virus	Chlorotic spotting and ringspots	El Salvador and USA	Waterworth et al. (1976)
	Hibiscus chlorotic ringspot virus	Chlorotic spotting and ringspots	Australia	Jones and Behncken (1980)
	Hibiscus chlorotic ringspot virus	Chlorotic spotting and ringspots	Singapore	Wong and Chng (1992)
	Hibiscus chlorotic ringspot virus	Chlorotic spotting and ringspots	Taiwan	Li and Chang (2002)
	Hibiscus chlorotic ringspot virus	Chlorotic spotting and ringspots	New Zealand	Tang et al. (2008)
	Alfalfa mosaic virus	Mosaic	Spain	Parrella et al. (2012)
	Hibiscus chlorotic ringspot virus	Chlorotic spotting and ringspots	Iran	Pourrahim et al. (2013)

(continued)

Table 8.1 (continued)

Genus	Virus	Symptoms	Geographical origin	References
	Hibiscus chlorotic ringspot virus IL	Chlorotic spotting and ringspots	Israel	Luria et al. (2013)
	Hibiscus chlorotic ringspot virus MGL1	Chlorotic spotting and ringspots	Turkey	Kamenova and Adkins (2004)
	Hibiscus chlorotic ringspot virus MGL2	Chlorotic spotting and ringspots	Turkey	Kamenova and Adkins (2004); Karanfil and Korkmaz (2017)
Begomovirus	Cotton leaf curl Multan virus	Leaf curl	China	Mao et al. (2008)
	Cotton leaf curl Burewala virus	Leaf curl	Pakistan	Akhtar et al. (2014)
	Cotton leaf curl Multan virus—01	Leaf curl	India	Srivastava et al. (2016)
	Cotton leaf curl Multan virus—02	Leaf curl	India	Srivastava et al. (2016)
	Cotton leaf curl Multan virus	Leaf curl	Philippines	She et al. (2017)
Cilevirus	Citrus leprosis virus C2	Green ringspots	Hawaii	Melzer et al. (2013)
	Hibiscus-infecting Cilevirus	Green ringspots	Florida	Avijit et al. (2018)
Ilarvirus	Tobacco streak virus	Necrotic streaks	Florida	Lewandowski et al. (2005)
	Hibiscus golden mosaic virus	Mosaic	Brazil	Allen et al. (2005)
Rhabdovirus	Eggplant mottled dwarf virus	Vein yellowing	Morocco	Lockhart (1987)
Tospovirus	Impatiens necrotic spot virus	Leaf chlorotic and necrotic	New Zealand	Elliot et al. (2009)
Hostuviroid	Hop stunt viroid	Upward curling leaf deformation	Italy	Luigi et al. (2013)

8.4 Detection of Viruses

8.4.1 Symptoms

Hibiscus plants shows viral symptoms annually on first young leaves during the February flush of growth when the plants are in full blooms (Wolfswinkel 1996) and basically exhibits systemic symptoms including dark and light green mosaic in young leaves, leaf puckering with malformation on older leaves, and significant stunting. *H. rosa-sinensis* with rod-shaped virus particles were isolated which express systemic symptoms. Symptoms of *Hibiscus chlorotic ringspot virus* (HCRSV)-infected plant leaves range from generalized mosaic, mottle, to chlorotic ringspots as well as vein-banding patterns, severe stunting, and flower distortion (Gao et al. 2012). Likewise, we mentioned all the viral symptoms on *H. rosa-sinensis* in Table 8.1, respectively.

Identification of virus on Hibiscus plant, based on symptomatology alone is not reliable diagnostic indicators for viral infection of *H. rosa-sinensis* due to vegetatively propagated crop over a time hibiscus can accumulate multiple viruses. Based on various surveys, probability of some of the plants sampled for one study could be co-infected with one or more additional viruses which complicate symptomatology as well. Simultaneously, to overcome this problem of symptomatology, we use multiple methods like electron microscopy, IC-RT-PCR, chemical analysis, and sequence analysis.

8.4.2 Microscopy

Electron microscopy (EM) is an important tool for revelation and analysis of virus replication. Electron microscopy depends on its ability i.e. to image the whole spectrum of interactions with resistance and non-reactions in the case when new virus isolates or species. EM can determine functional features of viruses along with bottom line mechanisms of interactions relevant in nature and in synthetic virology analysis. Electron microscopy (EM) can be of three different types: transmission electron microscopy (TEM), immunoelectron microscopy (IEM) (Milne and Lesemann 1984), and cryogenic electron microscopy (cryo-EM). These are basically used for plant viral detection. The electron microscopy which is the most powerful technique available to virologists today has versatility especially as a universal means of virus detection and describes its development from descriptive tool.

8.4.3 Serological and Molecular Detection

In the last decades, for detection of plant viruses, various techniques have been developed which include rapid and specific serological (enzyme-linked immunosorbent assay, ELISA) and molecular techniques (molecular hybridization and DNA amplification) for the detection of plant viruses. Serological techniques are generally used for detection of viruses to the species level; sometimes they allow discrimination between virus strains (serotypes) using monoclonal antibodies (Myrta et al. 2000; Permar et al. 1990; Sheveleva et al. 2018). Binding of viral proteins with antibodies known as ELISA (Clark and Adams 1977) but in molecular hybridization are basically used to detect virus species, but its detection level can be modified to a certain level by using different probes and hybridization conditions we can say that binding of viral nucleic acids with sequence-specific DNA or RNA probes, due to their sequence complementarities (Hull and Al-Hakim 1998).

The high sensitivity of these amplification techniques like in PCR can be a problem, contamination of reagents as well as instruments with amplicons from previous samples or sometimes cross-contamination between samples which we are using can produce false positives, reducing specificity. Simultaneously, universal detection in case of some plant viruses by ELISA with monoclonal antibodies failed for some plant virus isolates (Sheveleva et al. 2018).

8.4.4 High-Throughput Sequencing (HTS)

The most advanced and versatile technique since the nucleotide sequences which are frequently used in nowadays are HTS techniques that can be used for both to estimate the genetic variation along with structure of virus populations and to identify a virus sample at different taxonomic levels or we can say that discover new virus species, genera or families (Kreuze et al. 2009; Pecman et al. 2017; Verdin et al. 2017; Wu et al. 2015), according to their nucleotide or amino acid identity with known sequences in databases (Gen Bank) or the presence of sequence motifs. This is the most impressive technique for multiplex detection as it can identify and discover an unlimited number of viruses and virus variants within a plant (Jones et al. 2017).

8.4.5 RT-PCR

The amplification of multiple targets simultaneously in a single PCR reaction by using several primer pairs specific for each target is done in a multiplex PCR or RT-PCR technique. But the main problem with this RT- PCR is limitation in number of targets that can be amplified simultaneously since the more primers are used there will be the high probability of incompatibility between some of them. However, the number of products of different sizes that can be resolved by electrophoresis or the number of fluorescent dyes that can be used (Boonham et al. 2007) is also limited.

8.4.6 Features of Type of Molecular Technique for Plant Virus Detection and Diagnostics

The features of detection techniques that are costs, designability, versatility, multiplexing, sensitivity, specify, quantification, throughput screening i.e. number of samples analyzed simultaneously, rapidity, on-sited and easiness these features are not only during the application but also during the design or development as well. When we consider the sample for processing which is also a critical step and affects the rapidity, easiness and throughput of the detection process whereas if we take about versatility of a detection technique is a ability for different detection levels that means at family, genus, species, strain or isolate respectively. Multiplexing is an ability to perform parallel analysis and on-site is to detect viruses on field with portable devices is known as on-site. Moreover, while comparing all molecular detection techniques, it was found that RT-PCR is the best technique in case of quantification of samples, whereas lateral flow is the finest among all in terms of rapidity, on-site and easiness. HTS is supreme among all, as itis best in the features like designability, versatility, multiplexing, specifies and sensitivity features. Moreover, it is good in quantification as well (Rubio et al. 2020). But this high-throughput sequencing is still too expensive for most routine analyses, that's why it is necessary to develop rapid as well as accurate detection techniques for each virus, being PCR the easiest to develop. In contrast, the setup of molecular detection techniques is a directed process that should be cheap, fast, and versatile, enabling to address different detection levels by considering the genetic variability of virus populations.

8.4.7 Chemical Analysis

In this analysis, the effect of HCRSV infection on total titratable acidity, ascorbic acid, and total anthocyanin and flavonoid contents in sepals of Roselle plants has been evaluated (Shafie 2019). In this process, the acidity value, ascorbic acid, and flavonoids are detected high in Roselle sepals infected with HCRSV compared to sepals of healthy plants, whereas *Hibiscus chlorotic ringspot virus* infection significantly decreased the total anthocyanin content in sepals of infected plants compared with that of healthy ones, respectively. As a result, by applying the above analysis, we can detect the HCRSV infection.

8.5 Vectors of Hibiscus Virus

Plant virus affecting *H. rosa-sinensis* is transmitted by various means, which include transmission through vectors like insects, mites, fungi, nematode and parasitic plants, vegetative propagation, pollen, seed, and also by contact. Movement of viruses within a plant generally occurs through plasmodesmata for between-cell movement and via phloem for long-distance transport. Insect vectors specially whiteflies (*Bemisia tabaci*), a complex of morphologically indistinguishable species (Barro et al. 2011) are vectors of many plant viruses that includes several genera of these whitefly-transmitted plant viruses such as *Begomovirus, Carlavirus, Crinivirus, Ipomovirus and Torradovirus* include several hundred species of emerging along with economically significant pathogens of important food and fiber crops [reviewed by Lapidot et al. 2014; Castillo et al. 2011; King et al. 2012]. Such plant viruses do not replicate within its vector nevertheless they are moved readily by adult whitefly from plant to plant. For most of these plant viruses, whitefly feeding is necessary for acquisition and inoculation, while for others only probing is required. Among these viruses, many of them are unable to transmit themselves by other mode. Thus virus cultures is maintained, biological as well as molecular characterization (identification of host range and symptoms) (Cohen and Antignus 1994; Polston et al. 1993) ecology (Cohen 1990; Polston et al. 2006) all are require to show viruses can be transmitted to experimental hosts using the whitefly vector. New approaches to management, such as evaluation of cultural approaches or selection and development of resistant cultivars, new chemicals, or compounds, require the use of whiteflies for virus transmission.

8.6 Management

Considering efficient and durable control or management of viral diseases, it is necessary to consider the genetic diversity and evolution of virus populations and have specific, fast, and reliable diagnostic tools which should be necessary to implement.

Moreover, vectors are quite difficult to control in horticulture and agronomic production systems which were found to cause severe infection to green life. The severity rate of viral diseases is found higher in winter than in summer. Therefore, seed treatment is performed to control vectors in initial stage via systematic insecticides. The use of tolerant cultivars and suitable site selection are also important, to avoid mechanical transmission during pruning and propagation (Kamenova and Adkins 2004) implemented the everyday used of sanitation approach via pruning tool which was either applied for 1 min with 10% (wt/vol) NaOCl or 20% (wt/vol) NFDM to eradicate Hibiscus latent Fort Pierce virus or longer by using 2% (wt/vol) NaOH, 6% (vol/vol) NaOCl, 20% (wt/vol) TSP, may be effective procedures and has a practical significance for landscape maintenance workers, hibiscus growers, nurseries, and where frequent horticultural practices are means of virus spread and dissemination in the environment. Disease management should focus on virus reservoir, regulation of virus source, and control of load of vectors in the agroecosystem. Endemic and epidemics are risk factors for worldwide losses which include nature of weather, habitat, and agricultural practice which accelerate the development of population of vectors and reservoir in an area. Conventional approaches, such as cultural methods and biocide applications against plasmodiophorida vectors, nematode, and arthropod, have limited success at alleviating the impact of plant viruses. Initial planting of resistant cultivars is the most economical and effective way to control plant virus diseases. Natural sources of resistance have been utilized extensively to develop virus-resistant plants by conventional breeding. Non-conventional methods also give flourishing effects to confer virus resistance by transferring primarily virus-derived genes, including viral coat protein, movement protein, replicase, protease, defective interfering RNA, and non-coding RNA sequences, into susceptible plants. Non-viral genes (microRNAs, dsRNAse, R genes, ribosome-inactivating proteins, RNA-modifying enzyme protease inhibitors, and scFvs) have also been used successfully to engineer resistance to viruses in plants. MicroRNAs and its molecular background provide a completely different new scope for plant protection against viruses. This would be very promising for those viruses having ambience or negative polarity as this would accelerate reverse genetics. Very few genetically engineered and virus-resistant crops have been released for cultivation, and none is on hand in developing countries. However, a number of economically valued GEVR (genetically engineered virus-resistant) crops, transformed with viral genes, are of great interest in developing countries. The major issues encountering the yield and deregulation of GEVR crops in developing countries are chiefly socioeconomic matter related to bio-safety regulatory frameworks, intellectual property rights, expense to generate GE crops, and defiance by nongovernmental activists. Indication for satisfactory resolution of these of above mentioned factors, apparently leading to field tests and deregulation of GEVR crops in developing countries (Reddy et al. 2009).

Viruses exhibit resistance to external factors, and due to flexibility in their genome organization and expression mechanisms and also because of co-evolution and adaptability with host plant, the complexity of infection may lead to increase in number in future and become highly stable. These demonstrations can also contribute to expand host range. A greater understanding of co-evolution and recombination in viruses help in developing strategies for managing plant viral disease which could be done by in silico analysis (Morya et al. 2014). Thus, a good agronomical practice are required that minimize the activity of vectors that transmit virus and destroy infected host plants. Despite of this, there are still vital gaps and many unknown mechanisms regarding host interaction and host range and also the gene expression that needs to be understood. The advanced technology, multidisciplinary approaches generating integration data from transcriptomic, proteomics, metabolomics and genomic information invoke for various host plant may contribute further insights for sustainable diseases prevention in plants and suppressing virus reproduction.

8.7 Conclusion

Flowering plants have been widely used as curative agents for variety of ailments since the time immemorial. There, concentrated flowers as well as leaves extract can be found in various herbal preparations that are also available in the market today. *Hibiscus rosa-sinensis* L. is an important source of different types of compounds with diverse chemical structures along with pharmacological activities. Therefore, presence of such ample amount of chemical compounds indicates that *Hibiscus rosa-sinensis* L. could serve as a "prominent lead" for the development of unique agents having very good efficacy in various disorders in future. Further, more or less viruses are reported worldwide, out of which 32 viruses are housing in 8 genera summarized in this reveiw chapter which affects the economic aspects of *Hibiscus rosa-sinensis* L. generating huge yield loss. Moreover, *Hibiscus leaf curl, a begomovirus* disease, also causes yield losses of hibiscus flowers throughout India as well (*Anon* 1950; Mali 1980; Vasudeva et al. 1953). Although eight genera are emphasized here, a number of viruses associated with *Tobamovirus* and *Carmovirus* genera are profoundly studied for *Hibiscus* plant. Prominently these are transmitted by whiteflies especially in case of begomovirus. Although, damages caused by various pest have a considerable negative economic impact in agriculture and also ongoing pharmacological research have been affected, being emergent viral diseases particularly most important among them (Anderson et al. 2004; Mumford et al. 2016). However, the tremendous ability of viruses to evolve and generate molecular as well as biological variation is a considerable difficulty for virus detection and disease management. Presently, when a new virus-like disease appears, number of approaches (like ELISA, PCR, RPA, HTS, electron microscopy, etc.) has been developed in the last few decades which could be used for its detection. In addition, an effective pruning and propagation sanitization tool was performed by Kamenova and Adkins (2004) to avoid mechanical transmission of viruses, thus minimizing productivity loss. Control of plant viral infection becomes a challenge because of the lack of precise management implications with *Hibiscus rosa-sinensis*. From this review, we found that very limited attempts are made to manage *Hibiscus* plant from viral infection. Therefore, by employing various serological and molecular techniques may provide a new insight to combat with challenges by identifying and characterizing the new viruses that can help researcher to manage them accordingly and make the world continuously benefited by pharmacological values of these plant.

For better productivity and yield, a combination of various management strategies should be used to control plant virus diseases. The use of new technologies coupled with cultural and biological practices provides best measures for combating plant viruses. There is also a need for government and other agencies to introduce greatly improved methods.

References

Adkins S, Kamenova I, Achor D, Lewandowski DJ (2003) Biological and molecular characterization of a novel *tobamovirus* with a unique host range. Plant Dis 87:1190–1196

Adkins S, Kamenova I, Chiemsomhar P, Baker CA, Lewandowski DJ (2006) Tobamoviruses from Hibiscus in Florida and beyond. Virus Dis Ornam 11:65–70

Akhtar KP, Ullah R, Saeed M, Sarwar N, Mansoor S (2014) China rose (*Hibiscus rosa-sinensis)*: a new natural host of *cotton leaf curl Burewala virus* in Pakistan. J Plant Pathol 96:385–389

Allen JE, Kamenova I, Adkins S, Hanson SF (2005) First report of *Hibiscus latent Fort Pierce virus* in New Mexico. Plant Health Prog (Online)

Anderson PK, Cunningham AA, Patel NG, Morales FJ, Epstein PR, Daszak P (2004) Emerging infectious diseases of plants: pathogen pollution, climate change and agrotechnology drivers. Trends Ecol Evol 19:535–544

Anon (1950) Annual report of Division of Mycology & Plant Pathology for the year 1949–50. Indian Agricultural Research Institute, New Delhi

Avijit R, Andrew LS, Michael JM, Jonathan S (2018) Complete nucleotide sequence of a novel *Hibiscus-infecting Cilevirus* from Florida and its relationship with closely associated *Cileviruses*. Genome Announc 6(4):01521–01517

Barro DPJ, Liu SS, Boykin LM, Dinsdale AB (2011) *Bemisia tabaci*: a statement of species status. Annu Rev Entomol 56:1–19

Boonham N, Tomlinson J, Mumford R (2007) Microarrays for rapid identification of plant viruses. Annu Rev Phytopathol 45:307–328

Brunt AA, Barton RJ, Phillips S, Lana AO (1980) *Hibiscus latent ringspot virus*, a newly recognised virus from *Hibiscus rosa-sinensis* (Malvaceae) in western Nigeria. Ann Appl Biol 96(1):37–43

Castillo NJ, Olive FE, Sanchez-Campos S (2011) Emerging virus diseases transmitted by *whiteflies*. Annu Rev Phytopathol 49:219–248

Clark MF, Adams A (1977) Characteristics of the microplate method of enzyme-linked immunosorbent assay for the detection of plant viruses. J Gen Virol 34:475–483

Cohen S (1990) Epidemiology of whitefly-transmitted viruses. In: Gerling D (ed) Whiteflies: their bionomics, pest status and management. Intercept, Andover, pp 211–225

Cohen S, Antignus Y (1994) Tomato yellow leaf curl virus, a whitefly-borne geminivirus of tomatoes. Adv Dis Vector Res 10:259–288

Elliot DR, Lebas BSM, Ochoa-Corona FM, Tang J, Alexander BJR (2009) Investigation of *impatiens necrotic spot virus* outbreaks in New Zealand. Australas Plant Pathol 38(5):490–495

Gao R, Ng FKL, Liu P, Wong SM (2012) *Hibiscus chlorotic ringspot virus* coat protein up regulates sulfur metabolism genes for enhanced pathogen defense. Mol Plant Microbe Interact 25(12):1574–1583

Huang JG, Fan ZF, Li HF, Tian GZ, Hu JS (2004) First report of *tomato mosaic virus* on *Hibiscus rosa-sinensis* in China. Plant Dis 88:683–683

Hull R (2014) Plant virology, 5th edn. Elsevier, London; 854 p

Hull R, Al-Hakim A (1998) Nucleic acid hybridization in plant virus diagnosis and characterization. Trends Biotechnol 6:213–218

Jadhav VM, Thorat RM, Kadam VJ, Sathe NS (2009) Hibiscus rosa-sinensis Linn–"Rudrapuspa": a review. J Pharm Res 2(7):1168–1173

Jones DR, Behncken GM (1980) Hibiscus chlorotic ringspot, a widespread virus disease in the ornamental *Hibiscus rosa-sinensis*. Aust J Plant Pathol 9(1):4–5

Jones S, Baizan-Edge A, MacFarlane S, Torrance L (2017) Viral diagnostics in plants using next generation sequencing: computational analysis in practice. Front Plant Sci 8:1770

Kamenova I, Adkins S (2004) Transmission, in planta distribution, and management of *Hibiscus latent Fort Pierce virus*, a novel tobamovirus isolated from Florida hibiscus. Plant Dis 88:674–679

Karanfil A, Korkmaz S (2017) First report of *Hibiscus chlorotic ringspot virus* in Turkey. New Dis Rep 35:22

King AMQ, Adams MJ, Carstons EB, Lefkowitz EJ (2012) International Committee on Taxonomy of Viruses (ed) Virus taxonomy classification and nomenclature of viruses, vol 9. Elsevier, Amsterdam, pp 193–210

Kreuze JF, Perez A, Untiveros M, Quispe D, Fuentes S, Barker I (2009) Complete viral genome sequence and discovery of novel viruses by deep sequencing of small RNAs: a generic method for diagnosis, discovery and sequencing of viruses. Virology 388:1–7

Lapidot M, Legg JP, Wintermantel WM, Polston JE (2014) Management of whitefly-transmitted viruses in open-field production systems. In: Lobenstein G, Katis N (eds) Advances in virus research: control of plant virus diseases: seed-propagated crops. Elsevier, Oxford, pp 147–206

Lewandowski D, Afunian M, Adkins S (2005) Detection of *tobacco streak virus* from hibiscus in Florida. Phytopathology 95:S59

Li SC, Chang YC (2002) First report of *Hibiscus chlorotic ringspot virus* in Taiwan. Plant Pathol 51:803

Lockhart BEL (1987) Evidence for identity of plant *rhabdovirus* causing vein yellowing disease of tomato and *Hibiscus rosa-sinensis*. Plant Dis 71:731–733

Luigi M, Manglli A, Tomassoli L, Faggioli F (2013) First report of *Hop stunt viroid* in *Hibiscus rosa-sinensis* in Italy. New Dis Rep 27:14

Luria N, Reingold V, Lachman O, Dombrovsky A (2013) Full-genome sequence of *Hibiscus chlorotic ringspot virus* from Israel. Genome Announc 1(6):e01050–e01013

Mali VR (1980) Studies on leaf curl disease of China rose in Marathwada. Indian J Mycol Plant Pathol 9:119

Mao MJ, He ZF, Yu H, Li HP (2008) Molecular characterization of *cotton leaf curls Multan virus* and its satellite DNA that infects Hibiscus rosa-sinensis. Chin J Virol 24:64–68

Melzer MJ, Simbajon N, Carillo J, Borth WB, Freitas-Astua J, Kitajima EW et al (2013) A Cilevirus infects ornamental hibiscus in Hawaii. Arch Virol 158(11):2421–2424

Milne RG, Lesemann DE (1984) Immunosorbent electron microscopy in plant virus studies. Methods Virol 8:85–101

Mitrofanova IV, Zakubanskiy AV, Mitrofanova OV (2018) Viruses infecting main ornamental plants: an overview. Ornam Horticult 24(2):95–102

Morya VK, Singh BK, Thomas G (2014) Ecogenomic of geminivirus from India and neighbour countries: a *in-silico* analysis of recombination phenomenon. Interdiscip Sci Comput Life Sci 6:1–9

Mumford R, Macarthur R, Boonham N (2016) The role and challenges of new diagnostic technology in plant biosecurity. Food Security 8:103–109

Myrta A, Potere O, Crescenzi A, Nuzzaci M, Boscia D (2000) Properties of two monoclonal antibodies specific to the cherry strain of Plum pox virus. J Plant Pathol 82:95–101

Parizipour MHG, Keshavarz-Tohid V (2020) Identification and phylogenetic analysis of a tobamovirus causing hibiscus (*Hibiscus rosa-sinensis L.*) mosaic disease in Iran. J Plant Pathol 102:813–824

Parrella G, Fiallo-Olivé E, Navas-Castillo J (2012) First report of China rose (*Hibiscus rosa-sinensis*) as a host of *Alfalfa mosaic virus* in Spain. Plant Dis 96(3):462

Pecman A, Kutnjak G, Gutiérrez-Aguirre I, Adams I, Fox A, Boonham N (2017) Next generation sequencing for detection and discovery of plant viruses and viroids: comparison of two approaches. Front Microbiol 8:1998

Permar T, Garnsey S, Gumpf D, Lee R (1990) A monoclonal antibody that discriminates strains of citrus tristeza virus. Phytopathology 80:224–228

Polston JE, Hiebert E, McGovern RJ, Stansly PA, Schuster DJ (1993) Host range of *Tomato mottle virus*, a new geminivirus infecting tomato in Florida. Plant Dis 77:1181–1184

Polston JE, Cohen L, Sherwood TA, Ben-Joseph R, Lapidot M (2006) *Capsicum* species: symptomless hosts and reservoirs of *Tomato yellow leaf curl virus* (TYLCV). Phytopathology 96:447–452

Pourrahim R, Ghobakhlo A, Farzadfar S (2013) Biological and molecular detection of *Hibiscus chlorotic ringspot virus* infecting *Hibiscus rosa-sinensis* in Iran. Phytopathol Mediterr 52 (3):528–553

Quadros AFF, Silva JP, Xavier CDA, Zerbini FM, Boari AJ (2019) Two new begomoviruses infecting tomato and Hibiscus sp. *in the Amazon region of Brazil*. Arch Virol 164(7):1897–1901

Rajeshwari R, Reddy RVC, Maruthi MN, Colvin J, Seal SE, Muniyappa V (2005) Host range, vector relationships and sequence comparison of a begomovirus infecting hibiscus in India. Ann Appl Biol 147(1):15–25

Reddy DVR, Sudarshana MR, Fuchs M, Rao NC, Thottappilly G (2009) Genetically engineered virus-resistant plants in developing countries: current status and future prospects. Adv Virus Res 75:185–220

Regenmortel MHV, Fauquet CM, Bishop DHL, Carstens EB, Estes MK, Lemon SM et al (2000) Virus taxonomy: classification and nomenclature of viruses. The International Committee on Taxonomy of Viruses, vol 11. Academic, San Diego, p 1162

Rubio L, Galipienso L, Ferriol I (2020) Detection of plant viruses and disease management: relevance of genetic diversity and evolution. Front Plant Sci 11:1092

Shafie RM (2019) Hibiscus chlorotic ringspot virus (HCRSV) affecting roselle (Hibiscus sabdariffa L.) in Egypt. Middle East J Appl Sci 9(1):50–62

She XM, Tang YF, He ZF, Lan GB (2017) Molecular characterization of *Cotton Leaf Curl Multan Virus* and its associated betasatellite infecting Hibiscus rosa-sinensis in the Philippines. J Plant Pathol 99(3):765–768

Sheveleva A, Ivanov P, Gasanova T, Osipov G, Chirkov S (2018) Sequence analysis of Plum pox virus strain C isolates from Russia revealed prevalence of the D96E mutation in the universal epitope and interstrain recombination events. Viruses 10:450

Srinivasan KG, Narendra KR, Wong SM (2002) Hibiscus virus S is a new subgroup II *tobamovirus*: evidence from its unique coat protein and movement protein sequences. Arch Virol 147 (8):1585–1598

Srivastava A, Kumar S, Jaidi M, Raj SK (2016) Association of *Cotton leaf curl Multan virus* and its associated betasatellite with leaf curl disease of Hibiscus rosa-sinensis in India. New Dis Rep 33:4

Tang J, Elliott DR, Quinn BD, Clover GRG, Alexander BJR (2008) Occurrence of *Hibiscus chlorotic ringspot virus* in Hibiscus spp in New Zealand. Plant Dis 92:486

Vasudeva RS, Varma PM, Capoor SP (1953) Some whitefly (*Bemisia tabaci* Gen.) transmitted viruses in India. In: 6th international congress of microbiology, Rome, vol 5, pp 520–521

Verdin E, Wipf-Scheibel C, Gognalons P, Aller F, Jacquemond M, Tepfer M (2017) Sequencing viral siRNAs to identify previously undescribed viruses and viroids in a panel of ornamental plant samples structured as a matrix of pools. Virus Res 241:19–28

Waterworth HE, Lawson RH, Monroe RL (1976) Purification and properties of *Hibiscus chlorotic ringspot virus*. Phytopathology 64:570–575

Wolfswinkel LD (1996) A virus disease in *hibiscus rosa-sinensis.* L. Afr J Agric Sci Technol 9:483–486

Wong SM, Chng CG (1992) Occurrence of *Hibiscus chlorotic ringspot virus* in Singapore. Phytopathology 82:722

Wu Q, Ding S, Zhang Y, Zhu S (2015) Identification of viruses and viroids by next-generation sequencing and homology-dependent and homology-independent algorithms. Annu Rev Phytopathol 53:425–444

Yoshida T, Kitazawa Y, Komatsu K, Neriya Y, Ishikawa K, Fujita N (2014) Complete nucleotide sequence and genome structure of a Japanese isolate of hibiscus latent Fort Pierce virus, a unique *tobamovirus* that contains an internal poly (A) region in its 3′ end. Arch Virol 159 (11):3161–3165

Molecular Characterization of Viruses Occurring on Some Ornamental Shrubs Grown in India

9

Ashish Srivastava, Suwarna Verma, Tripti Singhal, and Shri Krishna Raj

Abstract

In this chapter, we describe the disease symptoms, molecular detection, and characterization of viruses naturally occurring on ornamental shrubs such as crotons, crape jasmine (*Tabernaemontana coronaria*), night-blooming jasmine (*Cestrum nocturnum*), hibiscus (*Hibiscus rosa-sinensis*), bimli jute (*Hibiscus cannabinus*), and some other ornamental shrub species that usually grow in Indian gardens and roadside for their beautiful flowers. The published work of various researchers all over the world has been summarized in this chapter.

Keywords

Shrubs ornamental plants · Virus disease symptoms · Virus characterization · *Begomovirus* · *Potyvirus* · *Tombusvirus* · *Tobamovirus* and *Varicosavirus*

9.1 Introduction

The occupancy of ornamental crops in the economy is increasing globally and become an important part of it. Their use is not limited as a flower itself, but they are extensively being utilized in the manufacturing of perfumes and other important oils. A number of ornamental crops such as crape jasmine, night-blooming jasmine, Indian mallow, *Clerodendrum*, honeysuckle, *Pseuderanthemum*, Eupatorium, croton, *Pedilanthus*, poinsettia, and common jasmine (such as mogra, chameli,

A. Srivastava · S. Verma · T. Singhal
Amity Institute of Virology and Immunology, Amity University, Uttar Pradesh, Noida, India

S. K. Raj (✉)
Plant Molecular Virology Laboratory, CSIR-National Botanical Research Institute, Lucknow, Uttar Pradesh, India

S. K. Raj et al. (eds.), *Virus Diseases of Ornamental Plants*,
https://doi.org/10.1007/978-981-16-3919-7_9

chandini, raatrani, etc.) are grown for their important value and many of them are for their exquisite scent. Many species of jasmine, viz., Arabian jasmine (*Jasminum sambac*), crape jasmine (*Tabernaemontana coronaria*), night-blooming jasmine (*Cestrum nocturnum*), jasmine (*J. officinale*), mullai (*J. auriculatum*), and (*J. grandiflorum*), and ornamental species of *Jatropha* are grown in India. The important growing states of jasmine are Tamil Nadu, West Bengal, Karnataka, Maharashtra, Andhra Pradesh, Odisha, Uttar Pradesh, Jammu and Kashmir, North East states, etc. (Prakash and Muniyandi 2014). *Cestrum nocturnum* is a species of *Cestrum* in the plant family Solanaceae, grown in subtropical regions as an ornamental plant for its flowers that released a powerful, sweet perfume at night. In traditional medicine, *C. nocturnum* have been reportedly used for wound healing and treatment of fever, pains, epilepsy, and dermatological conditions. In addition, *C. nocturnum* possess significant pharmacological and biological properties including analgesic, antitumor, anticancer, anti-inflammatory, antimicrobial, antiepileptic, antioxidant, and larvicidal and cytotoxic activities (Al-Reza et al. 2010). *Tabernaemontana coronaria*—an evergreen and common garden plant in tropical countries—has been used as a traditional medicine. *Tabernaemontana coronaria* is an aphrodisiac tonic, especially to the brain, liver, and spleen and purgative (Henriques et al. 1996). The other member *T. divaricata* is reported as antioxidant, anti-infectives, anticancer, analgesic and cholinergic in both peripheral and central nervous systems (Pratchayasakul et al. 2008).

Indian mallow (*Abutilon indicum*) of family Malvaceae are woody ornamental plants majorly grown in India in garden beds. These plants are also used for several medicinal plants. Ornamental species of *Jatropha, J. integerrima J. podagrica*, and *J. multifida* are grown as ornamental plants in the gardens in India as perennial plants. *J. integerrima* Jacq. is an upright decorative woody shrub and filled with beautiful flowers during summers (Sharma and Singh 2013). *J. podagrica* and *J. multifida* are also decorative woody shrubs and widely cultivated species throughout the tropics and are native from Mexico through Central America to Brazil (Bijan 1982). *Clerodendrum* of family Lamiaceae are native to the continent Africa and southern Asia and are well recognized for their attractive flowers. They are used as remedies for diabetes and high blood pressure in some parts of India. These eye-catching shrub species are also known to be host of many viruses (Bock et al. 1981; Narayana et al. 2006, 2007; Tewari et al. 2007; Raj et al. 2008a, b; Gao et al. 2010; Ramkat et al. 2011; Ramkat Rose et al. 2011; Snehi et al. 2011a, b, 2012); however, few of them showed more susceptibility towards the begomoviruses. We described about the various DNA or RNA viruses infecting these shrubs one by one.

9.2 Viruses Infecting Ornamental Shrubs

9.2.1 Begomoviruses Infecting *Jasmine* sp.

Jasmines are group of plants known as source of many exotic fragrances and beautiful flowers. These are the major sources noted in perfumes and oils. Here we are discussing about the Arabian jasmine (*J. sambac*), crape jasmine (*T. coronaria*), night-blooming jasmine (*C. nocturnum*), jasmine (*J. officinale*), and mullai (*J. auriculatum*). The perennial jasmine serves many viruses as alternate host during non-cropping seasons. Based on the high economic importance of these crops, the characterization and diversity studies are prime important. During the surveys of jasmine at Lucknow, New Delhi, Noida, and Bangalore, India, severe yellow mottling and leaf curl symptoms were observed (Fig. 9.1) in more than 90% plants. The association of a whitefly transmitting begomovirus, Pedilanthus leaf curl virus, with yellow mottling and leaf curl symptoms of jasmine was first reported in India (Srivastava et al. 2014). Another begomovirus, *Ageratum enation virus*, was reported from Laxmangarh and Rajasthan, India, on *J. sambac* (Marwal et al. 2013).

Fig. 9.1 Showing yellow mottling, blistering, and leaf deformation on crape jasmine in close view (**a**) and in a field (**b**) as compared with healthy plants (**c**). Severe leaf curl and deformation symptoms on night-blooming jasmine (**d**) compared with a healthy plant (**e**)

9.2.2 Begomoviruses Infecting *Jatropha* sp.

J. integerrima, *J. podagrica*, and *J. multifida* are the species of *Jatropha* which are grown as ornamental crops in the gardens in India. In these ornamental species, the infection of Cassava latent geminivirus from Kenya and Jatropha mosaic virus from Florida, USA, have been reported to cause mosaic disease of J. multifida (Bock et al. 1981; Polston et al. 2014). However, no virus has been reported hitherto from *J. integerrima* and *J. podagrica*. Srivastava et al. (2015a) first reported the infection of begomovirus among these species of *Jatropha* (Fig. 9.2). They described the association of a novel begomovirus with yellow mosaic symptoms of *J. integerrima*, *J. podagrica*, and *J. multifida* with sequence analyses of cloned complete DNA-A genome amplified by rolling circle amplification. The novel species of begomovirus was identified as "Jatropha mosaic Lucknow virus (JMLV)" (Srivastava et al. 2015b).

Snehi et al. (2014) have also reported three different begomovirus species, Jatropha mosaic India virus in *J. podagrica*; Tomato leaf curl Patna virus in *J. multifida*; and Papaya leaf curl virus in *J. integerrima*. However, we did not recover the begomoviruses species from ornamental *Jatropha* which are reported from *J. curcas* in India such as *Indian cassava mosaic virus* (Snehi et al. 2012), *Jatropha leaf yellow mosaic Katerniaghat virus* (Srivastava et al. 2015b), and *Jatropha leaf crumple India virus* (Srivastava et al. 2015c).

Fig. 9.2 Natural symptoms of severe yellow mosaic on *J. integerrima* (**a**), *J. multifida* (**b**), and *J. podagrica* (**c**), as compared to their respective healthy plants (**d**, **e**, **f**)

9.2.3 Begomoviruses Infecting Garden Croton

The garden croton (*Codiaeum variegatum*) of the family Euphorbiaceae was first included by Carl Linnaeus in 1753. The plants are widely grown in the garden of India, Indonesia, Malaysia, Australia, etc. Garden crotons are famous for their beautiful leaves which show yellow veins, yellow spots in lamina, bright yellowing, and colored spots in the dark green background. These natural leaf structures sometimes match with the natural virus symptoms in other croton wild species. Chenulu and Phatak (1965) first reported the whitefly-transmitted geminivirus associated with the yellow mosaic disease of croton.

Further, the whitefly transmission assay and the nucleic acid hybridization technique were used to detect the begomovirus: Acalypha yellow mosaic begomovirus causing yellowing and yellow spots symptoms (Raj et al. 1996).

During the surveys around Bengaluru, India, from February 2005 to July 2006, Mahesh et al. (2010) observed typical leaf curl symptoms of begomovirus on crotons, which were further detected by PCR. The partial coat protein was sequenced, and the infecting begomovirus was tentatively identified as croton leaf curl virus.

9.2.4 Croton Yellow Vein Mosaic Virus (CYVMV) on Croton

Croton (*Coidageum variegatum*; Euphorbiaceae) is a common ornamental plant valued for the variation of leaf morphology and bright tones of color shown by many cultivars. Leaves can be green with yellow spots or completely yellow. Some cultivars have yellow veins and/or spots that range from yellow to red. Although the color patterns on the leaves resemble symptoms caused by plant viruses, the general thought is that they are of genetic nature. A begomovirus, Croton yellow vein mosaic virus (CYVMV), and a putative rhabdovirus have been isolated from plants showing the virus-like symptoms described above (Bellardi 1991; Harrison et al. 1991). CYVMV was detected by molecular hybridization in several croton cultivars (Raj et al. 2006). The virus is transmitted by grafting and whiteflies (*B. tabaci*) to other croton plants and to several other plant species (Harrison et al. 1991). Although excessive yellowing and poor growth are associated with CYVMV infection on croton, differentiation between genetically inherited color patterns from those caused by the virus has not been well established (Raj et al. 2006). The croton rhabdovirus caused dwarfing, vein yellowing, and leaf malformation on croton (Bellardi 1991).

9.2.5 Clerodendron Yellow Mosaic Begomovirus Infecting Bougainvillea

Bougainvillea (*Bougainvillea peruviana*) is a common and popular ornamental plant of the family Nyctaginaceae commonly grown in the gardens of India. During a survey in the spring seasons of 2012 and 2013, bougainvillea plants throughout

Rajasthan State, India, exhibited typical symptoms for a begomovirus infection, including leaf curling and stunted growth. The presence of begomovirus DNA-A genomic components was detected in the plants exhibiting symptoms by PCR using begomovirus component-specific primers which produced amplicons of the expected size (~550 bp). PCR-positive samples were subjected to rolling circle amplification to obtain the full-length viral genomes. RCA products which showed single fragments of ~2.7 kbp were cloned and sequenced (GenBank Accession No. KF704391). Sequence analysis of DNA-A revealed 98% identity at nucleotide level with Clerodendron yellow mosaic virus (ClYMV) isolates suggesting that our isolate is an isolate of ClYMV. This was the first report of natural occurrence of ClYMV on *B. peruviana* in India (Nehra et al. 2014).

9.2.6 Clerondendron Yellow Mosaic Begomovirus Infecting Golden Dewdrop

Golden dewdrop (*Duranta erecta*, Verbenaceae) is a species of flowering shrub and is commonly grown as an ornamental hedge plant in gardens. During a survey in August 2013, yellow mosaic disease was observed on *D. erecta* growing in several public gardens/parks at Lucknow, India, with disease incidence of 80–90%. Infected plants exhibited severe yellow mosaic accompanied by upward leaf curl symptoms and stunting in severely infected plants as compared to healthy plants (Jaidi et al. 2015).

These symptoms reminiscent of a begomovirus infection reported previously on *D. erecta* in Pakistan, and hence a similar infection was suspected. Twenty leaf samples from infected and four from healthy plants were collected from Goldmohar Park at Jankipuram, Lucknow, India, and total DNA was isolated. PCR amplifications were performed with begomovirus-specific degenerate primers which resulted in the generation of amplicons of the anticipated size of ~1.2 kb in all samples from leaves showing disease symptoms and no amplicons from any of the four healthy samples suggestive of begomovirus in *D. erecta*.

To identify the begomovirus responsible for the infection, total DNA isolated from a representative sample was subjected to rolling circle amplification. Electrophoresis of product revealed a DNA fragment ~2.7 kb which was cloned, sequenced, and deposited in GenBank (Accession No. KR869857, MJGD1). Sequence analysis of isolate MJGD1 revealed 93–95% nucleotide identities with DNA-A of *Clerodendron yellow mosaic virus* (ClYMV): RKAS1 (KF704391) of *Bougainvillea peruviana*; IARI (EF408037) of *Clerodendron* spp. from India; and SA23 (HE863667) of *Croton* spp. from Pakistan. It also showed close phylogenetic relationships with these ClYMV isolates suggesting that begomovirus isolated from *D. erecta* is a new strain or species of ClYMV (Jaidi et al. 2015).

To assess the prevalence of ClYMV in *D. erecta*, total DNA was extracted from 28 symptomatic samples collected from 7 public gardens at Lucknow and tested by nucleic acid spot hybridization (NASH) tests using an α-32P radioactive labeled probe prepared from cloned ClYMV (KR869857). NASH resulted in positive signals from 27 out of 28 samples, indicating the prevalence of ClYMV in 96.4% *D. erecta* plants. Begomoviruses are associated with leaf curl disease in *D. erecta* (Irama et al. 2005); Catharanthus yellow mosaic virus in *D. repens*, and Tomato leaf curl New Delhi virus in *D. repens* from Pakistan. ClYMV has been reported on *B. peruviana* and *C. inerme* in India, and this was the first report of natural occurrence of ClYMV on *D. erecta* in India (Jaidi et al. 2015).

9.2.7 Begomoviruses Infecting China Rose

Hibiscus sp. (commonly known as China rose), of family Malvaceae, is a genus of around 220 species of flowering plants, and it is native to warm temperate, subtropical, and tropical regions throughout the world. Genus *Hibiscus* contains both annual and perennial woody ornamental shrubs and herbaceous plants (Fig. 9.3). They are also used medicinally for temperature and other urine disease while in some countries used for tea to treat constipation. Due to excessive presence of the polysaccharide content, mucilage, these plants are favorite for several viral vectors such as mealybugs, aphids, and whiteflies. The infection of viruses causes serious damages and bad flower quality of these plants and, therefore, causes loss of the ornamental qualities. Severe mosaic disease of China rose (*Hibiscus rosa-sinensis*) was also identified in association of cotton leaf curl Multan virus (Srivastava et al. 2016).

Fig. 9.3 Natural severe leaf curl symptoms of China rose (**a**) in comparison to healthy (**b**)

A new begomovirus species, *Hibiscus golden mosaic virus*, has been reported on *Hibiscus* sp. causing yellow mosaic symptoms in Brazil (Quadros et al. 2019).

9.2.8 *Hibiscus Chlorotic Ringspot Tombusvirus* Infecting *Hibiscus rosa-sinensis*

Due to perennial nature, these shrub plants serve as the potential alternative hosts for several DNA and RNA viruses during non-cropping seasons. Gao et al. (2012), Karanfil and Korkmaz (2017), and Ramos-González et al. (2020) have characterized the positive-sense RNA virus, Hibiscus chlorotic ringspot virus of family *Tombusviridae*, from vein banding and chlorotic ringspots disease of *H. rosa-sinensis*.

Hibiscus rosa-sinensis (rose of China) is an ornamental plant grown throughout the tropics and subtropics and under glass in more temperate areas. It is commonly used in landscape design. Hibiscus chlorotic ringspot virus (HCRSV) was reported for the first time in Turkey from *Hibiscus rosa-sinensis* by Karanfil and Korkmaz (2017). HCRSV is one of the pathogens causing diseases on *H. rosa-sinensis*. Symptoms of HCRSV vary from vein banding to chlorotic ringspots on leaves of *H. rosa-sinensis*. The virus belongs to the genus *Betacarmovirus* in the family *Tombusviridae*.

The virus was detected from infected leaves collected from eight different plants by RT-PCR using a pair of primers specific to parts of the coat protein (CP) gene of HCRSV. As expected, a 759 bp DNA fragment corresponding to the partial CP gene was amplified from all samples. Two PCR products were sequenced. BLAST analysis of these sequences confirmed similarity to HCRSV sequences. The identities at nucleotide and amino acid level between Turkish and other HCRSV isolates were from 94 to 98% and 86 to 97%, respectively. Phylogenetic relationship of different HCRSV isolates showed close relatedness with isolates from Iran and Israel. This was the first report of HCRSV in Turkey (Karanfil and Korkmaz 2017).

The complete genome sequence of the *Hibiscus chlorotic ringspot virus* (HCRSV) isolated from the Americas was determined by Ramos-González et al. (2020). The genome sequence of HCRSV isolate SBO1 was 3945 nucleotides long and showed 93.1% nucleotide sequence identity with HCRSV-Singapore (X86448), considered the type member of the species. These two viruses displayed a similar genomic organization and potentially encode seven open reading frames (ORFs). In addition, a phylogenetic analysis based on a fragment with 557 nts of p38, the coat protein gene, from a cohort of 14 samples collected in Brazil revealed a no clearly defined segregation of South American isolates from those previously detected in geographical areas outside this continent. The practical consequences of the use of p38 ORF-derived amplicons for detection and variability studies of HCRSV were discussed (Ramos-González et al. 2020).

9.2.9 *Tobacco Mosaic Virus/Tomato Mosaic Virus* (Tobamovirus) Infecting Jasmine

Jasminum multiflorum and *J. gracile* were reported hosts of a *Tobamovirus*, Tomato mosaic virus (ToMV) and the potyvirus, jasmine potyvirus T (JaVT-jasmine) and Jasmine yellow mosaic potyvirus (JaYMV), from the USA, India and China (Kamenova et al. 2006; Kaur et al. 2013).

Virus-like symptoms were observed on leaves of landscape and nursery downy and star jasmine (*Jasminum multiflorum*) and wax jasmine (*J. gracile*) in southeast Florida. Foliar symptoms included mottling, chlorotic ringspots, and chlorotic line patterns. An agent was mechanically transmitted with difficulty from symptomatic leaves of downy jasmine to *Nicotiana debneyi* and *N. tabacum* 'Xanthi' and subsequently from these hosts to *Chenopodium quinoa* and other herbaceous test plants. Virions were isolated from *N. tabacum* 'Xanthi.' Rod-shaped particles (297 × 18 nm) similar to tobamoviruses were observed in partially purified virus preparations and in leaf dip from symptomatic star jasmine and indicator plants (Kamenova et al. 2006).

Extraction of viral-associated double-stranded (ds) RNA revealed a profile consistent with that of a tobamovirus. Sodium dodecyl sulfate-polyacrylamide gel electrophoresis analysis showed that the virus contained one polypeptide with an approximate molecular weight (Mr) of 18 kDa. The virus reacted specifically with IgG for *Tobacco mosaic virus* (TMV) and related tobamoviruses (*Tomato mosaic virus*, ToMV) in double-antibody sandwich enzyme-linked immunosorbent assay. No reaction was observed with TMV-specific IgG (Kamenova et al. 2006).

Reverse transcription polymerase chain reaction with total RNA isolated from symptomatic jasmine leaves and infected *N. tabacum* 'Xanthi' using ToMV coat protein-specific primers amplified the expected product from downy and star (but not wax) jasmine and *N. tabacum* 'Xanthi.' The nucleotide and amino acid sequence of the products were 100% identical to the corresponding fragment of a Brazilian isolate of ToMV from *Impatiens*. This was the first report of ToMV in jasmine in the USA (Kamenova et al. 2006).

9.2.10 Potyviruses Infecting Jasmine

In India, Kaur et al. (2013) have reported the association of yellow mosaic symptoms on *J. sambac* with Jasmine yellow mosaic potyvirus (JaYMV-India). Association of Jasmine potyvirus with the yellow mosaic disease of *Jasminum sambac* L. was detected by reverse transcription polymerase chain reaction (RT-PCR) using potyvirus coat protein (CP) gene-specific primers and confirmed by Southern hybridization test with CP gene-specific probe prepared from a known potyvirus. The RT-PCR amplicons were cloned and sequenced and 616 nucleotide sequence data deposited in GenBank database (JN807771). The sequence data of jasmine virus isolate shared highest 94% identities during BLASTn, high amino acid sequence homology in multiple alignment analysis, and closest phylogenetic relationship with Taiwanese isolate of *J. potyvirus T.* infecting *J. sambac*; therefore, the virus under study was identified as an isolate of *J. potyvirus* (Kaur et al. 2013).

Further, the complete genome of jasmine virus T from *Jasminum sambac* was analyzed in China by Tang et al. (2016). The genome of a potyvirus (isolate JaVT_FZ) recovered from jasmine (*Jasminum sambac* L.) showing yellow ringspot symptoms in Fuzhou, China, was sequenced. JaVT_FZ is closely related to 7 other potyviruses with completely sequenced genomes, with which it shares 66–70% nucleotide and 52–56% amino acid sequence identity. However, the coat protein (CP) gene shares 82–92% nucleotide and 90–97% amino acid sequence identity with those of two partially sequenced potyviruses, named jasmine potyvirus T (JaVT-jasmine) and jasmine yellow mosaic potyvirus (JaYMV-India), respectively. This suggests that JaVT_FZ, JaVT-jasmine, and JaYMV-India should be regarded as members of a single potyvirus species, for which the name "Jasmine virus T" has priority (Tang et al. 2016).

9.2.11 Detection of a *Carlavirus* Species from Yellow Mosaic Jasmine Plants

Jasmine (*Jasminum sambac*) is widely grown as an ornamental plant and for extraction of essential oil in Taiwan. In 2003, a potyvirus, designated *Jasmine virus T* (JaVT), was isolated from jasmine exhibiting severe yellow mosaic symptoms. However, due to jasmine's vegetative propagation nature, confirmed virus-free jasmine plants were unavailable for inoculation tests. Determining whether JaVT could individually induce yellow mosaic symptom was therefore not testable. In 2010, another field survey was conducted and found that, in addition to plants showing a yellow mosaic symptom, there was also jasmine exhibiting mild mosaic, mottling symptoms, or even being nearly symptomless. Interestingly, JaVT could be detected from all of the jasmine plants exhibiting these diverse symptoms, suggesting JaVT might not be the only virus inducing the yellow mosaic disease (Chang et al. 2018).

In 2012, a *Carlavirus* was detected from yellow mosaic jasmine plants by RT-PCR using a degenerate *Carlavirus*-specific primer pair. Sequence analysis of these amplicons demonstrated the presence of a new species of *Carlavirus*, which was named *Jasmine virus C* (JaVC). Virus-specific ELISA and RT-PCR analysis confirmed that JaVC could only be detected in yellow mosaic jasmine, but not in those plants showing mild symptoms. Taken together, these results suggest that the yellow mosaic symptom in jasmine was induced by a mixed infection of JaVT and JaVC (Chang et al. 2018).

9.2.12 Camellia Yellow Mottle Virus Infecting *Camellia japonica*

Camellia yellow mottle virus (CYMoV) causes pale discoloration in camellia blossoms and leaves. CYMoV also causes irregular yellow mottling of camellia leaves and whitish blotches in the blossoms. The pattern of affected plant tissue is irregular on the whole plant; some branches show symptoms, and others do not. Some leaves can be almost entirely affected. Margins of discolored areas are not distinct and blur into the normal color of the plant part. CYMoV spreads by budding, grafting, or rooting cuttings from plants that are infected. The virus is sometimes deliberately introduced through grafting to produce leaf or flower variegation considered to be attractive. Once a plant becomes infected with virus, it usually remains infected throughout its life (http://ipm.ucanr.edu/PMG/GARDEN/PLANTS/DISEASES/camvirus.html).

It is also reported that CYMoV causes variegation of camellia flowers and occasional yellow mottle or ringspots on some leaves. Based on particle morphology and ultrastructure studies of infected cells, CYMoV has been proposed to be a tentative member of the genus *Varicosavirus* but is currently not recognized as such by the International Committee on Taxonomy of Viruses. Varicosaviruses consist of fragile, non-enveloped rods of about 320–360 nm in length with negative-sense ssRNA genome and are transmitted by the zoospores of the fungus *Olpidium brassicae*. Except for graft transmission, CYMoV has not been transmitted by other means. The host range of CYMoV was restricted to *Camellia* spp. (Valverde et al. 2012).

9.2.13 Association of a Complex Virome Containing Two Distinct Emaraviruses to Virus-Like Symptoms in *Camellia japonica*

In 2020, a complex virome that includes two distinct emaraviruses was recorded to be associated with virus-like symptoms in *Camellia japonica* by Peracchio et al. (2020) in Italy. *They described that Camellia japonica* plants manifesting a complex and variable spectrum of viral symptoms like chlorotic ringspots, necrotic rings, yellowing with necrotic rings, yellow mottle, leaves and petals deformations, and flower color breaking were studied since 1940 essentially through electron microscopic analyses; however, a strong correlation between symptoms and one or more well-characterized viruses was never verified.

In this work, samples collected from symptomatic plants were analyzed by NGS technique, and a complex virome composed by virus members of the *Betaflexiviridae* and *Fimoviridae* families was identified. In particular, the genomic fragments typical of the emaravirus group were organized in the genomes of two new emaraviruses species, tentatively named *Camellia japonica*-associated emaravirus 1 and 2. They were the first emaraviruses described in camellia plants and were always found solely in symptomatic plants. On the contrary, in both symptomatic and asymptomatic plants, they detected five betaflexivirus isolates that, based on aa identity comparisons, can be classified in two new putative species called *Camellia japonica*-associated betaflexivirus 1 and *Camellia japonica*-associated betaflexivirus 2. Together with other recently identified betaflexiviruses associated to *Camellia japonica* disease, the betaflexiviruses characterized in this study show an unusual hyper-conservation of the coat protein at amino acidic level (Peracchio et al. 2020).

9.2.14 Frangipani Mosaic Tobamoviruses Infecting Frangipani Tree (*Plumeria rubra*)

Frangipani mosaic virus (FrMV) is known to infect frangipani tree (*Plumeria rubra*) in India, but the virus had not been characterized at genomic level. An isolate of FrMV (FrMV-Ind-1) showing greenish mosaic and vein-banding symptoms *in P. rubra* in New Delhi was characterized based on host reactions, serology, and genome sequence. The virus isolate induced local symptoms on several new experimental host species: *Capsicum annuum*, *Nicotiana benthamiana*, *Solanum lycopersicum*, and *S. melongena*. The genome of FrMV-Ind-1 was 6643 (JN555602) nucleotides long with genome organization similar to tobamoviruses. The Indian isolate of FrMV shared a very close genome sequence identity (98.3%) with the lone isolate of FrMV-P from Australia. FrMV-Ind-1 together with FrMV-P formed a new phylogenetic group, i.e., Apocynaceae-infecting tobamovirus (Kumar et al. 2015).

The polyclonal antiserum generated through the purified virus preparation was successfully utilized to detect the virus in field samples of frangipani by ELISA. Of the eight different tobamoviruses tested, FrMV-Ind-1 shared distant serological relationships with only cucumber green mottle mosaic virus, tobacco mosaic virus, bell pepper mottle virus, and kyuri green mottle mosaic virus. RT-PCR based on coat protein gene primer successfully detected the virus in frangipani plants. This study was the first comprehensive description of FrMV occurring in India (Kumar et al. 2015).

9.3 Conclusion

It is concluded that various ornamental shrubs are attached and infected by mainly five types of viruses belonging to the *Begomovirus*, *Potyvirus*, *Tombusvirus*, *Tobamovirus*, and *Varicosavirus* group. Among them, most of the viruses were from the *Begomovirus* group followed by *Potyvirus* group. Some viruses were also detected that belong to *Tombusvirus*, *Tobamovirus*, and *Varicosavirus* groups.

Two *Begomovirus* species—*Pedilanthus leaf curl virus* on crape jasmine (*T. coronaria*) and night-blooming jasmine (*C. nocturnum*) and *Ageratum enation* viruses on Arabian jasmine (*J. sambac*)—were found from India. Three different *Begomovirus* species—Jatropha mosaic India virus, Tomato leaf curl Patna virus, and Papaya leaf curl virus—have been found in India on three species of ornamental Jatropha, viz., *J. podagrica*, *J. multifida*, and *J. integerrima*, respectively. Two begomoviruses—Acalypha yellow mosaic virus and Croton yellow vein mosaic virus—have been reported to cause yellowing, yellow spots and leaf curl symptoms in crotons (*Codiaeum variegatum*). A new begomovirus species, *Hibiscus golden mosaic virus*, has been reported on China rose (*Hibiscus* sp.) causing yellow mosaic symptoms in Brazil. The *Cotton leaf curl Multan begomovirus* was also found to be associated with severe mosaic disease of China rose (*Hibiscus rosa-sinensis*). The *Clerodendron yellow mosaic begomovirus* has been reported on bougainvillea (*Bougainvillea peruviana*) and golden dewdrop (*Duranta erecta*).

The *Potyvirus* species—*Jasmine potyvirus T* (JaVT-jasmine) and *Jasmine yellow mosaic potyvirus* (JaYMV-India) on *Jasmine multiflorum* and *J. gracile* and *Jasmine yellow mosaic potyvirus* (JaYMV-India) on *J. sambac*—were reported from India. A *Tobamovirus*, *Tomato mosaic virus* (ToMV), has been reported on *Jasmine multiflorum* and *J. gracile* from China and the USA. *Hibiscus chlorotic ringspot virus* of *Tombusvirus* group was found in *H. rosa-sinensis* causing vein banding and chlorotic ringspots disease. *Camellia yellow mottle virus of Varicosavirus* group on camellia (*Camellia japonica*) and *Frangipani mosaic virus* of *Tobamovirus* on frangipani tree (*Plumeria rubra*) have also been reported.

References

Al-Reza SM, Rahman A, Cho Y-S, Kang SC (2010) Chemical composition and antioxidant activity of essential oil and organic extracts of *Cestrum nocturnum* L. J Essent Oil Bearing Plants 13 (5):615–624

Bellardi MG (1991) First report of rhabdovirus-like particles in croton (*Codiaeum variegatum*). Plant Dis 75:862

Bijan D (1982) Comparative anatomy of the petiole and infrageneric relationships in Jatropha (*Euphorbiaceae*). Am J Bot 69(8):1283–1295

Bock KR, Guthrie EJ, Figueiredo G (1981) A strain of Cassava latent virus occurring in coastal districts of Kenya. Ann Appl Biol 99(2):151–159

Chang CA, Lin YC, Lin YY, Wang SM (2018) Etiological study of viruses associated with yellow mosaic disease in jasmine. Acta Hortic 1193:47–54. https://doi.org/10.17660/ActaHortic.2018.1193:7

Chenulu VV, Phatak HC (1965) Yellow mosaic of Acalypha indica L.—a new whitefly transmitted virus disease from India. Curr Sci 34:321–322

Gao SQ, Qu J, Chua NH, Ye J (2010) A new strain of Indian cassava mosaic virus causes a mosaic disease in the biodiesel crop *Jatropha curcas*. Arch Virol 155(4):607–612

Gao R, Ng FK, Liu P, Wong SM (2012) Hibiscus chlorotic ringspot virus coat protein upregulates sulfur metabolism genes for enhanced pathogen defense. Mol Plant-Microbe Interact 25 (12):1574–1583

Harrison BD, Muniyappa V, Swanson MM, Roberts IM, Robinson DJ (1991) Recognition and differentiation of seven whitefly transmitted geminiviruses from India and their relationship to African cassava mosaic and Thailand mung bean yellow mosaic viruses. Ann Appl Biol 118:299–308

Henriques AT, Melo AA, Morreno PRH (1996) Ervatamia coronaria: chemical constituents and some pharmacological activities J. Ethnopharmacol 50:19–25

Irama S, Amraoa L, Mansoora MS, Malika AH, Briddona RW, Zafara Y (2005) First report of a begomovirus associated with leaf curl disease of Duranta erecta in Pakistan. Plant Pathol 54:260

Jaidi M, Kumar S, Srivastava A, Raj SK (2015) First report of Clerodendron yellow mosaic virus on golden dewdrop (*Duranta erecta*) in India. New Dis Rep 32:12

Kamenova I, Adkins S, Achor D (2006) Identification of tomato mosaic virus infection in jasmine. Acta Hortic 722:277–284

Karanfil A, Korkmaz S (2017) First report of Hibiscus chlorotic ringspot virus in Turkey. New Dis Rep 35:22

Kaur C, Kumar S, Snehi SK, Raj SK (2013) Molecular detection of Jasmine potyvirus associated with yellow mosaic symptoms on Jasminum sambac L. in India. Arch Phytopathol Plant Protect 46(9):1102–1107

Kumar A, Solanki V, Verma HN, Mandal B (2015) Characterisation and diagnosis of frangipani mosaic virus from India. Virus Genes 51:310–314

Mahesh YS, Shankarappa KS, Rangaswamy KT, Prameela HA, Aswathanarayana DS, Divya BL, Nagesha N, Maruthi MN (2010) Detection and characterisation of a begomovirus associated with leaf curl disease of ornamental croton (*Codiaeum variegatum*). J Hortic Sci Biotechnol 85 (2):101–105

Marwal A, Sahu A, Prajapat R, Gaur RK (2013) First report of begomovirus infecting two ornamental plants: *Jasminum sambac* and *Millingtonia hortensis*. Indian Phytopathol 66:115–116

Narayana DSA, Shankarappa KS, Govindappa MR, Prameela HA, Gururaj Rao MR, Rangaswamy KT (2006) Natural occurrence of Jatropha mosaic virus disease in India. Curr Sci 91:584–586

Narayana DSA, Rangaswamy KT, Shankarappa KS, Maruthi MN, Lakshminarayana Reddy CN, Rekha AR, Keshava Murthy KV (2007) Distinct begomoviruses closely related to cassava mosaic viruses cause Indian jatropha mosaic disease. Int J Virol 3:1–11

Nehra C, Sahu AK, Marwal A, Gaur RK (2014) Natural occurrence of Clerodendron yellow mosaic virus on Bougainvillea in India. New Dis Rep 30:19

Peracchio C, Forgia M, Chiapello M, Vallino M, Turina M, Ciuffo M (2020) A complex virome that includes two distinct emaraviruses is associated to virus like symptoms in Camellia japonica. Virus Res 286:197964. https://doi.org/10.1016/j.virusres.2020.197964

Polston JE, London MA, Capobianco H (2014) The complete genome sequence of New World Jatropha mosaic virus. Arch Virol 159:3131–3136. https://doi.org/10.1007/s00705-014-2132-1

Prakash K, Muniyandi B (2014) Area, production of jasmine and rainfall of Madurai District in Tamil Nadu, South India. Int J Sci Res 3:1122–1125

Pratchayasakul W, Pongchaidecha A, Chattipakorn N, Chattipakorn S (2008) Ethnobotany & ethnopharmacology of Tabernaemontana divaricata. Indian J Med Res 127(4):317–335

Quadros AFF, Silva JP, Xavier CAD, Zerbini FM, Boari AJ (2019) Two new begomoviruses infecting tomato and Hibiscus sp. in the Amazon region of Brazil. Arch Virol 164(7):1897–1901

Raj SK, Sriavstava KM, Singh BP (1996) Evidence from nucleic acid hybridization tests for geminivirus infection of ornamental crotons in India. Eur J Plant Pathol 102:201–203

Raj SK, Srivastava KM, Singh B (2006) Evidence from nucleic acid hybridization tests for geminivirus infection of ornamental crotons in India. Eur J Plant Pathol 102:201–203

Raj SK, Snehi SK, Kumar S, Khan MS, Pathre U (2008a) First molecular identification of a distinct begomovirus causing mosaic disease on Jatropha curcas L. in India which is closely related to Indian cassava mosaic virus. Aust Plant Dis Notes 3:69–72

Raj SK, Kumar S, Snehi SK, Uday P (2008b) First report of Cucumber mosaic virus on *Jatropha curcas* in India. Plant Dis 92(1):171

Ramkat Rose C, Calari A, Maghuly F, Laimer M (2011) Biotechnological approaches to determine the impact of viruses in the energy crop plant *Jatropha curcas*. Virol J 8:386

Ramkat R, Calari A, Maghuly F, Laimer M (2011) Occurrence of African cassava mosaic virus (ACMV) and East African cassava mosaic virus-Uganda (EACMV-UG). In: Jatropha curcas. BMC proceedings, vol 5, no. Suppl 7, p P93

Ramos-González PL, da Costa-Rodrigues M, Potsclam-Barro M, Chabi-Jesus C, Banguela-Castillo A, Harakava R, Kitajima EW, Freitas-Astúa J (2020) First genome sequence of an isolate of hibiscus chlorotic ringspot virus from the Western hemisphere. Trop Plant Pathol 45:153–158

Sharma SK, Singh H (2013) Pharmacognostical standardisation of Jatropha integerrima Jacq. (Euphorbiaceae) roots. Pharm Lett 5(1):155–159

Snehi SK, Raj SK, Khan MS, Prasad V (2011a) Molecular identification of a new begomovirus species associated with yellow mosaic disease of *Jatropha gossypifolia* in India. Arch Virol 156:2303–2307

Snehi SK, Khan MS, Raj SK, Prasad V (2011b) Complete nucleotide sequence of Croton yellow vein mosaic virus and DNA-β associated with yellow vein mosaic disease of *Jatropha gossypifolia* in India. Virus Genes 43:93–101

Snehi SK, Srivastava A, Raj SK (2012) Biological characterization and complete genome sequence of a possible strain of Indian cassava mosaic virus from *Jatropha curcas* in India. J Phytopathol 160:547–553

Snehi SK, Raj SK, Prasad V, Singh V (2014) Molecular identification of begomoviruses affecting important ornamental plants in India. Int J Bacteriol Virol 1(1):1–8

Srivastava A, Kumar S, Raj SK (2014) Association of Pedilanthus leaf curl virus with yellow mottling and leaf curl symptoms in two jasmine species grown in India. J Gen Plant Pathol 80:370–373

Srivastava A, Kumar S, Jaidi M, Raj SK (2015a) Characterization of a novel begomovirus associated with yellow mosaic disease of three ornamental species of Jatropha grown in India. Virus Res 201:41–49

Srivastava A, Kumar S, Jaidi M, Raj SK (2015b) Molecular characterization of a new begomovirus associated with leaf yellow mosaic disease of Jatropha curcas in India. Arch Virol 160 (5):1359–1362

Srivastava A, Jaidi M, Kumar S, Raj SK (2015c) Molecular identification of a new begomovirus associated with leaf crumple disease of Jatropha curcas L. in India. Arch Virol 160(2):617–619

Srivastava A, Kumar S, Jaidi M, Raj SK (2016) Association of Cotton leaf curl Multan virus and its associated betasatellite with leaf curl disease of *Hibiscus rosa-sinensis* in India. New Dis Rep 33:4

Tang Y, Gao F, Yang Z, Wu Z, Yang L (2016) Complete genome analysis of jasmine virus T from Jasminum sambac in China. Arch Virol 161:2033–2036

Tewari JP, Dwivedi HD, Pathak M, Srivastava SK (2007) Incidence of a mosaic disease in Jatropha curcas L. from eastern Uttar Pradesh. Curr Sci 93(8):1048–1049

Valverde RA, Sabanadzovic S, Hammond J (2012) Viruses that enhance the aesthetics of some ornamental plants: beauty or beast? Plant Dis 96(5):600–611

Virus-Diseases Reported on *Calendula officinalis*

10

Akil A. Khan and Shoeb Ahmad

Abstract

Calendula officinalis L. of Asteraceae known as "pot marigold" is grown across Europe and Asia for its beautiful flowers and some medicinal properties. Plant is reported to be infected by various viruses like *Ageratum enation* virus, *Tomato leaf curl New Delhi virus*, *Bidens mottle virus*, *Cucumber mosaic virus*, *Papaya leaf curl virus*, etc. which reduced the quality and quantity of the flowers. In this review, we have discussed about the viruses and their diseases which invade *Calendula officinalis*.

Keywords

Calendula officinalis · *Ageratum enation virus* · *Tomato leaf curl New Delhi virus* · *Bidens mottle virus* · *Cucumber mosaic virus* · *Papaya leaf curl virus*

10.1 Introduction

Plant viruses can infect them and cause various diseases characterized by a combination of severe symptoms and high transmission rates (Anderson et al. 2004; Strange and Scott 2005). In natural ecosystems, plant viruses cause devastating diseases and often have wide host ranges (Strange and Scott 2005). While viruses can be visually impaired (Shang et al. 2011), a lack of visual symptoms (such as leaf mottling or malformation) is often believed to suggest a lack of virus infection (Cooper and Jones 2006).

We know very little about commercially important plant-virus interactions (e.g., seeds, pasture plants as well as horticultural varieties) (Nicaise 2014). Viral infection

A. A. Khan (✉) · S. Ahmad
Department of Botany, G.F. College, Shahjahanpur, UP, India

S. K. Raj et al. (eds.), *Virus Diseases of Ornamental Plants*,
https://doi.org/10.1007/978-981-16-3919-7_10

may decrease the growth of plant by depressing their photosynthesis, altering metabolism as well as changing their resource allocation (Radwan et al. 2007; Tecsi et al. 1996). Plant viruses around 70% are transmitted from plant to plant by invertebrate mobile vectors such as aphids, leafhoppers, whiteflies, mites, thrips, beetles, etc. Insect-mediated transmission is the most common mode of transmission of plant viruses in nature (Gray and Banerjee 1999). Aphids alone transmit about 290 plant viruses among all insect vectors. Some plant viruses are spread by aphids like *Cucumovirus*, *Potyvirus*, etc. (Ng and Perry 2004) in *Calendula officinalis* too.

Calendula officinalis, a medicinal plant of the *Calendula* genus of the Asteraceae family, is widely distributed throughout the tropical and temperate region. *Calendula officinalis* flower has many active substances, such as flavonoids, saponins, and carotenoids, which are antioxidants as well as important for immune system of plant (Breithaupt 2007). It also has many medicinal properties as suggested by Wang et al. (2017).

Calendula officinalis L. plant can develop many viral diseases which reduce its both medicinal and ornamental charm. Contaminated plants were stunted, producing few twisted flowers. Previous literature suggested that many viruses such as *Ageratum enation* virus (Jaidi et al. 2015), *Tomato leaf curl New Delhi virus* (Venkataravanappa et al. 2019, b), *Bidens mottle virus* (Huang and Jan 2011), *cucumber mosaic virus* (Milosevic et al. 2015), *Papaya leaf curl virus* (Venkataravanappa et al. 2019, b), *Tomato leaf curl Bangladesh virus* (Khan et al. 2005) are reported on *C. officinalis*. *Tomato leaf curl viruses* have bipartite genomes called DNA-A and DNA-B, which infect dicotyledon plants. Plants such as *C. officinalis* are widely distributed around the world and have high environmental adaptability and this may be the reason for occurrance of new viruses as well as considered as a reservoir of economically imperative viruses that are often overlooked in the study of diversity of viruses (Urbino et al. 2013). Several experimental studies have shown that such plants act as reservoirs or alternative hosts of begomovirus survival (Raj et al. 2008) as well as spread without major crops (Ilyas et al. 2013). Moriones et al. (2017) suggests that intimate molecular and cellular associations occur in hemolymph viruses between begomovirus genes and proteins, whitefly genes, and important protein receptors. Additional awareness of diversity and distribution is urgently needed. We addressed the viruses invading *Calendula officinalis* in terms of symptoms caused by them and ways of their transmission of causal viruses.

10.2 Viruses Infecting *Calendula officinalis*

Several viruses affect calendula, with symptoms such as yellow net symptoms, decreased leaf size, and number of flowers. Many viruses showed reduced flowers yield and plant quality. Padma et al. (1975) published on mosaic disease in *Calendula officinalis*. She mentioned that the disease was first reported from India that was transmissible to *Chenopodium amaranticolor* and *Nicotiana glutinosa*. Dhillon et al. observed, in their physiological studies on *Calendula officinalis*, leaves were found to be infected with *Potato yellow dwarf virus*. Full sugar, unreduced carbohydrate, starch, as well as the total carbohydrate reduced sugar along with total RNA content was found high in infected leaves. Lisa (1971) described two distinct *Calendula officinalis* viruses, Liguria. *Cucumber mosaic* and *turnip mosaic viruses* caused the infection separately or together.

Hristova et al. reported that indistinct mosaic symptoms were found on greenhouse-grown *Calendula officinalis* near the town of Pravets, Bulgaria. Based on host range studies and serological and EM examination, the virus isolate from *Calendula officinalis* was identified as *Tobacco mosaic tobamovirus*. Gupta and Verma (1983) describe a calendula rosette disease. The symptoms, host range, and transmission differ from those previously reported virus diseases of *Calendula officinalis*. Moreover, the calendula plant has also been found infected by *Tomato chlorosis virus* (ToCV) of the genus Crinivirus in the family *Closteroviridae*, which has broad host range as it includes 24 plant species in 7 plant families. Khan et al. (2005) reported natural infection of begomovirus on *Calendula officinalis* plant showed yellowing of the vein and stunted growth of whole plant. Descriptions of some viruses which regularly infect *Calendula officinalis* plant are as follows:

10.2.1 *Ageratum Enation Virus* (AEV)

It is a virus belonging to the genus *Begomovirus*. It is known to infect several hosts such as *Ageratum houstonianum*, *Trichosanthes dioica*, *Cleome gynandra*, *Crassocephalum crepidioides*, and *Amaranthus cruentus* including *Calendula officinalis*. It causes disease in *Calendula officinalis* and other hosts including leaf curl, leaf enation, and stunting symptoms. The highly infected *Calendula officinalis* plants showed yellow vein net symptoms on leaves as well as deformation in leaf as compared to the healthy plants. The highly infected plants remained stunted in growth and bore smaller sized as well as lesser flowers in number (Jaidi et al. 2015). Whitefly, *Bemisia tabaci*, spreads this virus from one host plant to another. *Begomovirus* was implicated in *Calendula officinalis* plants (Khan et al. 2005). Rana et al. (2016) suggested that midgut protein interacts with begomoviruses as well as plays a significant role in transmission of virus from the digestive tract to haemolymph. Polymerase chain reaction amplification with betasatellite-specific and begomovirus degenerate primers (Rojas et al. 1993; Briddon et al. 2002) is used for begomovirus molecular recognition where the full-length DNA genome was amplified by rolling circle amplification (Jaidi et al. 2015).

The genus *Begomovirus* of family Geminiviridae is a major constraint of tomato production, causing the most prevalent and economically significant disease affecting tomatoes in the Indian subcontinent as suggested by Moriones et al. (2017). Its one alternative host is *Calendula officinalis* in which it causes yellow vein disease (Singh et al. 2017). In this disease, younger leaves turn yellow and decrease in size, and the plant is stunted. Affected plants produce yellow or white fruits and are not marketable (Jose and Usha 2003). Srivastava et al. (2017) indicated that the average cell size in healthy leaves was approximately 125–130μm, which was decreased due

Fig. 10.1 *Calendula officinalis* diseased plant (**a**) growing in garden. The infected *Calendula officinalis* plant showing leaf deformation as well as yellow vein net symptoms on leaves (**a**, **b**) as compared to healthy plants (**c**)

to AEV infection to 77–110μm. In comparison to that in healthy leaves, the stomata were sunken and closed on the abaxial surface of infected plant's leaf, likely due to the extreme curl, but length of the stomata was almost same in both infected and healthy leaves (27–31μm). In addition, veins in infected plant's leaves often showed hyperplasia along with the wavy patterns as well as nonuniform thickening (some places between 400 and 500μm), but in healthy leaves, they remain straight, smooth as well as uniformly thickened (approximately about 420μm). Excessive staining in infected tissues (with trypan blue) shows the weakened plasma membranes, as well as also shows loss of integrity of the membrane which eventually causing loosen in their form and maintain blue stain, indicating the induced cell death in plants infected by AEV (Srivastava et al. 2017) (Fig. 10.1).

10.2.2 *Bidens Mottle Virus* (BiMoV)

Bidens mottle virus belongs to the *Potyviridae* family in the genus *Potyvirus*. Like other viruses in this family, sap and aphids spread *Bidens mottle virus* in a stylet-borne fashion. It infects many plants, including agricultural crops such as lettuce and escarole, or many ornamental plants, including *Calendula officinalis* (Huang and Jan 2011). It causes mild to extreme mottling, moderate to severe leaf distortion (Huang and Jan 2011), and stunting. In certain hosts, it can cause floral break symptoms and flower abortion, and at least one host has been identified that shows no noticeable symptoms. It also causes chlorotic and necrotic ringspots in *Calendula officinalis* (Desbiez et al. 2017). The virus causing the disease was identified as Bidens mottle virus, based on characteristic of cylindrical inclusions (Baker et al. 2001).

10.2.3 *Cucumber Mosaic Virus* (CMV)

Cucumber mosaic virus belongs to genus *Cucumovirus* of family Bromoviridae can spread from one plant to another via sap and aphids. Parasitic weeds will spread it, *Cuscuta* sp.; *Calendula officinalis* causes chlorotic mottle (Milosevic et al. 2015). Except this, its symptoms also include leaf mottling or mosaic, yellowing, stunting, ringspots as well as flower, fruit, and leaf distortion. *Calendula officinalis* represents a potential virus reservoir for this virus in Serbia (Milosevic et al. 2015). The presence of CMV in *Calendula officinalis* plants was detected by commercial DAS-ELISA sandwich kits and further confirmed by RT-PCR and sequencing using CMV CPfwd/CMVCPrev unique primers to amplify protein (CP) coat gene (Milosevic et al. 2015).

10.2.4 *Papaya Leaf Curl Virus* (PaLCV)

Calendula officinalis plants showed yellow vein symptoms in Karnataka state, India. PCR diagnosis and complete genome sequencing showed that the symptomatic *Calendula officinalis* plants are associated with PaLCV (Jaidi et al. 2015). Venkataravanappa et al. reported that PaLCV-infected calendula plants can serve as an alternative host for other commercially significant pathogens of plants (Jaidi et al. 2015). Whitefly, *Bemisia tabaci*, spreads this virus from one host plant to another. The main symptoms are severe thickening and curling of veins on leaves and reduced growth of the plants in size. Severely affected plants produced smaller, fewer as well as deformed flowers (Jaidi et al. 2015; Venkataravanappa et al. 2019, b).

10.2.5 *Tomato Leaf Curl Bangladesh Virus* (ToLCBV)

Bangladesh virus was described by Khan et al. (Milosevic et al. 2015). The disease symptoms in *Calendula officinalis* consisted of shortening and vein yellowing of leaves as well as petioles and stunting growth of plant (Milosevic et al. 2015). The disease was transmitted experimentally by whiteflies (Bemisia tabaci) (Khan et al. 2005) from infected *Calendula officinalis* to healthy seedlings.

10.2.6 *Tobacco Curly Shoot Virus* (TbCSV)

A monopartite begomovirus (*Begomovirus* genus, *Geminiviridae* family) is a significant threat to several crops as well as weeds in India and China. After infection, her host displays extreme curling and yellowing symptoms (Zhao et al. 2017). *Calendula officinalis* is naturally infected with this virus by whiteflies (*Bemisia tabaci*) and displays vein yellowing, leaves and petioles shortening, and plant body stunting (Milosevic et al. 2015).

10.2.7 *Potato Yellow Dwarf Virus* (PYDV)

It belongs to the *Mononegavirales*, Rhabdoviridae family, and *Alphanucleorhabdovirus*. The virus is noted for its various symptoms including dwarfing, stunted growth as well as apical yellowing. The virus also causes malformation and tuber cracking. *Agallia constricta* transmits the virus. After feeding on an infected plant, the insect becomes infected and then transmits the virus to every plant. In *Calendula officinalis*, this virus infects leaves and then spreads throughout the body of the plant and affects its physiology.

10.2.8 *Tobacco Streak Virus* (TSV)

Tobacco streak virus is a plant-pathogenic Ilavirus of family Bromoviridae which causes chlorosis and necrotic streaks on leaf veins. Temperature greatly affects symptoms (Gulati et al. 2016). Vemana and Jain (2010) observed localized (chlorotic and necrotic spots) and systemic (axillary shoot proliferation, necrotic spots, stunting, wilt as well as complete necrosis) symptoms in *Calendula officinalis* (Fig. 10.2). This virus is also transmitted through seed in case of several weeds such as *Datura stramonium*, *Chenopodium quinoa*, etc. and by thrips vectors (Beris and Vassilakos 2020).

10.2.9 *Tobacco Mosaic Virus* (TMV)

The infection spreads by direct contact to the neighboring cells. It also causes infection in *Calendula officinalis*, and the signs of infection are generally color loss or mottling, yellowing, stunting, and/or deformed leaves, flowers, or buds (Hosford 1967). It is transmitted by contact with infected plant or by vector like aphids (Sacristán et al. 2011).

Fig. 10.2 *Calendula officinalis* (**a** healthy, **b** necrotic spots regarding Tobacco streak virus)

10.2.10 *Calendula Yellow Net Virus* (CYNV)

On *Calendula officinalis*, this virus caused yellow net symptoms as observed by Naqvi and Samad (1985) in Kasimpur, Aligarh. In *Calendula officinalis*, the virus can be transmitted easily from infected plant to young healthy seedlings. Except this, the virus was also transmitted by two different species of aphid, *Myzus persicae* Sulz and *Aphis gossypii* Glov, in a regular manner. The symptoms of this virus were seen as a beautiful yellowing network on the infected leaves. Except this, the infected plants were shorter as well as have retarded growth (Afreen et al. 2010) (Table 10.1).

10.3 Conclusion

Calendula officinalis has a broad variety of phytochemicals and pharmacology practices, making it an excellent source of new medicines. It is also grown as an ornamental plant around the world, but both its medicinal properties and ornamental features are distorted by viral infection. So more research are required in future regarding viruses which affect this plant.

Table 10.1 Viruses, their symptoms and mode of transmission in *Calendula officinalis*

Virus name	Symptom	Transmitted by	References
Ageratum enation virus	On leaves yellow vein net symptom, deformation of leaf, leaf enation, and stunting growth with less number of flowers	Whitefly (*Bemisia tabaci*)	Khan et al. (2005), Jaidi et al. (2015), Rana et al. (2016)
Bidens mottle virus	Mottling, moderate to severe leaf distortion, stunting growth, chlorotic and necrotic ringspots	Aphids	Huang and Jan (2011), Desbiez et al. (2017)
Cucumber mosaic virus	Leaf mottling or mosaic, ringspots, yellowing, stunting, and flowers less in number as well as flower distortion seen	Aphids	Milosevic et al. (2015)
Papaya leaf curl virus	Severe curling and thickening of veins on leaves as well as low growth in plant. Severely affected plants produce smaller, fewer as well as deformed flowers	Whitefly (*Bemisia tabaci*)	Jaidi et al. (2015), Venkataravanappa et al. (2019, b)
Tomato leaf curl Bangladesh virus	Vein yellowing, shortening of leaves and petioles, and plant stunting	Whitefly (*Bemisia tabaci*)	Khan et al. (2005)
Tobacco curly shoot virus	Displays vein yellowing, leaves and petioles shortening, and plant body stunting	Whitefly (*Bemisia tabaci*)	Milosevic et al. (2015)
Potato yellow dwarf virus	Stunted growth, dwarfing, apical yellowing, and malformation of flowers	*Agallia constricta*	
Tobacco streak virus	Axillary shoot proliferation, necrotic spots, complete necrosis, wilt as well as stunting	By seed in case of several weeds such as *Datura stramonium*, *Chenopodium quinoa*, etc. and by thrips vectors	Vemana and Jain (2010), Gulati et al. (2016), Beris and Vassilakos (2020)
Tobacco mosaic virus	Generally color loss or mottling, yellowing, stunting, and/or deformed leaves, flowers, or buds	It is transmitted by contact with infected plant or by vector like aphids	Hosford (1967), Sacristán et al. (2011)
Calendula yellow net virus	Symptoms observed were a beautiful network of yellowing on the leaves, shorter and reduced growth of plant	*Aphis gossypii* Glov and *Myzus persicae*	Naqvi and Samad (1985), Afreen et al. (2010)

References

Afreen B, Baghel G, Khan AA, Naqvi QA (2010) Virus-vector relationship of yellow vein disease of calendula and the whitefly. J Plant Prot Sci 2(2):60–65

Anderson PK, Cunningham AA, Patel NG, Morales FJ, Epstein PR, Daszak P (2004) Emerging infectious diseases of plants: pathogen pollution, climate change and agrotechnology drivers. Trends Ecol Evol 19(10):535–544

Baker CA, Raid RN, Scully BT (2001) Natural infection of Vicia faba by Bidens mottle virus in Florida. Plant Dis 85(12):1290

Beris D, Vassilakos N (2020) Plant beneficial microbes: do they have a role as antiviral agents in agriculture? In: Sharma V, Salwan R, Al-Ani LKT (eds) Molecular aspects of plant beneficial microbes in agriculture. Elsevier, San Diego, pp 19–33

Breithaupt DE (2007) Modern application of xanthophylls in animal feeding—a review. Trends Food Sci Technol 18(10):501–506

Briddon RW, Bull SE, Mansoor S, Amin I, Markham PG (2002) Universal primers for the PCR-mediated amplification of DNA beta: a molecule associated with some monopartite begomoviruses. Mol Biotechnol 20(3):315–318

Cooper I, Jones RAC (2006) Wild plants and viruses: under-investigated ecosystems. Adv Virus Res 67:1–47

Desbiez C, Schoeny A, Maisonneuve B, Berthier K, Bornard I, Chandeysson C et al (2017) Molecular and biological characterization of two potyviruses infecting lettuce in southeastern France. Plant Pathol 66(6):970–979

Gray SM, Banerjee N (1999) Mechanisms of arthropod transmission of plant and animal viruses. Microbiol Mol Biol Rev 63(1):128–148

Gulati A, Alapati K, Murthy A, Savithri HS, Murthy MRN (2016) Structural studies on tobacco streak virus coat protein: insights into the pleomorphic nature of ilarviruses. J Struct Biol 193 (2):95–105

Gupta DM, Verma VS (1983) Calendula rosette disease [virus]. Gartenbauwissenschaft 48 (3):106–107

Hosford RM (1967) Transmission of plant viruses by dodder. Bot Rev 33(4):387–406

Huang C-H, Jan F-J (2011) First report of Bidens mottle virus infecting Calendula in Taiwan. Plant Dis 95(3):362

Ilyas M, Nawaz K, Shafiq M, Haider MS, Shahid AA (2013) Complete nucleotide sequences of two begomoviruses infecting Madagascar periwinkle (*Catharanthus roseus*) from Pakistan. Arch Virol 158(2):505–510

Jaidi M, Kumar S, Srivastava A, Raj SK (2015) First report of Ageratum enation virus and Ageratum leaf curl betasatellite infecting Calendula officinalis in India. New Dis Rep 32:6

Jose J, Usha R (2003) Bhendi yellow vein mosaic disease in India is caused by association of a DNA beta satellite with a begomovirus. Virology 305(2):310–317

Khan AA, Naqvi QA, Khan MS, Singh R, Raj SK (2005) First report of a begomovirus infecting calendula in India. Plant Pathol 54(4):569–569

Lisa V (1971) Two viruses naturally infecting cultivated Cardoon (Cynara cardunculus L.). Phytopathologiamediterranea 10(3):231–237

Milosevic D, Ignjatov M, Nikolic Z, Gvozdanovic-Varga J, Tamindzic G, Stankovic I et al (2015) The presence of cucumber mosaic virus in pot marigold (*Calendula officinalis* L.) in Serbia. Ratarpovrt 52(2):67–73

Moriones E, Praveen S, Chakraborty S (2017) Tomato leaf curl New Delhi virus: an emerging virus complex threatening vegetable and fiber crops. Viruses 9(10):264

Naqvi QA, Samad A (1985) Purification and properties of Calendula yellow net virus. Indian J Virol 1:143–146

Ng JCK, Perry KL (2004) Transmission of plant viruses by aphid vectors. Mol Plant Pathol 5 (5):505–511

Nicaise V (2014) Crop immunity against viruses: outcomes and future challenges. Front Plant Sci 5:660

Padma R, Verma VS, Singh S (1975) Observations on the occurrence of a mosaic disease on *Calendula officinalis* L. [Marigold]. Gartenbauwissenschaft 40:133–134

Radwan DEM, Fayez KA, Mahmoud SY, Hamad A, Lu G (2007) Physiological and metabolic changes of Cucurbita pepo leaves in response to zucchini yellow mosaic virus (ZYMV) infection and salicylic acid treatments. Plant Physiol Biochem 45(6–7):480–489

Raj SK, Snehi SK, Kumar S, Khan MS, Pathre U (2008) First molecular identification of a begomovirus in India that is closely related to Cassava mosaic virus and causes mosaic and stunting of *Jatropha curcas* L. Australas Plant Pathol 3(1):69–72

Rana VS, Popli S, Saurav GK, Raina HS, Chaubey R, Ramamurthy VV et al (2016) A Bemisia tabaci midgut protein interacts with begomoviruses and plays a role in virus transmission: begomovirus-bemisiatabaci interaction. Cell Microbiol 18(5):663–678

Rojas MR, Gilbertson RL, Russell DR, Maxwell DP (1993) Use of degenerate primers in the polymerase chain reaction to detect whitefly-transmitted geminiviruses. Plant Dis 77:340–347

Sacristán S, Díaz M, Fraile A, García-Arenal F (2011) Contact transmission of tobacco mosaic virus: a quantitative analysis of parameters relevant for virus evolution. J Virol 85 (10):4974–4981

Shang H, Xie Y, Zhou X, Qian Y, Wu J (2011) Monoclonal antibody-based serological methods for detection of Cucumber green mottle mosaic virus. Virol J 8(1):228

Singh J, Khanna S, Ranjan K, Pant RP, Kumar P, Sirohi A et al (2017) Evidence of association of begomovirus with the yellow vein disease of an ornamental plant pot marigold (*Calendula officinalis*) from western Uttar Pradesh. J Pure Appl Microbiol 11(3):1609–1615

Srivastava A, Agrawal L, Raj R, Jaidi M, Raj SK, Gupta S et al (2017) Ageratum enation virus infection induces programmed cell death and alters metabolite biosynthesis in Papaver somniferum. Front Plant Sci 8:1172

Strange RN, Scott PR (2005) Plant disease: a threat to global food security. Annu Rev Phytopathol 43(1):83–116

Tecsi LI, Smith AM, Maule AJ, Leegood RC (1996) A spatial analysis of physiological changes associated with infection of cotyledons of marrow plants with cucumber mosaic virus. Plant Physiol 111(4):975–985

Urbino C, Gutiérrez S, Antolik A, Bouazza N, Doumayrou J, Granier M et al (2013) Within-host dynamics of the emergence of tomato yellow leaf curl virus recombinants. PLoS One 8(3): e58375

Vemana K, Jain RK (2010) New experimental hosts of tobacco streak virus and absence of true seed transmission in leguminous hosts. Indian J Virol 21(2):117–127

Venkataravanappa V, Sonavane P, Reddy CNL, Reddy MK (2019) Associations of papaya leaf curl virus and betasatellite with yellow vein disease of pot marigold (Calendula officinalis L.). Arch Phytopathol Pflanzenschutz 52(19–20):1289–1297

Wang S, Zhang L, Li J, Cong J, Gao F, Zhou G (2017) Effects of dietary marigold extract supplementation on growth performance, pigmentation, antioxidant capacity and meat quality in broiler chickens. Asian Australas J Anim Sci 30(1):71–77

Zhao LL, Ding M, Zhang XY, Yin YY, Li TT, Zhang ZK (2017) First report of Tobacco Curly Shoot Virus (TbCSV) and its associated satellites from watermelon in China. J Plant Pathol 99 (3):761–764

Viruses and Satellites Infecting Hollyhock (*Alcea rosea*) Plants: An Update

11

Manish Kumar and Supriya Chakraborty

Abstract

Hollyhock (*Alcea rosea*) is an annual ornamental plant grown at several places of the world. Several distinct pathogens including viruses (DNA and RNA), fungus, bacteria, and nematodes infect hollyhock. However, geminiviruses (family: *Geminiviridae*), ssDNA viruses which infect many crops, vegetables, medicinal, and ornamental plants, are among the most devastating pathogens causing huge economic loss. In the past decades, the emergence of begomoviruses and associated satellite molecules is more prevalent in Asian countries (India, China, Pakistan), USA, Africa, and Europe (Spain), which can partly be attributed to increase in international trades. Plant DNA or RNA viruses have a high potential of posing genetic variability into their genome, and the process has accomplished various ways like recombination, nucleotide substitution, mutation, and recurrent pseudo-recombination. However, genome-wide variability and changes incorporated into their viral genome lead to the emergence of a novel virus. The spread and emergence of plant viruses also rely on the availability of viral genomic components, occupancy of vector populations, favorable environmental factors, and permissive host plants. The emergence of novel viruses, their expanding host range, and a widespread distribution make management of plant disease and achieving food security globally challenging. An integrated disease and pest management (IPM) approach and development of disease-resistant plant using molecular methods should be prioritized to limit economic losses.

M. Kumar · S. Chakraborty (✉)
Molecular Virology Laboratory, School of Life Sciences, Jawaharlal Nehru University, New Delhi, India

S. K. Raj et al. (eds.), *Virus Diseases of Ornamental Plants*,
https://doi.org/10.1007/978-981-16-3919-7_11

Keywords

Hollyhock · Begomoviruses · RNA viruses · Host range · Genome map · Evolutionary relatedness · Emergence · IPM

11.1 Introduction

Alcea rosea (family: *Malvaceae*) is an annual ornamental plant. More than 60 species are reported in the genus *Alcea*, which belongs to the family *Malvaceae*. The word "hollyhock" derives from the term "holly" (meaning absolved healing power) and "hock" (meaning mellow plant). True to its name, *Alcea rosea* possess anti-inflammatory, astringent, demulcent, and diuretic properties. It is used for the remedial in healing bleeding gums, gastrointestinal disorders, ulcers, peptic ulceration, renal dysfunction, kidney disorders, cough, arthritis, inflammatory conditions, and asthma. Hollyhock is widely distributed and grows well at the Mediterranean region (Italy, Spain) to Asian countries (India, China, Pakistan). In the recent years, the predominance of begomovirus molecules and their associated satellites have increased considerably in ornamental plants causing huge economic losses (Mitrofanova et al. 2018; Marwal and Gaur 2020). Other than DNA viruses, hollyhock plants also suffer from ssRNA viruses like Zucchini yellow mosaic virus (ZYMV), Malva vein clearing mosaic virus (MVCM), and Alcea rosea virus 1 (ArV1). Countries like USA have mandatory plant quarantine protocols to screen plants for various plant pathogens. However, the high degree of interspecific variation among the pathogens allows them to affirm specific host range. The current chapter discussed the presence of diverse viruses of a medicinally important ornamental plant (*Alcea rosea*, hollyhock), etiology, and the current management practices.

11.2 Habitat

Alcea rosea is cultivated under a wide range of climatic and environmental conditions. The genus of this species is distributed throughout Asia and also commonly grown in gardens in several tropical, subtropical, and temperate countries including India, Turkey, Taiwan, Greece, Pakistan, and China.

11.3 Diseases and Yield Loss

Two major diseases, leaf curl disease (LCD) and yellow mosaic disease (YMD), were reported on hollyhock plants (Kumar et al. 2020), which were caused by the circular, ssDNA viruses (family: *Geminiviridae;* genus: Begomovirus). The typical symptoms of LCD and YMD include yellowing of veins, leaf curling, leaf distortion, vein thickening, and stunted plant growth (Fig. 11.1). Besides, RNA viruses also

Fig. 11.1 Diversity of symptoms induced by begomoviruses on hollyhock plants: (**a**) hollyhock plant exhibiting yellow mosaic symptoms, (**b**) leaf curl symptoms, (**c**) severely infected hollyhock plants showing leaf curl symptoms under field condition, and (**d**) healthy hollyhock plant. Images (**a**, **b**, **d**) were from the authors itself, and image (**c**) was reproduced from Ashwathappappa et al. (2020)

infect hollyhock causing leaf blistering and yellowing of leaves. Ornamental plants are grown for decorative and utility purposes in several metropolitan landscapes, gardens, and the nursery, and they are prevalent and economically relevant crops worldwide. As per the Union Fleurs (an international flower trade association) annual report, the global export of ornamental plants is grown by more than 10% annually and is expected to reach 57.5 billion US$ till 2030. The increase in trade will further enhance requirement of cultivation of virus-free plants worldwide (https://unionfleurs.org/).

11.4 Host Ranges

Begomoviruses infects several economically important plants such as tomato, chili, okra, and various ornamental and medicinal plants including *Alcea rosea* (hollyhock), *Catharanthus roseus* (Madagascar periwinkle), *Hibiscus rosa-sinensis* (China rose), *Chrysanthemum indicum*, *Solanum capsicastrum* (Jerusalem cherry), *Alternanthera sessilis* (Matikaduri), *Mentha piperita* (peppermint), *Hibiscus sabdariffa* (Roselle), *Mirabilis jalapa* (4 o'clock flower), *Ludwigia parviflora* (water primrose), *Helianthus annuus* (wild sunflower), *Amaranthus viridis* (green amaranth), *Cestrum nocturnum* (Jessamine), and *Bougainvillea peruviana* (bougainvillea). More than 35 begomovirus species are known to be associated with ornamental and weed plants in India (Marwal and Gaur 2020).

11.5 Insect Vector and Genome Organization

The insect vector whitefly (*Bemisia tabaci*) transmits the begomoviruses from an infected plant to healthy plants via their tiny stylets. These insect-transmissible plant-infecting viruses are encapsidated within a twinned (geminate) icosahedral particle-containing circular, non-enveloped ssDNA of 2.5–5.2 kb in size (Navas-Castillo et al. 2011). Based on genome organization, nucleotide sequence identity, and insect-vector transmission, it is further classified into nine genera as *Becurtovirus, Begomovirus, Capulavirus, Curtovirus, Eragrovirus, Grablovirus, Mastrevirus, Topocuvirus,* and *Turncurtovirus* (Zerbini et al. 2017). *Begomovirus* comprises the largest genus with more than 400 virus species which infect several economically important crops, weeds, and various ornamental plants worldwide (ICTV master species list, 2019). These whitefly-transmitted viruses are further subdivided into bipartite viruses containing DNA-A (approximately 2.8 kb in size) and DNA-B (approximately 2.8 kb in size) genomes and monopartite viruses with only DNA-A-like genome. DNA-A of the begomoviruses has six to seven ORFs, four to five in the complementary (AC1, AC2, AC3, AC4, AC5) or antisense strand, and two in the virion or sense strand (AV1 and AV2). AC1 encodes replication protein (Rep), AC2 is transcriptional activator protein (TrAP), AC3 codes for replication enhancer protein (REn), AC4 acts as a silencing suppressor, and AC5 function is not fully characterized (Hanley-Bowdoin et al. 2013). Rep is the only protein which is indispensable for viral replication to occur, and it regulates early events of begomovirus pathogenesis in a permissive host (George et al. 2014; Ruhel and Chakraborty 2019). In virion strand, AV1 encodes coat protein (CP), responsible for insect transmission, whereas AV2 encodes the pre-coat protein that interferes with RNA silencing (Basu et al. 2018). The DNA-B genome has two ORFs, BC1 encodes for the movement protein (MP), and BV1 encodes for the nuclear shuttle protein (NSP), both of which are necessary for movement of viral genome for short distance- and long-distance spread (Sanderfoot and Lazarowitz 1996) (Table 11.1). Bipartite begomoviruses are predominantly grouped in the "New World" (NW), whereas the other diversified monopartite begomoviruses are widespread geographically in the "Old World" (OW). However, OW begomoviruses are generally associated with their satellites of approximately 1.3–1.5 kb in size, categorized as alphasatellite (family *Alphasatellitidae*; subfamilies G*emini alphasatellitinae*), betasatellite, and a recently discovered deltasatellite (both betasatellite and deltasatellite belong to the family *Tolecusatellitidae*) (Dry et al. 1997; Alabi et al. 2020). Alphasatellite has only one ORF encoding alpha-rep protein. However, betasatellite contains one ORF in the complementary strand; βC1 and the other in anticomplementary or virion strand; and βV1 in addition to having the satellite-conserved region (SCR) (George et al. 2014; Ruhel and Chakraborty 2019; Basu et al. 2018). Betasatellite serves as pathogenicity determinant and also have multitasking roles such as suppressor of gene silencing, induces disease symptoms, and assists in weakening of plant defense response in begomovirus pathogenesis (Hu et al. 2020; Gnanasekaran and Chakraborty 2018; Bhattacharyya et al. 2015; Zhou 2013) (Fig. 11.2a).

Table 11.1 Gene products of hollyhock-infecting viruses and their functions

	Functions	References
Begomovirus encoded genes and their products		
Replication protein/ AC1/C1	Virus replication, reprogramming of cell cycle, role in autophagy, and involved in epigenetic modifications	Kushwaha et al. (2017), George et al. (2014), Ruhel and Chakraborty (2019)
Replication enhancer/AC2/C2	Trans-activation of viral genome and posttranscriptional gene silencing	Kumar et al. (2015)
Transcriptional activator/AC3/C3	Virus replication and reprogramming of cell cycle	Pasumarthy et al. (2011)
AC4/C4	Virus movement, target hormone signaling, and HR response	Fondong et al. (2007)
Coat protein/AV1/ V1	Encapsidation, vector transmission, and trafficking of viral genome	Ohnesorge and Bejarano (2009)
Pre-coat protein/ AV2/V2	PTGS and TGS gene silencing	Basu et al. (2018)
Movement protein/ BC1	Cell to cell movement	Hou et al. (2000)
Nuclear shuttle protein/BV1	Nuclear-cytoplasmic trafficking of viral genome, suppression of host defense, and HR response	Hou et al. (2000)
βC1	Autophagy, symptom determinant, suppression of host defense, and photosynthesis impairment	Gnanasekaran et al. (2019a, b)
α-Rep	Gene silencing and suppressor of PTGS and TGS	Vinoth Kumar et al. (2017)
Potyvirus-encoded genes and their products		
Viral protein genome-linked (VpG)	Translation, virus movement, and replication	Revers and García (2015)
First protein proteinase (P1-Pro)	Translation and modulator of replication	Nigam et al. (2019)
Helper component-proteinase (HC-Pro)	Silencing suppression and aphid transmission	Valli et al. (2018)
P3N-pretty interesting potyviridae ORF (P3-PIPO)	Cell-to-cell movement	Vijayapalani et al. (2012)
Cylindrical inclusion (CI)	Helicase involved in virus movement and replication	Deng et al. (2015)
Nuclear inclusion a (NIa)	Polyprotein processing	Revers and García (2015)
Nuclear inclusion a (NIb)	RNA-dependent RNA polymerase	Revers and García (2015)
Coat protein (CP)	Virus movement, virion formation, and aphid transmission	Carbonell et al. (2013)
6 kDa (6K1 and 6K2)	Formation of replication vesicles	Revers and García (2015)
Closterovirus encoded genes and their products		

(continued)

Table 11.1 (continued)

	Functions	References
Leader proteinases (L-pro)	Papain-like cysteine protease	Dolja and Koonin (2013)
Methyltransferase (Met)	Methyltransferase	Erokhina et al. (2001)
Helicase (hel)	Helicase	Dolja and Koonin (2013)
RNA-directed RNA polymerases (RdRP)	RNA-directed RNA polymerase	Dolja and Koonin (2013)
Coat protein (CP)	Capsid protein	Chen et al. (2011)
Minor coat protein (CPm)	Minor capsid protein and initiate virion assembly	Alzhanova et al. (2007)
Heat shock protein (Hsp70)	Heat shock protein 70, required for proper virion tail assembly, and long-distance transport	Prokhnevsky et al. (2002)
64 kDa protein (p64)	Cell-to-cell movement	Dolja and Koonin (2013)
20 kDa protein (p20)	Long-distance transport through phloem	Dolja and Koonin (2013)
21 kDa protein (p21)	Suppressor of RNAi	Chapman et al. (2004)

11.6 Emergence and Spread

Plant viruses are widely spread among the noncultivated crop such as weeds and ornamental plants. Like other plant viruses, begomoviruses use these plants as an alternative host and potential reservoirs for infection to occur. It might be one reason why the Indian subcontinent is considered as the hot-spot region for begomoviruses. Evolution and spread of ssDNA viruses also depend on the diversity of insect vectors, availability of begomoviral genomic components, environmental factors, and abundance of permissive host plants. Several ornamental plants and weeds are infected with begomovirus species, including eight known species associated with hollyhock plants. Hollyhock leaf curl virus (HoLCV), Hollyhock yellow vein mosaic virus (HoYVMV), Hollyhock leaf crumple virus (HLCrV), Cotton leaf curl Multan virus (CLCuMuV), Tomato leaf curl Gujarat virus (ToLCGuV), Tomato leaf curl Karnataka virus (ToLCKV), Cotton leaf curl Gezira virus (CLCuGV), and Tomato yellow leaf curl virus (TYLCV) are associated with hollyhock plants which further suggest that ornamental plants are latent repositories for begomoviruses. Phylogenetic analysis of DNA-A molecules indicated that the begomoviruses which infect hollyhock also infect other plants of the *Malvaceae* families as *Hibiscus rosa sinensis, Abelmoschus esculentus, Malva parviflora, Malvastrum coromandelianum,* and *Gossypium hirsutum* (Fig. 11.3).

Plant viruses display genetic variation in their genomes primarily because of the error-prone polymerases employed for viral replication. To further understand the extent of variation at the genomic level, molecular epidemiologists begin to understand the factors responsible for genomic variation such as recombination, mutation, and nucleotide substitution rate. However, ssDNA viruses such as begomoviruses

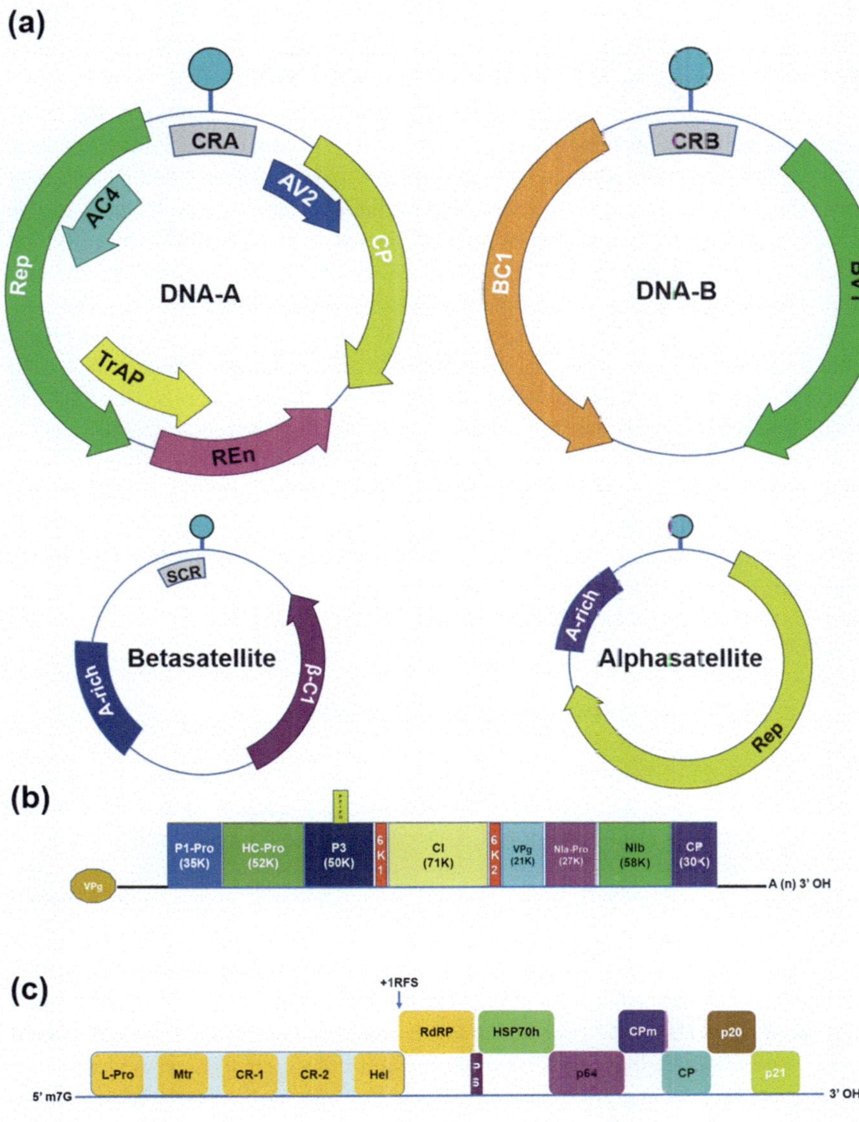

Fig. 11.2 Genome organization of viruses infecting hollyhock plants: (**a**) bipartite begomoviruses; DNA-A, and its associated satellites like alphasatellite and betasatellite. AC1, replication-associated protein (Rep); AC2, transcriptional activator protein (TrAP); AC3, replication enhancer protein (REn); AV1, coat protein; AV2, pre-coat protein; BC1, movement protein (MP); and BV1, nuclear shuttle protein (NSP). CR, and SCR, attributes to the conserved region and short-intergenic region, respectively. (**b**) Genomic map of a representative member of the genus *Potyvirus*. A straight line depicted monopartite ssRNA genome, and their polypeptide ORF have been shown by square-shaped colored boxes. A covalently attached viral protein genome-linked (VPg) at the 5′ terminal, followed by the protein 1 proteases (P1-Pro); helper component protease (HC-Pro), P3; pretty interesting Potyviridae ORF (PIPO); nuclear inclusion A and B protease (NIa-Pro and NIb); cytoplasmic inclusion (CI); RNA-dependent RNA polymerase (RdRP); and Coat protein (CP) are shown. (**c**) A representative genome organization of genus closterovirus exhibiting the relative position of the ORFs and their gene products. Leader proteinase (L-Pro), methyltransferase (Mtr),

also showed a high level of genetic diversity even though they utilize DNA polymerases. A well-studied example is the Tomato yellow leaf curl virus (TYLCV). The mean substitution rate estimated for TYLCV was 2.88×10^{-4} nucleotide substitutions/site/year (Duffy and Holmes 2008). However, the exact mechanism behind the substitution dynamics is still unclear. Genetic material exchange through recombination concedes plant viruses to evolve rapidly and may extend the host range and virulence. Recombination in viruses leading to the emergence of novel begomoviruses is well known. A recent study suggests that CLCuMuV which is identified in hollyhock-associated begomoviruses were recombinant. The hot-spot region of the putative parents was identified at AC1, AV1, and AV2 region (DNA-A), Rep region (in alphasatellites), and satellite-conserved region (in betasatellite) (Kumar et al. 2020). Potential recombination events at AC1 and AC2/AC3 regions were also detected in HoYVMV infecting hollyhock plant in India. These data suggest that begomoviruses molecules are recombination-prone, and it might contribute significantly to their evolutionary dynamics. Further, it is important to note that simple sequence repeats (SSR) are known to occur on geminiviral genomes, and nonrandom distribution of compound SSRs (cSSRs) have been found to be linked with recombination hot spots on geminivirus genomes (George et al. 2012, 2015).

11.7 Hollyhock-Associated Begomoviruses

In India, leaf curling and yellowing of leaves in *Alcea rosea* was reported in early 1970s (Srivastava et al. 2020) at the National Botanical Garden, Lucknow. Concurrently, Hollyhock leaf crumple virus (HLCrV) was reported to infect hollyhock plant at Cairo, Egypt. From Cairo, Egypt, the infected plants showed severe leaf crumpling symptoms, cupping of leaves, and vein thickening (Abdel-Salam et al. 1998). In 2001, a novel virus species named Althea rosea enation virus (AREV) (name changed to HLCrV as per the ICTV rectification; https://talk.ictvonline.org/ictv/proposals/Ratification_2004.pdf) (Mayo 2005) was observed to infect hollyhock plants in Egypt (Bigarré et al. 2001). It indicates that whitefly-transmitted begomoviruses induce leaf enation symptoms and induce changes at the histopathological level such as alteration of cambial structure and higher pectin-like accumulation as an intracellular fibrous material in the infected tissues. Furthermore, the association of a monopartite begomovirus, Hollyhock yellow vein mosaic virus (HoYVMV), and Ludwigia leaf distortion betasatellite (LuLDB) was reported from diseased hollyhock plants in India (Srivastava et al. 2014) suggesting presence of satellite molecules with the diseased plants (Table 11.2). In another case, a novel species of Hollyhock leaf curl virus-Malva (HoLCV-Mal), which shows 92.5%

Fig. 11.2 (continued) helicase (Hel), RNA-directed RNA polymerase (RdRP), heat shock protein-70 homolog (HSP70h), coat protein (CP), and minor coat protein (CPm)

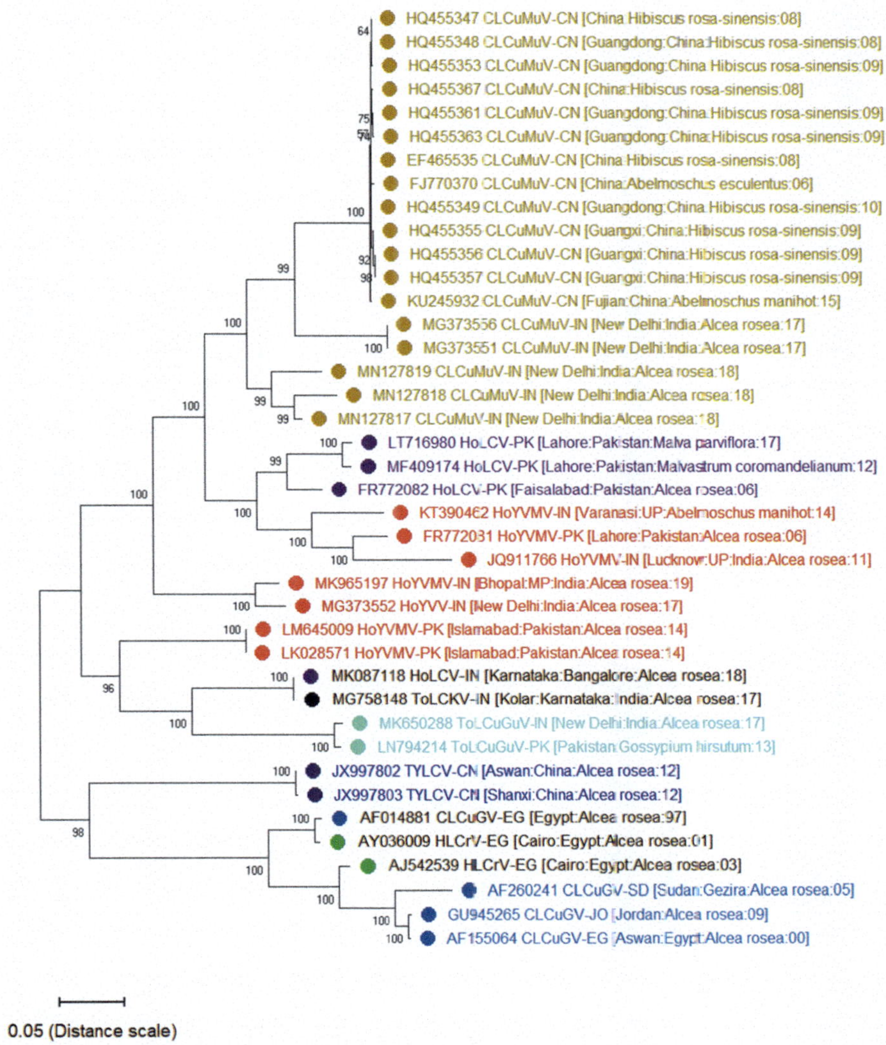

Fig. 11.3 Phylogenetic dendrogram displaying genetic relatedness of hollyhock-infecting begomoviruses The nucleotide sequences were aligned using Muscle algorithm, Jukes-Cantor model with 1000 bootstrap replicates, and the tree was generated via maximum likelihood method. The distance scale bar represents the number of substitutions rate/sites/year. Each colored code indicates a single representative species. There are eight species (isolates numbers are in bracket): Hollyhock leaf curl virus (HoLCV; 04), Hollyhock yellow vein mosaic virus (HoYVMV; 07), Hollyhock leaf Crumple virus (HLCrV; 02), Cotton leaf curl Multan virus (CLCuMuV; 18), Tomato leaf curl Gujrat virus (ToLCGuV; 02), Tomato leaf curl Karnataka virus (ToLCKV; 01), Cotton leaf curl Gezira virus (CLCuGV; 04), and Tomato yellow leaf curl virus (TYLCV; 02)

nucleotide identity with previously identified infect *Alcea rosea* infecting HoLCV-PK, was associated with *Malva parviflora* (family, *Malvaceae*) in Pakistan. Further, analysis suggests the presence of Kenaf leaf curl betasatellite (KLCuB) and

Table 11.2 List of viruses and satellites associated with hollyhock plants

Name	Abbreviation	Accession no.	Associated betasatellite	Associated alphasatellite	Species	References
Hollyhock leaf curl virus	HoLCV-PK [Lahore:Pakistan:17]	LT716980	KLCuB (LT716981)	ACSLA (LT716982)	*Malva parviflora*	Sattar et al. (2017)
				AYVIA (LT716983)		
	HoLCV-PK [Lahore:Pakistan:12]	MF409174	KLCuB (MF409173)	HoLCA (MF409175)	*Malvastrum coromandelianum*	Sattar et al. (2017)
	HoLCV-IN [Karnataka: Bangalore:18]	MK087118	CLCuBaB (MK087119)		*Alcea rosea*	Ashwathappappa et al. (2020)
	HoLCV-PK [Faisalabad:Pakistan:06]	FR772082			*Alcea rosea*	
Hollyhock yellow vein mosaic virus	HoYVMV-IN [Bhopal: MP:India:19]	MK965197			*Alcea rosea*	
	HoYVMV-PK [Islamabad:Pakistan:14]	LM645009			*Alcea rosea*	
	HoYVMV-IN [Lucknow: UP:India:11]	JQ911766	LuLDB (JQ408216)		*Alcea rosea*	Srivastava et al. (2014)
	HoYVMV-PK [Islamabad:Pakistan:14]	LK028571			*Alcea rosea*	
	HoYVMV-PK [Lahore:Pakistan:06]	FR772081			*Alcea rosea*	
	HoYVMV-IN [Varanasi: UP:14]	KT390462			*Abelmoschus manihot*	
	HoYVMV-IN [New Delhi:India:17]	MG373552	LuLDB (MG373553)	CLCuMuA (MG373558)	*Alcea rosea*	Kumar et al. (2020)
				GDarSLA (MG373559)		
		AY036009			*Alcea rosea*	

Hollyhock leaf Crumple virus	HLCrV-EG [Cairo: Egypt:01]					
	HLCrV-EG [Cairo: Egypt:03]	AJ542539			*Alcea rosea*	
Cotton leaf curl Multan virus	CLCuMuV-IN [New Delhi:India:18]	MN127819			*Alcea rosea*	
	CLCuMuV-CN [China:06]	FJ770370			*Abelmoschus esculentus*	
	CLCuMuV-CN [Fujian:China:15]	KU245932			*Abelmoschus manihot*	
	CLCuMuV-IN [New Delhi:India:18]	MN127818			*Alcea rosea*	
	CLCuMuV-IN [New Delhi:India:18]	MN127817			*Alcea rosea*	
	CLCuMuV-IN [New Delhi:India:17]	MG373556			*Alcea rosea*	Kumar et al. (2020)
	CLCuMuV-IN [New Delhi:India:17]	MG373551			*Alcea rosea*	Kumar et al. (2020)
	CLCuMuV-CN [China:08]	EF465535			*Hibiscus rosa-sinensis*	
	CLCuMuV-CN [China:08]	HQ455347			*Hibiscus rosa-sinensis*	
	CLCuMuV-CN [Guangdong:China:08]	HQ455348			*Hibiscus rosa-sinensis*	
	CLCuMuV-CN [Guangdong:China:10]	HQ455349			*Hibiscus rosa-sinensis*	
	CLCuMuV-CN [Guangdong:China:09]	HQ455353			*Hibiscus rosa-sinensis*	

(continued)

Table 11.2 (continued)

Name	Abbreviation	Accession no.	Associated betasatellite	Associated alphasatellite	Species	References
	CLCuMuV-CN [Guangxi:China:09]	HQ455355			*Hibiscus rosa-sinensis*	
	CLCuMuV-CN [Guangxi:China:09]	HQ455356			*Hibiscus rosa-sinensis*	
	CLCuMuV-CN [Guangxi:China:09]	HQ455357			*Hibiscus rosa-sinensis*	
	CLCuMuV-CN [Guangdong:China:09]	HQ455361			*Hibiscus rosa-sinensis*	
	CLCuMuV-CN [Guangdong:China:09]	HQ455363			*Hibiscus rosa-sinensis*	
	CLCuMuV-CN [China:08]	HQ455367			*Hibiscus rosa-sinensis*	
Tomato leaf curl Gujrat virus	ToLCuGuV-IN [New Delhi:India:17]	MK650288	LuLDB MG373553	CLCuMuA (MG373558)	*Alcea rosea*	Kumar et al. (2020)
				MG373559 (GDarSLA)		
	ToLCuGuV-PK [Pakistan:13]	LN794214			*Gossypium hirsutum*	
Tomato leaf curl Karnataka virus	ToLCKV-IN [Kolar: Karnataka:India:17]	MG758148			*Alcea rosea*	
Cotton leaf curl Gezira virus	CLCuGV-SD [Sudan: Gezira:05]	AF260241			*Alcea rosea*	
	CLCuGV-EG [Aswan: Egypt:00]	AF155064			*Alcea rosea*	
	CLCuGV-EG [Egypt:97]	AF014881			*Alcea rosea*	
	CLCuGV-JO [Jordan:09]	GU945265			*Alcea rosea*	

Tomato yellow leaf curl virus	TYLCV-CN [Aswan:China:12]	JX997802			*Alcea rosea*	
	TYLCV-CN [Shanxi:China:12]	JX997803			*Alcea rosea*	
Zucchini yellow mosaic virus (Potyvirus)	ZYVMV-SK [South Korea: Seoul:02]	AB098081			*Alcea rosea*	Desbiez and Lecoq (1997)
Malva vein clearing virus (Potyvirus)	MCMV-GR [Germany:09]	GQ856544			*Alcea rosea*	Menzel et al. (2010)
Alcea rosea virus 1	ArV1-CN [China:16]	MK107841			*Alcea rosea*	Wang et al. (2020)

Ageratum conyzoides symptomless alphasatellite (ACSLA) along with DNA-A molecules with HoLCV-Mal (Sattar et al. 2017).

Molecular characterization of begomoviruses complex associated with other malvaceous plants gives an insight into trans-replication abilities of HoLCV. A recent study suggests Tomato leaf curl virus (ToLCV) can also infect *Alcea rosea* plants in southern India. Furthermore, recombination analysis showed that the recombinant fragments might have originated from begomoviruses infecting tomato and ageratum (Ashwathappappa et al. 2020). This observation is further strengthened by the sequence analysis which shows that ToLCV infecting hollyhock from Karnataka shares 91% sequence identity with ToLCV (GenBank accession nos. KP178730 and AY754812) isolated from tomato plants. Apart from the DNA-A molecules, an association of betasatellite, Cotton leaf curl Bangalore betasatellite (CLCuBaB), was also confirmed in hollyhock plants. An association of bipartite begomoviruses with hollyhock was reported first in India (Kumar et al. 2020). Hollyhock plants showed typical leaf curling vein thickening, leaf crinkling, venial chlorosis, or yellow mosaic symptoms. Further, investigation of the collected samples revealed the presence of several monopartite and bipartite begomovirus along with satellite molecules, including a novel begomovirus: Hollyhock yellow vein mosaic New Delhi virus (HoYVMV-IN). Moreover, an ornamental plant harboring begomoviral complexes during the off-season could serve as a new source of virus inoculum in the subsequent seasons. The evolution of cotton- and tomato-associated begomoviruses to hollyhock might prompt a potential threat to its cultivation. Taken together, one can assume that the possible spread of begomoviruses might be because of the extended host range, availability of susceptible host, and viruliferous whitefly population throughout the year.

11.8 Hollyhock-Associated RNA Viruses: ssRNA, Positive Strand, Monopartite Genome

11.8.1 Potyvirus

The family *Potyviridae* consists of more than 160 species with positive-sense ssRNA viruses of approximately 8–11 kb in size that infect monocot and dicot plants worldwide. They are transmitted by aphids in a nonpersistent and non-circulative manner. The genomic RNA is transcribed into the polyprotein, which is further processed by the proteolytic digestion and produced several proteins that are required for successful pathogenesis and transmission between host plants by insect vectors: the first protein proteinase (P1) for translation; Helper component-proteinase (HC-Pro) as a silencing suppressor and helper component protease; the third protein proteinase (P3) to assist in replication and movement (P3N-pretty interesting potyviridae ORF) of P3N-PIPO during the cellular movement; 6K1 for vesicles formation; viral protein genome-linked (VPg) having role in viral protein translation, movement, and viral replication; Nuclear inclusion protein a (NIa-Pro) for processing of polypeptides; Nuclear inclusion protein b (NIb) as RNA-dependent

RNA polymerase (RdRP); and CP (coat protein) to help in movement of virus and aphid transmission (Nigam et al. 2019) (Fig. 11.2 b).

Zucchini yellow mosaic virus (ZYMV) (family *Potyviridae*) was reported to infect hollyhock and Zucchini squash (family: *Cucurbitaceae*) plants. Plants inoculated with ZYMV showed typical leaf blistering and yellowing leaves (Desbiez and Lecoq 1997). Inoculation of coat protein (CP) of ZYMV isolated from hollyhock (nonhost) showed severe symptom when compared to the ZYMV isolated from the cucumber (natural host). Another ssRNA virus, Malva vein clearing virus (MVCV) (family: *Potyviridae*) was identified infecting *Alcea rosea* and *C. quinoa* in Germany (Menzel et al. 2010). Presence of virus has been further confirmed by RT-PCR (reverse transcription PCR) followed by the partial sequencing. Further, MVCV was mechanically inoculated on several other herbaceous plant species (*Chenopodium quinoa, Chenopodium foliosum, Chenopodiastrum murale, Chenopodium amaranticolor, Datura stramonium, Nicotiana benthamiana, Nicotiana hesperis, Petunia hybrida*, and *Solanum lycopersicum*) to investigate the host range. However, only *C. quinoa* plant showed disease symptoms at 2 weeks postinoculation.

11.8.2 Closteroviruses

The genus *Closterovirus* comprises of 16 species having positive strand, ssRNA with monopartite genome of 14.5–19.3 kb in size. Larger genome size and the presence of minor coat protein (CPm) gene at the upstream region of the coat protein (CP) gene is a distinct feature of closteroviruses. It is transmitted by the sap-sucking insect vector, aphids. Members of the genus encodes proteins: a larger polypeptide with conserved domain of papain-like proteases (P-Pro), methyltransferase (Met), variable central region (CR-1 and CR-2), helicase (Hel) domain, RNA-directed RNA-polymerases (RdRP), 6 kDa membrane binding hydrophobic protein (P6), cellular heat shock protein-70 (HSP70) protein, and several structural and nonstructural proteins such as p64, CPm, CP, p20, and p21 with RNA silencing suppressor activity (Fig. 11.2c) (Gaur et al. 2016).

A recent study suggests that ssRNA viruses (positive strand), having filamentous particles, can infect ornamental plants. It is transmitted by several species of aphids and whiteflies. Recently, *Alcea rosea* plants showing yellow mosaic and vein clearing symptoms were reported in China (Wang et al. 2020). Small-RNA deep sequencing technologies and in silico analysis of infected leaf samples confirmed the occupancy of watermelon mosaic virus (WMV), Malva vein clearing virus (MVCV), and a novel closteroviral genome, Alcea rosea virus 1 (ArV1) (family: *Closteroviridae*), in infected tissues. Infectivity of these viruses needs to be ascertained in order to confirm the etiology. Polarity analysis suggests that ArV1-vsiRNAs (virus-derived small-interfering RNAs) were heterogeneously distributed throughout the ArV1 genome. This vsiRNAs profiling gives an insight into region of the ArV1 genome targeted by the host RNA silencing pathway, and this information can be further used to generate resistance against ArV1.

11.9 Management

Viruses can efficiently infect their hosts and extend their host range, and plants evolve in order to overcome infection. To further maintain crop productivity and economic interest, protective measures need to be taken. Integrated pest management (IPM) is the ecosystem-based approach, focused primarily on the long-term prevention of pest and their destruction. It includes combinations of biological control and chemical control in a coordinated manner. Several strategies have been designed for plant DNA viruses, such as developing the resistant plant, antibody-based system, and artificially synthesized peptide aptamers (Beam and Ascencio-Ibáñez 2020). However, it is difficult for plants to defend against mixed begomoviruses infections, a more common scenario. Genomic approaches have ushered to identify and understand the roles of host factors that regulate host resistance to plant viruses (Sharma et al. 2015a; Prakash et al. 2020). Further, transgenic as well as RNA vaccination approaches have been successfully used to control both RNA and DNA viruses (Alam et al. 2019; Sharma et al. 2015b; Namgial et al. 2019); hence, such strategies may be employed controlling hollyhock viruses too. This also emphasizes the need for standardization of transformation technique for hollyhock as has been done in the past for other recalcitrant crop plants (Kumar et al. 2012).

Therefore, a combinatorial approach that targets both vector and virus might be a more immeasurable chance of impairing the disease. The insect vector transmits the begomoviruses; therefore, spraying insecticide to control the vector population is widely used, although the excessive use of chemical pesticides always increases concerns about its deleterious effect on human health and the ecosystem. Other nonchemical approaches are the resistance breeding program that primarily relies on identifying resistance/tolerance variety or germplasm. Furthermore, genome-wide deep sequencing and advancement in RNAi technology might provide an insight view of the etiology of the diseased plant (Chakraborty and Kumar 2020). RNA viruses are recognized by either protein-based assay such as ELISA and Western blotting or RNA-based assays such as RT-PCR with specific CP primers (Mahuku et al. 2015). Similarly, for detecting closteroviruses, PCR (targeting conserved HSP70 coding region) and ELISA (German-Retana et al. 1999) are often used. Timely detection of plant viruses and application of integrated pest and disease management strategies hold the key for maintaining future plant health.

References

Abdel-Salam AM, El-Shazly M, Thouvenel JC (1998) Biological, biochemical and serological studies on hollyhock leaf crumple virus (HLCrV): a newly discovered whitefly transmitted geminivirus. Arab J Biotechnol 1:41–58

Alabi OJ, Diaz-Lara A, Al RM (2020) Satellite nucleic acids and viruses. In: Reference module in life sciences. Elsevier, New York. http://www.sciencedirect.com/science/article/pii/B9780128096338213083

Alam CM, Jain G, Kausar A, Singh AK, Mandal B, Varma A et al (2019) Dicer 1 of Candida albicans cleaves plant viral dsRNA in vitro and provides tolerance in plants against virus infection. Virus Dis 30(2):237–244. https://doi.org/10.1007/s13337-019-00520-x

Alzhanova DV, Prokhnevsky AI, Peremyslov VV, Dolja VV (2007) Virion tails of Beet yellows virus: coordinated assembly by three structural proteins. Virology 359(1):220–226

Ashwathappappa KV, Venkataravanappa V, Lakshminarayana Reddy CN, Krishna RM (2020) Molecular characterization of Tomato leaf curl virus infecting hollyhock (Alcea rosea L.) in India. Indian Phytopathol 73(2):339–347. https://doi.org/10.1007/s42360-020-00221-2

Basu S, Kumar Kushwaha N, Kumar Singh A, Pankaj Sahu P, Vinoth Kumar R, Chakraborty S (2018) Dynamics of a geminivirus-encoded pre-coat protein and host RNA-dependent RNA polymerase 1 in regulating symptom recovery in tobacco. J Exp Bot 69(8):2085–2102

Beam K, Ascencio-Ibáñez JT (2020) Geminivirus resistance: a minireview. Front Plant Sci 11:1–9

Bhattacharyya D, Gnanasekaran P, Kumar RK, Kushwaha NK, Sharma VK, Yusuf MA et al (2015) A geminivirus betasatellite damages the structural and functional integrity of chloroplasts leading to symptom formation and inhibition of photosynthesis. J Exp Bot 66(19):5881–5895. https://pubmed.ncbi.nlm.nih.gov/26113193

Bigarré L, Chazly M, Salah M, Ibrahim M, Padidam M, Nicole M et al (2001) Characterization of a new begomovirus from Egypt infecting hollyhock (Althea rosea). Eur J Plant Pathol 107 (7):701–711

Carbonell A, Maliogka VI, De Jesús PJ, Salvador B, San León D, García JA et al (2013) Diverse amino acid changes at specific positions in the N-terminal region of the coat protein allow plum pox virus to adapt to new hosts. Mol Plant-Microbe Interact 26(10):1211–1224

Chakraborty S, Kumar M (2020) Tomato Leaf Curl New Delhi virus (Geminiviridae). Reference module in life sciences. Elsevier, Amsterdam, pp 1–14. https://doi.org/10.1016/B978-0-12-809633-8.21561-6

Chapman EJ, Prokhnevsky AI, Gopinath K, Dolja VV, Carrington JC (2004) Viral RNA silencing suppressors inhibit the microRNA pathway at an intermediate step. Genes Dev 18 (10):1179–1186

Chen AYS, Walker GP, Carter D, Ng JCK (2011) A virus capsid component mediates virion retention and transmission by its insect vector. Proc Natl Acad Sci U S A 108(40):16777–16782

Deng P, Wu Z, Wang A (2015) The multifunctional protein CI of potyviruses plays interlinked and distinct roles in viral genome replication and intercellular movement. Virol J 12(1):1–11. https://doi.org/10.1186/s12985-015-0369-2

Desbiez C, Lecoq H (1997) Zucchini yellow mosaic virus. Plant Pathol 46:809–829

Dolja VV, Koonin EV (2013) The closterovirus-derived gene expression and RNA interference vectors as tools for research and plant biotechnology. Front Microbiol 4:1–10

Dry IB, Krake LR, Rigden JE, Rezaian MA (1997) A novel subviral agent associated with a geminivirus: the first report of a DNA satellite. Proc Natl Acad Sci U S A 94(13):7088–7093. http://www.ncbi.nlm.nih.gov/pubmed/9192696

Duffy S, Holmes EC (2008) Phylogenetic evidence for rapid rates of molecular evolution in the single-stranded DNA begomovirus tomato yellow leaf curl virus. J Virol 82(2):957–965

Erokhina TN, Vitushkina MV, Zinovkin RA, Lesemann DE, Jelkmann W, Koonin EV et al (2001) Ultrastructural localization and epitope mapping of the methyltransferase-like and helicase-like proteins of Beet yellows virus. J Gen Virol 82(8):1983–1994

Fondong VN, Reddy RVC, Lu C, Hankoua B, Felton C, Czymmek K et al (2007) The consensus N-myristoylation motif of a geminivirus AC4 protein is required for membrane binding and pathogenicity. Mol Plant-Microbe Interact 20(4):380–391

Gaur RK, Petrov NM, Patil BL, Stoyanova MI (eds) (2016) Plant viruses: evolution and management. Springer, pp 231–252

George B, Mashhood Alam C, Jain SK, Sharfuddin C, Chakraborty S (2012) Differential distribution and occurrence of simple sequence repeats in diverse geminivirus genomes. Virus Genes 45 (3):556–566. https://doi.org/10.1007/s11262-012-0802-1

George B, Ruhel R, Mazumder M, Sharma VK, Jain SK, Gourinath S et al (2014) Mutational analysis of the helicase domain of a replication initiator protein reveals critical roles of Lys 272 of the B′ motif and Lys 289 of the β-hairpin loop in geminivirus replication. J Gen Virol 95 (7):1591–1602. https://www.microbiologyresearch.org/content/journal/jgv/10.1099/vir.0.064923-0

George B, Alam CM, Kumar RV, Gnanasekaran P, Chakraborty S (2015) Potential linkage between compound microsatellites and recombination in geminiviruses: evidence from comparative analysis. Virology 482:41–50. http://www.sciencedirect.com/science/article/pii/S0042682215001397

German-Retana S, Candresse T, Martelli G (1999) Closteroviruses (Closteroviridae). In: Granoff A, Webster RG (eds) Encyclopedia of virology, 2nd edn. Elsevier, Oxford, pp 266–273. http://www.sciencedirect.com/science/article/pii/B0122270304000534

Gnanasekaran P, Chakraborty S (2018) Biology of viral satellites and their role in pathogenesis. Curr Opin Virol 33:96–105. http://www.sciencedirect.com/science/article/pii/S1879625718300361

Gnanasekaran P, KishoreKumar R, Bhattacharyya D, Vinoth Kumar R, Chakraborty S (2019a) Multifaceted role of geminivirus associated betasatellite in pathogenesis. Mol Plant Pathol 20 (7):1019–1033

Gnanasekaran P, Ponnusamy K, Chakraborty S (2019b) A geminivirus betasatellite encoded βC1 protein interacts with PsbP and subverts PsbP-mediated antiviral defence in plants. Mol Plant Pathol 20(7):943–960

Hanley-Bowdoin L, Bejarano ER, Robertson D, Mansoor S (2013) Geminiviruses: masters at redirecting and reprogramming plant processes. Nat Rev Microbiol 11(11):777–788. https://doi.org/10.1038/nrmicro3117

Hou YM, Sanders R, Ursin VM, Gilbertson RL (2000) Transgenic plants expressing geminivirus movement proteins: abnormal phenotypes and delayed infection by Tomato mottle virus in transgenic tomatoes expressing the Bean dwarf mosaic virus BV1 or BC1 proteins. Mol Plant-Microbe Interact 13(3):297–308

Hu T, Song Y, Wang Y, Zhou X (2020) Functional analysis of a novel βV1 gene identified in a geminivirus betasatellite. Sci China Life Sci 63(5):688–696

Kumar RV, Sharma VK, Chattopadhyay B, Chakraborty S (2012) An improved plant regeneration and Agrobacterium-mediated transformation of red pepper (Capsicum annuum L.). Physiol Mol Biol Plants 18(4):357–364. https://doi.org/10.1007/s12298-012-0132-8

Kumar V, Mishra SK, Rahman J, Taneja J, Sundaresan G, Mishra NS et al (2015) Mungbean yellow mosaic Indian virus encoded AC2 protein suppresses RNA silencing by inhibiting Arabidopsis RDR6 and AGO1 activities. Virology 486:158–172. https://doi.org/10.1016/j.virol.2015.08.015

Kumar M, Vinoth Kumar R, Chakraborty S (2020) Association of a begomovirus-satellite complex with yellow vein and leaf curl disease of hollyhock (Alcea rosea) in India. Arch Virol 165 (9):2099–2103. https://doi.org/10.1007/s00705-020-04696-2

Kushwaha NK, Bhardwaj M, Chakraborty S (2017) The replication initiator protein of a geminivirus interacts with host monoubiquitination machinery and stimulates transcription of the viral genome. PLoS Pathogens 13:1–41

Mahuku G, Lockhart BE, Wanjala B, Jones MW, Kimunye JN, Stewart LR et al (2015) Maize lethal necrosis (MLN), an emerging threat to maize-based food security in sub-Saharan Africa. Phytopathology 105(7):956–965

Marwal A, Gaur RK (2020) Molecular diversity of begomoviruses and DNA satellite molecules infecting ornamental plants in India. In: Applied plant virology. Academic Press, pp 475–491. https://doi.org/10.1016/B978-0-12-818654-1.00034-7

Mayo MA (2005) Changes to virus taxonomy 2004. Arch Virol 150(1):189–198

Menzel W, Winter S, Richert-Pöggeler KR (2010) First report of Malva vein clearing virus naturally occurring in hollyhock in Germany. Plant Dis 94(2):276. https://doi.org/10.1094/PDIS-94-2-0276B

Mitrofanova IV, Zakubanskiy AV, Mitrofanova OV (2018) Viruses infecting main ornamental plants: an overview. Ornam Hortic 24(2):95–102

Namgial T, Kaldis A, Chakraborty S, Voloudakis A (2019) Topical application of double-stranded RNA molecules containing sequences of Tomato leaf curl virus and Cucumber mosaic virus confers protection against the cognate viruses. Physiol Mol Plant Pathol 108:101432. http://www.sciencedirect.com/science/article/pii/S0885576519300499

Navas-Castillo J, Fiallo-Olivé E, Sánchez-Campos S (2011) Emerging virus diseases transmitted by whiteflies. Annu Rev Phytopathol 49:219–248

Nigam D, LaTourrette K, Souza PFN, Garcia-Ruiz H (2019) Genome-wide variation in potyviruses. Front Plant Sci 10:1–28

Ohnesorge S, Bejarano ER (2009) Begomovirus coat protein interacts with a small heat-shock protein of its transmission vector (Bemisia tabaci). Insect Mol Biol 18(6):693–703

Pasumarthy KK, Mukherjee SK, Choudhury NR (2011) The presence of tomato leaf curl Kerala virus AC3 protein enhances viral DNA replication and modulates virus induced gene-silencing mechanism in tomato plants. Virol J 8:1–14

Prakash V, Singh A, Singh AK, Dalmay T, Chakraborty S (2020) Tobacco RNA-dependent RNA polymerase 1 affects the expression of defence-related genes in Nicotiana benthamiana upon Tomato leaf curl Gujarat virus infection. Planta 252(1):11. https://doi.org/10.1007/s00425-020-03417-y

Prokhnevsky AI, Peremyslov VV, Napuli AJ, Dolja VV (2002) Interaction between long-distance transport factor and Hsp70-related movement protein of Beet Yellows Virus. J Virol 76 (21):11003–11011

Revers F, García JA (2015) Molecular biology of potyviruses, Advances in virus research, vol 92, 1st edn. Elsevier, Amsterdam, pp 101–199. https://doi.org/10.1016/bs.aivir.2014.11.006

Ruhel R, Chakraborty S (2019) Multifunctional roles of geminivirus encoded replication initiator protein. Virus Dis 30(1):66–73. https://doi.org/10.1007/s13337-018-0458-0

Sanderfoot AA, Lazarowitz SG (1996) Getting it together in plant virus movement: cooperative interactions between bipartite geminivirus movement proteins. Trends Cell Biol 6(9):353–358. http://www.sciencedirect.com/science/article/pii/0962892496100313

Sattar MN, Qurashi F, Iqbal Z, Saleem Haider M (2017) Molecular characterization of Hollyhock leaf curl virus and associated DNA-satellites infecting Malva parviflora in Pakistan. Can J Plant Pathol 39(2):229–234. https://doi.org/10.1080/07060661.2017.1336490

Sharma VK, Kushwaha N, Basu S, Singh AK, Chakraborty S (2015a) Identification of siRNA generating hot spots in multiple viral suppressors to generate broad-spectrum antiviral resistance in plants. Physiol Mol Biol Plants 21(1):9–18. https://doi.org/10.1007/s12298-014-0264-0

Sharma VK, Basu S, Chakraborty S (2015b) RNAi mediated broad-spectrum transgenic resistance in Nicotiana benthamiana to chilli-infecting begomoviruses. Plant Cell Rep 34(8):1389–1399. https://doi.org/10.1007/s00299-015-1795-8

Srivastava A, Kumar S, Raj SK, Pande SS (2014) Association of a distinct strain of hollyhock yellow vein mosaic virus and Ludwigia leaf distortion betasatellite with yellow vein mosaic disease of hollyhock (Alcea rosea) in India. Arch Virol 159(10):2711–2715

Srivastava KM, Gupta RP, Singh BP (2020) Occurrence of a complex viral infection on Althaea Rosea L. Curr Sci 46(18):644–646. https://www.jstor.org/stable/24215775

Valli AA, Gallo A, Rodamilans B, López-Moya JJ, García JA (2018) The HCPro from the Potyviridae family: an enviable multitasking Helper Component that every virus would like to have. Mol Plant Pathol 19(3):744–763

Vijayapalani P, Maeshima M, Nagasaki-Takekuchi N, Miller WA (2012) Interaction of the trans-frame potyvirus protein P3N-PIPO with host protein PCaP1 facilitates potyvirus movement. PLoS Pathog 8(4):e1002639

Vinoth Kumar R, Singh D, Singh AK, Chakraborty S (2017) Molecular diversity, recombination and population structure of alphasatellites associated with begomovirus disease complexes. Infect Genet Evol 49:39–47. https://doi.org/10.1016/j.meegid.2017.01.001

Wang D, Cui L, Yu P, Qi Y, Wang B, Niu Y (2020) A new closterovirus in Alcea rosea identified by small RNA deep sequencing. J Plant Pathol 102(1):29–37

Zerbini FM, Briddon RW, Idris A, Martin DP, Moriones E, Navas-Castillo J et al (2017) ICTV virus taxonomy profile: Geminiviridae. J Gen Virol 98(2):131–133

Zhou X (2013) Advances in understanding begomovirus satellites. Annu Rev Phytopathol 51 (1):357–381. https://doi.org/10.1146/annurev-phyto-082712-102234

Major Viral Diseases of *Salvia* spp., Diagnosis, and Their Management Strategies

12

Soumya Sinha and Abdul Samad

Abstract

Salvia is a well-known genus of plants distributed throughout the world including Central and South America, Central Asia and Mediterranean, as well as Eastern Asia. It is widely cultivated as ornamental, medicinal, and aromatic plants. A wide range of shapes, sizes, and colors of flowers makes it attractive and suitable to different climates, landscape, and gardening requirements. Also, characteristic aromas increase its value in addition to colors. Viral diseases provide a major challenge to twenty-first-century agriculture worldwide. Climate change and human population pressures are driving rapid alterations in agricultural practices and cropping systems that favor destructive viral disease outbreaks. Modern agriculture must provide sufficient food to feed the world's growing population, which is projected to increase from 7.5 billion in 2018 to at least 9.8 billion by 2050. This goal is made even more challenging because of crop loss to diseases. Incidence of virus diseases and their damage caused to economically important crops are reportedly increasing every year particularly in the countries embedding tropical and subtropical conditions. Integrated management approaches involving utilization of virus resistant crops and efficient management of insect vectors can reduce this disastrous problem. Agricultural globalization and international trade are spreading viruses and their vectors to new geographical regions with unexpected consequences for food production and natural ecosystems. Advances in scientific understanding of virus pathosystems, rapid technological innovation, innovative communication strategies, and global scientific networks provide opportunities to build epidemiologic intelligence of virus threats to crop production and global food security. A paradigm shift toward deploying integrated,

S. Sinha · A. Samad (✉)
Department of Plant Pathology, CSIR- Central Institute of Medicinal and Aromatic Plants, Lucknow, India

S. K. Raj et al. (eds.), *Virus Diseases of Ornamental Plants*,
https://doi.org/10.1007/978-981-16-3919-7_12

smart, and eco-friendly strategies is required to advance virus disease management in diverse agricultural cropping systems.

Keywords

Plant virus · Salvia · Vectors · Diagnosis · Management

12.1 Introduction

Salvia, commonly referred to as sage, is the largest genus of plants that belongs to the family Lamiaceae, within the order Lamiales and subfamily Nepetoideae. The name Salvia derives from the Latin word salvere which means "to feel well and healthy." The genus consists of around 1000 species which is distributed all over the world including Central and South America (~600 species), Central Asia and Mediterranean (~250 species), and Eastern Asia (~90 species) (Walker et al. 2004). Its species consists of annual, biennial, or perennial herbs. The stems are angled; leaves can be entire, toothed, or pinnately divided. Some of the species contain small bracts, which are ornamental and attractive in appearance. The flowers are produced in racemes, or panicles, with different colors ranging from blue, white, yellow to red. The calyx is generally tubular or bell shaped, and corolla is claw shaped with two lips. The upper lip is entire, and the lower lip has two lobes. The stamens are reduced to two short structures with two-celled anthers. The fruits are smooth ovoid or oblong nutlets and have a mucilaginous coating in different species (Bailey 1949). Salvia species are widely cultivated as medicinal, aromatic, and ornamental plants. Terpenoids and phenolic compounds, such as flavonoids and caffeic acid derivatives, are known to be present in aerial parts of Salvia species. Hundreds of secondary metabolites have been isolated from different species of *Salvia* and have been tested for a wide range of biological activities (Wu et al. 2012). Glandular and nonglandular trichomes are present on the surface of stem, leaves, and flowers. Glandular gland is of two types: capitates and peltate one. It is a source of essential oil that has high demand in cosmetic, perfumery, and pharmaceutical industries. The essence of long-lasting aroma due to the presence of volatile compounds like monoterpenes, diaterpenes, and sesquiterpenes increases its value as an ornamental plant.

12.1.1 Some of the Well-Known Ornamental Spp. of Salvia

12.1.1.1 *S. officinalis*

S. officinalis is a perennial, evergreen plant, commonly known as garden sage, common sage, or culinary sage. It is native to Mediterranean region and North Africa and naturalized throughout the world in different places. The plant flowers in late spring or summer which is blue to purplish in color. The leaves are oblong, gray green, rugose on the upper side, and nearly white underneath due to the many short soft hairs. It is widely used as an ornamental plant in border garden as well as rock garden. Apart from this, it is widely used as a spice in the food industry and also in the pharmaceutical industries because it possesses anticancer, anti-inflammatory, antinociceptive, antioxidant, antimicrobial, antimutagenic, and antidementia activities (Ghorbani and Esmaeilizadeh 2017).

12.1.1.2 *S. sclarea*

Salvia sclarea (commonly known as clary sage) is a perennial herb. It is native to Southern Europe and Central Asia mainly distributed in Mediterranean countries including Italy, southern France, Morocco, and India mostly/primarily found in Kashmir Valley. It grows to a height of about 4 ft, and stem is thick, covered with hairs, and quadrangular in shape. Leaves are simple, opposite, and ovate cordate in shape, and surface is rugose and covered with many glandular and nonglandular trichomes. Flowers are white, zygomorphic, and symmetrical; are in verticils, with two to six flowers in each verticil; and are held in large colorful bracts that range in color from pale mauve to lilac or white to pink with a pink mark on the edge giving showy appearance. Because of its attractive characters, it has high ornamental value and have high demand in trade market. More than 150 botanic gardens currently hold accessions of this plant under ex situ conservation (https://bgci.org/plant_search.php). Beside this, clary sage seeds are rich in fatty acids and contained high levels of antioxidant and antiradical activities making them ideal for use as nutraceuticals. It represents one of the known vegetable sources for omega 3-linoleic acid to the functional food industry. It bears important antioxidant compounds, polyphenolic acids like rosmarinic acid, a caffeic acid derivative, and carnosic acid and flavonoids being mostly present as flavones, flavonols, and their glycosides (Lu and Foo 2002). Its essential oils have diverse role, used in perfumery and food industries and also known to have anti-inflammatory, antioxidant, antifungal, and antibacterial activities (Kuźma et al. 2009). It is also classified as Pb hyperaccumulator and Cd and Zn accumulators and showed promising results in the phytoremediation of soil contaminated from heavy metals (Angelova et al. 2016).

12.1.1.3 *S. uliginosa*

S. uliginosa is a perennial herb, commonly called as bog sage. It is native to Southern Brazil, Uruguay, and Argentina. It grows up to a height of 3–6 ft (0.91–1.83 m) and has multiple thin stems and yellow-green lance-shaped leaves with serrated edges. The bright azure-blue flowers are 0.5 in. (1.3 cm) long with a white beeline in the throat pointing toward the nectar and pollen. They grow in whorls beginning in summer until fall, with many flowers coming into bloom at the same time. The flowers of *S. uliginosa* contain a pigment called as cyanosalvianin which imparts blue color (Mori et al. 2008). *S. uliginosa* was introduced into horticulture in 1912 and has become popular in gardens and public landscapes for its azure-blue flowers, ability to grow under various conditions, and its pollinator habitat attributes. This plant has been awarded by the Royal Horticultural Society (2017a).

12.1.1.4 *S. splendens*

S. splendens is commonly known as scarlet sage. It is a tropical perennial plant, native to Brazil. The wild form, rarely seen in cultivation, reaches 1.3 m (4.3 ft) tall. Smaller cultivars are very popular as bedding plants, seen in shopping malls and public gardens all over the world. Its leaves are in even, elliptical arrangements, with dentate margin. Flowers are in erect spikes that sprout from the center of the plant in groups of two to six together in each leaf node; scarlet, tubular, or bell shaped; and 35 mm long, with two lobes toward the apex. The upper lobe is 13 mm long. It flowers a good part of summer and autumn. It is widely grown as an ornamental plant, with a large number of cultivars selected by different colors from white to dark purple (Clebsch 2003).

12.1.1.5 *S. leucantha*

S. leucantha, commonly known as the Mexican bush sage, is a herbaceous perennial plant that is native to subtropical and tropical conifer forests in central and eastern Mexico. The flowers are usually white, emerging from colored bracts. It is not frost hardy but is often grown in warmer latitudes for its prominent arching velvety blue or purple inflorescences. It grows up to 1.3 m (4.3 ft) high and 2 m (6.6 ft) wide, with numerous erect stems, often arching at their tips, and with long inflorescences. The linear-lanceolate leaves are a soft mid-green, with whitish, hairy undersides. Its flowers are white with lavender-blue calyces, blooming nearly the entire year (Clebsch 2003). This plant has ornamental value in the garden and has gained the Royal Horticultural Society's Award of Garden Merit (Royal Horticultural Society 2017b).

12.1.2 Viruses Globally Infecting *Salvia* spp.

As plants are not mobile, viruses use insect, arthropods, nematode, or fungal vectors to reach their hosts and have been shown to manipulate the plant physiology and/or the vector behavior to enhance their transmission (Blanc and Michalakis 2016). Different groups of plant pathogenic viruses are infecting *Salvia* spp. globally and documented in the literature as:

12.1.2.1 Cucumber Mosaic Virus (CMV)

CMV is an icosahedral virus approximately 28–30 nm in diameter and belongs to the genus Cucumovirus in the family Bromoviridae. It was first reported by Doolittle and Jagger on cucumber in 1916 and till date known to have very broad host range, infecting more than 1200 crops including both monocots and dicots. Scholfold reported that CMV occupied fourth position among the top ten plant viruses on the basis of its scientific/economic importance (Scholthof et al. 2011). It is widely distributed all along the tropical as well as temperate areas (Hord et al. 2001). Based on the biological, serological, and molecular properties, CMV isolates are grouped into subgroups I and II, and subgroup I further divided into IA and IB on the basis of diversity of CP gene and 5′ untranslated (UTR) sequence. The virus is rapidly transmitted by more than 80 aphid species in a nonpersistent manner (Palukaitis et al. 1992). CMV genome is made up of tripartite, linear, single-stranded plus sense RNAs designated as RNA1, RNA2, and RNA3 (Lot et al. 1974) and also have two subgenomic RNAs named as RNA4 and RNA4a (Zitter and Murphy 2009; De Wispelaere and Rao 2009). Each RNA segments have methylated nucleotide cap at their 5′ terminus and tRNA-like structure at their 3′ terminus (Dreher 1999, 2009).

In 2014, on the basis of DAC-ELISA results, Dikova reported an association of CMV with infected plants *S. sclarea* exhibiting yellow spotted and dwarfed plants (Dikova 2014). Earlier, infection of CMV on *S. sclarea* was established by Pisi and Vicchi (1989) in Italy. Typical viral symptoms like chlorotic line patterns and ring spots were observed on the leaves of *S. uliginosa* at Burden Research plantation in Baton Rouge. Based on symptoms, agar double diffusion assay, and particle morphology (spherical particles of ~30 nm), CMV was found to be associated with the disease (Holcomb and Valverde 1998).

Joshi and Dubey reported infestation of CMV on *S. splendens* KER-GAWL plants at Gorakhpur, India, in 1972. The affected plants showed abnormal light-green spotting on the leaves. The disease starts as vein clearing when plants are few weeks old, followed by the appearance of light-green spots which afterward enlarge to form bigger patches and finally giving the leaf a mosaic pattern. In more severe stages, leaves are distorted. Such affected plants are weak, are dwarfed, and bear a few small unattractive flowers. The symptoms produced on differential hosts and physical properties of the virus under study suggested the infection of CMV (Joshi and Dubey 1972).

Later in 2012, Ali et al. also reported association of CMV with disease of the *S. splendens* from India. Infected plants showed severe mosaic, mottling, and distortion of leaves and retarded growth. The virus was transmitted to the healthy plants of Salvia spp. as well as many other hosts by mechanical inoculation. *Myzus persicae* Sulzer and *Aphis gossypii* Glover transmit the virus in nonpersistent manner. Purified sample in EM showed spherical particles ~28 nm in diameter. RT-PCR was performed using total RNA as templates, and CMV coat protein (CP) gene-specific primer gel was electrophoresed on 1% agarose. An amplification of expected size 650 bp fragment was obtained only in the infected sample (Ali et al. 2012). Beside these hosts, earlier, CMV infection was also observed on *S. patens* in 1951 (Faan and Johnson 1951) and *S. lyrata* in 1959 (Anderson 1959).

12.1.2.2 Alfalfa Mosaic Virus (AMV)

It is a worldwide distributed pathogen that belongs to the genus Alphamovirus of Bromoviridae family (Jaspars and Bos 1980). Its particles are icosahedral in shape measuring 30–57 nm in length and ~18 nm in diameter. It is one of the most biologically variable plant viruses with numerous natural variants having different pathogenicity (Crill et al. 1970; Hajimorad and Francki 1988; Hiruki and Miczynski 1987). It is transmitted by different aphid species in a nonpersistent manner (Jeffries 1998). The genome of AMV consists of three single-stranded, plus-senses RNA molecules designated as RNA1, RNA2, and RNA3 in decreasing order of molecular size (Jaspars 1985). RNAs 1 and 2 encode the 1a and 2a subunits of the replicase proteins (Bol 1999). RNA 3 is bicistronic and codes for a movement protein (Langereis et al. 1986) and the coat protein, which is translated from a subgenomic messenger, RNA 4 (Bol 1999; Tenllado and Bol 2000). RNAs 1, 2, 3 and 4 are, respectively, ~3.65 kb, ~2.6 kb, ~2.2 kb, and ~0.8 kb nucleotides long.

AMV on clary sage was earlier proven by Bellardi et al. (1999) in Italy. In 2014, infection of AMV on *S. sclarea* showing yellow spot and retarded growth was reported from Bulgaria on the basis of DAC-ELISA results (Dikova 2014). Later, in 2016, AMV infection was also detected in wild growing populations of *S. sclarea* exhibiting bright yellow mosaic symptoms near Lake Velence at Middle-West Hungary. Initial detection was carried out by mechanical transmission assay which was further confirmed by sequencing of RT-PCR product (by using AMV-CP-specific primers) that showed 99% similarity with other AMV isolates (Salamon et al. 2018).

12.1.2.3 Begomoviruses

Begomovirus is the largest genus of plant viruses that belongs to the family Geminiviridae The virus is obligately transmitted by an insect vector, which can be the whitefly *Bemisia tabaci* or can be other whiteflies in a persistent, circulative, non-propagative manner. Begomoviruses are divided into old and new world where monopartite begomoviruses are mostly old world (OW) with few bipartite viruses, while new world (NW) begomoviruses are all bipartites. Old world are more diverse in comparison to new world as AV2 gene is present in OW but absent in NW begomoviruses (Rahman et al. 2005). Genome of begomovirus is ss circular DNA, either bipartite (DNA-A and DNA-B) or monopartite (DNA-A). In monopartite begomovirus, all genes are present on the DNA-A molecule itself for replication while in bipartite begomoviruses, Rep and CP gene are only present on DNA-A molecule and gene for movement MP and NSP are present on DNA-B (Kumar et al. 2010; Maruthi et al. 2007). The monopartite genome is often assisted by satellite molecules, betasatellite or alphasatellite. These satellite molecules are of the half size of genome not related with the pathogenicity of the virus but depend on the helper virus (Borah and Dasgupta 2012) to enhance symptom/severity of the disease. During the survey, chlorosis was observed on the leaves of *S. splendens* in Gazipur, Bangladesh. On the basis of indicator plant assay and indirect ELISA, the associated disease was identified as Okra Yellow Vein clearing Mosaic Virus (OYVCMV) (Ara et al. 2012).

During a survey at a local nursery in Baton Rouge, Louisiana, *S. splendens* cv. "Dancing Flame" were found to be exhibited symptoms like bright yellow mosaic, yellow vein, and mild leaf distortion. DNA was successfully amplified from total DNA extracts of infected samples but not from healthy plants of the same cultivar. The obtained RCA product was digested with Pst I. Three DNAs of approximately 2.7, 2.3, and 0.4 kb were visualized in 6% PAGE. Based on the size and number of the DNA that resulted from RCA, it was suspected that a bipartite geminivirus was present in the sample. The Pst I-digested DNA generated by RCA was cloned and sequenced. The sequence analysis revealed that DNA-A is 2777 nt in size and, as reported for members of the begomovirus genus, and contains seven open reading frames. The bicistronic DNA-B consists of 2729 nt. Blast analysis showed that the virus associated with *S. splendens* is Clerodendrum golden mosaic China virus (ClGMCNV) (Valverde et al. 2012).

S. hispanica plants showed leaf deformation, chlorosis, and dwarfing in the production fields located in the north of Salta. Infected samples were analyzed by DAS-ELISA as well as PCR by using PAL1v1978/PAR1c496 which amplifies a 1100–1300 nt fragment corresponding to the 5′ region of the Rep gene, the entire common region, and the 5′end of the CP gene of begomoviruses. The comparison of the complete nucleotide sequences showed the presence of two viral species with the typical genome organization of bipartite new world begomovirus, identified as Sida mosaic Bolivia virus 2 and Tomato yellow spot virus (Celli et al. 2014).

12.1.2.4 Tobacco Mosaic Virus (TMV)

TMV is a positive-sense single-stranded RNA virus species in the genus *Tobamovirus*. It has a rodlike appearance. The TMV genome consists of a 6.3–6.5 kb single-stranded (ss) RNA. The 3′ terminus has a tRNA-like structure, and the 5′ terminus has a methylated nucleotide cap. (m7G5′pppG). The genome encodes four open reading frames (ORFs), two of which produce a single protein due to ribosomal read through a leaky UAG stop codon. The four genes encode a replicase (with methyltransferase [MT] and RNA helicase [Hel] domains), an RNA-dependent RNA polymerase, a so-called movement protein (MP) and a capsid protein (CP) (Klug 1999).

Earlier, in 1946, four species of Salvia were found as host of TMV (HOLMES 1946). Later, based on ELISA results, it was found that yellow spotted and dwarfed plants of *S. sclarea* were found to be associated with TMV in Bulgaria (Dikova 2014).

12.1.2.5 Broad Bean Wilt Virus (BBWV)

BBWV is the type species of the genus *Fabavirus* in the family *Comoviridae* (Taylor and Stubbs 1972). It has wide host range and mainly transmitted by aphids (Benner et al. 1985; Brunt et al. 1996). On the basis of serological and molecular studies, BBWV isolates are divided into two groups: *Broad bean wilt virus 1* (BBWV1) and *Broad bean wilt virus 2* (BBWV2) (Kobayashi et al. 2003; Uyemoto and Provvidenti 1974). The genome is composed of two single-stranded positive-sense RNA molecules, RNA-1 and RNA-2, that encapsidated separately into icosahedral virions (Lisa and Boccardo 1996). The nucleotide (nt) sequence identity between both of them is 39–67%.

In 2002, *S. officinalis* leaves showing twisting and distortion were subjected to ELISA, and an isolate BBWV-2 was identified as a causal pathogen (Mumford et al. 2006). That was the first report of BBWV-2 in the UK. Later, in 2014, BBWV was also reported on *S. sclarea* from Bulgaria (Dikova 2014).

12.1.2.6 Tomato Spotted Wilt Virus (TSWV)

TSWV belongs to the genus Tospovirus in the family tospoviridae and ranked second among the top ten economically important viruses in the world (Scholthof et al. 2011; Mandal 2012). Its particles are spherical in shape having diameter between 80 and 110 nm. Its genome is made up of single-stranded RNA. Bronzing of old leaves of *S. sclarea* were observed, and the pathogen associated with disease was found to be TSWV on the basis of ELISA result (Dikova 2014).

12.1.2.7 Potato Virus Y (PVY)

PVY is the type species of the genus *Potyvirus* in the family *Potyviridae*. It is naturally transmitted by vegetatively propagated material and by different species of aphids in a nonpersistent manner. The virion is non-enveloped with a flexuous and filamentous nucleocapsid, 680–900 nanometers (nm) long, and 11–20 nm in diameter. The nucleocapsid contains ~2000 copies of the capsid protein. The symmetry of the nucleocapsid is helical with a pitch of 3.4 nm. Its genome consists of single-stranded, positive-sense RNA genome of approximately 9.7 kb (Brunt 1992; Dougherty and Carrington 1988; Hollings and Brunt 1981). Genomic RNA is translated into two polyproteins that require proteolytic processing to produce ten mature and one fusion protein essential for replication and movement: P1 (translation, modulator of replication), helper component proteinase HC-Pro (silencing suppression and aphid transmission), P3 (virus replication and movement), P3N-PIPO (cell-to-cell movement), 6K1 (formation of replication vesicles), cytoplasmic inclusion protein (CI, helicase involved in virus movement and replication), 6K2 (formation of replication vesicles), genome-linked protein VPg (translation, virus movement, and replication), NIa-Pro (polyprotein processing), NIb (RNA-dependent RNA polymerase), and CP (virus movement, virion formation, and aphid transmission) (Riechmann et al. 1992; Chrzanowska 1991). In 2014, based on ELISA results, Dikova reported infestation of Potyvirus on *S. sclarea* plants showing yellow spots and dwarfing (Dikova 2014).

12.1.2.8 Clerodendrum Chlorotic Spot Virus (CCSV)

Clerodendrum chlorotic spot virus (ClCSV) belongs to the genus Dichorhavirus in the family Rhabdoviridae (Dietzgen et al. 2014). Most common symptom produced by this virus is chlorotic spots (Kitajima et al. 2008). In C1CSV-infected cells, rod-shaped viral particles (40 × 100–110 nm) are found in nuclear electron-lucent inclusions called viroplasms and also adjacent to the membranous systems of both the nuclear envelope and the endoplasmic reticulum (Kitajima et al. 2008, 2003). ClCSV is naturally transmitted by Brevipalpus phoenicis sensu lato mites that belong to the genus Brevipalpus (Acari: Tenuipalpidae) (Kitajima et al. 2008). It can be mechanically propagated to a selected group of experimental hosts, and in some of them, the infection becomes systemic when plants are kept at 28–29 °C. Its genome consists of negative-sense single-stranded RNA molecules that encode six proteins known as the nucleocapsid protein (gene N/ORF1), phosphoprotein (gene P/ORF2), movement protein (gene MP/ORF3), matrix protein (gene M/ORF4), glycoprotein (gene G/ORF5), and RNA-dependent RNA polymerase (gene L/ORF6) (Dietzgen et al. 2017).

In 2008, chlorotic spots were observed on the leaves of infected *S. leucantha* plants which was found to be caused by Clerodendrum chlorotic spot virus (ClCSV)

(Kitajima et al. 2008). *S. leucantha* was also found infected by an unidentified C-BTV, i.e., cytoplasmic type *Brevipalpus*-transmitted virus in a residential garden at Piracicaba, SP, showing green spots on senescent leaves (Kitajima et al. 2003).

12.1.3 Management

Management of plant virus diseases is a matter of vital importance and concern to farmers, horticulturists, foresters, manufacturers, as well as consumers. It is well established that virus diseases in different crops cause enormous losses all over the world in terms of quantity and/or quality of products. Although it's very difficult to put a clear figure on the financial impact of plant viruses in agriculture, the yield losses that can be ascribed to plant viruses are estimated to cost worldwide more than $30 billion annually (Sastry and Zitter 2014).

Sustainable plant disease management requires a multidimensional consideration of the impacts of management approaches on economics, sociology, and ecology by fully understanding the mechanisms of plant disease epidemics, the functioning of healthy agroecosystems, and individual and collective roles of approaches on disease management. Such type of actions helps not only to increase agricultural productivity and improve food quality but also to protect the ecological environment and natural resources. The availability of diagnostic measure would provide a reliable virus management strategy. Some of the most common techniques are discussed below.

12.1.3.1 Cultural Practices

Continuous cultural practices are found to be effective providing management against virus diseases. Regular weeding, host-free periods, eradication of source plants, use of reflective mulches and physical barriers, intercropping or delayed sowing time to divert vectors, and use of virus-free transplants are some of the approaches found to be beneficial in reducing the incidence of viruses.

12.1.3.2 Vector Management

Major transmission of the viruses is carried out by vectors. These vectors can be aphids, whiteflies, thrips, etc. Insecticides are commonly used to manage these vectors, but it has harmful effects on human health and on environment. Also, frequent use leads to the resistance in the vectors. Therefore, identification and use of natural parasites are found to be effective for the management of vectors. For example, *Encarsia* spp. and *Eretmocerus* spp. have been widely observed (Cudjoe et al. 2005; Kamau et al. 2005) for the management of *B. tabaci* of cassava in Africa. The feeding behavior of ladybird (*Coccinella transversalis*) predator of green and black aphids has been observed on chrysanthemum plants that caused reduction in the aphid population (Raj et al. 2005).

12.1.3.3 Virus Reservoirs

Weeds and wild plants are found to be the alternative host of many viruses. *A. conyzoides* in Cameroon is recorded a host to a complex consisting of Ageratum leaf curl Cameroon virus (ALCCMV), Ageratum leaf curl betasatellite (ALCCMB), and Ageratum leaf curl Cameroon alpha satellite (ALCCMA) (Leke et al. 2012). *Salvia plebeia* R. BR., a perennial weed, grows in abundance adjacent to the gardens of Gorakhpur. This plant is noticed to harbor cucumber mosaic virus throughout the year and acting as potential source of infection of *S. splendens* (Joshi and Dubey 1972). Therefore, identification of reservoir or alternate host is very crucial step for the proper management.

12.1.3.4 Plant Extracts

Botanical extracts have been known to combat bacterial and fungal diseases, thus seeking attraction in crop protection (Chakraborty and Chakraborty 2010). Endogenous viral inhibitors are proteinaceous antiviral substances extracted from plants and used against plant viral infections (Madhusudhan et al. 2011; Varma and Malathi 2003; Verma et al. 1992). Such antiviral substances have been reported in the extracts of *Bougainvillea spectabilis*, *Azadirachta indica*, *Pongamia glabra*, *Clerodendrum aculeatum*, *Phytolacca americana*, *Dianthus caryophyllus*, *Mirabilis jalapa*, *Boerhaavia diffusa*, etc., against serious pathogenic viruses such as TMV, PVX, PVY, begomoviruses, etc. (Al-mazra'awi and Ateyyat 2009; Bellotti and Arias 2001; Pandey et al. 2010). These extracts reduce the infectivity of the viruses to 60–80% and thus reduce the number of local lesions. Likewise, extracts of *Annona muricata* and *Datura stramonium* were most effective in suppressing disease. The plants' medicinal extracts were found to have increased the plant height and total weight of the plant, fruit amount, and fruit weight (Hamidson et al. 2018).

12.1.3.5 Training Programs

Losses caused by pathogens are growing day by day. Therefore, use of proper management strategies at right time is needed. Farmers are often confused about the pesticides and fungicides that are used to combat the diseases. This often causes environmental pollutions as these chemicals are very hazardous for their surroundings. Knowledge about the virus transmission, vector, and their infection cycle is important. Farmers should be informed about the major symptoms associated with the fungal and viral diseases. Farmers should recommend attending workshops related to agricultural practices to increase their skills.

12.1.3.6 Virus-Free Plants

Using planting material from which all infected plants have been rouged, applying heat therapy (35–54 °C), use of meristem tip cultures, cold treatment, and chemotherapy are other means suggested for obtaining virus-free plants. It is possible to produce thousands of virus-free planting materials within a relatively short period of time through apical meristem and tip culture technology (Verma et al. 2005; Fraga et al. 2004).

12.1.3.7 Molecular-Based Approaches

The majority of approaches currently being used for virus management studies rely entirely on molecular methods. Genetic engineering has been documented as most effective approach for controlling virus diseases in a wide range of crops grown worldwide. Pathogen-derived resistance has been observed to be mediated either by the protein encoded by the transgene (protein mediated) or by the transcript produced from the transgene (RNA mediated) also known as posttranscriptional gene silencing (PTGS) or both (Varma et al. 2002). The coat protein-mediated protection has been reported to be successful against numerous RNA viruses; however, few results of engineered resistance against begomoviruses have also been reported. Kunik et al. (1994) extended the idea of capsid protein-mediated protection that could be applied successfully with many RNA viruses to the DNA begomovirus. RNA silencing refers to small interfering (si) RNAs- or micro (mi) RNAs-mediated sequence-specific gene silencing mechanisms, which play important role in antiviral defense, development, and maintenance of genome integrity (Ding 2000; Qu et al. 2007; Chen 2012; Chen et al. 2016). Based on antiviral silencing mechanisms, it is categorized into four groups: sense gene-induced posttranscriptional gene silencing (S-PTGS), hairpin RNA-induced PTGS (hp-PTGS), artificial miRNA-induced PTGS (AMIR), and trans-acting siRNA-induced PTGS (TAS). Many researches have been carried out by using these approaches. RNA silencing should not only be viewed as an antiviral mechanism that must be counteracted. In fact, many viruses encode weak or transiently active suppressors and probably use RNA silencing to limit their accumulation and prevent catastrophic damage to their hosts, as evidenced in the case of symptom recovery (Ghoshal and Sanfaçon 2015).

Virus-derived small RNAs are the by-products of RNA silencing that guide ARGONAUTE proteins to target RNAs, and they occasionally share sequence complementarily with plant mRNAs, providing viruses with another tool to fine-tune plant gene expression (Adkar-Purushothama et al. 2015). Characterizing the impact of the downregulation of plant mRNAs by viral small RNAs on the outcome of infection is likely to be a focus of attention in the future.

12.1.3.8 Gene Pyramiding Strategy

The concept of transferring several characterized resistance genes into one plant is called "Gene Pyramiding." The dogma behind this strategy is that the probability of a pathogen mutating to "virulence against all resistance genes in the pyramid would be the product of the probabilities for each gene singly" (Mundt 2014), thus making the probability of a virulent pathotype arising highly unlikely. Hence, pyramiding relies on resistance genes that have been previously characterized singly and whose functions are combined within the same plant. This strategy has been successfully applied to plant-virus interactions (Shi et al. 2009) and could bring the opportunity to associate different types of antiviral plant resistances targeting different virus-derived molecules (proteins or nucleic acids) and various processes during virus cycle (replication/translation, plasmodesmata crossing, systemic colonization), with the possible consequence to reduce significantly the probability of resistance breaking by new virus variants (Quenouille et al. 2013).

12.1.3.9 The TILLING and ecoTILLING Revolution

The early 2000s have seen the emergence of the TILLING (Targeting Induced Local Lesions IN Genomes) method that consists of a classical mutagenesis step followed by the targeted search for plants carrying a mutation in a gene of interest (Szarejko et al. 2017). By creating artificial polymorphism directly into crops, this technique allows potentially (1) to replace a resistance allele overcome by viral strains, (2) to generate novel resistances with a broader spectrum, or (3) to create a new resistance gene based on knowledge acquired in heterologous systems. Such a strategy increases the natural allelic diversity by the identification of novel artificial alleles. Although this strategy requires the previous characterization of the gene conferring the resistance, the main advantage of TILLING is that it can be applied to any plant species, regardless of its genome size, ploidy level or method of propagation, and without introducing heterologous DNA as for GM plants. Similarly, the TILLING natural alternative (using natural germplasms collections instead of EMS mutants collections) is calling ecoTILLING and consists of exploiting the whole natural variability of a plant species (including wild-related and cultivated genotypes) (Nicaise 2014). Initially developed in *A. thaliana*, both TILLING and ecoTILLING have spread rapidly to other model plants (*Medicago truncatula*, *Lotus japonicus*, *Brachypodium distachyon*) and major crops (e.g., maize, soybean, sorghum, tomato, pepper, cucumber, pea, wheat, banana, bean, rice, barley, *Brassica napus*). It seems now clear that these strategies are emerging as major crop improvement tools, with especially successful examples of recent applications to antiviral protection (Piron et al. 2010).

12.1.3.10 Resistance Gene Enrichment Sequencing (RenSeq)

RenSeq is a comparative genomics tool used to identify nucleotide-binding leucine-rich repeats (NLR) gene family members that play a role in disease resistance. During RenSeq, a library of NLR gene-specific DNA fragments are used as baits to capture and enrich for genomic DNA fragments encoding NLRs from resistant germplasms. The enriched NLR-encoding DNA is sequenced and compared with the reference genome to identify the polymorphisms in the NLR genes that potentially account for the observed disease resistance in the resistant germplasm of interest (Jupe et al. 2013).

Recent metagenomic studies have revealed the large diversity of viruses found in wild plants. Perhaps not surprisingly, many natural infections do not have detrimental effects on the host. Some are even beneficial (Roossinck 2015). These studies have highlighted the delicate equilibrium between plant viruses and their natural hosts that arose from long-term coevolution and that can be broken by large-scale monocultures. Thus, characterizing the impact of virus infection in natural ecosystems may reveal new beneficial uses of plant viruses and help develop more informed agricultural practices.

12.1.3.11 CRISPER/Cas9

CRISPR/Cas9 technology could be a new ray of hope to engineer resistance against single and multiple geminivirus infections in plants (Zaidi et al. 2016). Clustered regularly interspaced short palindromic repeats (CRISPRs)/CRISPR-associated 9 (Cas9) is a prokaryotic molecular immunity system against invading viruses and has been harnessed as a powerful tool for targeted genomic editing. During subsequent infections, spacers are transcribed as part of the CRISPR array; after transcription and maturation, CRISPR RNA guides the Cas9 endonuclease to scan invading DNA and cleave the target sequence (Wright et al. 2016; Zhang et al. 2018) at a site preceding the protospacer associated motif (PAM), a trinucleotide sequence that is recognized by Cas9 and necessary for its binding to target DNA. Ali et al. (2015a, b) showed that CRISPR/Cas9 technology could impart molecular immunity against three geminiviruses, i.e., Tomato yellow leaf curl virus (TYLCV), Beet curly top virus (BCTV), and Merremia mosaic virus (MeMV) in *N. benthamiana* plants (Zhang et al. 2018), and revealed that a sgRNA designed to target a conserved sequence (TAATATTAC) in the viral intergenic region could be used to target multiple geminiviruses simultaneously (Zhang et al. 2018). Zhang et al. expressed the CRISPR/Cas9 system from *F. novicida* in plants was found to be effective against Cucumber mosaic virus (CMV) and Tobacco mosaic virus (TMV). Chandrasekaran et al. (2016) showed that disruption of the *eIF4E* gene in cucumber by CRISPR/Cas9 sgRNA led to the development of virus-resistant plants (Chandrasekaran et al. 2016). Harnessing the CRISPR/Cas9 machinery to engineer plant resistance to viral pathogens also opens the possibility of addressing basic questions in virus infection and plant host resistance. For example, the CRISPR/Cas9 platform could be used to investigate the evolution of the viral genome to counteract plant immunity by examining the genomes of viruses that escape recognition by the CRISPR/Cas9 system. The CRISPR/Cas9 platform could also be used for targeted mutagenesis to identify host factors that control plant resistance and susceptibility to viral infection. Thus, CRISPR/Cas9 technology offers a promising approach for understanding and engineering resistance to single and multiple viral infections in plants.

12.2 Conclusion

Huge economic losses are being reported continuously every year due to plant virus diseases especially in tropics and subtropics. Intensifying spread and prevalence of virus disease epidemics have been reported in many previous studies. Viruses cause epidemics on all major cultures of agronomic importance, representing a serious threat to global food security as well as horticulture, fruits, ornamental, and essential oil-bearing plants. As strict intracellular pathogens, they cannot be controlled chemically, and prophylactic measures consist mainly in the destruction of infected plants and excessive pesticide applications to limit the population of vector organisms. A powerful alternative frequently employed in agriculture relies on the use of crop genetic resistances, approach that depends on mechanisms governing plant-virus interactions.

Salvia is one of the largest genuses of plants that have diverse economic importance. It has been widely cultivated as medicinal, aromatic, and ornamental plants throughout the world and contributes in economic growth. Growing incidence of viral diseases could be a limiting factor for its commercial cultivation. Among different species of Salvia, *S. sclarea* was found to be most prone against different viruses. It has been found infected by CMV, AMV, PVY, BBSW, and TMV. CMV has been reported on majority of the *Salvia* spp. as reported earlier that it is known to have largest host range. It might be possible because of its convenient transmission mode and rapidly evolving genomic structure. Therefore, there is a need for a better understanding of the factors that have led to the increase in the vector populations in diverse cropping systems. To date, there is no report regarding the management of *Salvia* spp., and modern molecular approaches like RNAi and CRISPER/cas9 technology can be employed for better management of Salvia. CRISPR/Cas9 technology offers a promising approach for understanding and engineering resistance to single and multiple viral infections in plants. These approaches could also be used for targeted mutagenesis to identify host factors that control plant resistance and susceptibility to viral infection.

Therefore, knowledge related to the molecular bases of viral infections and crop resistances is key to face viral attacks in fields. Over the past 80 years, great advances have been made on our understanding of plant immunity against viruses. Although most of the known natural resistance genes have long been dominant R genes (encoding NBS-LRR proteins), a vast number of crop recessive resistance genes were cloned in the last decade, emphasizing another effective strategy to block viruses. In addition, the discovery of RNA interference pathways highlighted a very efficient antiviral system targeting the infectious agent at the nucleic acid level. Insidiously, plant viruses evolve and often acquire the ability to overcome the resistances employed by breeders. The development of efficient and durable resistances able to withstand the extreme genetic plasticity of viruses therefore represents a major challenge for the coming years. Here, we have described some of the most devastating diseases caused by viruses on *Salvia* spp. and summarizes current knowledge about plant-virus interactions, focusing on resistance mechanisms that prevent or limit viral infection in plants. In addition, the current outcomes of the actions are employed to control viral diseases in fields and the future investigations that need to be undertaken to develop sustainable broad-spectrum crop resistances against viruses. In parallel, advancements in molecular stacking and targeted gene insertion through genome editing are expected to play a major role in generating broad-spectrum resistance against both viral and nonviral pathogens in the near future.

In a coordinated manner with technological advances, fundamental research needs to explore new scientific leads, deciphering more and more thoroughly the intricate molecular dialog between a plant virus and its host. Hence, future challenges associated to the management of crop viral diseases will rely mainly on integrated research actions with a view to translating fundamental understanding toward applied programs and thus reducing the gap existing between the laboratory and the field (Fig. 12.1, Table 12.1).

Fig. 12.1 (**a**) Infected plant of *S. sclarea* showing yellow mosaic. (**b**) Healthy plants of *S. sclarea*

Table 12.1 List of viruses infecting *Salvia* spp.

S. No.	Virus	Natural host	Country	Year	References
1	CMV	*S. sclarea*	Italy	1989	Pisi and Vicchi (1989)
2	CMV	*S. sclarea*	Bulgaria	2014	Dikova (2014)
3	CMV	*S. uliginosa*	Baton Rouge, Louisiana	1998	Holcomb and Valverde (1998)
4	CMV	*S. splendens* KER-GAWL	India	1972	Joshi and Dubey (1972)
5	CMV	*S. splendens*	India	2012	Ali et al. (2012)
6	CMV	*S. patens*	–	1951	Faan and Johnson (1951)
7	CMV	*S. lyrata*	–	1959	Anderson (1959)
8	AMV	*S. sclarea*	Italy	1999	Bellardi et al. (1999)
9	AMV	*S. sclarea*	Bulgaria	2014	Dikova (2014)
10	AMV	*S. sclarea*	Hungary	2016	Salamon et al. (2018)
11	OYVCMV	*S. splendens*	Bangladesh	2012	Ara et al. (2012)
12	ClGMCNV	*S. splendens* cv. Dancing Flame	Baton Rouge, Louisiana	2011	Valverde et al. (2012)
13	SMBV2	*S. hispanica*	Salta	2013	Celli et al. (2014)
14	TYSV	*S. hispanica*	Salta	2013	Celli et al. (2014)
15	TMV	*S. sclarea*	Bulgaria	2014	Dikova (2014)
16	BBWV	*S. officinalis*	UK	2002	Mumford et al. (2006)
17	BBWV	*S. sclarea*	Bulgaria	2014	Dikova (2014)
18	TSWV	*S. sclarea*	Bulgaria	2014	Dikova (2014)
19	PVY	*S. sclarea*	Bulgaria	2014	Dikova (2014)
20	ClCSV	*S. leucantha*	–	2008	Kitajima et al. (2008)
21	C-BTV	*S. leucantha*	–	2013	Kitajima et al. (2003)

References

Adkar-Purushothama CR, Brosseau C, Giguère T, Sano T, Moffett P, Perreault JP (2015) Small RNA derived from the virulence modulating region of the potato spindle tuber viroid silences callose synthase genes of tomato plants. Plant Cell 27(8):2178–2194

Ali Z, Abulfaraj A, Idris A, Ali S, Tashkandi M, Mahfouz MM (2015a) CRISPR/Cas9-mediated viral interference in plants. Genome Biol 16(1):1–1

Ali Z, Abul-Faraj A, Li L, Ghosh N, Piatek M, Mahjoub A, Aouida M, Piatek A, Baltes NJ, Voytas DF, Dinesh-Kumar S (2015b) Efficient virus-mediated genome editing in plants using the CRISPR/Cas9 system. Mol Plant 8(8):1288–1291

Ali S, Akhtar M, Singh KS, Naqvi QA (2012) RT-PCR and CP gene based molecular characterization of a cucumber mosaic cucumovirus from Aligarh, UP, India. Agric Sci 3:971

Al-mazra'awi MS, Ateyyat M (2009) Insecticidal and repellent activities of medicinal plant extracts against the sweet potato whitefly, *Bemisia tabaci* (Hom.: Aleyrodidae) and its parasitoid Eretmocerus mundus (Hym.: Aphelinidae). J Pest Sci 82(2):149–154

Anderson CW (1959) A study of field sources and spread of five viruses of Peppers in central Florida. Phytopathology 49(2):97

Angelova VR, Ivanova RV, Todorov GM, Ivanov KI (2016) Potential of *Salvia sclarea* L. for phytoremediation of soils contaminated with heavy metals. Int J Agric Biosyst Eng 10 (12):780–790

Ara MR, Masud MM, Akanda AM (2012) Detection of plant viruses in some ornamental plants that act as alternate hosts. Agriculturists 10(2):46–54

Bailey LH. Manual of cultivated plants. 1949. (Rev. Edn.).

Bellardi MG, Autonell CR, Bertaccini A (1999) Virus and phytoplasma infections on 'officinal' plants in Emilia-Romagna. III report. Informat Fitopatol 49(6):47–53

Bellotti AC, Arias B (2001) Host plant resistance to whiteflies with emphasis on cassava as a case study. Crop Prot 20(9):813–823

Benner CP, Kuhn CW, Demski JW, Dobson JW, Colditz P (1985) Identification and incidence of pepper viruses in northeastern Georgia. Plant Dis 69(11):999–1001

Blanc S, Michalakis Y (2016) Manipulation of hosts and vectors by plant viruses and impact of the environment. Curr Opin Insect Sci 16:36–43

Bol JF (1999) Alfalfa mosaic virus and ilarviruses: involvement of coat protein in multiple steps of the replication cycle. J Gen Virol 80(5):1089–1102

Borah BK, Dasgupta I (2012) Begomovirus research in India: a critical appraisal and the way ahead. J Biosci 37(4):791–806

Brunt AA (1992) The general properties of potyviruses. Arch Virol Suppl 5:3–16

Brunt A, Crabtree K, Dallwitz MJ, Gibbs AJ, Watson L (1996) Viruses of plants. In: Descriptions and lists from the VIDE database. CAB International, Wallingford

Celli MG, Perotto MC, Martino JA, Flores CR, Conci VC, Pardina PR (2014) Detection and identification of the first viruses in chia (*Salvia hispanica*). Viruses 6(9):3450–3457

Chakraborty D, Chakraborty S (2010) Bioassay guided isolation and identification of antibacterial and antifungal components from methanolic extract of green tea leaves (*Camellia sinensis*). Res J Phytochem 4:78–86

Chandrasekaran J, Brumin M, Wolf D, Leibman D, Klap C, Pearlsman M, Sherman A, Arazi T, Gal-On A (2016) Development of broad virus resistance in non-transgenic cucumber using CRISPR/Cas9 technology. Mol Plant Pathol 17:1140–1153

Chen X (2012) Small RNAs in development–insights from plants. Curr Opin Genet Dev 22:361–367

Chen L, Cheng X, Cai J, Zhan L, Wu X, Liu Q, Wu X (2016) Multiple virus resistance using artificial trans-acting siRNAs. J Virol Methods 228:16–20

Chrzanowska M (1991) New isolates of the necrotic strain of potato virus Y (PVY N) found recently in Poland. Potato Res 34(2):179–182

Clebsch B (2003) New book of salvias. Timber Press, Portland, OR

Crill P, Hagedorn DJ, Hanson EW (1970) Alfalfa mosaic, the disease and its virus incitant. Univ Wisconsin Agric Sci Res Bull 280:40

Cudjoe A, Gyamenah J, James B (2005) Ghana. In: Anderson PK, Morales FJ (eds) Whitefly and whitefly borne viruses in the tropics. Building a knowledge base for global action. CIAT, Cali, pp 24–29

De Wispelaere M, Rao AL (2009) Production of cucumber mosaic virus RNA5 and its role in recombination. Virology 384(1):179–191

Dietzgen RG, Kuhn JH, Clawson AN, Freitas-Astúa J, Goodin MM, Kitajima EW, Kondo H, Wetzel T, Whitfield AE (2014) Dichorhavirus: a proposed new genus for Brevipalpus mite-transmitted, nuclear, bacilliform, bipartite, negative-strand RNA plant viruses. Arch Virol 159 (3):607–619

Dietzgen RG, Kondo H, Goodin MM, Kurath G, Vasilakis N (2017) The family Rhabdoviridae: mono-and bipartite negative-sense RNA viruses with diverse genome organization and common evolutionary origins. Virus Res 227:158–170

Dikova B (2014) Establishment of virus pathogens on the medicinal plant *Salvia sclarea*. Sci Technol 4:15–22

Ding SW (2000) RNA silencing. Curr Opin Biotechnol 11:152–156

Dougherty WG, Carrington JC (1988) Expression and function of potyviral gene products. Annu Rev Phytopathol 26(1):123–143

Dreher TW (1999) Functions of the 3′-untranslated regions of positive strand RNA viral genomes. Annu Rev Phytopathol 37(1):151–174

Dreher TW (2009) Role of tRNA-like structures in controlling plant virus replication. Virus Res 139(2):217–229

Faan HC, Johnson JA (1951) The overwintering of the Cucumber mosaic virus. Phytopathology 41 (11):1001

Fraga M, Alonso M, Borja M (2004) Shoot regeneration rates of perennial Phlox are dependent on cultivar and explant type. HortScience 39(6):1373–1377

Ghorbani A, Esmaeilizadeh M (2017) Pharmacological properties of *Salvia officinalis* and its components. J Tradit Complement Med 7(4):433–440

Ghoshal B, Sanfaçon H (2015) Symptom recovery in virus-infected plants: revisiting the role of RNA silencing mechanisms. Virology 479:167–179

Hajimorad MR, Francki RI (1988) Alfalfa mosaic virus isolates from lucerne in South Australia: biological variability and antigenic similarity. Ann Appl Biol 113(1):45–54

Hamidson H, Damiri N, Angraini E (2018) Effect of medicinal plants extracts on the incidence of mosaic disease caused by cucumber mosaic virus and growth of chili. In: IOP Conference Series: Earth and Environmental Science, vol 102, No. 1. IOP Publishing, Bristol, p 012062

Hiruki C, Miczynski KA (1987) Severe isolate of alfalfa mosaic virus and its impact on alfalfa cultivars grown in Alberta. Plant Dis 71(11):1014–1018

Holcomb GE, Valverde RA (1998) Natural Infection of *Salvia uliginosa* with Cucumber mosaic cucumovirus. HortScience 33(7):1215–1216

Hollings M, Brunt AA (1981) Potyvirus group. CMI/AAB Descript Plant Viruses 245(7)

Hord MJ, Garcia A, Villalobos H, Rivera C, Macaya G, Roossinck MJ (2001) Field surveys of Cucumber mosaic virus subgroups I and II in crop plants in Costa Rica. Plant Dis 85(9):952–954

Jaspars EM (1985) Molecular plant virology. In: Virus structure and assembly and nucleic acid-protein interactions, vol 1. CRC Press, Boca Raton, FL, pp 155–221

Jaspars EM, Bos L (1980) Alfalfa mosaic virus. No. 229. In: Descriptions of plant viruses. Commonwealth Mycological Institute, The Association of Applied Biologists, Kew

Jeffries CJ (1998) FAO/IPGRI technical guidelines for the safe movement of Germplasm: potato No. 19. Food and Agriculture Organization of the United Nations/International Plant Genetic Resources Institute, Rome

Joshi RD, Dubey LN (1972) Observations on a virus disease of *Salvia splendens* occurring in Gorakhpur. In: Phyton-Annales Rei Botanicae, vol 14, No. 3–4. Ferdinand Berger Soehne, Honr, p 309

Jupe F, Witek K, Verweij W, Śliwka J, Pritchard L, Etherington GJ, Maclean D, Cock PJ, Leggett RM, Bryan GJ, Cardle L (2013) Resistance gene enrichment sequencing (R en S eq) enables reannotation of the NB-LRR gene family from sequenced plant genomes and rapid mapping of resistance loci in segregating populations. Plant J 76(3):530–544

Kamau J, Sseruwagi P, Aritua V (2005) Whitefly and whitefly-borne viruses in the tropics. In: Anderson PK, Morales FJ (eds) Building a knowledge base for global action. CIAT, Cali, pp 54–60

Kitajima EW, Chagas CM, Rodrigues JC (2003) Brevipalpus-transmitted plant virus and virus-like diseases: cytopathology and some recent cases. Exp Appl Acarol 30(1):135–160

Kitajima EW, Kubo KS, Ferreira PD, Alcântara BK, Boari AJ, Gomes RT, Freitas-Astua J, Rezende JA, Morais GJ, Salaroli RB (2008) Chlorotic spots on Clerodendrum, a disease caused by a nuclear type of Brevipalpus (Acari: Tenuipalpidae) transmitted virus. Sci Agric 65(1):36–49

Klug A (1999) The tobacco mosaic virus particle: structure and assembly. Philos Trans R Soc Lond B Biol Sci 354(1383):531–535

Kobayashi YO, Kobayashi A, Nakano M, Hagiwara K, Honda Y, Omura T (2003) Analysis of genetic relations between Broad bean wilt virus 1 and Broad bean wilt virus 2. J Gen Plant Pathol 69(5):320–326

Kumar A, Kumar J, Khan JA (2010) Sequence characterization of cotton leaf curl virus from Rajasthan: phylogenetic relationship with other members of geminiviruses and detection of recombination. Virus Genes 40(2):282–289

Kunik T, Salomon R, Zamir D, Navot N, Zeidan M, Michelson I, Gafni Y, Czosnek H (1994) Transgenic tomato plants expressing the tomato yellow leaf curl virus capsid protein are resistant to the virus. Biotechnol J 12(5):500–504

Kuźma Ł, Kalemba D, Różalski M, Różalska B, Więckowska-Szakiel M, Krajewska U, Wysokińska H (2009) Chemical composition and biological activities of essential oil from *Salvia sclarea* plants regenerated in vitro. Molecules 14(4):1438–1447

Langereis K, Mugnier MA, Cornelissen BJ, Pinck L, Bol JF (1986) Variable repeats and poly (A)-stretches in the leader sequence of alfalfa mosaic virus RNA 3. Virology 154(2):409–414

Leke WN, Brown JK, Ligthart ME, Sattar N, Njualem DK, Kvarnheden A (2012) *Ageratum conyzoides*: a host to a unique begomovirus disease complex in Cameroon. Virus Res 163 (1):229–237

Lisa V, Boccardo G (1996) *Fabaviruses:* broad beans wilt and allied viruses. In: Harrison BD, Murant AF (eds) The plant viruses, vol. 5: Polyhedral virions and bipartite RNA genomes. Plenum Press, New York, NY, pp 229–250

Lot H, Marchoux G, Marrou J, Kaper JM, West CK, Van Vloten-Doting L, Hull R (1974) Evidence for three functional RNA species in several strains of cucumber mosaic virus. J Gen Virol 22 (1):81–93

Lu Y, Foo LY (2002) Polyphenolics of Salvia—a review. Phytochemistry 59(2):117–140

Madhusudhan KN, Vinayarani G, Deepak SA, Niranjana SR, Prakash HS, Singh GP, Sinha AK, Prasad BC (2011) Antiviral activity of plant extracts and other inducers against tobamoviruses infection in bell pepper and tomato plants. Int J Plant Pathol 2(1):35–42

Mandal SK (2012) Field evaluation of alternate use of insecticides against chilli thrips, Scirtothrips dorsalis (Hood). Ann Plant Protect Sci 20(1):59–62

Maruthi MN, Rekha AR, Muniyappa V (2007) Pumpkin yellow vein mosaic disease is caused by two distinct begomoviruses: complete viral sequences and comparative transmission by an indigenous *Bemisia tabaci* and the introduced B-biotype. EPPO Bull 37(2):412–419

Mori M, Kondo T, Yoshida K (2008) Cyanosalvianin, a supramolecular blue metalloanthocyanin, from petals of *Salvia uliginosa*. Phytochemistry 69(18):3151–3158

Mumford RA, Jarvis B, Harju V, Boonham N, Skelton A (2006) First report of Broad bean wilt virus 2 in the UK: findings in foxglove and salvia. Plant Pathol 55(6):819

Mundt CC (2014) Durable resistance: a key to sustainable management of pathogens and pests. Infect Genet Evol 27:446–455

Nicaise V (2014) Crop immunity against viruses: outcomes and future challenges. Front Plant Sci 5:660

Palukaitis P, Roossinck MJ, Dietzgen RG, Francki RI (1992) Cucumber mosaic virus. Adv Virus Res 41:281–348

Pandey SK, Mathur AC, Manisha S (2010) Management of leaf curl disease of chilli (*Capsicum annuum* L.). Int J Virol 6(4):246–250

Piron F, Nicolaï M, Minoïa S, Piednoir E, Moretti A, Salgues A, Zamir D, Caranta C, Bendahmane A (2010) An induced mutation in tomato eIF4E leads to immunity to two potyviruses. PLoS One 5(6):e11313

Pisi A, Vicchi V (1989) Infection by Cucumber mosaic virus (CMV) on *Salvia sclarea* in Italy. Informa Fitopathol 39:49–51

Qu J, Ye J, Fang R (2007) Artificial microRNA-mediated virus resistance in plants. J Virol 81:6690–6699

Quenouille J, Montarry J, Palloix A, Moury B (2013) Farther, slower, stronger: how the plant genetic background protects a major resistance gene from breakdown. Mol Plant Pathol 14 (2):109–118

Rahman M, Hussain D, Malik TA, Zafar Y (2005) Genetics of resistance to cotton leaf curl disease in *Gossypium hirsutum*. Plant Pathol 54(6):764–772

Raj SK, Singh R, Khan MS, Kumar S, Pratap D, Snehi SK (2005) Eco-friendly approaches for the management of plant virus diseases. In: Third International Conference on Plants and Environmental pollution. NBRI, Lucknow, pp 58–59

Riechmann JL, Lain S, García JA (1992) Highlights and prospects of potyvirus molecular biology. J Gen Virol 73(1):1–6

Roossinck MJ (2015) Plants, viruses and the environment: ecology and mutualism. Virology 479:271–277

Royal Horticultural Society (2017a) AGM plants-ornamental. Royal Horticultural Society, London, p 95. Accessed 14 Oct 2018

Royal Horticultural Society (2017b) AGM plants-ornamental. Royal Horticultural Society, London, p 94. Accessed 12 Oct 2018

Salamon P, Sos-Hegedus A, Gyula P, Szittya G (2018) First report of the infection of alfalfa mosaic virus in *Salvia sclarea* in Hungary. J Plant Pathol 100(3):607

Sastry KS, Zitter TA (2014) Management of virus and viroid diseases of crops in the tropics. In: Plant virus and viroid diseases in the tropics. Springer, Dordrecht, pp 149–480

Scholthof KB, Adkins S, Czosnek H, Palukaitis P, Jacquot E, Hohn T, Hohn B, Saunders K, Candresse T, Ahlquist P, Hemenway C (2011) Top 10 plant viruses in molecular plant pathology. Mol Plant Pathol 12(9):938–954

Shi Y, Klutstein M, Simon I, Mitchell T, Bar-Joseph Z (2009) A combined expression-interaction model for inferring the temporal activity of transcription factors. J Comput Biol 16 (8):1035–1049

Szarejko I, Szurman-Zubrzycka M, Nawrot M, Marzec M, Gruszka D, Kurowska M, Chmielewska B, Zbieszczyk J, Jelonek J, Maluszynski M (2017) Creation of a TILLING population in barley after chemical mutagenesis with sodium azide and MNU. In: Biotechnologies for plant mutation breeding. Springer, Cham, pp 91–111

Taylor RH, Stubbs LL (1972) Broad bean wilt virus. In: Description of plant viruses. Commonwealth Mycological Institute, The Association of Applied Biologists, Kew, p 81

Tenllado F, Bol JF (2000) Genetic dissection of the multiple functions of alfalfa mosaic virus coat protein in viral RNA replication, encapsidation, and movement. Virology 268(1):29–40

Uyemoto JK, Provvidenti R (1974) Isolation and identification of two serotypes of broad bean wilt virus. Phytopathology 64(12):1547–1548

Valverde RA, Singh R, Sabanadzovic S (2012) Detection and identification of Clerodendron golden mosaic China virus in *Salvia splendens*. Eur J Plant Pathol 133(3):499–503

Varma A, Malathi VG (2003) Emerging geminivirus problems: a serious threat to crop production. Ann Appl Biol 142(2):145–164

Varma A, Jain RK, Bhat AI (2002) Virus resistant transgenic plants for environmentally safe management of viral diseases. Indian J Biotechnol 1:73–86
Verma A, Dhar AK, Mandal B (1992) MYMV transmission and control in India. AVRDC, Taiwan
Verma N, Ram R, Zaidi AA (2005) In vitro production of Prunus necrotic ringspot virus-free begonias through chemo-and thermotherapy. Sci Hortic 103(2):239–247
Walker JB, Sytsma KJ, Treutlein J, Wink M (2004) Salvia (Lamiaceae) is not monophyletic: implications for the systematics, radiation, and ecological specializations of *Salvia* and tribe Mentheae. Am J Bot 91(7):1115–1125
Wright AV, Nuñez JK, Doudna JA (2016) Biology and applications of CRISPR systems: harnessing nature's toolbox for genome engineering. Cell 164(1–2):29–44
Wu YB, Ni ZY, Shi QW, Dong M, Kiyota H, Gu YC, Cong B (2012) Constituents from *Salvia* species and their biological activities. Chem Rev 112(11):5967–6026
Zaidi SS, Mansoor S, Ali Z, Tashkandi M, Mahfouz MM (2016) Engineering plants for geminivirus resistance with CRISPR/Cas9 system. Trends Plant Sci 21(4):279–281
Zhang T, Zheng Q, Yi X, An H, Zhao Y, Ma S, Zhou G (2018) Establishing RNA virus resistance in plants by harnessing CRISPR immune system. Plant Biotechnol J 16:1415–1423
Zitter TA, Murphy JF (2009) Cucumber mosaic. Plant Health Instruct 10:516–518

A Current Overview of Viruses of Chrysanthemum in India: Perspective and Prospective

13

Susheel Kumar, Karmveer Kumar Gautam, and Shri Krishna Raj

Abstract

This chapter deals with an overview on the diagnosis and characterization of viruses infecting chrysanthemums. This also describes about the virus management of chrysanthemum through in vitro chemotherapy to eliminate the virus (es) infecting chrysanthemum explants for virus-free plants and development of virus-resistant transgenic plants through genetic engineering. The published work by researchers all over the world and the work done by virology group at CSIR-NBRI, Lucknow, India, have been summarized in this chapter to conclude about the perspectives and prospectives of chrysanthemum, an ornamental plant important for floriculture trade.

Keywords

Chrysanthemums · Virus diagnostics · Virus characterization · Virus-free plants · Virus-resistant transgenic plants

13.1 Introduction

Chrysanthemum (*Chrysanthemum morifolium* Ramat. of family Asteraceae) called as "Queen of East Asia" is a native to northern hemisphere mainly Asia and northeastern Europe. Chrysanthemum is an important cut flower and pampered pot

S. Kumar · K. K. Gautam · S. K. Raj (✉)
Plant Molecular Virology Laboratory, CSIR-National Botanical Research Institute, Lucknow, Uttar Pradesh, India

S. K. Raj et al. (eds.), *Virus Diseases of Ornamental Plants*, https://doi.org/10.1007/978-981-16-3919-7_13

plant, cultivated worldwide. It ranks third, after rose and carnations, in cut flower sale and has gained immense commercial importance under cut flower trade due to its beautiful blooms of various shapes, sizes, and colors and long self-vase life.

Besides their floriculture importance, chrysanthemums have other economic values. The leaves and flower heads are consumed as vegetable. The boiled flowers are used to make a drink called "chrysanthemum tea" in some parts of Asia used to treat headaches, fever, hepatitis, detoxification, and influenza and improve eyesight.

Since chrysanthemums are propagated by cuttings/suckers and a few major propagators supply most of the plants to growers over very large geographic areas, these conditions make the ornamental crop production industry conducive to the proliferation and distribution of diseases. Viruses and viroids frequently infecting chrysanthemum and jeopardizing its quality and production are *Cucumber mosaic virus* and *Tomato aspermy virus* (*Bromoviridae*), *Chrysanthemum virus B* (*Flexiviridae*), *Tomato spotted wilt virus* (*Tospoviridae*), *Impatiens necrotic spot virus* (*Bunyaviridae*), and *Chrysanthemum chlorotic mottle viroid* (Avsunviroidae) and *Chrysanthemum stunt viroid* (Pospiviroidae) (Bouwen and van Zaayen 1995). Two *Begomoviruses*, *Clerodendron yellow mosaic virus* and *Tomato leaf curl New Delhi virus*, have also been reported from India on chrysanthemum (Marwal et al. 2013; Ashwathappa et al. 2020) (Table 13.1).

Among all the viruses, cucumoviruses have been very frequently reported in chrysanthemums; hence, they are considered as the most important ones. There are three type members of genus *Cucumovirus*: *Cucumber mosaic virus* (CMV), *Tomato aspermy virus* (TAV), and *Peanut stunt virus* (PSV). They are aphid transmittable, are single-stranded, and have plus sense messenger RNA genome. The frequent natural occurrence of CMV and TAV on chrysanthemum cultivars affected the commercial production in terms of the yield and quality of chrysanthemum flowers worldwide (Raj et al. 2008).

In this chapter, we have described the overview on the diagnosis and characterization of viruses infecting chrysanthemums. We have also described about the virus management of chrysanthemum through in vitro chemotherapy for elimination of virus infection and development of virus-resistant transgenic plants through genetic engineering.

13.2 Characterization of Viruses Infecting Chrysanthemums

13.2.1 *Cucumber Mosaic Virus* (Cucumoviruses)

Srivastava et al. in 1992 (Srivastava et al. 1992) reported CMV from India that infected *C. morifolium*, where it caused diffuse chlorotic dots near the veins, severe chlorosis accompanied by necrosis, and stunting of the plants. The virus isolate was purified and characterized as chrysanthemum strain of CMV (CMV-CI) based on host reaction, in vitro properties, particle morphology (29 nm), molecular weight of the coat protein subunits (24.5 kDa), nucleic acid, and serological relationships with other cucumoviruses (Srivastava et al. 1992). The association of CMV with mosaic

Table 13.1 Viruses infecting chrysanthemums worldwide

Virus genus	Country	Symptomatology	Citations
Cucumovirus			
CMV	India	Diffuse chlorotic dots near the veins, severe chlorosis, necrosis, and stunting in *C. morifolium*	Srivastava et al. (1992)
CMV	Japan	Mosaic symptoms	Nakamura et al. (1994)
CMV	–	–	Bouwen and van Zaayen (1995)
CMV	India	Chlorotic spots on leaves and stunting of plants in *C. morifolium*	Verma et al. (2004)
CMV	India	Vein yellowing, yellow necrotic spots, and severe mosaic in *C. morifolium*	Kumar et al. (2005), Susheel (2009)
TAV	USA	Symptomless appearance with flower distortion	Brierley (1958)
TAV	USA	Slight dwarfing of blooms with waving and curling of ray florets	Brierley (1955)
TAV	–	Color break, dwarfing, and discoloration of flowers	Hollings (1955)
TAV	India	Chlorotic mottle of *C. morifolium*	Singh and Gupta (1978)
TAV	–	Mottling following deformed flowers with color break	Chuyan and Krylov (1979)
TAV	India	Distortion and depression on ray florets in *C. morifolium*	Gupta and Singh (1981)
TAV	India	Chlorotic ring mosaic on leaves of *C. morifolium*	Raj et al. (1991)
TAV	Taiwan	–	Lin et al. (2004)
TAV	India	Mottling and deformed inflorescence in *C. morifolium*	Verma et al. (2004)
TAV	India	Ringspot and flower deformation	Raj et al. (2008)
Carlavirus			
CVB	Japan	–	Nakamura et al. (1994)
CVB	Taiwan	–	Lin et al. (2004)
CVB	India	–	Verma et al. (2003)
CVB	Taiwan	–	Lin et al. (2005)
CVB	India	–	Singh et al. (2011)
Potyvirus			
Potyvirus	–	Chlorotic spots on the leaves	Bertaccini et al. (1994)
TuMV	Japan	Albinic mosaic disease	Chen et al. (1999)
TuMV	Japan	Mosaic and mottling of garland chrysanthemum	Chen et al. (2000)
CVMV	India	–	Mehra et al. (2009)

(continued)

Table 13.1 (continued)

Virus genus	Country	Symptomatology	Citations
BMV	Taiwan	Mosaic and deformation in leaves and stunting	Chen and Lee (2012)
PVY	China	Yellowing and mottled leaves	Liu et al. (2014)
ZYMV	China	Foliar yellowing and mottle	Niu et al. (2015)
Nepovirus			
Arabis mosaic virus	Iran	Mosaic, leaf chlorosis, small necrotic lesions, and leaf malformation	Ghotbi and Shahraeen (2005)
Tospovirus			
TSWV	Japan	Chlorotic spots on top leaves followed by wilting	Matsuura et al. (2002), Matsuura et al. (2007)
CSNV	Brazil	Necrotic lesions surrounded by yellow areas on leaves followed by necrosis on stems	Bezerra et al. (1999)
CSNV	–	Stem necrosis	Dullemans et al. (2015)
INSV	Japan	Chlorotic mottle and necrosis	Kondo et al. (2011)
Geminivirideae			
Cl YMV	India	Yellowing of leaf vein disease	Marwal et al. (2013)
TLCNDV	India	Mosaic and leaf curl disease	Ashwathappa et al. (2020)

TLCNDV tomato leaf curl New Delhi virus, ClYMV clerodendron yellow mosaic virus, ZYMV zucchini yellow mosaic virus, PVY potato virus Y, BMV bidens mottle potyvirus, CVMV chilli veinal mottle virus, TuMV turnip mosaic virus, INSV impatiens necrotic spot virus

symptoms of *Chrysanthemum frutescens* was also reported in Japan based on mechanical inoculation on test plants and observed by EM (Nakamura et al. 1994).

Verma et al. in 2004 (Verma et al. 2004) reported infection of a viral disease showing characteristic chlorotic spots on leaves and stunting in plants of chrysanthemum cv. Regol Time. The infecting virus was identified as *Cucumber mosaic virus* (CMV) on the basis of mechanical inoculation on *Chenopodium amaranticolor* and *Cucumis sativus*, DAS-ELISA, and RT-PCR. They also eliminated CMV from chrysanthemum plants using meristem tip culture method (Verma et al. 2004).

Molecular detection of *Cucumber mosaic virus* in chrysanthemum cultivars had been done in India by Kumar et al. in 2005 (Kumar et al. 2005). Molecular detection of *Cucumber mosaic virus* in various chrysanthemum cultivars was attempted by RT-PCR and Southern hybridization tests using CMV-specific primers and cloned probes, respectively. A pair of primers was designed from a conserved region of the *Cucumber mosaic virus* coat protein (CMV-CP) gene, capable of amplifying a product of ~650 bp from various CMV strains. RT-PCR using the total nucleic acid from infected leaf samples and the specific primers resulted in positive amplification of an expected size band of ~650 bp in most of the samples. The identity of the PCR amplicons was checked by Southern hybridization using the a32P-labeled

Fig. 13.1 Naturally infected chrysanthemums showing mosaic and ring spots (indicated by arrows) (**a**), chlorotic dots (**b**), purplish tinge (**c**), green mosaic (**d**), and vein clearing of leaf lamina (**e**) (Susheel 2009)

DNA probes prepared from the cloned coat protein gene of a well-identified strain of CMV isolated from *Amaranthus*. Positive signal of hybridization of PCR products and CMV probes confirmed the identity of PCR amplicons as a fragment derived from the CMV genome in infected chrysanthemum samples (Kumar et al. 2005).

During surveys in two subsequent years (2004 and 2006), several chrysanthemum cultivars were found to be affected by typical mosaic and ring spot disease in Lucknow, India (Susheel 2009). The naturally infected chrysanthemum plants exhibited severe mosaic and ring spots, chlorotic dots, purplish tinge, and yellow veins on leaves (Fig. 13.1). Infected plants also showed stunted growth, bearing small and deformed flower heads.

The presence of CMV in naturally infected chrysanthemum was detected by Ouchterlony double diffusion test (ODDT) using antisera of viruses (CMV=PVAS 242a, TAV=PVAS 24, CVB=PVAS 349, *Potyvirus*=PVAS 268). The test resulted in a sharp line of precipitin band with PVAS 242a and a faint precipitin line with TAV antiserum, but not with CVB and *Potyvirus* antisera confirming the presence of CMV in chrysanthemum samples.

Further, molecular identification of the CMV strain was done by amplification of the complete RNA 3 genome by RT-PCR using coat protein gene, movement protein gene, and UTR primers. The resulting amplicons were cloned and sequenced and sequence data were assembled to obtain the full-length RNA 3 genome (GenBank accession no. EF153733). BLAST analysis of complete RNA 3 component of CMV-Ch isolate under study revealed highest 99% nucleotide sequence identity with CMV isolates of *Amaranthus*, *Jatropha*, *Rauvolfia*, and *Datura*, earlier established strains in India. However, banana and tomato strains of Indian CMV showed only 95% and 93% sequence identities, respectively. CMV-Ch showed 93% sequence identity with Tfn and Nt9 strains of CMV of subgroup IB and 90% identity with Le02 and Fny strains of CMV of subgroup IA (Susheel 2009; Kumar et al. 2016).

The nucleotide identities of virus isolate at CP and MP levels were 88–99% and 84–100%, respectively, with several IB strains of CMV reported from all over the world. It was highest 95–99% and 95–100% at CP and MP with Indian strains of CMV belonging to IB subgroup. It also shared 86–90% and 72–91% nucleotide identities at CP and MP level with the members of IA subgroup of CMV. The subgroup II members of CMV showed 71–87% and 73–75% identities at CP and MP, respectively. The nucleotide identities at 5′UTR, 3′UTR, and interdomain region (IR) of CMV-Ch isolate with the selected Indian strains of CMV revealed 99% and 100%. The maximum identity may be because these regions contain the core promoters and domains essentially required for virus replication and maintenance.

The phylogenetic analysis of the complete RNA 3 genome sequences of CMV-Ch isolate under study (EF153733) with reported strains of CMV belonging to subgroups IA, IB, and II was done using MEGA v4.1 tool. RNA 3 genome-based phylogenetic analysis resulted in noncongruent tree, clearly dividing all CMV strains into three clusters of subgroups IA, IB, and II (Fig. 13.2). CMV chrysanthemum isolate under study clustered within subgroup IB and showed close relationships with CMV isolates of *Datura*, *Amaranthus*, *Musa* (banana), *Jatropha*, and *Rauvolfia*, which is reported from India. All Indian strains of CMV clustered within subgroup IB and formed a separate clade, indicating that Indian strains are closely related to each other grouped within subgroup IB, with exception of CMV-Ts and CMV-lily strain which showed closeness with subgroup IA strains and clustered with the isolates of subgroup IA. Subgroup II strains showed distant relationships with the chrysanthemum isolate under study (Fig. 13.2).

This was the first CMV chrysanthemum strain of subgroup IB from India which had been characterized at complete RNA 3 (its ORFs and non-translated regions) that is represented in GenBank database from India (Susheel 2009; Kumar et al. 2016).

13.2.2 *Tomato Aspermy Virus* (Cucumoviruses)

It is generally observed that *Tomato aspermy virus* (TAV) and *Cucumber mosaic virus* (CMV) have many common properties, and both have been assigned to the cucumovirus (Harrison et al. 1971). TAV-V and CMV-Q strains are reported as having identical size, morphology, sedimentation rate, RNA base ratio, and buoyant density (Habili and Francki 1975). But the amino acid compositions of proteins from the two similar viruses are distinguishable (Habili and Francki 1975). It is reported that they are distantly related serologically (Habili and Francki 1975). Rao et al. in 1982 (Rao et al. 1982) reported distant antigenic relationship between Australian strain of TAV and CMV. Similarly, TAV chrysanthemum isolate, which showed identical nucleotide sequence to TAV-V strain and the coat protein gene region, was also serologically distinct from CMV-pepper (Choi et al. 1999).

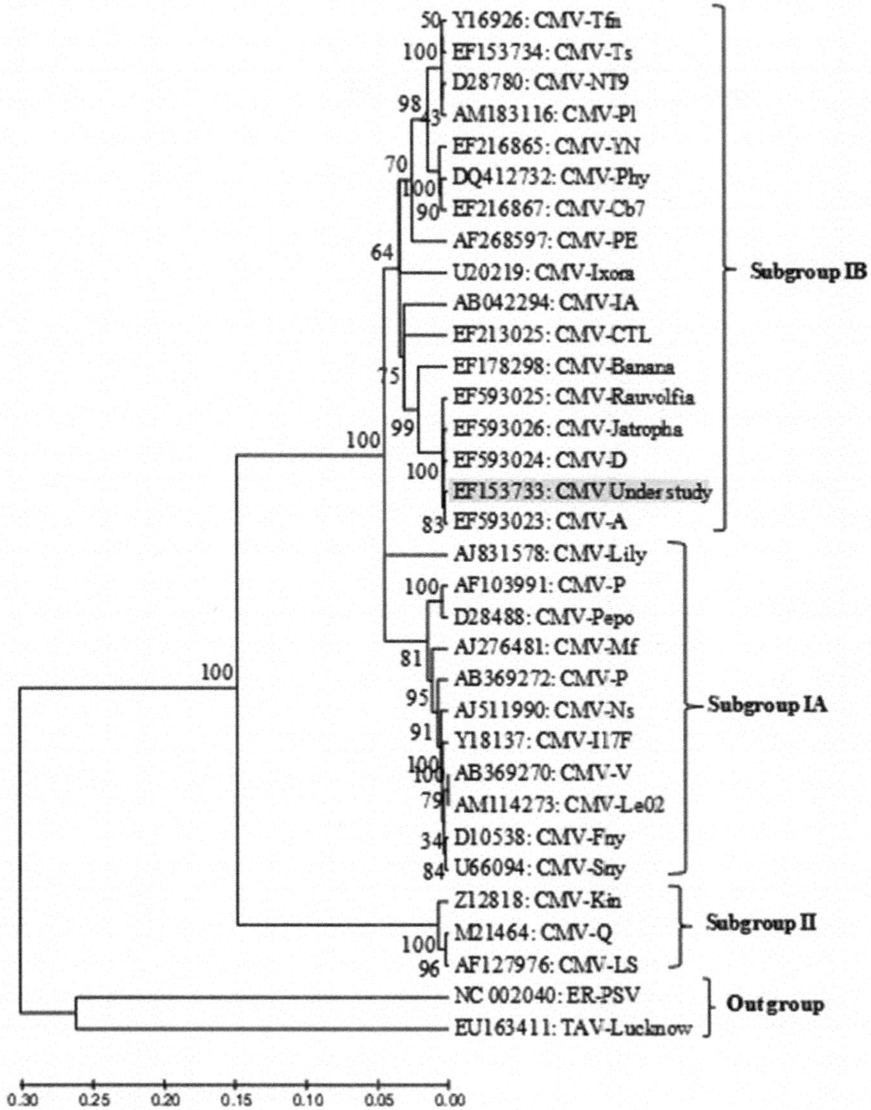

Fig. 13.2 Complete RNA-based phylogenetic analysis of CMV chrysanthemum isolate at nucleotide by MEGA v4.1. The evolutionary history was inferred using the neighbor-joining method with 100 replicate bootstraps (Susheel 2009)

Tomato aspermy virus (TAV) has been first reported on *Chrysanthemum morifolium* from England. TAV rarely produced leaf symptoms, but sometimes caused slight dwarfing of blooms together with waving and curling of the ray florets (Brierley 1955, 1958); on the other hand, Hollings in 1955 (Hollings 1955) reported that TAV causes of severe breakage, dwarfing, and distortion of the flowers in chrysanthemum, while in most varieties, the plant remains symptomless. TAV causing deformed flowers with color break and mottled leaves has been reported on chrysanthemum cultivars (Sastry 1964; Chuyan and Krylov 1979). A strain of *Chrysanthemum aspermy virus* (CAV) on chrysanthemum has been reported for the first time from India by Sastry in 1964 (Sastry 1964). Then after, studies on CAV on chrysanthemum were published by Gupta and Singh in 1981 (Gupta and Singh 1981).

The complete genome sequences of two TAV strains, V-TAV from Australia (Moriones et al. 1991; Moreno et al. 1997a) and KC-TAV from South Korea (Choi et al. 2002), and RNA 3 sequences of five TAV strains, SPN, C, P, I, and B (Moreno et al. 1997b; Salanki et al. 1994), are available in NCBI database. The 3′ termini of the V-TAV and N strain of TAV have also been reported by (Wilson and Symons 1981).

Raj et al., in 1991 (Raj et al. 1991), reported chlorotic ring mosaic of chrysanthemum caused by a strain of *Chrysanthemum aspermy virus* in India. The causal virus was identified as a strain of Chrysanthemum aspermy virus-CR strain on the basis of particle morphology (28 nm), transmission by aphids in nonpersistent manner, molecular weight of coat protein subunits as 24.5 kDa, and presence of five species of single-stranded RNA encapsidated in virus particles. The isolate designated as Chrysanthemum aspermy virus-CR strain did not show any serological relationship with Cucumber mosaic virus strains, viz., CMV-C, CMV-D, CMV-S, CMV-L, and CMV-T, in agar gel double diffusion tests. A positive serological relationship has been established with Tomato aspermy virus-N and *Chrysanthemum aspermy virus* (Raj et al. 1991).

Lin et al., in 2004 (Lin et al. 2004), performed the molecular and serological detection of *Tomato aspermy virus* infecting chrysanthemum in Taiwan. In this study, they described that Tomato aspermy virus (TAV) is known to induce malformation of floral parts on sensitive chrysanthemum cultivars jeopardizing the quality and yield. They designed a set of primer from TAV's coat protein (CP) gene sequence (accession No. D01015) and successfully amplified a 780 bp DNA product by RT-PCR from chrysanthemum specimens collected from Changhua, the major chrysanthemum production area in Taiwan. The product was cloned and sequenced. It was found to consist of 776 nucleotides (nts) corresponding to the genome organization of the 3′-terminal region of RNA 3 of cucumoviruses. The only open reading frame deducing in this sequence contains 220 amino acid residues coinciding to the size of reported TAV-CP. Compared with the known sequence of TAV-CP (D01015), the percent identities of nucleotide and amino acid sequences are 92% and 93%, respectively. This result indicates the sequence amplified from chrysanthemum is originated from a strain of TAV. To our knowledge, this is the first report of the occurrence of TAV in Taiwan (Lin et al. 2004).

Verma et al. (Verma et al. 2006) did detection and molecular characterization of a *Tomato aspermy virus* isolate infecting chrysanthemums in India. *Tomato aspermy virus* (TAV) was detected by double-antibody sandwich enzyme-linked immunosorbent assay (DAS-ELISA) from chrysanthemums exhibiting mottling and deformed inflorescence grown in various states of India. Out of 15 cvs tested,

Fig. 13.3 Naturally infected *Chrysanthemum morifolium* plant showing yellow mosaic, growth stunting, and deformation in flowers (Raj et al. 2007)

11 (73.3%) were found to be positive for TAV. *Myzus persicae* and *Aphis gossypii* transmitted the virus nonpersistently. Electron microscopy of partially purified viral preparations revealed polyhedral virions (ca. 29 nm dia). Slot blot hybridization was performed to detect the virus in various chrysanthemum cultivars. The virus has confirmed positivity by reverse transcription polymerase chain reaction (RT-PCR) and nucleotide sequencing of coat protein gene (CP). The amplified 657 bp fragment was about 97% identical with respect to CP sequences of other TAV isolates available in the database. In terms of amino acid sequence similarity, the homology value was 99%. High genetic similarities of CP gene of TAV Indian isolates with that of other isolates of TAV indicated their probable common ancestry (Verma et al. 2006).

Raj et al. (Raj et al. 2007) reported *Tomato aspermy virus* as the cause of yellow mosaic and flower deformation of *Chrysanthemum morifolium* cultivars in India (Fig. 13.3). They detected TAV by ELISA and RT-PCR using TAV-specific primers and identified it based on the coat protein gene sequence analysis. In this study, the total RNA was isolated from healthy and naturally infected leaf tissue of chrysanthemum plants, and RT-PCR was done using TAV-specific primers. An amplicon of the expected size (~650 bp) was obtained which was cloned and sequenced (GenBank accession number DQ191798). The isolate was 97% identical at the nucleotide level and 95% at amino acid level with seven other TAV isolates from

India and elsewhere in the world. Therefore, it is concluded that the cause of yellow mosaic and flower deformation in chrysanthemums was TAV. This was the first report of the molecular characterization of a TAV isolate infecting chrysanthemums in India (Raj et al. 2007).

Further, Raj et al., in 2009 (Raj et al. 2009), published the biological and molecular characterization of three isolates of *Tomato aspermy virus* (TAV) infecting chrysanthemums in India. In this study, severe mosaic, chlorotic ringspots, and flower deformation were observed during the winter of November 2006–February 2007 on chrysanthemums (*Chrysanthemum morifolium*) at three locations in India: Lucknow (UP), Dhanbad (MP), and Kolkata (WB). These TAV isolates were mechanically transmitted to test plant species and also by aphids (*Aphis gossypii*) to tomato (*Lycopersicon esculentum*).

The complete RNA 3 of each TAV isolate was cloned and sequenced and determined to be 2386 nucleotides (nt) long and encodes two open reading frames (ORFs): the movement protein (MP) of 741 nt and the coat protein (CP) of 657 nt translating into 246 and 218 amino acids (aa), respectively. When RNA 3 sequences of the Indian isolates were multiple aligned with seven other strains of TAV occurring worldwide, Indian isolates shared 98–99% identities among themselves and with the KC, V, P, B, I, and C strains of TAV. In phylogenetic analysis, the Lucknow and Kolkata isolates of TAV clustered together and showed a close relationship with the KC-TAV strain from South Korea, whereas the Dhanbad isolate formed an independent cluster and showed closeness with the V-TAV strains from Spain and Australia. Recombination events were also observed in the CP region of the Dhanbad isolate, supporting its diverse behavior. This was the first report of the complete RNA 3 sequence of these three Indian TAV isolates (Raj et al. 2009).

During the survey from 2004 to 2007, the infected chrysanthemum samples were collected from 15 different locations in India: Kolkata (West Bengal); Dhanbad (Jharkhand); Patna (Bihar); Jammu and Katra (Jammu and Kashmir); Baijnath Dham and Chintpurni Devi (Himachal Pradesh); Chandigarh; New Delhi; Lakhimpur, Lucknow, and Kanpur (Uttar Pradesh); Hyderabad (Andhra Pradesh); Mysore (Karnataka); and Ooty (Tamil Nadu). Naturally infected chrysanthemum plants exhibited various symptoms: yellow mosaic, marginal yellowing, vein yellowing, vein banding, stunting of whole plant, and flower deformations (Fig. 13.4) (Susheel 2009).

The presence of TAV was confirmed by direct antigen coating enzyme-linked immunosorbent assay (DAC-ELISA) using TAV-specific antibodies (PVAS 24), and sap of infected plants was used as antigen. A total 18 of samples collected from various locations and one healthy symptomless chrysanthemum (as negative control) were screened. During DAC-ELISA, the maximum mean absorbance value was observed as 0.916, 0.801, 0.620, and 0.323 in naturally infected chrysanthemums collected from Lucknow, Kolkata, Katra, and Dhanbad samples, respectively, as compared to 0.043 in case of the negative control (healthy chrysanthemum) plant. However, TAV infection was detected from 15 out of 18 samples collected from

Fig. 13.4 Flower deformations in large bloom varieties as compared to the healthy ones (right)

various locations which indicate the distribution of TAV in India (Fig. 13.5) (Susheel 2009).

For molecular identification of the TAV isolates, the full-length RNA 3 genome was amplified by RT-PCR using the RNA 3-specific primers designed for TAV isolates. The expected size (~2.4 kb) amplicons were obtained at all the temperatures, but amplicon's intensity and quality as a single band were obtained at 66 °C. The full-length amplicons of three samples were purified, cloned in a pGEM-T easy vector system, and sequenced. The sequence obtained from the three clones were assembled and analyzed for the consensus sequence in its entirety with no remaining ambiguities. The analysis of sequence data resulted in 2386 nucleotide complete genome of RNA 3. The complete RNA 3 (2386 nucleotide) sequences of three samples were deposited to GenBank: Dhanbad isolate (EU163410), Lucknow isolate (EU163411), and Kolkata isolate (EF153735).

Data analyses of RNA 3 genome of Kolkata, Dhanbad, and Lucknow-TAV isolates revealed that complete length of RNA 3 is 2386 nucleotide and consisted of two ORFs. The first in frame ATG(AUG) (start codon) in the ORF was at position 192–194 nucleotide and had termination-TAG(UAG)-codon at 930–932 nucleotide position and identified as MP gene of 741 nucleotide translating 246 amino acid residues. The second ORF started (ATG/AUG) at position 1228–1230 nucleotide and terminated (TAG/UAG) at 1882–1884, and identified as CP gene of 657 nucleotide translating 218 amino acid residues. MP and CP ORFs were separated by 295 nucleotide long intergenic region and flanked by 5′ un-translated region (UTR) and 3′UTR of 191 and 502 nucleotide, respectively (Susheel 2009).

Sequence alignment of complete RNA 3 components of TAV isolate of Lucknow with the selected TAV strains revealed that Lucknow isolate shared common features, viz., conserved TG tract in 5′UTR; size of CP (657 nt) equal to all TAV strains, except B-TAV (654 nucleotide, Acc S72468) and C-TAV (690 nucleotide,

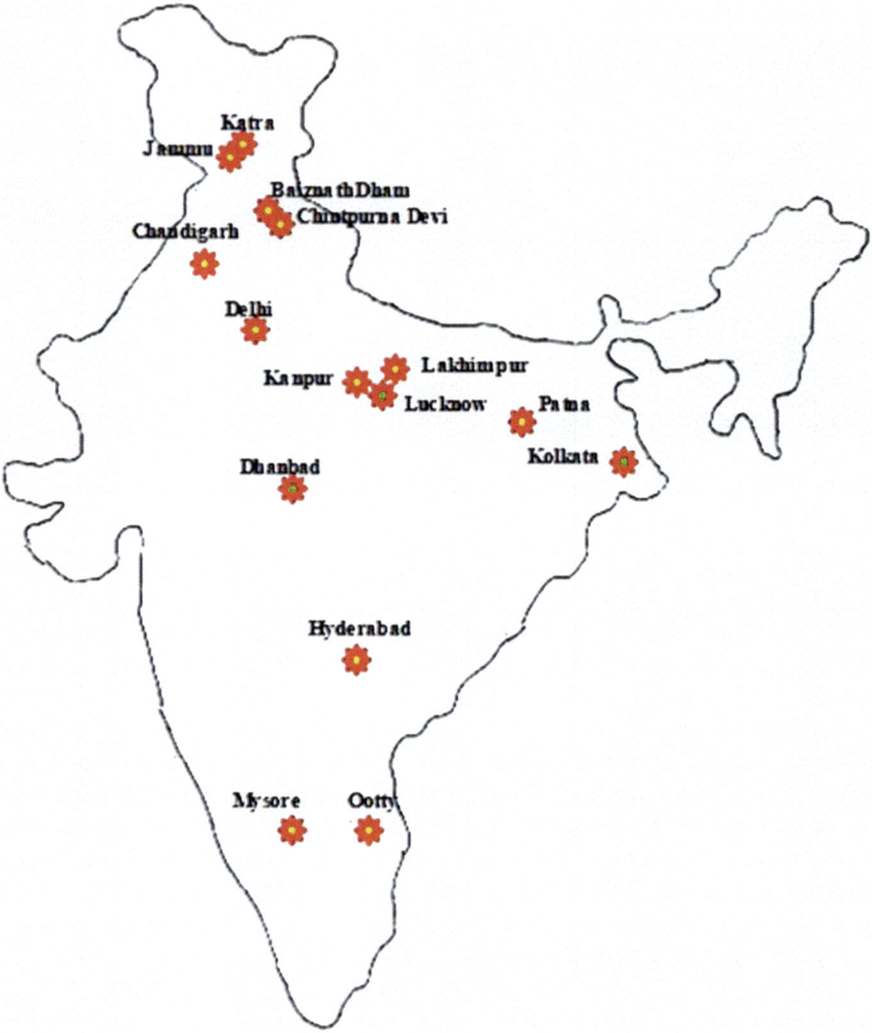

Fig. 13.5 Distribution of TAV in different locations in India based on ELISA tests conducted in various chrysanthemum samples collected during survey during 2004–2007 (Susheel 2009)

Acc D01015), and presence of the motif "GTTCAATTCC" and ICR-2-like conserved motif and a highly conserved region of 40 nucleotides "GAA CGG GTT GTC CAT CCA GCT AAC GGC TAA AAT GGT CAGT" at 3′UTR, the unique features found in all cucumoviruses sequenced to date. Around 116 additional nucleotides were present only in I-TAV isolate (reported from the USA).

BLAST analyses of complete RNA 3 gene of TAV Lucknow isolate revealed highest nucleotide identity (99%) with Kolkata, V, and Dhanbad strains of TAV. Ninety-eight percent identity was obtained with KC, C, and B-TAV strains, whereas a minimum of 97% identity was obtained with I and P-TAV. During *Genomatix DiAlign* analysis, complete RNA 3 component of virus isolates shared 98–99% nucleotide identities among themselves (Lucknow, Kolkata, and Dhanbad) and with V, KC, C, P, I, and B strains of TAV reported from other parts of the world, when analyzed by multiple sequence alignment. The minimum 48% and 42% identities were obtained with *Peanut stunt virus* (Acc AY775057) and *Cucumber mosaic virus* (EF153734), members of genus *Cucumovirus* used for reference as out-group.

CP gene-based analysis of Lucknow, Kolkata, and Dhanbad isolates showed highest nucleotide identity (96–99%), whereas the amino acid identities were 96–100%. A maximum nucleotide identities (89%) were recorded with the V and KC strains of TAV. However, highest amino acid identities (95%) were obtained with V and KC strains of TAV. The identities of other gene as MP gene, 5′UTR, 3′UTR, and intergenic regions at nt level were 95–99%, 98–99%, 94–100%, 94–99%, and 98–100%, respectively, with the selected strains of TAV considered for the study.

Phylogenetic analysis of complete RNA 3 component of TAV isolates resulted in noncongruent dendrogram. Among Indian strains, Kolkata and Lucknow isolates clustered together, whereas Dhanbad isolate clustered discretely. However, all the three isolate showed close relationship with KC strain, whereas distant relationships were observed with C, P, I, V, and B-TAV strains (Fig. 13.6). Spanish and Australian V-TAV strains fall in one cluster, whereas I, P, B, and C-TAV strains clustered together. Phylogenetic analysis indicated that Indian isolates have some unique genetic features, which lead them to clade distinctly. PSV strain was found closer to TAV strains as compared to CMV.

To find the putative cause of this diverge behavior of Dhanbad isolate, recombination analysis was also carried out with recombination detection program (RDP) v1.08 using GENECONV method. The analysis showed maximum (seven) recombination events in initial 450 bp region of CP gene of Dhanbad isolate that could explain its divergent behavior. Two recombination events were also observed in Lucknow isolate, with the remaining two Indian strains under study as well as other TAV strains reported from abroad (Fig. 13.7). However, Kolkata isolate did not show any recombination events. The result obtained revealed that this recombination is the cause of this diverge behavior of Dhanbad isolate.

Biological assay with Dhanbad isolate also indicated clear-cut differentiation between Dhanbad isolate and other two Indian strains under study, supporting the recombination in CP region of Dhanbad isolate.

The sequence data of RNA 3 genome of TAV isolate of Kolkata generated in this study will help in the understanding of the taxonomic status of Indian strains of TAV with respect to TAV strains established all over the world. The CP clones may be utilized for development of TAV-resistant transgenic chrysanthemum plants. ELISA, RT-PCR, and nucleic acid probe-based protocols developed out of the project for detection of TAV may be used for virus indexing of chrysanthemum cultivars and searching of virus-free propagating materials for the floriculture industries (Susheel 2009).

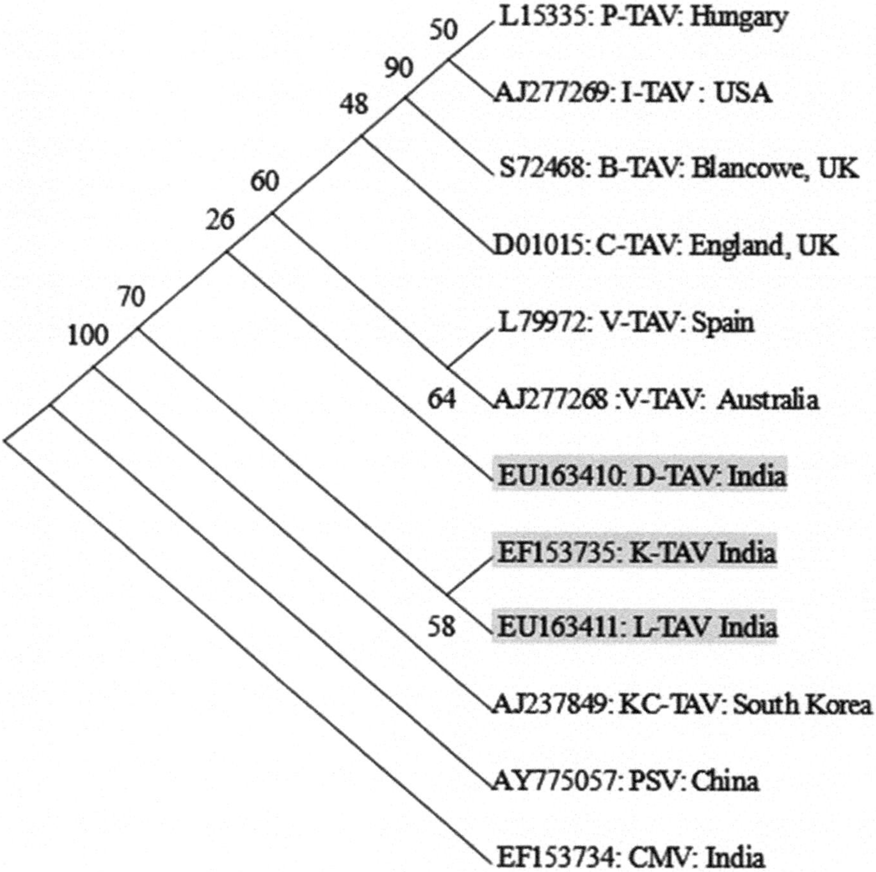

Fig. 13.6 Complete RNA 3 gene-based phylogenetic analysis of Indian isolates of TAV (in gray) by MEGA v 4.1. Dendrogram was generated using neighbor-joining method with 100 replicate bootstraps (Susheel 2009)

13.2.3 *Tomato Spotted Wilt Virus* (*Tospovirus*)

Bezerra et al. (Bezerra et al. 1999) reported the increase of tospoviral diversity in Brazil with the identification of two new *Tospovirus* species, one from chrysanthemum and one from zucchini. During a survey conducted in several different regions of Brazil, two unique tospoviruses were isolated and characterized, one from chrysanthemum and the other from zucchini. The chrysanthemum virus displayed a broad host range, whereas the virus from zucchini was restricted mainly to the family Cucurbitaceae. Double-antibody sandwich enzyme-linked immunosorbent assay and western immunoblot analyses demonstrated that both viruses were serologically distinct from all reported tospovirus species including the recently proposed peanut yellow spot virus and iris yellow spot virus (IYSV) species.

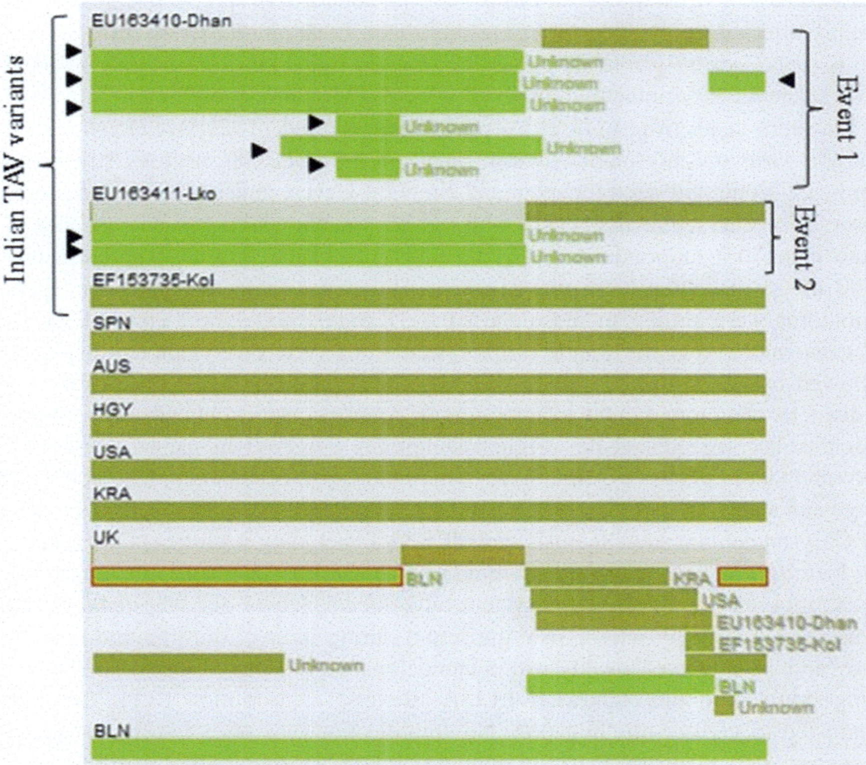

Fig. 13.7 Analysis of recombination events using RDP v1.082 showing two independent recombination events, identified by GENECONV method (Susheel 2009)

The nucleotide sequences of the nucleocapsid (N) genes of both viruses contain 780 nucleotides encoding for deduced proteins of 260 amino acids. The N proteins of these two viruses displayed amino acid sequence similarities with the previously described tospovirus species ranging from 20% to 75%, but they were more closely related to each other (80%). Based on the biological and molecular features, these viruses are proposed as two new tospovirus species, designated as chrysanthemum stem necrosis virus (CSNV) and zucchini lethal chlorosis virus (ZLCV). With the identification of CSNV and ZLCV, in addition to tomato spotted wilt virus, groundnut ringspot virus, tomato chlorotic spot virus, and IYSV, Brazil harbors the broadest spectrum of tospovirus species reported (Bezerra et al. 1999).

Matsuura et al. (Matsuura et al. 2002) reported the occurrence of *Tomato spotted wilt virus* in chrysanthemum fields. DAS-ELISA proved to be reliable enough to detect a latent infection by *Tomato spotted wilt virus* (TSWV) in asymptomatic stock plants of chrysanthemum. A high density of thrips (*Frankliniella occidentalis*), the predominant vector, in the presence of latently infected stock plants resulted in a high incidence of disease in the chrysanthemum production field. The incidence of disease was low when the vector thrips were not abundant in spite of the presence of latently infected stock plants. These results suggest that an infestation of the vector thrips causes severe secondary spread of TSWV originating from latently infected stock plants in chrysanthemum production fields (Matsuura et al. 2002).

Matsuura et al. (Matsuura et al. 2007) first reported of *Chrysanthemum stem necrosis virus* on chrysanthemums in Japan. In August 2006, necrotic streaks on stems, chlorotic and necrotic spots and rings on leaves, and leaf distortions were observed on chrysanthemum cvs. Jimba and Seinotama, with a disease incidence of more than 70% (approximately 30,000 plants), which represents approximately 1000 m^2 of greenhouses of one grower in Hiroshima Prefecture, western Japan. Symptoms were similar to those caused by *Tomato spotted wilt virus* (TSWV). *F. occidentalis* was the major thrips species observed on symptomatic plants, followed by *F. intonsa*. Tospovirus-like spherical particles that were 80–100 nm in diameter were found in the infected leaves. After mechanical inoculation, a single lesion isolate reproduced the original symptoms observed in nature on healthy chrysanthemum plants (cv. Jimba). As determined by mechanical inoculation, host range and symptomatology of the isolate were similar to those described previously for Chrysanthemum stem necrosis virus (CSNV), including necrotic spots on *Petunia hybrida*. The isolate caused stunting, severe necrotic lesions on stems, necrotic spots, rings, and vein necrosis on systemically infected leaves of *L. esculentum*. The virus reacted strongly with CSNV antiserum during indirect dot immuno-binding assay and cross-reacted weakly with a monoclonal antibody to N protein of TSWV using double-antibody sandwich ELISA. Reverse transcription (RT)-PCR was conducted to verify virus infection. No amplification was observed from extracts of symptomatic plants ($n = 10$) by multiplex RT-PCR using TSWV and Impatiens necrotic spot virus-specific primer sets, indicating that the diseased chrysanthemums were not doubly infected with these viruses. However, a DNA fragment of approximately 450 bp was amplified in samples by RT-PCR using tospovirus universal primers. The nucleotide sequence of the amplified fragment had 98.1% identity with the corresponding region of the CSNV nucleocapsid protein gene (GenBank accession no. AF067068). The results indicated that the virus associated with a stem necrosis disease of chrysanthemums in Hiroshima was an isolate of CSNV (Matsuura et al. 2007).

Kondo et al. (Kondo et al. 2011) first reported *Impatiens necrotic spot virus* (*Tospovirus*) infecting chrysanthemum (*Chrysanthemum morifolium*) in Japan. During a survey in 2009, chlorotic mottle and necrosis were observed on chrysanthemums (cv. Jimba) in Aomori Prefecture, Japan. A virus was isolated from the chrysanthemum plants by serial local-lesion transfer. The symptoms exhibited by the test plants, the particle morphology, the features of the protein, and the potential for transmission by thrips were similar to those for *Impatiens necrotic spot virus* (INSV). The partial nucleotide sequences of the nucleocapsid protein gene and the 3′-untranslated sequence of the S RNA shared 99% identity with that of an INSV isolate. This was the first report of INSV infection of chrysanthemums in Japan (Kondo et al. 2011).

Dullemans et al. (Dullemans et al. 2015) investigated the complete nucleotide sequence of *Chrysanthemum stem necrosis virus*. The complete genome sequence of chrysanthemum stem necrosis virus (CSNV) was determined using Roche 454 next-generation sequencing. CSNV is a tentative member of the genus *Tospovirus* within the family *Bunyaviridae*, whose members are arthropod-borne. This is the first report of the entire RNA genome sequence of a CSNV isolate. The large RNA of CSNV is 8955 nucleotides (nt) in size and contains a single open reading frame of 8625 nt in the antisense arrangement, coding for the putative RNA-dependent RNA polymerase (L protein) of 2874 aa with a predicted Mr of 331 kDa. Two untranslated regions of 397 and 33 nt are present at the 5′ and 3′ termini, respectively. The medium (M) and small (S) RNAs are 4830 and 2947 nt in size, respectively, and show 99% identity to the corresponding genomic segments of previously partially characterized CSNV genomes. Protein sequences for the precursor of the Gn/Gc proteins, N and NSs, are identical in length in all of the analyzed CSNV isolates (Dullemans et al. 2015).

13.2.4 *Arabis Mosaic Virus* (*Nepovirus*)

Arabis mosaic virus (ArMV), a *Nepovirus* infection in plants of *Chrysanthemum* spp. was reported from Iran (Ghotbi and Shahraeen 2005). The virus-infected chrysanthemum plants exhibit symptoms of mosaic, leaf chlorosis, small necrotic lesions, and leaf malformation and deformation. This virus is transmitted by nematode vectors (*Xiphinema* spp.) in a nonpersistent manner, and also by mechanical sap inoculation. Viral-like symptoms were observed on ornamental plants including Chrysanthemum sp. in the main cultivation region in Iran. Samples of different ornamentals from glasshouses, mainly in Markazi and Tehran provinces, with symptoms of mosaic, leaf chlorosis, small necrotic lesions, and leaf malformation and deformation were collected. Arabis mosaic virus (ArMV) infections of samples and that of indicator test plants were confirmed by DAS-ELISA and tissue blotting immunoassay with ArMV-specific polyclonal antibody. Results confirmed that samples of Chrysanthemum sp. from Tehran area were infected with ArMV. This was thought to be the first report on the occurrence of ArMV on Chrysanthemum sp. from Iran (Ghotbi and Shahraeen 2005).

13.2.5 *Chrysanthemum Virus B* (Carlaviruses)

Verma et al. in 2003 (Verma et al. 2003) performed the detection and identification of and studied the incidence of *Chrysanthemum B carlavirus* in chrysanthemum in India. During the study, symptoms were observed on a variety of chrysanthemum grown in Himachal Pradesh (India). The disease incidence ranged between 40.62% and 94.66%. Out of the 36 cvs. surveyed, seven were found to be negative for CBV in DAS-ELISA. The virus has a very narrow host range infecting only *N. clevelandii*, *N. glutinosa*, *N. rustica*, *Petunia hybrida*, and *Vicia faba*. In cytopathology, infected cells showed absence of specific inclusion body(ies) and presence of virus particles in large numbers along the outer envelope of chloroplast with abnormalities like electron-dense matrix, loss of chloroplast envelope, and extensive invigilation of cytoplasm. In immune electron microscopy, there was heavy decoration and trapping of virus particle with the antiserum specific to CBV. The particle length was ca. 680 × 12 nm. The causal virus was found to be *Chrysanthemum B carlavirus* on the basis of host range, DAS-ELISA, immune electron microscopy, and cytopathology. This may have practical benefits for producing virus-free stocks of chrysanthemum for growers (Verma et al. 2003).

Lin et al., in 2005 (Lin et al. 2005), reported the occurrence of *Chrysanthemum virus B* in Taiwan and preparation of its antibody against coat protein expressed in bacteria. In this study, they conducted an extensive survey from the year 2000–2005, for the occurrence of CVB in different chrysanthemum plantations and cutting producing nurseries in Changhua, Taiwan. Among 504 chrysanthemum leaf samples collected from nine different locations, 394 of them were detected with CVB infection in indirect ELISA. This was the first evidence showing the existence of CVB in chrysanthemum in Taiwan. To characterize CVB isolates from Taiwan, their CP gene was amplified by reverse transcriptase polymerase chain reaction (RT-PCR) using a set of primer (CVBup, CVB-dw). A DNA product of 1028 bp was amplified, cloned, and sequenced. It was found to contain an open reading frame with 316 amino acid residues corresponding to the size of reported CVB CPs, and it shared 78.4–88.0% and 83.1–93.7% identities in nucleotide and amino acid sequence, respectively, with 20 known CVB CP gene sequences (Lin et al. 2005). In order to produce specific antiserum against CVB for its detection, we expressed the cloned CVB CP gene in bacteria culture and used bacteria-expressed CP as immunogen for antiserum preparation. An antiserum (#107) was prepared against the expressed CVB CP and shown to be useful in ELISA to detect CVB in chrysanthemum plants. By comparing it with a commercialized CVB antiserum (Agdia Inc., USA), the reactivity in terms of ELISA readings of antiserum #107 to the same dilution of infected chrysanthemum tissue was always higher than that of Agdia's CVB antiserum (Lin et al. 2005).

Further, Singh et al. (Singh et al. 2011) studied the intermolecular interactions of *Chrysanthemum virus B* coat protein and implications for capsid assembly. They described that *Chrysanthemum virus* B coat protein constitutes the viral capsid which, besides other functions, encapsulates and protects the viral nucleic acid. They demonstrated homotypic interaction of the coat protein subunits, essentially important for dimer formation, which is the first step during capsid assembly in vivo and in vitro. Interaction capacity of full-length and truncated protein had been investigated, and important regions had been identified through protein-protein interaction in yeast and by co-immunoprecipitation assays. Complete coat protein was found to interact strongly with similar subunits. Constructs with 102 amino acids from the N-terminal and 64 amino acids from the C-terminal were found to be inconsequential for dimer formation as they did not show any interaction with similar subunits or with full-length protein when analyzed for β-galactosidase or histidine prototrophy. Results suggest that the region of 98–184 amino acids from the middle plays an important role in the process, probably without the involvement of N- and C-terminals (Singh et al. 2011).

13.2.6 *Potyviruses* Infecting Chrysanthemums

Bertaccini et al. in 1994 (Bertaccini et al. 1994) reported a *Potyvirus* infecting *Chrysanthemum frutescens*. In this work, symptomatology of probable viral origin was observed in some Ligurian cultivation of *Chrysanthemum frutescens*. The flowers and the growth of the affected plants were normal, but the presence of marked chlorotic spots on the leaves reduced the commercial value of the cut flowers. Samples of leaves from symptomatic and asymptomatic *C. frutescens* were used in mechanical inoculation on herbaceous hosts, leaf dip preparation, ultrastructural observations of embedded tissues, and immune electron microscopy (IEM) using the "decoration" tests. Of the 15 species inoculated belonging to six families, only *Chenopodium amaranticolor*, *C. quinoa*, and *C. frutescens* are infected with a filamentous virus (*Chrysanthemum spot virus*, ChSV) which, in leaf dip preparations, was ~750 nm. The observations of ultrathin sections of both *C. amaranticolor* and *C. frutescens* revealed the presence of cylindrical inclusions (pinwheels, tubes, and scrolls) together with filamentous viruslike particles in the cytoplasm of parenchymatic cells. The IEM tests were carried out on crude sap of *C. amaranticolor*-infected leaves using 22 antisera belonging to the potyvirus group. None of these were completely homologous to ChSV; the sera to Bidens mosaic virus (BMV), potato virus Y (PVY), Iris severe mosaic virus (ISMV), and carnation vein mottle virus (CVMV) gave a partial decoration of the viral particles (Bertaccini et al. 1994).

Chen et al., in 1999 (Chen et al. 1999), studied the host reactions, cytological characteristics, and serological properties of a *Turnip mosaic virus* isolate causing albinic mosaic disease of garland chrysanthemum. Chen et al., in 2000 (Chen et al. 2000), attempted characterization of the coat protein gene of a *Turnip mosaic virus* isolate infecting garland chrysanthemum. In this study, the causal agent that induced albinic mosaic disease of garland chrysanthemum (*Chrysanthemum coronarium*) was identified as an isolate of turnip mosaic virus (TuMV) and designated as TuMV-GC. Previous studies indicated that host reactions of TuMV-GC are similar to those of TuMV-C5 strain, and it is indistinguishable from other TuMV strains from Taiwan by cytological characteristics and serological properties. In order to further characterize the virus at the molecular level, the coat protein (CP) gene of this isolate and Taiwanese isolates of TuMV-C4 and TuMV-C5 strains were cloned and sequenced. Comparison the CP genes of the three isolates showed that they share nucleotide identities of 97.92–99.88% and amino acid identities of 98.61–100%. The 3′ noncoding regions of the three isolates have nucleotide identities of 98.56–99.52%. Furthermore, CP genes of TuMV-GC and TuMV-C5 isolates share an amino acid identity of 100% and the 3′ non-coding regions a nucleotide identity of 99.52%. This molecular evidence substantiates that TuMV-GC is a variant of TuMV-C5 strain. Further comparison with other reported TuMV isolates from different areas of the world revealed that the CP genes of the TuMV-GC and TuMV-C5 isolates from Taiwan share higher amino acid identities of 98.96–99.65% with the MV, MUT, and RN strains from the UK, and the TuMV-C4 isolate shares the highest identity of 99.31% with the NIAP strain from Korea. In addition, the 3′ noncoding regions of the three TuMV isolates from Taiwan share higher nucleotide identities of 98.56–99.04% with the CAPPI strain from Canada. These results indicated that the degrees of the homology of the CP genes were probably related to the major differences in host specificity, but not to geographic distribution. The nucleotide sequences reported were deposited under GenBank accession numbers AF233888–AF233890 (Chen et al. 2000).

Mehra et al., in 2009 (Mehra et al. 2009), reported a new chrysanthemum potyvirus based on molecular evidence. In the study, a previously unknown potyvirus infecting chrysanthemum plants was detected using techniques such as ELISA, RT-PCR, and hybridization. The ELISA-positive samples were amplified using a potyvirus group-specific primer which gave an amplification of ~850 bp. The amplified product was cloned and sequenced and showed 72–73% homology with known potyviruses that infect chrysanthemums such as *Potato virus Y*, *Soybean mosaic virus*, and *Turnip mosaic virus* when compared to the sequence available in the database. However, the present potyvirus isolates show 93% homology with *Chilli veinal mottle virus* and *Pepper vein banding virus*. The results were further confirmed by Northern hybridization. This was the first report of a potyvirus similar to *Chilli veinal mottle virus* and *Pepper vein banding virus* infecting chrysanthemums (Mehra et al. 2009).

The first report of *Bidens mottle potyvirus* causing mosaic and leaf deformation in garland chrysanthemum and lettuce in Taiwan was investigated by Chen and Lee (2012). Symptoms of mosaic and deformation in leaves and stunting of plants have been observed in chrysanthemum and lettuce, respectively. Filamentous virus particles (approximately 780 × 13 nm) in the crude sap and pinwheel inclusions in infected cells were observed in the preparations of both diseased hosts with electron microscopy. Virus cultures were isolated and inoculated to their original host individually, and all caused symptoms similar to that observed in the field, indicating their pathogenicity. A cDNA fragment consisting of partial nuclear inclusion (NIb) and coat protein (CP) genes were amplified with potyvirus degenerate primers. The amplified fragments were cloned and sequenced, and the combined cDNA sequences were deposited under GenBank accession numbers AB491763 (chrysanthemum) and AB491764 (lettuce). Sequence analysis showed that both cloned sequences shared more than 97% nucleotide similarity to that of BiMoV. The amino acid sequence of the CP of both isolates shared a 99.3% identity and a 98.9–99.3% identity to that of other BiMoV isolates (Chen and Lee 2012).

Liu et al., in 2014 (Liu et al. 2014), reported *Potato virus Y* strain N-Wilga infecting chrysanthemum for the first time in China. The typical viral symptoms were observed in field-grown chrysanthemum with leaf yellowing and mottled leaves in Wenjiang District, Sichuan Province, of China. The incidence of these symptoms in the field was 12.3%. Sixteen symptomatic leaves and 12 non-symptomatic leaves were collected and tested for *Potato virus Y* (PVY) presence using commercial PVY-specific DAS-ELISA kits. Six samples were found positive for PVY. RT-PCR tests using potyvirus-specific primers for CP gene amplified a single, expected 218 bp fragment from chrysanthemum extracts from all six samples positive for PVY in ELISA. These six PCR fragments were sequenced and found 100% identical to each other. The sequence (GenBank accession no. KJ174515) shared 99% identity with corresponding sequences of several PVY isolates. The same six positive samples were subjected to a multiplex RT-PCR assay to identify the PVY strain type, and all six PVY samples from Sichuan were found to belong to the PVY$^{\text{N-Wi}}$ strain (Liu et al. 2014).

Niu et al., in 2015 (Niu et al. 2015), reported *Zucchini yellow mosaic virus* in chrysanthemum for the first time from China. In July 2014, viruslike symptoms such as foliar yellowing and mottle were observed on chrysanthemum plants on farms in Wenxi, Shanxi Province, China. The incidence of this disease was more than 65%. Leaf samples were collected from eight symptomatic plants in a suburban farm of Wenxi for further analysis. These samples were first tested by enzyme-linked immunosorbent assay (ELISA) using monoclonal antibodies specific for *Tobacco mosaic virus*, *Cucumber mosaic virus*, *Chrysanthemum virus B*, and *Zucchini yellow mosaic virus* (ZYMV). All eight samples were found only positive for ZYMV. To further verify the presence of this virus, double-stranded RNAs (dsRNA) were extracted from ELISA-positive samples and used as templates in reverse transcription using degenerate primer M4-T. The cDNAs were then used in PCR using a pair of ZYMV-specific primers. A 971 bp DNA fragment was amplified and cloned into pUCm-T Vector. Six clones were sequenced, and their nucleotide sequences were 100% identical. The sequence (KP742962) had 99.2% identity with a ZYMV isolate (AY597207), confirming the infection of ZYMV in the diseased chrysanthemum plants. This was the first report anywhere of ZYMV infecting chrysanthemum (Niu et al. 2015).

13.2.7 *Clerodendron Yellow Mosaic Virus* and *Tomato Leaf Curl Virus* (Begomoviruses)

Marwal et al., in 2013 (Marwal et al. 2013), first reported the association of a begomovirus with yellowing of leaf vein disease of *Chrysanthemum indicum* in India. Infected leaf samples of *C. indicum* showing yellowing of leaf veins were collected from gardens of New Delhi (India). An expected PCR product of size ~500 bp was amplified from total DNA extracts of symptomatic leaf samples with universal primers on the gene of coat protein region of begomovirus DNA-A component. The presence of begomoviruses was also confirmed by Southern blot analysis using control cloned DNA-A probe of *Cotton leaf curl virus*. Sequence analysis of the virus infecting *Chrysanthemum indicum* showed 99% nucleotide sequence identity with *Clerodendron yellow mosaic virus* (Marwal et al. 2013).

Ashwathappa et al., in 2020 (Ashwathappa et al. 2020), reported association of *Tomato leaf curl New Delhi virus* with mosaic and leaf curl disease of *Chrysanthemum*. During the study, the leaf samples of *chrysanthemum* plants showing typical begomovirus symptoms of mosaic, mottling, and downward leaf curl were collected from farmer's field in Karnataka state of India. The *begomovirus* infection was confirmed by PCR using specific primers in 22 infected chrysanthemum samples, and no amplification was observed in 22 healthy chrysanthemum samples. The partial begomovirus genome (1.2 kb) sequencing indicated 22 infected chrysanthemum samples were infected with begomovirus (*Tomato leaf curl New Delhi virus*, ToLCNDV). Therefore, one representative sample was selected for complete genome amplification using RCA method. The complete genome amplified was cloned and sequenced. The Species Demarcation Tool (SDT) analysis of full genome of ToLCNDV showed maximum nucleotide (nt) identity of 89.2–97.6% (DNA A) and 87.7–96.3% (DNA B) with ToLCNDV infecting cucurbits. Based on begomovirus species demarcation criteria (91% nucleotide sequence identity), the virus infecting chrysanthemum is considered as a strain of ToLCNDV. The betasatellite amplified from the begomovirus-infected chrysanthemum showed maximum nt identity of 89% with Tomato leaf curl Bangladesh betasatellite (ToLCBDB) infecting papaya in India. Further, break point recombination analysis showed that genome and betasatellite of ToLCNDV infecting chrysanthemum are recombinant with already known begomoviruses. The whitefly cryptic species predominantly present in the begomovirus-infected chrysanthemum fields was identified as Asia-II-5 group (Ashwathappa et al. 2020).

13.3 Management of Viruses Infecting Chrysanthemums

The management of viruses and viroids in chrysanthemums has been suggested as follows:

(a) since no chemical treatment is available to eliminate a virus or viroid from an infected plant, the disease control can only be achieved by use of virus-tested planting stock in conjunction with strict hygienic practices, (b) the movement of insects, such as thrips and aphids, should be controlled as these can vector virus diseases, (c) propagating tools and knives must be sterilized after use, and (d) washing of hands with hot soapy water before handling chrysanthemum plants helps safeguard against virus diseases that are transmitted by contact.

Conventional and nonconventional methods are used for a long time for management of viruses in many crops. These are discussed as follows:

13.3.1 Sanitation of Cropping Field and Planting Practices

Cleaning and eradication of infected plant material together with the potential reservoir of cucumovirus from the fields was found to be a very effective method. Weeds have been identified as key sources of viral infections for a long time. Eradication of perennial weeds from around greenhouses, gardens, and fields to eliminate possible sources of virus may therefore prove helpful (Agrios 1978). For control of CMV, weed hosts harboring cultivated fields have been found successful in reducing the incidences of virus in crop fields of cucumber and celery (Rist and Lorbeer 1989; Raj et al. 2008). Along with this, the practices such as early plantation and plant spacing use of silver- or white-colored mulches were found effective in reducing disease incidence and obtaining maximum yield (Pratap et al. 2006).

13.3.2 Spray of Insecticide for Virus Vector Control

Application of insecticides, viz., Malathion and Rogor (0.1%) solution either by spraying or drenching soil, will effectively reduce the virus burden on commercial crops (Khan et al. 2006). The regular use of insecticide should not be in practice because it adversely affects the environment and may diminish the quality of crops. Three sprayings of Malathion insecticide (0.2%) at every 21 days of intervals were found to be effective for management and minimizing the population of both insect- and vector-borne diseases (Khan et al. 2006).

13.3.3 Biological Control of Virus Transmitting Vectors

Biological control of aphids (the virus vector) capable of transmitting various types of viruses, especially the cucumo-(CMV and TAV) and potyviruses (BYMV) in several plant species, has been attempted. The feeding behavior of ladybird (*Coccinella transversalis*) predator of green and black aphids has been observed on chrysanthemum plants. Different larval stages as well as adult ladybird predator have been exploited for minimizing the aphid population (Raj et al. 2008).

13.3.4 Reliable and Sensitive Virus Diagnostics for Virus Indexing

To date, highly reliable and sensitive diagnostic techniques are available to accomplish this need. To control the virus infection in chrysanthemum cultivation, it is essential to ensure that the source plants used for propagation are virus-free (Raj et al. 2008). In a study, ELISA protocol was standardized for the detection of CMV and TAV. ELISA using CMV and TAV antibodies could detect CMV and TAV in leaf samples of various cultivars of chrysanthemum. The 1:500 and 1:1000 dilutions of antigen and antibodies respectively, were found ideal for optimum ELISA reaction for both TAV and CMV (Susheel 2009).

Further, the RT-PCR was standardized for the sensitive detection of CMV and TAV. The primers of the CP region of CMV and TAV were chosen for general diagnosis of both viruses in chrysanthemums. Optimization of RT-PCR conditions for diagnosis of CMV and TAV in chrysanthemums was done by considering the PCR conditions and using CP primers of CMV and TAV with variable $MgCl_2$ concentrations (1–5 mM) and different annealing temperatures (52–64 °C). The conditions 3.0 mM concentration of $MgCl_2$ and 58 °C annealing temperature were found optimum for TAV, whereas the 52 °C annealing temperature with 3.0 mM concentration of $MgCl_2$ was found to be ideal for CMV. Therefore, all the RT-PCR reactions were performed with optimum conditions for CMV and TAV for indexing of chrysanthemum cultivars/samples (Susheel 2009).

In this study, the detection limit of PCR using TAV-CP primers at different dilutions of cDNA template, 3, 2, 1, 0.1, 0.01, 0.001, and 0.0001μL, was also optimized. The 3.0μL cDNA was equivalent to RNA extracted from 2.25 mg leaf tissue. The RNA extracted from 100 mg tissue, suspended in 40μL water and 6μL, was used for cDNA synthesis. Detection limit by RT-PCR was obtained at 0.001μL which is equivalent to 0.00075 mg leaf tissue. The detection limit for CMV was similar as in case of TAV. The presence of CMV and TAV was successfully detected by the standardized TR-PCR in infected chrysanthemum plants and aphids. These tests developed were useful for virus indexing of chrysanthemum cultivars as well as the virus diagnosis in aphid vectors for epidemiological studies of the CMV and TAV. RT-PCR was found more sensitive than ELISA; however, ELISA has been considered as a feasible methodology for routine screening of CMV and TAV or for indexing of bulk material of chrysanthemums (Susheel 2009).

13.3.5 In Vitro Regeneration and Production of Virus-Free Chrysanthemums

Plant biotechnology offers an opportunity to develop new germplasms, conservation and development of variable traits in chrysanthemum, and regeneration of virus-free chrysanthemum plants. Datta et al. in 2011 (Datta et al. 2001) established a protocol using direct shoot regeneration system from ray florets of 28 genotypes in which regeneration frequency and average number of shoots per explant varied among the cultivars, among Japanese chrysanthemum (Datta et al. 2001).

In vitro production of TAV-free (Singh and Gupta 1978), TSWV-free (Balukiewicz and Kryczynski 2001), and CMV-free (Verma et al. 2004) chrysanthemums has been successfully achieved by shoot meristem culture alone or in combination with chemo- and/or thermotherapy. Verma et al. (Verma et al. 2004) eliminated CMV from chrysanthemum plants using meristem tip culture method. MS medium amended with BAP (2 mg/L) and IBA (0.05 mg/L) was used for shoot proliferation and IBA (0.05 mg/L) for rooting of the plants. Virus-free plants (84%) were obtained from the optimum size (0.3 mm) of meristem tips (as indexed by DAS-ELISA). Of these, 72% plants were found negative for CMV as indexed by RT-PCR. Virus indexing by RT-PCR was found a reliable method (Verma et al. 2004).

Kumar et al., in 2009 (Kumar et al. 2009), attempted the elimination of mixed infection of *Cucumber mosaic* virus (CMV) and *Tomato aspermy virus* (TAV) from *Chrysanthemum morifolium* Ramat. cv. Pooja by shoot meristem culture. In this study, mixed infection of CMV and TAV was detected in *C. morifolium* cv. Pooja plants by direct antigen coating enzyme-linked immunosorbent assay (DAC-ELISA) and reverse transcription polymerase chain reaction (RT-PCR). The elimination of CMV and TAV was achieved by in vitro culturing 0.3 mm long shoot meristem of infected plants on MS medium supplemented with 3.0 mg/L BAP and 0.5 mg/L NAA. The regenerated plants were indexed by DAC-ELISA and confirmed by RT-PCR. A total of 78.1% CMV and TAV-free shootlets were obtained from the regenerated shoot meristem as indexed by DAC-ELISA, of which only 65.6% were found truly virus-free when confirmed by RT-PCR. Virus-free shootlets were rooted on half MS medium and acclimatized under glasshouse. These plants showed better growth and quality of blooms as compared to diseased ones (Kumar et al. 2009).

13.3.6 Genetic Transformation of Chrysanthemum for Virus Resistance

Kumar et al., in 2012 (Kumar et al. 2012), attempted the genetic transformation and development of *Cucumber mosaic virus*-resistant transgenic plants of *Chrysanthemum morifolium* cv. Kundan. In this study, *Chrysanthemum morifolium* cv. Kundan had been found susceptible to infections of CMV which drastically affects the quality and quantity of blooms and posed significant constraints in commercial cultivation of chrysanthemum in India. Therefore, development of inbuilt resistance in *C. morifolium* against CMV seemed to be essential. *Agrobacterium*-mediated transformation of petiole explants of *C. morifolium* was attempted using pRoK2 binary vector harboring coat protein (CP) gene of CMV under the control of CaMV 35S promoter.

A total of 257 explants were transformed and 73 putative transgenic plants from seven independent co-cultivation events were obtained with 6% transformation efficiency. Molecular analysis of these plants confirmed the successful integration of CP transgene in 63% plants, of which 12.3% plants were able to transcript and translate the transgene. Expression of coat protein did not evoke any abnormal phenotype. The regenerated chrysanthemum cv. Kundan plants showed similar plant morphology, leaf size, growth, and color of blooms as shown by mother Kundan plants. Transgenic plants showed delayed resistance when challenged by CMV-chrysanthemum strain which produced good quality blooms as compared to the susceptible ones (Kumar et al. 2012).

The transgenic chrysanthemum plants containing genes of interest have been successfully obtained for TSWV resistance (Urban et al. 1994; Yepes et al. 1995), while there is a single report of utilization of virus nucleocapsid gene of TWSV by Urban et al. (1994) and developed a high-frequency transformation protocol for commercial chrysanthemum cultivar Iridon. Although no report is available in literature showing the use of any gene of CMV strain or its gene fragments for chrysanthemum transformation, the present study would also for the first time use nucleocapsid gene of CMV.

13.4 Conclusion

The study will delineate about the frequency of virus infection in ornamental chrysanthemums and their biological, serological, and molecular characterization. It also enlightens the molecular constitution and phylogenetic relationships of the present isolates with the cucumoviruses reported from India and abroad which provides better understanding of the taxonomy of virus(es). The work will also delineate how the present CMV and TAV isolates have relationships to the indigenous strains of local geographic region of the same agroclimatic conditions. Beside this, the work also explores the regeneration and transformation methodologies evaluated for a particular chrysanthemum crop, in terms of high-frequency and disease (virus)-free chrysanthemum regeneration and transformation of chrysanthemum for novel traits.

The use of disease-resistant cultivars is common for controlling plant viruses in many crop species. By using genetic engineering technology, the application of pathogen-derived resistance (PDR) in improving host resistance overcomes the source limitation of natural resistance to viral infections and transforms the novel transgene constructs to improve the phenotype. Though successful to limit the crop losses, this has not always led to virus-free propagation of plants as they are tolerant rather than resistant to viral diseases. Hence, novel approaches that are generally applicable in breeding for resistance to viruses in ornamental plants are the necessity of the time. To date, many pathogen-derived resistance approaches have been successfully developed for many crop plants; however, in India, no such study ever before has been done in transforming the chrysanthemum, either for disease resistance or for phenotype improvement traits.

Such a study would be of practical value that will help in extending the genetically engineered resistance to plant viruses infecting chrysanthemums, an important plant for floricultural trade, and also to other economical crops grown in the country.

References

Agrios GN (1978) Plant pathology, 2nd edn. Academic Press, Inc., San Diego, CA, pp 466–470

Ashwathappa KV, Venkataravanappa V, Reddy CNL, Reddy MK (2020) Association of Tomato leaf curl New Delhi virus with mosaic and leaf curl disease of Chrysanthemum and its whitefly cryptic species. Indian Phytopathol 73:533–542

Balukiewicz A, Kryczynski S (2001) Attempts to eliminate TWSV from Chrysanthemum plants by meristem tip culture. Phytopathol Pol 21:101–108

Bertaccini A, Bellardi MG, Marani F, Rabiti A (1994) A Potyvirus infecting *Chrysanthemum frutescens*. Acta Hortic 377:107–114

Bezerra IC, de Resende RO, Pozzer L, Nagata T, Kormelink R, de Avila AC (1999) Increase of tospoviral diversity in Brazil with the identification of two new *Tospovirus* species, one from chrysanthemum and one from zucchini. Phytopathology 89:823–830

Bouwen I, van Zaayen AZ (1995) Chrysanthemum. In: Loebenstein G, Lawwon RH, Brunt AA (eds) Virus and virus-like diseases of bulb and flower crops. John Wiley and Sons, New York, NY, pp 396–408

Brierley P (1955) Symptoms induced in chrysanthemum on inoculation with the viruses of mosaic, aspermy and flower distortion. Phytopathology 45:2–7

Brierley P (1958) Tomato aspermy in chrysanthemum from Asia and Europe. Plant Dis Rep 42:61–62

Chen YK, Lee JY (2012) First report of *Bidens mottle potyvirus* causing mosaic and leaf deformation in garland chrysanthemum and lettuce in Taiwan. Plant Dis 96:464

Chen TM, Chen YK, Chen MJ, Yeh SD (1999) Host reactions, cytological characteristics and serological properties of a *Turnip mosaic virus* isolate causing albinic mosaic disease of garland chrysanthemum. Plant Pathol Bull 8:73–82

Chen TM, Chen MJ, Yeh SD (2000) Characterization of the coat protein gene of a *Turnip mosaic virus* isolate infecting garland chrysanthemum. Plant Protect Bull Taipei 42:83–96

Choi SK, Choi JK, Park WM, Ryn KH (1999) RT-PCR detection and identification of three species of Cucumoviruses with a genus-specific single pair of primers. J Virol Methods 83:67–73

Choi SK, Choi JK, Ryn KH, Park WM (2002) generation of infectious cDNA clones of a Korean strain of *Tomato aspermy virus*. Mol Cell 13:52–60

Chuyan AKH, Krylov AV (1979) Properties of *Tomato aspermy virus* from chrysanthemum and its host range in the Priomore region. Byulleten Glamogo Botaniches Kogo Soda 114:84–92

Datta SK, Chakrabarti D, Saxena M, Mandal AKA, Biswas AKA (2001) Direct shoot regeneration from florets of chrysanthemum cultivars. Indian J Gent Plant Breed 61:373–376

Dullemans AM, Verhoeven JTJ, Kormelink R, van der Vlugt RAA (2015) The complete nucleotide sequence of *Chrysanthemum stem necrosis virus*. Arch Virol 160:605–608

Ghotbi T, Shahraeen N (2005) First report on incidence of *Arabis mosaic virus* (ArMV, Nepovirus) on ornamental plants in Iran. Iranian J Plant Pathol 41:305–306

Gupta RP, Singh BP (1981) Studies on Chrysanthemum aspermy virus on Chrysanthemum. Indian J Mycol Pathol 11:158–160

Habili N, Francki RIB (1975) Comparative studies on tomato aspermy and Cucumber mosaic virus. IV. Immunogenic and serological properties. Virology 64:421

Harrison BD, Finch JT, Gibbs AJ, Hollongs M, Shepherd RJ, Valenta V, Wetter C (1971) Sixteen groups of plant viruses. Virology 45:356–363

Hollings M (1955) Investigation of Chrysanthemum viruses I. Aspermy flower deformation. Ann Appl Biol 43:83–102

Khan MS, Raj SK, Bano T, Garg VK (2006) Incidence and management of mosaic and leaf curl disease in cultivars of chilli (*Capsicum annuum*). J Food Agric Environ 44:99–102

Kondo T, Yamashita K, Sugiyama S (2011) First report of *Impatiens necrotic spot virus* infecting chrysanthemum (*Chrysanthemum morifolium*) in Japan. J Gen Plant Pathol 77(4):263–265

Kumar S, Srivastava A, Raj SK (2005) Molecular detection of Cucumber mosaic virus in Chrysanthemum cultivars. Acta Phytopathol Entomol Hungarica 40(3):205–211

Kumar S, Khan MS, Raj SK, Sharma AK (2009) Elimination of mixed infection of Cucumber mosaic and Tomato aspermy virus from Chrysanthemum morifolium Ramat. cv. Pooja by shoot meristem culture. Sci Hortic 119:108–112

Kumar S, Raj SK, Sharma AK, Varma HN (2012) Genetic transformation and development of *Cucumber mosaic virus* resistant transgenic plants of *Chrysanthemum morifolium* cv. Kundan. Sci Hortic 134:4045

Kumar S, Raj SK, Verma HN (2016) Cucumoviruses causing diseases in chrysanthemum and their management. Scholar's Press, Omni Scriptum, Saarbrücken. ISBN 978-3-659-84149-1

Lin MJ, Chang CA, Chen CC, Cheng YH (2004) Molecular and serological detection of *Tomato aspermy virus* infecting chrysanthemum in Taiwan. Plant Pathol Bull 13:291–208

Lin MJ, Chang CA, Chen CC, Cheng YH (2005) Occurrence of *Chrysanthemum virus B* in Taiwan and preparation of its antibody against coat protein expressed in bacteria. Plant Pathol Bull 14:191–202

Liu XL, Wei Q, Hong B, Zhao XT (2014) First report of *Potato virus Y* strain N-wilga infecting chrysanthemum in China. Plant Dis 98:1589

Marwal A, Sahu AK, Kgaur R (2013) First report on the association of a begomovirus with chrysanthemum indicum exhibiting yellowing of leaf Vein disease characterized by molecular studies. J Hortic Res 21(2):17–21
Matsuura S, Hoshino S, Hayashi H, Kohguchi T, Hagiwara K, Omura T (2002) Effects of latent infection of stock plants and abundance of thrips on the occurrence of Tomato spotted wilt virus in chrysanthemum fields. J Gen Plant Pathol 68:99–102
Matsuura S, Kubota K, Okuda M (2007) First report of Chrysanthemum stem necrosis virus on chrysanthemums in Japan. Plant Dis 91:468
Mehra A, Jabeen N, Singh AK, Hallan V, Zaidi AA (2009) A new chrysanthemum potyvirus: molecular evidence. Achiv Phytopathol Plant Protect 42:436–441
Moreno IM, Bernal JJ, de Blass BG, Rodrigue-Cerezo E, Garcia-Arenal F (1997a) The expression level of the 3a movement protein determine differences in severity of symptoms between two strains of tomato aspermy cucumovirus. Mol Plant-Microbe Interact 10:171–179
Moreno IM, Bernal JJ, de Blass BG, Rodrigue-Cerezo E, Garcia-Arenal F (1997b) The expression level of the 3a movement protein determine differences in sevreoty of symptoms between two strains of tomato aspermy cucumovirus. Mol Plant-Microbe Interact 10:171–179
Moriones E, Roosink MJ, Garci-Arenal F (1991) Nucleotide sequence of *Tomato aspermy virus* RNA 2. J Gen Virol 72:779–783
Nakamura Y, Fugisawa I, Lee KH, Uematsu (1994) Cucumber mosaic virus and Chrysanthemum virus B isolated from margnerate, Chrysanthemum morifolium L. Proc Kauto-tosao Plant Protect Soc 41:177–179
Niu EB, Chen LJ, Niu YB (2015) First report of *Zucchini yellow mosaic virus* in chrysanthemum. Plant Dis 99:1289
Pratap D, Kumar S, Raj SK (2006) molecular characterization of the virus causing shoestring in Tomato. In: XVI Annual Convention and International symposium on management of vector-borne viruses. 7–10 February, ICRISAT, Hyderabad, India, pp 8–06
Raj SK, Srivastava KM, Raizada RK, Singh BP (1991) Chlorotic ring mosaic of Chrysanthemum caused by a strain of *Chrysanthemum aspermy virus* in India. J Plant Dis Protect 98:640–646
Raj SK, Kumar S, Choudhari S (2007) Identification of *Tomato aspermy virus* as the cause of yellow mosaic and flower deformation on chrysanthemums in India. Aust Plant Dis Notes 2:1–2
Raj SK, Kumar S, Pratap D, Singh BP (2008) Virus affecting ornamental plants. In: Rao GP, Bragrd C, Lebas BSM (eds) Characterization, diagnosis & management of plant viruses, Vol. 4: Grain, crops and ornamentals. Studium Press LLC, Houston, TX, pp 1–29
Raj SK, Kumar S, Choudhari S, Verma DK (2009) Biological and molecular characterization of three isolates of tomato aspermy virus infecting chrysanthemums in India. J Phytopathol 157 (2):117–125
Rao ALN, Hatta T, Francki RIB (1982) Comparative studies on tomato aspermy and Cucumber mosaic viruses. VII: Serological relationships reinvestigated. Virology 116:318–326
Rist DL, Lorbeer JW (1989) Occurrence and overwintering of *Cucumber mosaic virus* and broad bean wilt virus in weeds growing near commercial lettuce fields in New York. Phytopathology 79:65–69
Salanki K, Balazs E, Burgyan J (1994) Nucleotide sequence and infectious in vivo transcripts of RNA 3 of *Tomato aspermy virus* pepper isolate. Virus Res 33:281–289
Sastry KSM (1964) A new strain of CAV from India. Indian J Microbiol 4:37–40
Singh BP, Gupta RP (1978) Control of Chrysanthemum aspermy virus by heat therapy. Phytopathology 47:909–910
Singh L, Hallan V, Zaidi AA (2011) Intermolecular interactions of chrysanthemum virus B coat protein: implications for capsid assembly. Indian J Virol 22:111–116
Srivastava KM, Raj SK, Singh BP (1992) Properties of a *Cucumber mosaic virus* strain naturally infecting Chrysanthemum in India. Plant Dis 76:474–477
Susheel K (2009) Molecular characterization of Cucumoviruses causing severe mosaic and ring sport diseases in Chrysanthemums and development of their management strategies. PhD thesis, Lucknow University, Lucknow, India

Urban LA, Sherman JM, Moyer JW, Daub ME (1994) High frequency shoot regeneration and Agrobacterium-mediated transformation of chrysanthemum (*Dendranthema grandiflora*). Plant Sci 98:69–79

Verma N, Sharma A, Raja R, Hallan V, Zaidi AA, Garg ID (2003) Detection, identification and incidence of Chrysanthemum B carlavirus in chrysanthemum in India. Crop Prot 22:425–429

Verma N, Ram R, Hallan V, Zaidi AA (2004) Production of Cucumber mosaic virus-free chrysanthemums by meristem tip culture. Crop Prot 23(5):469–473

Verma N, Hallan V, Ram R, Kumar K, Garg ID, Zaidi AA (2006) Detection and molecular characterization of a *Tomato aspermy virus* isolate infecting chrysanthemums in India. Acta Hortic 722:41–53

Wilson PA, Symons RH (1981) The RNAa of cucumoviruses: 3'-terminal sequences of two strains of *Tomato aspermy virus*. Virology 112:342–345

Yepes LC, Mittak V, Pang S-Z, Gonsalves C, Slightom JL, Gonsalves D (1995) Biolistic transformation of chrysanthemum with the nucleocapsid gene of *Tomato spotted wilt virus*. Plant Cell Rep 14:694–698

Chrysanthemum Production, Viral Diseases and Their Management

14

Ruchi Gupta, Ashutosh Gupta, Shampi Jain, Doomar Singh, and Neeraj Verma

Abstract

Chrysanthemum is the next only to rose in importance among the flower crops in the world. In India it has been recognised as fifth commercially important potent flower crop. It is versatile, it can be planted in pots, used for garland making and also as cut flower for flower arrangement. In India, large flowered varieties are grown for exhibition purpose while small flowered varieties are grown for cut flower, making garland, wreaths, veni and religious offerings. Apart from their ornamental flower, Chrysanthemum is a good source of nutrients, vitamins and minerals. The most important chemical extracts of Chrysanthemum include flavonoids, betaine, choline and vitamin B1. Because elite cultivars are typically propagated vegetatively, the risk of disseminating fungal, bacterial, phytoplasmal and viral pathogens is high. A number of viruses are known to infect chrysanthemum that cause benign, stunt, chlorosis, mosaic symptoms, and mottle. Consequently, this has led to qualitative and quantitative losses, resulting in serious problems in chrysanthemum production worldwide. Some viruses which affect the production of chrysanthemum are: - Chrysanthemum virus B (CVB), Chrysanthemum stem necrosis virus (CSNV), Chrysanthemum stunt viroid (CSVd), Cucumber mosaic virus (CMV), Tomato aspermy virus (TAV), Tomato spotted wilt virus (TSWV), Potato virus X and Begomovirus. Management of viral diseases is generally similar to those used for other pathogens, except that as yet chemicals find little application in viral diseases, although they may be used for vectors. Moreover, vectors involved in viral diseases have also complicated the problem in their management. The production of flowers and ornamental plants has been continuously growing due to improvement in market structure,

R. Gupta · A. Gupta · S. Jain · D. Singh · N. Verma (✉)
Department of Agriculture Science, Faculty of Agriculture Science and Technology, AKS University, Satna, MP, India

S. K. Raj et al. (eds.), *Virus Diseases of Ornamental Plants*,
https://doi.org/10.1007/978-981-16-3919-7_14

growth in population's and producers' purchasing power, diversification of species, diffusion of new production technologies and professionalization of the members of the Brazilian production chain. Factors, such as, number of plants, genetics and environmental, are determinant for the production of floral stems with quality. This quality can be evaluated using the length, stiffness and healthy floral stem, diameter and degree of inflorescence opening.

Keywords

Chrysanthemum · Plant virus · Management · Production and marketing

14.1 Introduction

Chrysanthemum is next only to rose in importance among the flower crops in the world. Chrysanthemum is a woody herbaceous perennial, producing blooms. The genus, belonging to the family Asteraceae, includes over 200 species of annuals, herbaceous perennials and subshrubs. It is distributed over Asia (mainly China), Africa, America and Europe. In Japan, it underwent a great transformation resulting in evolution of a wide range of types varying from perfectly regular blooms of the most irregular forms. Chrysanthemum has its origin from northern hemisphere, mainly Europe and Asia. Many authorities claim that it originated in China. In England, the popularity of chrysanthemum as an exhibition flower and as a commercial crop is as its peak. In the USA, it is the number one "dollar earner" flower and the most reliable. It is very difficult to say with certainty when chrysanthemum culture began in India. Sant Gyaneshwar has mentioned it in his famous Marathi exposition of Gita "Gyaneshwari" written in 1290 AD The Hindi name Guldaudi (flower of Daud) suggests that it must have been grown during the Mughal Period in India. British, American and Australian origin improved cultivars are grown in the northern parts of this country. The yellow flowered cultivars of chrysanthemum are grown in South-West and Central India on a field scale.

In India, it has been recognized as the fifth commercially important potent flower crop by the All India Coordinated Floriculture Improvement Project (AICFIP) of the Indian Council of Agriculture Research. It is versatile, it can be planted in pots, and it is used for garland making and also as cut flower for flower arrangement. Therefore, the modern techniques based on scientific research and by evolving more attractive colours have immensely increased its potential as a commercial crop, which can be raised year-round.

The Netherlands, Italy, Colombia, Spain, Germany and the USA are the important countries where chrysanthemum is mainly grown under greenhouse conditions. In different states of India, it is known with different names, Guldaudi in Hindi belt, Chandramallika in the eastern states, Samanti in the southern states and Shevanti in the western states. It is grown in an area of about 4000 ha. In India, large flowered varieties are grown for exhibition purposes, while small flowered varieties are grown for cut flower production; making garland, wreaths, and veni, and for religious offerings. The species of chrysanthemum have fibrous root system (shallow rooted plant). Chrysanthemum is an herbaceous perennial plant growing up to 50–150 cm tall, with deeply lobed leaves and large flower heads, in white, yellow or pink colour.

Chrysanthemums are highly prized for its varied colour, form, size, shape and use. This diversity is combined with wide variation in growth habit, vase life and amenability to various growth regulatory practices that has made this flower popular among flower growers. It is grown both for its aesthetic and commercial value. In North India, various hues of red, yellow, white and purple chrysanthemums are grown in abundance for decorating the landscape either in the ground or in pots. But in South India, mostly the yellow-coloured flowers are highly preferred and grown as loose flowers for trade (Prabhu et al. 2018).

Chrysanthemum is one of the most popular cut flowers and commercially important floral crops with hundreds of millions of its stems sold as cut flower every year in North America and Europe alone. It is grown throughout the world as cut flowers, loose flowers and potted plants besides being used as border plants in the gardens. The wide variation exhibited by its large number of cultivars in respect of growth habit, size, colour and shape of blooms makes it suitable for every place and every purpose conceivable for an excellent flower crop. Its erect and tall growing cultivars are suitable for background planting in borders or as cut flowers. It covers nearly 17% (3752 ha) of the total area under floriculture in the country. In India, Karnataka is the most prominent chrysanthemum growing state with an area of 1445 ha followed by the state of Tamil Nadu, Andhra Pradesh, Maharashtra, Rajasthan and Bihar where it is also grown on commercial scale. Chrysanthemum owes its popularity to a wide range of forms and colours in its flowers and its excellent keeping quality and availability throughout the year.

Chrysanthemum is a plant of vast potential for landscaping, besides being used as a cut flower. The state of Jammu and Kashmir is ideally suited for its cultivation almost all through the year, particularly being traditionally regarded as an autumn flower because of its ability to develop buds only at a day length of 13.5 h, and it can be used for off-season flower production in Kashmir. Garden chrysanthemum also given the name as 'Queen of the East' is a highly versatile and accommodating ornamental with a wide range of type, size and colour in its flowers and is too responsive to horticultural manipulations that plants can be 'tamed' to assure any form. Further, it possesses a wide range of adaptability and does remarkably well in the climate of Kashmir.

14.1.1 Chrysanthemum Can Be Used in Many Ways

- Tall growing plants are suitable for background planting in borders and beds.
- The dwarf and compact growing ones are suitable for front row planting in borders and also can be grown in pots.
- Long stems with attractive flowers have large vase life suitable for cut flowers.
- The decorative and fluffy bloomed small flowered cultivars are used for loose flowers and for garland making and hair decorations.

- The extra-large bloomed cultivars serve as exhibition material and bouquets.
- Flowers of spray-type varieties are highly suitable as loose flowers for making garlands and floral arrangements.

14.1.2 Classification

The flowers are classified by the kind and arrangement of florets (National Chrysanthemum Society of England) into five broad groups (Ahmad 2011).

1. Single: have one or more outer pistillate flowers (ray) with disc florets at the centre.
2. Anemones: similar to the singles except the disc florets are elongated and tubular forming a cushion. Disc flowers may be the same or a different colour from the ray flowers.
3. Pompons: a globular head formed by short uniform ray flowers, the shape is considered formal, and disc flowers are not apparent, which is further classified into three distinct sizes:
   (a) Small buttons (4 cm or less in diameter)
   (b) Intermediate (4–6 cm diameter)
   (c) Large (6–10 cm in diameter).
4. Decorative: have a floret arrangement similar to pompons since they are composed mainly of ray flowers, but the outer rows are longer than the central flower giving the flowers a flatter appearance or informal shape and size and are mostly intermediate and large.
5. Large flower: blooms are greater than 10 cm and are classified in many shapes, and disc florets are not apparent in most of these forms:
   (a) In curved double: Globose and formal with ray flowers similar in size to the disc flowers and that curve inwards and towards the top.
   (b) Reflexed double: Less formal and globose than the incurved double with overlapping ray flowers curved downwards except for the ray flowers.
   (c) Tubular ray flowers:
      - Spider: Ray flowers tubular and elongated into the outer rows but short in the centre. The dropping outer row ray flowers are sometimes hooked on the ends.
      - Fuji: Similar to the spider except the ray flowers may be shorter, dropless and lack hooks on the ends.
      - Quill: Tubular ray flowers, long on the outside and short near the centre, resembling feather quills. The ends of flowers are open and not flattened.
      - Spoon: Similar to the quill except the outer row flowers are open and are flattened, resembling a spoon.
6. Miscellaneous: Novelty types consisting of feathery plume, like of hairy ray florists.

14.1.3 Medicinal Use of Chrysanthemum

Apart from being an ornamental flower, chrysanthemum is a good source of nutrients, vitamins and minerals. Consuming 51 g of chrysanthemum offers 90 μg of vitamin B9, 0.481 mg of manganese, 17 mg of iron, 0.07 mg of copper, 0.09 mg of vitamin B6, 48 μg of vitamin A, 289 mg of potassium and 60 mg of calcium. Chrysanthemum has been used for hundreds of years in Chinese medicine. People use it to treat respiratory problems, high blood pressure and hyperthyroidism. Fans of the flower also say it can reduce inflammation and calm your nerves. Wild chrysanthemums itself and its compound preparations can be used to treat infectious diseases of the skin and respiratory system (Shun et al. 2005).

A tangy aromatic tea is made from its flowers or flower petals, commonly called chrysanthemum tea, and is drunk without or with little sugar or honey. Chrysanthemum tea acts as a coolant and helps in dealing minor rashes on the skin due to heat. You can apply the powdered form of this tea externally on the impacted area or drink it two to three times a day till symptoms disappear. It has amazing anti-inflammatory qualities that treat swelling due to bee stings, broken muscles and allergic reactions too. Chrysanthemum tea is also beneficial for heat rashes. According to Chinese doctors, chemical imbalances in the body are the main cause of heat rash. Chrysanthemum tea is one of the best cures for heat rash symptoms. For people with unstable vision, chrysanthemum tea improves the sharpness of vision. Chrysanthemum tea is also quite beneficial for different eye problems, like blurriness of vision, watering of eyes and spotty vision. Apart from that, cataract disease can also be treated with the help of chrysanthemum. To get rid of tired eyes, drink chrysanthemum tea or apply it on your eyes, by sprinkling it into the eyes directly. The Memorial Sloan Kettering Cancer Center claims that due to consumption of chrysanthemum tea, the passage of blood in the coronary arteries is improved. Certain health problems like dizziness, insomnia and headaches are cured by the use of chrysanthemum tea. According to the City University of New York, chrysanthemum can also be used to remove toxins from the blood. Hypertension (abnormal blood pressure level) can also be treated with the herbal tonics which include chrysanthemum. In traditional Chinese medicine, chrysanthemum is a cold herb which helps dispel pathogenic heat, helps the liver, improves eyesight and aids detoxification.

The most important chemical extracts of chrysanthemum include flavonoids, betaine, choline and vitamin B1. Thirteen important compounds of chrysanthemum flowers are acacetin-7-0-beta-D-glucopyranoside, luteolin, luteolin-7-*O*-beta-D-glucopyranoside, acaciin, acacetin 7-0-(6″-0-alpha-Lrhamnopyranosyl)-beta-sophoroside, 3-0-caffeoylquinic acid, syringaresinol 0-beta-D-glucopyranoside, 5,7-dihydroxychromone, uracil, *p*-hydroxybenzoic acid, 4-0-beta-D-glucopyranosyloxybenzoic acid, boscialin and blumenol A.

14.1.4 Insecticidal Properties of Chrysanthemum

Pyrethroids found in chrysanthemum were introduced by a team of Rothamsted Research scientists in the 1960s and 1970s following the elucidation of the structures of pyrethrin I and II by Hermann Staudinger and Leopold Ružička in the 1920s (Staudinger and Ruzicka 1924).

The pyrethroids represented a major advancement in the chemistry that would synthesize the analogue of the natural version found in pyrethrum. Its insecticidal activity has relatively low mammalian toxicity and an unusually fast biodegradation. Their development coincided with the identification of problems with DDT use. Their work consisted firstly of identifying the most active components of pyrethrum, extracted from East African chrysanthemum flowers and long known to have insecticidal properties. Pyrethrum rapidly knocks down flying insects but has negligible persistence—which is good for the environment but gives poor efficacy when applied in the field. Pyrethroids are essentially chemically stabilized forms of natural pyrethrum and belong to IRAC MoA group 3 (they interfere with sodium transport in insect nerve cells) (Haddi et al. 2012).

The pyrethrins are a class of organic compounds normally derived from *Chrysanthemum cinerariifolium* that have potent insecticidal activity by targeting the nervous systems of insects. Pyrethrin naturally occurs in chrysanthemum flowers and is often considered an organic insecticide when it is not combined with piperonyl butoxide or other synthetic adjuvants. Their insecticidal and insect repellent properties have been known and used for thousands of years.

Pyrethrin is most commonly used as an insecticide and has been used for this purpose since the 1900s (Metcalf 2000). In the 1800s, pyrethrin was known as 'Persian powder', 'Persian pellitory' and 'zacherlin'. Pyrethrins delay the closure of voltage-gated sodium channels in the nerve cells of insects, resulting in repeated and extended nerve firings. This hyperexcitation causes the death of the insect due to loss of motor coordination and paralysis. Pyrethrins are effective insecticides because they selectively target insects rather than mammals due to higher insect nerve sensitivity, smaller insect body size, lower mammalian skin absorption and more efficient mammalian hepatic metabolism (Bradberry et al. 2005).

Although pyrethrin is a potent insecticide, it also functions as an insect repellent at lower concentrations. Observations in food establishments demonstrate that flies are not immediately killed, but are found more often on windowsills or near doorways. This suggests, due to the low dosage applied, that insects are driven to leave the area before dying (Todd et al. 2001). Because of their insecticide and insect repellent effect, pyrethrins have been very successful in reducing insect pest populations that affect humans, crops, livestock and pets, such as ants, spiders and lice, as well as potentially disease-carrying mosquitoes, fleas and ticks.

14.1.5 Viral Diseases of Chrysanthemum

Because elite cultivars are typically propagated vegetatively, the risk of disseminating fungal, bacterial, phytoplasmal and viral pathogens is high. A number of viruses are known to infect chrysanthemum that cause benign, stunt, chlorosis, and mosaic symptoms and mottle. Consequently, this has led to qualitative and quantitative losses, resulting in serious problems in chrysanthemum production worldwide. Some viruses which affect the production of chrysanthemum are Chrysanthemum virus B (CVB) (Guan et al. 2017; Verma et al. 2007), Chrysanthemum stem necrosis virus (CSNV) (Boben et al. 2007); Chrysanthemum stunt viroid (CSVd) (Matsushita 2013; Yoon et al. 2012), Cucumber mosaic virus (CMV) (Srivastava et al. 1992; Verma et al. 2004, 2007; Choi et al. 2015), Tomato aspermy virus (TAV) (Verma et al. 2007; Liu et al. 2010), Tomato spotted wilt virus (TSWV) (Fukuta et al. 2004; Choi et al. 2015), Potato virus X (Choi et al. 2015) and Begomovirus (Marwal et al. 2013).

The most problematic pathogen of chrysanthemum is typically considered Chrysanthemum virus B (CVB). It is a member of the genus *Carlavirus* (Wetter and Milne 1981), is a single-stranded RNA virus and is the causal agent of some severe disease in chrysanthemum. The virus particles of CVB are slightly flexuous, rod-shaped and 685 nm long and 12 nm in diameter. They consist of multiple copies of a coat protein (Mr 37,000) encapsulating the viral RNA (Hollings and Stone 1972). CVB is transmitted by aphids in a non-persistent manner and by inoculation with sap. CVB is widespread throughout the world in cultivated varieties of chrysanthemum. The length of the genomic RNA of CVB is 8000–9000 nucleotides, excluding the poly tail, containing six open reading frames. Mild to severe symptoms are produced by CVB in chrysanthemums. Usually, the flowers become malformed in heavily infected plants. However, occasionally, no symptoms are seen. Thus, it is a serious potential threat to the floriculture industry worldwide.

Chrysanthemum stunt was first described by A. W. Dimock in 1945 of Cornell University. The disease became generally prevalent in the USA and Canada in 1946 and 1947. By that time, the disease was familiar to most producers of florists' chrysanthemums, but its nature remained a mystery. The symptoms and the rapid extension of the stunt disease immediately suggested a virus as the causal agent, but growers who had not previously contended with any virus disease of major importance were slow to accept that explanation. Chrysanthemum stunt viroid (CSVd) is a small, single-stranded, infectious RNA forming a circular secondary structure, which belongs to the Pospiviroidae family and mainly infects certain species and cultivars of Compositae and Solanaceae. The symptoms are overall reduction in plant size, foliage become pale in colour, flower may open prematurely and flower colour bleaching. However, the expression of the symptoms depends on the chrysanthemum cultivars. CSVd is known to be readily transmitted by sap, grafting, flower cutting shears and dodder connection, but the infection rate and incubation period were observed to differ according to the varieties. Some CSVd-resistant cultivars have also been reported, and the resistance is heritable in crosses between a CSVd-resistant chrysanthemum cultivars and CSVd-susceptible cultivars

(Matsushita 2013). Strains have been described for many viroids; these have not been described as such specifically for CSVd, despite the determination of 117 sequence variants of CSVd in 16 countries. Tomato has been described as an experimental host, and most of these studies were done with single isolates of CSVd and one cultivar of tomato plants. Biological activities of three isolates of CSVd, one each from the USA, China and Australia, varying in their nucleotide sequence were compared directly. These isolates showed no differences in symptoms induced on either chrysanthemum or tomato plants, suggesting that much of the biological variation reported previously may have been due to cultivar differences and/or environmental effects (Yoon et al. 2012).

Tomato aspermy virus (TAV) was found first on tomato (*Lycopersicum esculentum* Mill.) by Blencowe and Caldwell (1949) and was later identified as the cause of flower distortion in chrysanthemums (*Chrysanthemum indicum* L.) (Prentice 1952; Hollings 1955; Govier 1957). It is a ssRNA virus which has spherical particles of ca. 24 nm in diameter, with a central hollow core belonging to CMV. In chrysanthemum, symptoms of TAV infection generally appear as a diffuse mottling on young leaves, with distortion and colour break of flowers. Symptoms vary in flowers, not only in different cultivars but also between flowers of the same cultivar. Infection also causes varying degrees of stunting, depending on the cultivars and cultural practices. When the flowers produced are green instead of the normal colour for the variety, aster yellow is clearly present. Sometimes the upper branches of a flowering stem are thin, pale or yellowish and more upright than usual. Again, a number of thin, weak shoots bearing tiny leaves may arise from the base of the plant. Aster yellow appears in outdoor chrysanthemums and also in the greenhouse when plants are brought in for propagation or for flowering. TAV is transmitted in the non-persistent manner by four species of aphids, the foxglove aphid, the green peach aphid, the black chrysanthemum aphid and the green chrysanthemum aphid. Cucumber mosaic virus (CMV) is also a problematic virus for chrysanthemum growers. It is a member of *Bromoviridae* family which has isometric particles of ca. 29 nm in diameter with hollow centre and chrysanthemum strain of cucumber mosaic virus is similar to the aspermy virus (a member of *Bromoviridae*) in many respects, but is serologically related to the cucumber mosaic virus. In chrysanthemums, it is associated with distortion and changes of colour in the flower. Aspermy and the chrysanthemum strain of cucumber mosaic may be merely two names for the same virus. The symptom of CMV in chrysanthemum may be yellow mosaic with green veins, severe mosaic, yellowing of veins and yellow to necrotic spots. This variability in symptoms is a strong indication of the existence of CMV strains in the natural infection of chrysanthemum (Srivastava et al. 1992; Verma et al. 2004, 2007).

Tomato spotted wilt virus (TSWV) is the type member of the *Tospovirus* genus in the family *Bunyaviridae*. This virus has one of the widest host ranges of any plant virus, infecting over 600 species of plants in more than 50 families, including both monocots and dicots (Best 1968; De Avila et al. 1990). TSWV has several unusual characteristics for a plant virus, including an ambisense genome organization and quasispheroidal enveloped virions. It is transmitted by the western flower thrips

(*Frankliniella occidentalis*) and mechanically by sap. The particles of TSWV are spherical with ca. 85 nm in diameter. Symptomatology is highly variable and cultivar dependent. Most commonly, chlorotic spots and foliar necrosis are observed, with leaf bronzing, chlorotic ring and line patterns, internal necrosis, stem cankers, black streaks along the stem, wilting of plant parts, necrosis of axillary shoots, withering of leaves and stunting which occurs less frequently. Symptomless infection may occur and is of importance because stock plants are customarily maintained for long periods as sources of cuttings.

Chrysanthemum stem necrosis virus (CSNV), an RNA virus, belongs to the genus *Tospovirus* and family *Bunyaviridae*. CSNV was first reported in Slovenia in 2001. The disease symptoms on host plants cannot be distinguished from those caused by closely related viruses of the same genus, such as Tomato spotted wilt virus (TSWV). The disease symptoms may vary between host plants and can be quite severe.

14.1.6 Management

Management of viral diseases is generally similar to those used for other pathogens, except that as yet chemicals find little application in viral diseases, although they may be used for vectors. Moreover, vectors involved in viral diseases have also complicated the problem in their management. Viruses are the obligatory intracellular parasites infecting microbes, plants, animals and humans. They are dead outside the host cell but can take over the host's cell machinery as soon as they are into it. Several studies on inhibitor compounds have been done for animal viruses including those that are affecting humans, but there is inadequacy in terms of research and literature for plant viruses that are responsible for losses in crop yield and quality loss all across the globe. This could be focal point to study plant viruses, and their transmission and pathogenicity, and to establish widely used, effective and advanced approaches for their control. Plant viral diseases are not possible to control due to their intracellular habit in living cells, but these plant viral diseases and losses caused by them can be minimized through integration of management practices. Avoidance, exclusion and control of the pathogens are the principles of plant disease management. Integration of these three principles is fit for viral diseases because there is no chemical developed till now in the whole world which can kill the virus inside the living host cell. Many chrysanthemum growers, gardeners and florists have been using management practices since the ancient time.

Chrysanthemum is commercially propagated vegetatively through root suckers and terminal cuttings which carry the virus in new fields; therefore, it is necessary to follow quarantine measures to check the introduction of virus in new areas. However, the existence of symptomless hosts, the incubation period after inoculation and the absence of obvious symptoms in propagating material make quarantine difficult. Virus-free propagating material is the best option to eliminate the viral diseases of chrysanthemum; therefore, an extra care has to be taken by the growers to eliminate this problem making sure that the propagating material of chrysanthemum purchased is free from pathogens (viral, fungal, bacterial, etc.). It can be practised through the identification of virus in propagating material by ELISA method. Micropropagation (through meristem tip culture) is another method to obtain virus-free propagating material, but it is an expensive method for small and marginal chrysanthemum growers. Verma et al. (2003) obtained 84% CMV-free chrysanthemum plants through meristem tip culture. CVB-free plants of chrysanthemum were obtained by meristem culture following ribavirin or 3 weeks of heat treatment (Budiarto et al. 2011). It is also advisable not to try to directly import chrysanthemum propagating materials from any foreign country without phytosanitary certificate from a competent authority. Also avoid bringing or importing chrysanthemum from a country where viral diseases are more prevalent for many years. It has seen that large flower size and high yielding chrysanthemums are more susceptible to viral diseases as compared to local, small and medium flower size varieties. Eradication and roughing of infected plants after confirming virus infection without any greed is a good practice to get rid of the viral disease problem. Since chrysanthemum is generally produced by vegetative propagation, it is important to maintain free mother plants which prevent virus infection in the field. Accordingly, management of viral diseases basically involves preventing mechanical transmission by hands, agricultural tools and equipment. To date, several chemical agents, including sodium hypochlorite (NaOCl) and sodium hydroxide plus formaldehyde, have reportedly been effective in disinfecting viroid and virus contaminated tools. It is also advised to avoid staff moving from areas of virus-infected plants to healthy planting areas.

Cucumber mosaic virus is brought into crops by airborne aphid vectors. When vertical, sticky, yellow polyethylene sheets are erected along the edges of chrysanthemum crop, a considerable number of aphids are attracted to and stick to the plastic. This is done primarily to trap and monitor incoming insects, but to some extent, it also reduces the amount of virus inoculum reaching the crop. However, if reflectant aluminium or black, whitish-grey or coloured polyethylene sheets are used as mulches between the plants or rows in the field, incoming aphids, thrips and possibly other insect vectors are repelled and misguide away from the field. As a result, fewer virus-carrying vectors land on the plants and fewer plants become infected with the virus. Reflectant mulches, however, cease to function as soon as the crop canopy covers them. Black polyethylene sheets are also helpful in reducing weeds which give shelter to virus vectors. Viruses which infect chrysanthemum can also be managed by hot water treatment of nursery stocks.

Chrysanthemums may be protected against some viruses by protecting them against the vectors. Management of insect vectors and removing the weeds which serve as hosts (for vectors) may help in viral disease management. Although non-selective weedicides cannot be used as post-emergence weedicide in the chrysanthemum crop, few selective post-emergence weedicides can protect against vector hosts. Some pre-emergence broad-spectrum herbicides can also be used before planting the chrysanthemums to eradicate the vectors of chrysanthemum viruses, but this is only a partial solution of weed hosts due to short period persistence of any herbicide/weedicide. Therefore, application of selective insecticides against aphids and thrips may help to check the transmission of viruses from infected to healthy plants through these insect vectors. Some insecticides such as imidacloprid 17.8 SL, acetamiprid 20 SP, indoxacarb 14.5% + acetamiprid 7.7% SC, fipronil 5 SC, spinosad 45 SC, Cyazypyr 10 OD and afidopyropen (Inscalis™) 50 g/L DC can be used to control aphids and thrips.

Unlike humans and animals, plants lack an antibody-producing system and cannot be immunized by vaccination the way humans can, e.g. Covid-19 vaccine. Through genetic engineering, however, scientists have introduced and expressed in plants genes from mice coding for the production of antibodies against certain plant viruses, with which the mice had been injected artificially. Although plants so engineered do produce antibodies, called plantibodies, against specific plant virus, it is not yet known whether such plantibodies will effectively protect the plant from becoming diseased by that pathogen. In some host-virus combinations, the disease caused by severe strains of the virus can be avoided if the plants are inoculated first with a mild strain of the same virus, which then protects the plant from infection by the severe strain of the virus. Systemic acquired resistance can also be induced in plants against a variety of diverse viruses by treating the plants with certain chemical compounds, such as salicylic acid and 2,6-dichloroisonicotinic acid (INA), and certain benzothiazoles. By far the most common improvement of host resistance to almost any virus, however, is brought about by improving the genetic resistance of the host, i.e. by breeding and using resistant varieties. In the mid-1990s, genetic engineering technology made possible the isolation of individual resistance (R) genes from resistant plants and the transfer of such genes into susceptible plants in which they induce the hypersensitive (resistant) response. It is expected that this approach for improving resistance in susceptible plants, combined with conventional plant breeding, will provide one of the most effective tools for controlling plant diseases. It is apparent now that gene silencing is part of the normal defence system of plants against foreign nucleic acids; therefore, for viruses to become established and cause infection, they must overcome this defence.

14.2 Production and Marketing of Floriculture Products

The production of flowers and ornamental plants has been continuously growing due to improvement in market structure, growth in population's and producers' purchasing power, diversification of species, diffusion of new production technologies and professionalization of the members of the Brazilian production chain. The market totalled 5.22 billion reais (US$ 2.21 billion). Chrysanthemum is the second main species of cut flower produced in Brazil (15%), second only to roses, which represent 30% of the market. Other species like lisianthus (12%), lily (7%) and gerbera (6%) complement the main cut flowers in the national market. Chrysanthemum also stands out as flower pots, considering the main species: orchids (14%), lily (7.5%), chrysanthemum (7%), kalanchoe (6.4%), violet (6%) and bromeliad (4.5%). In floriculture, quality is defined as the set of attributes which makes the product saleable. Consumer acceptability of these products is evaluated by visual characteristics, such as size and shape, health conditions, turgor and maturity.

Factors, such as number of plants, genetics and environment, are determinant for the production of floral stems with quality. This quality can be evaluated using the length, stiffness and healthy floral stem and diameter and degree of inflorescence opening. The Instituto Brasileiro de Floricultural (Brazilian Institute of Floriculture) (IBRAFLOR) follows the quality criteria proposed by the Department de Qualidade (Quality Department) at Veiling Holambra Cooperative. The classification criterion of chrysanthemum is used to separate the products according to the standard and quality. The standard is measurable characteristics of the product containing 95% uniformity, such as the stem length (which ranges from 45 to 80 cm), bunch weight (1.1 or 1.3 kg with 5% tolerance), point of maturation and soft stems (stems that have a slope when held by the end of the base, acceptable up to 30° of slope). The quality category is defined as absence of defects and characterizes the lot quality. The quality should be established according to tolerance limits for severe damage (rust damage, botrytis and pests, mechanical damage, yellow and/or dry leaves and lack of total leaves) and minor defects (burning by phytotoxicity and chemical residue) (Zandonadi et al. 2018).

14.2.1 Quality Requirements of Chrysanthemum for International Markets

The cut flowers are graded according to stem length, colour and diameter of flowers. Grading in the USA is done according to the standards of the Society of American Florist (SAF), while in Europe the grades suggested by the European Economic Community (EEC) are followed. The metric grade classification for sprays, which works well for bulk packing, is given below:

Stem length should not be less than 66 cm for the British markets and those less than 51 cm should be marketed 'Shorts'.

14.2.2 Export Standards for Chrysanthemum

Parameters	Standard	Spray	Dwarf
Stem length (cm)	88–100	75–88	25–38
Weight (g)	30/stem of 90 cm	30 g/stem of 85 cm	15 g/stem of 30 cm
Number of flowers	Only one flower with five buds	Ten flowers with five to eight buds	10–12 flowers
Diameter	60–80 mm	35 mm for half bloom	30 mm
		45 mm for full bloom	

14.2.3 Society of American Florists (SAF) Grades for Standard Chrysanthemum

Quality parameters	Blue	Red	Green	Yellow
Min. stem length (cm)	75	75	60	60
Min. flower diameter (cm)	15	12.5	10	–
Stem strength	Strong	Strong	Strong	–

14.2.4 Polish Grades for Cut Chrysanthemum

Parameters	I	II
Flower diameter (cm)	13	8
Sterr. length (cm)	13	45
Permissible faults	–	Slight stem deformities

SAF specification standards for chrysanthemum (Kumar et al. 2007):

1. Bright, clean and healthy foliage and flowers
2. Lowers with similar varietal characteristics bunched together
3. Shape and size of flowers with underdeveloped centre
4. Flowers and foliage free from any defect, injury and dirt or any foreign material
5. Free from discoloration, nutrition and chemical or mechanical abnormalities
6. Fairly tight flower with underdeveloped centre
7. Fairly straight, stiff stems capable of supporting the flower in an upright position
8. Foliage stripped off from not more than one third of the stem
9. Stem length according to the requirement of grade

14.3 Farmers Are Facing the Problem in Marketing of Chrysanthemums

Colourful flowers with pleasant fragrance have been a source of attraction to mankind. Flowers provide pleasure through enlightening colours and spreading fragrance. Therefore, man has always taken the support of flowers as a token of expression for different kinds of sentiments on a number of occasions, and consequently, the ever-increasing demand of flowers has made the floriculture of paramount importance for conducting economic evaluation and marketing investigation. Ornamental plants (floriculture) can be found in the form of ornamental plants of pots and cut flowers. Cut flowers are one of the agricultural commodities that can help improve the income and welfare of farmers. One of the cut flowers having high economic value is chrysanthemum. Chrysanthemum is one of the floriculture commodities. The existence of chrysanthemum as an ornamental plant of commercial cut flowers is increasingly popular in various countries. It is not surprising that much research is being done to increase chrysanthemum production so that income also increases.

14.4 Cut Chrysanthemum Production in Two Soilless Systems

Chrysanthemum production is possible by soilless culture system. The chrysanthemum growers encountered soil-borne diseases, nematodes and accumulation of salinity when production in the same area was practised continuously. Soilless culture was a cultivation technique independent of soil condition. The purpose of the research was to determine the growth of chrysanthemum grown in two soilless systems including tray system and trough system. The results are the growth and quality of flowers produced in the tray system and the trough system were similar. No significant differences in flower characteristics were observed between the two systems except for flower colour. Chrysanthemum produced in both soilless systems received the same price for grade A as soil-grown chrysanthemum. Chrysanthemum became one of the types of floriculture plants that is in demand by the people of Indonesia in the past 7 years. Chrysanthemum production is dominated from Java Island with a total production of 432,827,108 stems. Chrysanthemum production in Tomohon continuously increases. Decorative flower in Tomohon is part of its culture farmer did not have capability in accessing source potential including capital and technology that increasingly develop to obtain feasible income (Lagana et al. 2018).

Acknowledgement Authors are thankful to Mr. B.P. Soni, Chancellor, and Er. A.K. Soni, Prochancellor, for providing facilities and encouragement in this project. Authors are also thankful to their family members for their moral support.

References

Ahmad S.S. 2011. Chrysanthemums. Department of Floriculture Kashmir, pp. 1-10.

Best RJ (1968) Tomato spotted wilt virus. Adv Virus Res 13:65–146

Blencowe JW, Caldwell J (1949) Aspermy - a new virus disease of the tomato. Ann Appl Biol 36:320–326

Boben J, Mehle N, Pirc M, Pleško IM, Ravnikar M (2007) New molecular diagnostic methods for detection of Chrysanthemum stem necrosis virus (CSNV). Acta Biol Slovenica 50(1):41–51

Bradberry SM, Cage SA, Proudfoot AT, Vale JA (2005) Poisoning due to pyrethroids. Toxicol Rev 24(2):93–106

Budiarto K, Marwoto B, Sanjaya L, Soedarjo M, Rahardjo IB (2011) Elimination of cvb (Chrysanthemum virus B) from a range of chrysanthemum varieties by apical meristem culture following antiviral agent and heat treatments. Biotropia 18(2):94–101

Choi H, Jo Y, Lian S, Jo KM, Chu H, Yoon JY, Choi SK, Kim KH, Cho WK (2015) Comparative analysis of chrysanthemum transcriptome in response to three RNA viruses: Cucumber mosaic virus, Tomato spotted wilt virus and Potato virus X. Plant Mol Biol 88(3):233–248

De Avila AC, Huguenot C, de Oliveira Resende R, Kitajima EW, Goldbach RW, Peters D (1990) Serological differentiation of 20 isolates of tomato spotted wilt virus. J Gen Virol 71:2801–2807

Fukuta S, Ohishi K, Yoshida K, Mizukami Y, Ishida A, Kanbe M (2004) Development of immunocapture reverse transcription loop-mediated isothermal amplification for the detection of tomato spotted wilt virus from chrysanthemum. J Virol Methods 121:49–55

Govier DA (1957) The properties of tomato asperrny virus and its relationship to cucumber mosaic virus. Ann Appl Biol 45:62–73

Guan Z, Wu D, Song A, Chen F, Chen S, Fang W (2017) A highly sensitive method for the detection of Chrysanthemum virus B 2017. Electron J Biotechnol 26:64–68

Haddi K, Berger M, Bielza P, Cifuentes D, Field LM, Gorman K, Rapisarda C, Williamson MS, Bass C (2012) Identification of mutations associated with pyrethroid resistance in the voltage-gated sodium channel of the tomato leaf miner (Tutaabsoluta). Insect Biochem Mol Biol 42(7):506–513

Hollings M (1955) Investigation of chrysanthemum viruses. 1. Aspermy flower distortion. Ann Appl Biol 43:86–102

Hollings M, Stone OM (1972) Chrysanthemum virus B. In: CMI/AAB descriptions of plant viruses, vol 110

Kumar R, De LC, Baiswar P (2007) Production of chrysanthemum under greenhouse condition. In: MTC on "Advanced techniques for production of commercial flower crops under green house conditions". In: Training manual on advanced technologies for production of commercial flower crops under green house condition, organized by NRC for orchids, Pakyong, Sikkim, pp 40–44

Lagana CA, Darwanto DH, Masyhuri (2018) Feasibility of chrysanthemum farming in Tomohon city. Agro Ekonomi 29:118–131

Liu J, Lin HC, Yi WZ, Hai ZX, Qin WY (2010) Detection of Tomato aspermy virus infecting chrysanthemums by LAMP. Sci Agric Sin 43:1288–1294

Marwal A, Sahu AK, Gaur RK (2013) First report on the association of a begomovirus with chrysanthemum indicum exhibiting yellowing of leaf vein disease characterized by molecular studies. J Hortic Res 21(2):17–21

Matsushita Y (2013) Chrysanthemum stunt viroid. JARQ 47(3):237–242

Metcalf RL (2000) Insect control. In: Ullmann's encyclopedia of industrial chemistry. https://doi.org/10.1002/14356007.a14_263. ISBN 978-3-527-30673-2

Prabhu G, Thamaraiselvi SP, Aruna P, Sudhakar R (2018) Evaluation of chrysanthemum (Dendranthemagrandiflora Tzelev.) genotypes for loose flower production under Coimbatore conditions. Int J Chem Stud 6(4):1618–1621

Prentice TW (1952) Chrysanthemum flower distortion survey, 1951. Plant Pathol 1:77–80. https://doi.org/10.1111/j.1365-3059.1952.tb00033x

Shun YZ, Yang Y, Huaidong Y (2005) Chemical composition and antimicrobial activity of the essential oils of Chrysanthemum indicum. J Ethnopharmacol 96(1):151–158

Srivastava KM, Raj SK, Singh BP (1992) Properties of a Cucumber mosaic virus strain spontaneously infecting Chrysanthemum in India. Plant Dis 76:474–477

Staudinger H, Ruzicka L (1924) Insektentötende Stoffe I. ÜberIsolierung und Konstitution des wirksamenTeiles des dalmatinischenInsektenpulvers [Insecticidal substances I. On isolation and constitution of the active part of the Dalmatian insect powder]. Helvetica Chimica Acta 7 (1):177–201. https://doi.org/10.1002/hlca.19240070124

Todd GD, Wohlers D, Citra M (2001) Toxicological profile for pyrethrins and pyrethroids. ATSDR, Atlanta, GA. Web. 26. 2015

Verma N, Sharma A, Rama R, Hallana V, Zaidia AA, Garg ID (2003) Detection, identification and incidence of Chrysanthemum B carlavirus in chrysanthemum in India. Crop Prot 22:425–429

Verma N, Ram R, Hallan V, Kumar K, Zaidi AA (2004) Production of Cucumber mosaic virus-free chrysanthemums by meristem tip culture. Crop Prot 23:469–473

Verma N, Mehra A, Singh L, Hallan V, Singh AK, Jabeen N, Singh MK, Ram R, Zaidi AA (2007) Screening for viruses infecting chrysanthemum cultivars in India. Sci Hortic 111:260–265

Wetter C, Milne RG (1981) Carlaviruses. In: Krustak E (ed) Handbook of plant virus infectious and comparative diagnosis. Elsevier, North Holand and Biochemical Press, pp 695–730

Yoon JY, Baek E, Palukaitis P (2012) Are there strains of chrysanthemum stunt viroid. J Plant Pathol 94(3):697–701

Zandonadi AS, Maia C, Barbosa JG, Finger FL, Grosi JAS (2018) Influence of long days on the production of cut chrysanthemum cultivars. Hortic Bras 36:33–39

Viruses Infecting Bulbous Ornamental Plants and Their Diagnosis and Management

15

S. Madhavan, V. Balasubramanian, and R. Selvarajan

Abstract

Floriculture flourished as an important fast-emerging sector of horticulture in the last few decades and is being viewed as a high growth industry in the world. Rapid urbanization, increased income levels, changes in lifestyles and social values resulted in increase in domestic and world market for flower crops significantly. Flower crops are propagated using various seed materials and modern methods, and the crops like gladiolus, tuberose and lily are mainly propagated by bulbs. Viruses are the major limiting factor and a major concern in the production of bulbous ornamentals. Bulbs, the propagating material usually harbour many viruses and act as the primary source of inoculum to cause diseases in the main crop. The viruses present in the bulbs infect the main crop, spread to healthy plants secondarily through various sap feeding insect vectors, such as aphids. Among the plant viruses, the single stranded RNA genome viruses like CMV, BYMV, TuMMoV, LSV, LMoV, TBV, TNV, TRV are considered as major constraints of bulbous ornamental flower production of the bulbous flower crops. Various methods such as electron microscopy, ELISA and PCR based technologies have been successfully developed and are being employed in the diagnosis and detection of the important viruses infecting bulbous ornamental crops. Management of these economically important viruses is successful when integration of multiple methods such as use of virus free certified planting materials and timely management of vectors in the field. In this chapter, we described the symptoms, transmission, characterization, diagnosis, and the

S. Madhavan
ICAR-Directorate of Floricultural Research, Regional Station, Vemagiri, Andhra Pradesh, India

V. Balasubramanian · R. Selvarajan (✉)
ICAR-National Research Centre for Banana, Tiruchirapalli, Tamil Nadu, India

S. K. Raj et al. (eds.), *Virus Diseases of Ornamental Plants*,
https://doi.org/10.1007/978-981-16-3919-7_15

management aspects of various taxonomic groups of viruses infecting above mentioned ornamentals.

Keywords

Gladiolus · Tuberose · Lily · *Potyvirus* · *Cucumovirus* · *Bromovirus* · *Tospovirus* · RNA viruses · CMV · BYMV · TuMMoV · LmoV · LSV · Lateral flow assay · RT-PCR · LAMP

15.1 Introduction

Floriculture is a fast-emerging venture throughout the world and is being viewed as a high growth industry and becoming important from the export angle. Worldwide, the area under cultivation in floriculture occupies more than 3 lakhs hectares and the worth of this industry is estimated to be about US$ 70 billion. Cut flower crops viz. rose, chrysanthemum, gerbera, orchids, tulips, lilium and anthurium are widely grown throughout the world and they contribute a significant share in the world floriculture trade. Ornamentals crops are known to be severely affected by plant pathogens, and among the diseases, viral diseases are considered very serious as they cause severe yield loss and the aesthetic value of flower is lost. Most of the ornamental plants are propagated by corms and cormels or rhizomes. These propagules harbour viruses if the plants are infected with viruses. Integrated management of these viruses is possible by employing multiple methods such as use of virus-free certified planting materials and timely management of vectors in the field. In this chapter, viral diseases affecting four important bulbous ornamental crops, namely, gladiolus, lily, tuberose and tulips, are dealt in detail. The symptoms, transmission, virus characterization and management aspects are reviewed from the published literature.

15.2 Gladiolus

Gladiolus is known as an important bulbous flower crop because of the aesthetic nature of its flower and is considered as one among the top ten cut flowers in the global floriculture industry. Though the value of gladiolus is high in the global floriculture trade, its production has declined tremendously both in quality and quantity due to diseases caused by fungal and viral pathogens. Gladiolus is commercially propagated through underground vegetative organ (corms). Continuous vegetative propagation of gladiolus through corms often results in harbouring of many corm borne viruses which may become major threat to commercial gladiolus production. Many viruses infect gladiolus but the major loss has occurred in its production because of the diseases caused by Cucumber mosaic virus (CMV) and Bean yellow mosaic virus (BYMV). These two viruses are the most prevalent viruses infecting gladioli worldwide. The symptoms, transmission, biology and management of Cucumber mosaic virus and Bean yellow mosaic virus are furnished hereunder.

15.2.1 Cucumber Mosaic Virus

15.2.1.1 Symptoms

CMV induce the symptoms on gladiolus at the time of flowering. Appearance of white, light grey or yellowish blotches can be seen in flowers; hence, the disease is commonly called as white break mosaic. Discoloured flowers are accompanied by crinkling, shrinking and often malformation. Opening of affected flowers occurs slowly and imperfectly. Sometimes, early withering of florets occur after yellowing in the later stages of disease development. The CMV-infected plants are stunted. On foliage, grey, yellow, brown or reddish small chlorotic square spots appeared between two leaf veins. In severe cases, the entire plant becomes yellow. Infected plants of some varieties produce warty corms (Fig. 15.1) (Univ. of Illinois Extension 1983; Stein 1995).

15.2.1.2 Characterization

CMV is an omnipresent and deleterious virus spreading worldwide (Brierley 1952; Berkeley 1953; McWhorter 1957; Wade 1984; Stein 1995; Navalinskiene and Samuitiene 2001; Dubey et al. 2008; Duraisamy and Pokorny 2009) which infects more than 1200 plant species in a systemic manner (Zitter and Murphy 2009; Dubey et al. 2010). CMV is the type member of the genus *Cucumovirus* of the family *Bromoviridae*. The virus is tripartite and isometric in shape and is 30 nm dia in size (Fig. 15.2); its genome consists of three positive single-stranded RNAs, viz. RNA1, RNA2 and RNA3, and a subgenomic particle, RNA4, encoded by RNA3, and these particles are encapsidated by a coat protein (Suzuki et al. 1991; Palukaitis et al. 1992; Raj et al. 2002). On the basis of serology, sequence homology and nucleic acid hybridization, isolates of CMV infecting different host plants are divided into two subgroups, I and II (Owen and Palukaitis 1988; Wahyuni et al. 1992). Again, subgroup I is divided into two subgroups, IA and IB, which are distinguished by their nucleotide sequences, and the CMV that infects gladiolus is grouped under subgroup I A (Roossinck 2002; Dubey et al. 2010). The phylogenetic analysis using coat protein gene and amino acid sequences of CMV strains from different hosts including gladiolus revealed that the CMV strain from gladiolus is distinct from the Indian CMV isolates which are infecting different host plants and has very close relation with CMV Fyn strain (Dubey et al. 2010).

15.2.1.3 Transmission

Transmission of CMV occurs mechanically and primarily through sap and secondarily by many aphid species in a non-persistent manner. *Macrosiphum euphorbiae* is the most efficient vector in transmitting CMV with an acquisition feeding period of 3–5 min and with overnight inoculation feeding period than other aphid species, *Aphis gossypii*, *A. citricola* and *Myzus persicae*. The CMV transmission efficiency

Fig. 15.1 Symptom induced by CMV in gladiolus. (**a**) Yellow streaking; (**b**) Yellow or grey or brown spots; (**c**, **d**) Flower colour breaking; (**e**) Deformed Corms. (Kathy Kamo. USDA)

of other aphid species is relatively close with *M. euphorbiae* (Aly et al. 1986). The CMV can also be transmitted mechanically by cutting knife used in preparation of corms for planting (Brierley 1962). Despite these methods of transmission, infected plants and viruliferous aphids are necessary for the field level spread.

15.2.1.4 Diagnosis

Diagnosis of CMV can be done by applying serological and other nucleic acid- and protein-based molecular methods. ELISA kits have been developed for diagnosis of

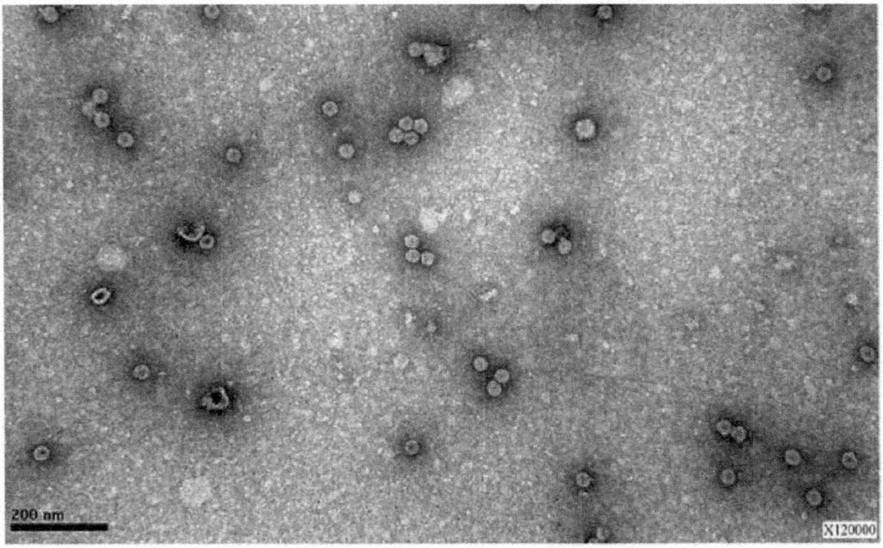

Fig. 15.2 Electron micrographs of negatively stained virus particles (~30 nm in diameter) with 2% uranyl acetate

viruses infecting gladiolus by different companies in the USA (www.agdia.com/testing-services/Gladiolus.cfm), Switzerland (http://www.bioreba.ch/files/Tecnical_Info/TS_Broschuere_2011_e_CHF.pdf) and Neogen Europe, Ltd., Auchincruive, Scotland, UK, manufactures. Raj et al. (2002) developed highly sensitive reverse transcription polymerase chain reaction (RT-PCR) and southern hybridization methods to detect the CMV from leaves and corms of gladiolus. These methods can be used to detect the presence of virus in the samples either in latent form or at very low concentration.

15.2.1.5 Management

Since gladiolus is mainly propagated through corms, it is advised to use resistant cultivars like Trader Horn and Eurovision (Aly et al. 1986) as mother plants for production of planting materials. Initial transmission of CMV occurs mainly through infected corms and cutting knives; hence, use disinfected cutting knives for preparation of corms and use certified corms for planting. Avoid growing gladiolus near bean, cucumber, tomato, clover and melon field. Mulching of gladiolus plants with aluminium foil repels the aphids with an efficiency of 90–95% and showed a good restriction on the spread of CMV in the field (Bing and Johnson 1971). Controlling of aphid population by using fine mesh coated with mineral oil or by direct spraying of mineral oil on the plants helps in restricting the movement of aphid vectors and thus reducing the outbreak of CMV (Aly et al. 1986). Development of transgenic gladiolus expressing pathogen-derived genes for conferring resistance could be exploited in the management strategy for CMV and for increasing the floral quality and productivity of gladiolus (Kamo et al. 2010; Dubey et al. 2015).

15.2.2 Bean Yellow Mosaic Virus

15.2.2.1 Symptoms

Like CMV, BYMV also produce mosaic, leaf and flower colour breaking symptoms in gladiolus (Fig. 15.3a, b). Mottling symptoms appear in leaves and flower stems. BYMV develop faint, inconspicuous pencil-stripe break patterns on the flowers. These symptoms appeared on the flowers when the plants grow rapidly with mild temperature conditions, and the symptoms get masked when the temperature is high. The infected plants become severely stunted and often produce less cormels (Srivastava et al. 1983; Zaidi et al. 1993; Selvarajan et al. 1999; Park et al. 1998; Katoch et al. 2003; Dubey et al. 2009; Duraisamy et al. 2011; Hemachandra Reddy et al. 2019).

15.2.2.2 Characterization

BYMV belongs to the genus *Potyvirus* of the family *Potyviridae* with a wide host range worldwide. BYMV is flexuous rod-shaped and is about 750 nm in length and 11 nm in dia (Fig. 15.3c). The genome of BYMV is a positive-sense single-stranded RNA with 3×10^6 MW. Complete genome sequence of CK-GL2, an isolate from *Gladiolus dalenii* cv. Sylvia, consists of 9532 nucleotides containing 5′-UTR, a large ORF, a putatively small overlapping ORF, coding for PIPO protein and 3′-UTR regions. The type of genome organization present in other BYMV isolates of different hosts is also found in the CK-GL2 isolate infecting gladiolus. Besides that, both ORFs encoded gene products, viz. P1 proteinase; helper component proteinase (HC-Pro); P3 protein; 6K1 protein; cylindrical inclusion (CI) protein; 6K2 protein; nuclear inclusion a (NIa) protein, a polyprotein that is further processed into the viral protein genome linked (VPg) and NIa proteinase (NIa-Pro); nuclear inclusion b (NIb) protein, the viral RNA-dependent RNA polymerase; and coat protein (CP), and an additional gene product "P3N-PIPO" was also found in the study undertaken by Kaur et al. (2015). Kehoe et al. (2014) demonstrated the presence of IX phylogenetic groups of BYMV isolates infecting monocots and dicots based on the phylogenetic analysis of their complete genome sequences. BYMV isolate infecting gladiolus belongs to phylogenetic group III as shown in the study of Kehoe et al. (2014) which was further supported by Kaur et al. (2015).

15.2.2.3 Transmission

BYMV is present in any one of its cultivated and wild hosts from which it is transmitted by many aphid species to the crop in a non-persistent manner (Agrios 2005). Kaur et al. (2015) demonstrated the transmission of BYMV by *Aphis craccivora* from gladiolus to *Vicia faba*. The transmission of BYMV is also reported to occur through contaminated tools used when harvesting flowers and corms (Brierley 1962).

Fig. 15.3 Symptom induced by BYMV in Gladiolus. (**a**) Dark green patches in venial and light green to yellowish interveinal region irregular chlorotic spots with yellow streaking and unevenly spread bright yellowish interveinal region; (**b**) Flower colour breaking symptoms; (**c**) Electron micrograph showing flexuous, filamentous virus particle (Magnification ×120,000) (Bar –120 nm) (Selvarajan and Gupta 1996)

15.2.2.4 Diagnosis

Presence of BYMV in the samples can be detected by using serological, nucleic acid, protein-based methods and electron microscopic observation of inclusion bodies (Kaur et al. 2015). However, ELISA kits that have been developed commercially and made available to the public by many companies which are mentioned early in the diagnosis of gladiolus CMV can be used in the diagnosis of BYMV infecting gladiolus. Although electron microscopy and commercial diagnostic kits are available in the market for detection of viruses present in the samples with high concentration, rapid, reliable and quantifiable diagnostic methods are very essential nowadays for indexing of BYMV present in different tissues of gladiolus plants at low concentration level. Rosner et al. (1994) demonstrated that use of additional amplification of the PCR products by transcription, using T7 RNA polymerase (PCR/T) in the detection of BYMV, presents low concentration level in the in vitro-cultured gladiolus. Duraisamy et al. (2011) developed real-time RT-PCR (rt-RT-PCR) and immunocapture real-time RT-PCR (IC-rt-RT-PCR) for the sensitive detection of BYMV in all tissues of gladiolus at very low concentration.

15.2.2.5 Management

Use of virus-free certified corms/plantlets for planting is considered as an initial step of virus disease management strategy. Seventy percent alcohol should be used to disinfect the harvesting tools between cuts to prevent cross-contamination. Callus developed from central bud of cormels infected with BYMV and CMV was subjected to virazole (ribavirin, 1-b-D-ribofuranosyl-1,2,4-triazole-3-carboxamide) treatment at 20, 40 and 60 mg/L for 6–8 weeks, and then callus is differentiated into plantlets, and tested for the presence of viruses. The results of the experiment revealed that virazole at 40 and 60 mg/L concentration levels completely eliminated the viruses from the callus tissue and corms established from the treated callus tissues certified as virus-free (Singh et al. 2007). Transgenic gladiolus plants transformed with the BYMV coat protein gene in either sense or antisense orientation were found to have delayed infection of BYMV and did not show the resistance against the eventual infection with BYMV (Kamo et al. 2005). Thermotherapy combined with meristem tip culture greatly improves the BYMV elimination efficiency from infected gladiolus corms, resulting in the production of BYMV-free gladiolus plants (Nezamabad et al. 2015). Thermotherapy, chemotherapy, electrotherapy and combination of these therapies have been successfully employed in the production of virus-free cormels/plantlets in gladiolus against BYMV (Selvarajan et al. 1999; Singh et al. 2007; Kaur et al. 2019). Since BYMV has wide host range and transmission occurs through aphids, it is advised to avoid growing of gladiolus near tomato, bean, clover, peas and infected gladiolus field. Virus-free gladiolus should be grown at least 1 km away from virus-infected gladiolus and leguminous plants. The inoculum of virus is present in the weed host and harbour aphids during the absence of main host. Hence, removal of the all nearby weeds (groundcherry, milkweed, pokeweed, pigweed, clover, henbane, jimsonweed, wild cucumber, bryony, shepherd's purse, goosefoot, thistle, chickweed, nettle, deadnettle and sowthistle) (Univ. Illinois Extension 1983) helps in the spread of the disease. The insect vectors, aphids, are the main cause of transmitting BYMV from reservoirs, to build up greatest secondary inoculum; thus, aphids contribute for the development of epiphytotics. The management practices for controlling aphids as described for CMV will avoid the spread of BYMV and restrict development of epiphytotics in gladiolus.

15.3 Tuberose

Tuberose is a bulbous ornamental flower crop belonging to the family Agavaceae, which native to Central America. This crop is growing worldwide for cut and loose flower trade and for extraction of essential oil. Tuberose mild mottle virus occurs worldwide and causes economic damage to the tuberose production. Symptoms and the management aspects are described hereunder.

15.3.1 Tuberose Mild Mottle Virus

15.3.1.1 Symptoms

Infected plants show mosaic and mottling symptoms on leaves. Mosaic also appears on flower peduncles and leads to reduction of the quality of flowers. Infected plants become stunted with reduced number of tillers, thereby reducing yield.

15.3.1.2 Characterization

The virus infecting tuberose showed mosaic and mottling symptoms and was first reported in New Zealand by Pearson and Horner in 1986. Subsequently, the presence of a similar disease on tuberose was reported from China (Lin et al. 2004), India (Kulshrestha et al. 2005) and the USA (Dey et al. 2017). The disease is caused by a potyvirus, and they have tentatively named it as Tuberose mild mottle virus (TuMMoV) (Lin et al. 2004). TuMMoV belongs to the genus *Potyvirus* of the family *Potyviridae*. TuMMoV is a flexuous rod-shaped virus with a mean length of 750 nm and 12 nm diameter. It produces cytoplasmic cylindrical inclusions, pinwheel and laminated virus aggregates in the infected leaves of tuberose (Horner and Pearson 1988). The mechanical inoculation of TuMMoV to a wide range of plants including tuberose showed that it could infect and produce characteristic symptoms only on the tuberose and not on the other inoculated plants. The presence of the TuMMoV on inoculated tuberose plants was further confirmed by ELISA and PCR (Horner and Pearson 1988; Lin et al. 2004; Kulshrestha et al. 2005; Krishnareddy et al. 2007). This study implies that the TuMMoV potyvirus could have only one host which infects tuberose. Lin et al. (2004) used potyvirus-specific primers to amplify TuMMoV from tuberose. RNA isolated from Indian isolate of TuMMoV from tuberose gave ~750 bp fragment when amplified with potyvirus group-specific primers. The CP gene nucleotide and its amino acid sequence alignment of Indian isolate of TuMMoV with the corresponding sequences of other potyvirus group showed that 97% sequence homology with TuMMoV was reported from China at nucleotide level and 100% homologous at amino acid level (Kulshrestha et al. 2005). TuMMoV cause mosaic disease in tuberose from Bangalore region identified as a distinct strain of TuMMoV which has differences in its sequences of coat protein gene and 3′ UTR with other isolates, TuMMoV-Hangzhou and TuMMoV-Palampur (Krishnareddy et al. 2007).

15.3.1.3 Transmission

This virus reported to be transmitted by green peach aphid, *Myzus persicae*, in a non-persistent manner (Kulshrestha et al. 2005).

15.3.1.4 Diagnosis

Initially, potyvirus-specific primers and antibodies were used in PCR and ELISA, respectively, to detect the presence of TuMMoV in tuberose (Kulshrestha et al. 2005; Krishnareddy et al. 2007). Lin et al. (2004) specifically designed primers which amplify a 600 bp from the 5′ end of CP gene coding region of TuMMoV. Dey et al. (2017) reported RT-PCR-specific primers to amplify 470 bp product to confirm the presence of TuMMoV in the samples.

15.3.1.5 Management

Tuberose is mainly propagated using corms; hence, certified corms can be used for plantings to avoid the spread of TuMMoV-infected corms to growing plants. Since there is no report available for the management of TuMMoV and the virus transmission was reported to be carried out by aphid, the control measures described for aphids in the management of gladiolus CMV and BYMV can also be applied in TuMMoV management in tuberose.

15.4 Lily

Lily is one of the important ornamental crops growing worldwide as cut flower and also as potted flowers. Lily is one among the top ten ranked flowers in the global export market. Commercially, three important lily species, Easter lily (*Lilium longiflorum*), Asiatic and Oriental hybrids, are grown in the world. *Lilium davidii* var. *unicolor* is being grown in China for its edible and medicinal value. The diseases caused by the viruses are the major concern, and it greatly influences the quality and quantity of lily production. Lily symptomless virus (LSV), Cucumber mosaic virus (CMV) and Lily mottle virus (LMoV) are the most commonly occurring viral pathogens of many cultivars of lily (Derks 1995; Asjes 2000; Sharma et al. 2005; Kwon et al. 2013; Lim et al. 2016).

15.4.1 Lily Mottle Virus

15.4.1.1 Symptoms

Symptoms in the lily plants infected with LMoV vary with the cultivars. Infected plants of some cultivars of lily show vein clearing; mosaic, mottle, chlorotic and yellow streaking; and curling of leaves (Fig. 15.4a). Colour breaking and malformations and asymmetry of the flowers may produce in other cultivars. Premature falling of buds and flowers may occur and the vase life of flowers is also reduced (Aravintharaj et al. 2017; Asjes 1997; Derks 1988; Lee et al. 1996; Sharma et al. 2005).

15.4.1.2 Characterization

LMoV belongs to the genus *Potyvirus* (family *Potyviridae*); it's a flexuous, non-enveloped and rod-shaped particle which is 11–15 nm wide and 680–900 nm long (King et al. 2011). It consists of a positive-sense single-stranded RNA genome with 9644 nt and encodes a 351.0 kDa polyprotein of 3095 amino acids. The sequences of LMoV isolates from lily are very closely related with Tulip breaking virus and Tulip breaking virus lily strain and had only 45.1–54.4% identity to other completely sequenced potyviruses (Alper et al. 1982; Zheng et al. 2003).

15.4.1.3 Transmission

Aphid species, *Macrosiphum euphorbiae*, *Myzus persicae* and *Aphis gossypii*, act as vectors to transmit the LMoV in a non-persistent manner (Asjes 2000; Conijn 2014). Among the aphid species, *M. persicae* efficiently transmits the LMoV within 2–3 min of inoculation, and the plants used for the virus transmission produce symptoms within 27–30 days of inoculation (Sharma et al. 2005).

15.4.1.4 Diagnosis

The oligonucleotide primers, U341 and D341, designed by Langeveld et al. (1991) were used to amplify the coat protein gene of LMoV infecting lily and identified the

Fig. 15.4 (**a**) Mosaic, mottle, chlorotic and yellow streaking symptom induced by Lily Mottle virus. (**b**) Mild, pale vein-clearing resembling micronutrient deficiency induced by Lily Symptomless Virus. (John Fisher, Ohio Department of Agriculture, Bugwood.org)

LMoV as a different virus from Tulip breaking virus which causes mottle in lily and flower breaking disease in tulip based on the sequence variation (Dekker et al. 1993). The virus can be detected by using ELISA, immunoblot hybridization, dot-blot hybridization and immunosorbent electron microscopy (Derks 1988; Derks et al. 1994). The virus present in the samples at low concentration could not be detected, but at the same time, a strong reaction was observed between the samples collected from the plants during flowering stage and LMoV antisera in DAS-ELISA (Sharma et al. 2005). PCR-based methods have been used extensively worldwide in diagnosing and indexing plants for LMoV, and these methods counteract the problems of ELISA in detection when the virus is present in the samples even at very low concentration (Niimi et al. 2003; Sharma et al. 2005; Zhang et al. 2017; Zhao et al. 2018; Zhang et al. 2020).

15.4.1.5 Management

Planting of virus-free certified propagative material and controlling the vectors by spraying with mineral oil and synthetic pyrethroids help in production of disease-free healthy crop. Growing of some of the disease-resistant Asiatic hybrids helps in good crop return to the farmers by reducing the cost of crop protection measures, thus in turn reducing the development of resistance in the insects against chemicals and also environmental hazards (Asjes 2000; Chastagner et al. 2018).

15.4.2 Lily Symptomless Virus

15.4.2.1 Symptoms

LSV infections occur and reside in the lily plants symptomlessly and cause symptoms in certain genotypes under certain environmental conditions. It causes symptoms like mild, pale vein clearing and mottling of leaves (Fig. 15.4b) in the glasshouse-grown lily plants (Zhang et al. 2017; Chastagner et al. 2018). Leaves of infected plants show yellowing quickly (Boontjes 1983). Quantitative and qualitative loss of flower and bulb production occurs in lily infection with LSV (Asjes 2000). Growth reduction, production of small size flowers and lower bulb yield are the symptoms produced in lily when it has combined infection with LSV and CMV (Asjes 2000).

15.4.2.2 Characterization

LSV is a species of the genus *Carlavirus* of the family *Flexiviridae*. It is a filamentous particle measuring 17–18 nm wide and 640 nm long (Allen 1972). The genome of LSV is monopartite, consists of a positive-sense single-stranded RNA genome constituted with 8394 nt and contains six open reading frames (ORFs). Single type of 32 kDa coat protein encapsulates the genomic RNA of LSV (Choi and Ryu 2003; Zheng et al. 2003).

15.4.2.3 Transmission

Myzus persicae, *Macrosiphum euphorbiae*, *Aulacorthum solani* and *Aphis fabae* transmit the LSV in a non-persistent manner. *M. persicae* is a more efficient vector than other species for transmitting LSV (Mowat and Stefanac 1974; Asjes 2000; Sharma et al. 2005).

15.4.2.4 Diagnosis

Filamentous particles of LSV are detected in the specimens stained with 1% uranyl acetate using a transmission electron microscope (Wang et al. 2007). The virus present at higher concentration in the samples collected during flowering stage is detected based on the reaction observed between LSV antiserum and antigen in DAS-ELISA (Sharma et al. 2005). Several other highly sensitive methods such as real-time reverse transcription polymerase chain reaction (RT-PCR) (Wei et al. 2012), quantitative RT-PCR (Nesi et al. 2013; Sun et al. 2014), RT-PCR combined with enzyme-linked immunosorbent assay (ELISA) (Sharma et al. 2005), immunocapture RT-PCR combined with indirect ELISA (Kim et al. 2012), microsphere-based fluorescent immune assay (Zou et al. 2011), dot- or tissue-blot immune enzyme assay (Makkouk et al. 1993) and loop-mediated isothermal amplification (He et al. 2016) have been employed to detect the LSV in the samples. The immunochromatographic strip (ICS) test with polyclonal antibodies developed against LSV showed 98.6% specificity and 100% sensitivity for LSV in field samples (Zhang et al. 2015).

15.4.2.5 Management

Using of virus-free planting materials and spraying with mineral oil and pyrethroids help in preventing the spread of LSV.

15.4.3 Cucumber Mosaic Virus

15.4.3.1 Symptoms

Chlorotic and yellow spots appear initially and later grey or necrotic spots may be produced in most of the cultivars of lily when infected with CMV. Flower colour breaking and curling or malformation of leaf and petal may be seen in some cultivars (Derks et al. 1997; Ryu et al. 2002). Asiatic hybrid lily infection with CMV produces symptoms such as mild mosaic, ring spot on leaves and transient vein yellowing. Growth deformities and flower colour breaking also occur occasionally (Ram et al. 1999).

15.4.3.2 Characterization

Although the CMV isolates were reported to have wide host range worldwide, the CMV isolate collected from Asiatic lilies grown in Kangra Valley, Himachal Pradesh, India are able to infect only few plant species and cause disease (Ram et al. 1999). This CMV isolate differs from other CMV isolates prevailing in Kangra Valley. Isolates of CMV collected from different *Lilium* species showed high similarity with CMV isolates of different crop species in their RNA 3 sequences. Restriction pattern of RT-PCR products of CP gene revealed that lily isolates belong to subgroup I of CMV (Jung et al. 2000; Chen et al. 2001; Ryu et al. 2002). In contrast, analysis of the CP gene sequence showed that CMV lily isolate collected from Himachal Pradesh has close affinity with CMV subgroup II. Though the sequences of the CMV lily isolates have high similarity with other CMV isolates, their pathogenicity and its symptom expression patterns are highly different from others (Choi et al. 2004; Lee et al. 2007).

15.4.3.3 Detection and Diagnosis

Several methods such as bioassay, double antibody sandwich enzyme-linked immunosorbent assay (DAS-ELISA), electron microscopy, immunosorbent electron microscopy, transmission electron microscopy, cytopathology, reverse transcription polymerase chain reaction (RT-PCR) and sequencing have been used for the detection and characterization of lily CMV. Zhang et al. (2017) developed triplex immunocapture reverse transcriptase PCR which has 81.4% sensitivity of detecting CMV of lilies.

15.4.3.4 Transmission

Myzus persicae, *Macrosiphum euphorbiae* and *Aphid fabae* are the common vectors of lily viruses. *M. persicae* is an efficient vector among other aphid species in transmitting CMV in a non-persistent manner (Sharma et al. 2005).

15.4.3.5 Management

Growing of virus-free certified planting materials and fallowing of land for some period combined with clean cultivation and removal of volunteer lily plants help to reduce the viral inoculum build-up and enhance the quality and quantity production of bulb and flower (Ram et al. 1999). Spraying of mineral oil and pyrethroid compounds on lily plants helps to prevent the spread of virus by aphids.

15.5 Tulips

Tulip is one of the most important bulbous ornamental crops which is being cultivated for cut flower and as potted plant and garden plant and used for landscaping in temperate regions of most countries. Diseases caused by the viruses led to quantitative and qualitative loss of tulip flower and bulb production. Twenty-two types of viruses have been reported worldwide that cause infection and damage in tulips (Sochacki 2013). Among the viruses, Tulip breaking virus (TBV), Tobacco necrosis virus (TNV), LSV, CMV and Tobacco rattle virus (TRV) are considered as the most dangerous viruses which cause severe loss in tulips (Deligoz and Sevik 2019). The characters of these important tulip viruses and its symptoms and transmission are given in Table 15.1 (Fig. 15.5).

Table 15.1 List of viruses, symptom, transmission and diagnosis in tulip

Virus, genus	Characters	Symptoms	Transmission	Diagnosis
TBV, *Potyvirus*	Filamentous, non-enveloped flexuous rods, 680–900 nm long, 12–15 nm in diameter, with helical symmetry, and contain non-segmented single- positive sense ssRNA (Khan and Dijkstra 2002)	Mosaic and chlorosis on leaves, and striping of the leaves and irregular flame, feather-like patterns on the flowers and production of abnormal flowers (Fig. 15.5) (Dekker et al. 1993; Lesnaw and Ghabrial 2000; Agoston et al. 2020)	Mechanical sap transmission and by aphid species including *Myzus persicae* and *Aphis gossypii* (Sochacki 2013)	DAS-ELISA (Deligoz and Sevik 2019), RT-PCR, cloning and sequence analysis, optical sensor technique and use of IoT with deep convolutional neural network-based detection (Derks et al. 1982; Dekker et al. 1993; Polder et al. 2010; Deligoz and Sevik 2019; Agoston et al. 2020)
TRV, *Tobravirus*	Rigid rod-shaped particles and bipartite ssRNA genome	Appearance of light green, grey lozenge-shaped discoloured spots or irregular bands on the leaves. Scattered appearance of lozenge-shaped light or dark-coloured spots in the petals of the flowers. Stunted growth of plants and flower malformation (Asjes 1994; Engelmann and Hamacher 2008)	Mechanical and by stubby root Nematodes, *Paratrichodorus* spp. and *Trichodorus* spp. (Engelmann and Hamacher 2008)	Immunoelectron microscopy, cDNA hybridization assay, DAS-ELISA, RT-PCR and sequence analysis (Van der Vlugt et al. 1988; Asjes et al. 1992; Pearson et al. 2009; Sochacki 2013)
LSV, *Carlavirus*	Filamentous particle, 17–18 nm wide and 640 nm long (Allen 1972). The genome of LSV is monopartite, and the genome is	Darkening of petal veins in the pink- and red-flowering cultivars (Derks and Asjes 1975), colour breaking in flowers and	Mechanical and aphids, *M. persicae*, *Macrosiphum euphorbiae* and *A. gossypii* (Derks and Asjes 1975)	Quantitative RT-PCR (Nesi et al. 2013; Sun et al. 2014), RT-PCR combined with enzyme-linked immunosorbent assay (ELISA)

(continued)

Table 15.1 (continued)

Virus, genus	Characters	Symptoms	Transmission	Diagnosis
	ssRNA with 8394 nt and contains six open reading frames (ORFs)	leaf mottling (Horst 2008)		(Sharma et al. 2005), immunocapture RT-PCR (Kim et al. 2012), microsphere-based fluorescent immune assay (Zou et al. 2011), dot- or tissue-blot immune enzyme assay (Makkouk et al. 1993) and loop-mediated isothermal amplification (He et al. 2016)
TNV, *Necrovirus*	The genome of TNV is unsegmented, consists of a single-stranded linear positive-sense RNA of 3.8 kb that lacks a 30 poly-A tail and replicates itself with the aid of its own RNA-dependent RNA polymerase (Fang and Coutts 2013; Newburn et al. 2014). Two separate species of TNV such as TNV-A and TNV-D have been reported to cause diseases in different crops (Rubino and Martelli 2010; Sit and Lommel 2010; King et al. 2011).	Plants become stunted and distorted when symptoms appear early and after emergence of flowers in tulips. Chlorotic, brown elliptical or oval or round necrotic streaks appear in the leaves when severe infection occurs. Streaks and malformation of flowers. Pre-mature death of plants (Asjes 1994; Engelmann and Hamacher 2008)	Mechanical and by the fungus *Olpidium brassicae* (Asjes 1994; Engelmann and Hamacher 2008; Sochacki 2013)	DAS-ELISA and RT-qPCR (Sochacki 2013; Bacso et al. 2016)

(continued)

Table 15.1 (continued)

Virus, genus	Characters	Symptoms	Transmission	Diagnosis
CMV, *Cucumovirus*	CMV is a multipartite virus with three isometric particles and its genome consists of three positive single-stranded RNAs, viz. RNA1, RNA2, RNA3, and a subgenomic particle RNA4	Appearance of chlorotic or brown or grey necrotic streaks or spots or bands in the leaves. Darkening of petal edges of the flowers. Sunken brown spots or rings in bulbs which produce distorted and stunted plants, colour breaking in flowers (Yamamoto 1971; Asjes 1994	Mechanical; and more than 60 aphid spp. are reported to transmit the virus	DAS-ELISA (Duraisamy and Pokorny 2009; Pearson et al. 2009; Sochacki 2013) and tissue blotting immunoassay (Kim et al. 1995)

15.5.1 Management of Tulip-Infecting Viruses

Growing of virus-free certified planting materials and TBV-resistant cultivars such as Cantata, Juan, Madame Lefeber and Princeps (Romanow et al. 1990), clean cultivation and removal of volunteer plants and weeds help to reduce the viral inoculum build-up and enhance the quality and quantity production of bulb and flowers of tulips. Disinfection of cutting tools, containers and equipment prevents the spread of virus inoculum. Application of spent mushroom in the furrows during planting and subsequent cultivation of the field with fodder radish reduced the infection percentage in tulip under favourable conditions for TRV infection (Zoon et al. 2002). Late planting of tulips, that is, after the first week of December, and crop rotation with barley had a high impact on the TRV carrying nematode population and resulted in the control of TRV infection in tulips (Asjes 1974, 1994). The spread of TRV was more in the sandy soil when the tulip plants are grown with a spacing of 15 cm than with 7.5 cm (Decraemer and Geraert 2006). Use an appropriate soil fumigants and soil solarization practice to reduce the fungal and nematode vectors of TNV and TRV. Spraying of mineral oil combined with pyrethroid compounds reduced the spread by aphids of TBV and LSV in lily (Asjes 1991).

Fig. 15.5 Tulip breaking virus. (**a**, **b**) Striping of leaves; (**c**, **d**) Flower colour breaking (Agoston et al. 2020); (**e**) Color breaking symptoms on petals. (John Fisher, Ohio Department of Agriculture, Bugwood.org)

References

Agoston J, Almasi A, Salanki K, Palkovics L (2020) Genetic diversity of potyviruses associated with tulip breaking syndrome. Plants 9(12):1807. https://doi.org/10.3390/plants9121807

Agrios GN (2005) Plant pathology, 5th edn. Elsevier Academic Press, Amsterdam, p 767

Allen TC (1972) Control of viruses in lilies. In: Commonwealth Mycological Institute/Association of Applied Biologists. Descriptions of Plant Virus, vol 96. Institute of Horticulture Research, Warwich

Alper M, Koenig R, Lesemann DE, Loebenstein G (1982) Mechanical transmission of a strain of tulip breaking virus from *Lilium longiflorum* to Chenopodium spp. Phytoparasitica 10:193–199

Aly R, Stein A, Levy S, Raccah B, Loebenstein G (1986) Spread and control of Cucumber mosaic virus in gladiolus. Phytoparasitica 14:205–217

Aravintharaj R, Balaji CG, Nagendran K, Priyanka R, Karthikeyan G (2017) First report of lily mottle virus on lily (Lilium sp.) in southern India. Virus Dis 28(2):222–223

Asjes CJ (1974) Soil-borne virus diseases in ornamental bulbous crops and their control in the Netherlands. Agric Environ 1:303–315

Asjes CJ (1991) Control of air-borne field spread of tulip breaking virus, lily symptomless virus and lily virus X in lilies by mineral oils, synthetic pyrethroids, and a nematicide in the Netherlands. Neth J Plant Pathol 97(30):129–138

Asjes CJ (1994) Viruses in tulip in the Netherlands. Acta Hortic 377:289–300

Asjes CJ (1997) Virus in bloembollen in kaart gebracht. Bloembollencultuur 108(4):50

Asjes CJ (2000) Control of aphid-borne Lily symptomless virus and Lily mottle virus in Lilium in the Netherlands. Virus Res 71:23–32

Asjes CJ, Van Schadewijk AR, Segers LC (1992) Detection of tobacco rattle virus by ELISA and test plants in main sprouts of tulip during storage at different temperatures. Neth J Pl Path 98:213–219

Bacso R, Kunstler A, Kiraly L (2016) *Tobacco necrosis virus* replication and spread in *Arabidopsis thaliana* ecotype Columbia: a potential system for studying plant defense reactions to symptomless virus infections. Acta Physiol Plant 38:139

Berkeley GH (1953) Some viruses affecting Gladiolus. Phytopathology 43:111–115

Bing A, Johnson GV (1971) Aluminum foil mulch reduces aphid transmission of Cucumber mosaic virus in gladiolus. Acta Hortic 23:286–291

Boontjes J (1983) Virusvrije lelies zijn groter areaal waard. Vakblad voor de Bloemisterij 38 (18):36–39

Brierley P (1952) Evidence on the significance of Cucumber mosaic and Tobacco ringspot viruses in Gladiolus. Gladiolus Mag 16(28–29):36–37

Brierley P (1962) Transmission of some Gladiolus viruses on tools in harvesting flowers and corms. Plant Dis Rep 46:505

Chastagner GA, van Tuyl JM, Verbeek M, Miller WB, Westerdahl BB (2018) Diseases of lily. In: McGovern RJ, Elmer WH (eds) Handbook of florists' crops diseases. Springer International Publishing, pp 1255–1256

Chen YK, Derks AFLM, Langeveld S, Goldbach R, Prins M (2001) High sequence conservation among cucumber mosaic virus isolates from lily. Arch Virol 146:1631–1636

Choi SA, Ryu KH (2003) The complete nucleotide sequence of the genome RNA of lily symptomless virus and its comparison with that of other carlaviruses. Arch Virol 148:1943–1955

Choi SK, Ahn HI, Kim MJ, Choi JK, Ryu KH (2004) Symptom determinant as RNA3 of lily isolates of Cucumber mosaic virus on Zucchini Squash. Plant Pathol J 20(3):212–219

Conijn CGM (2014) Developments in the control of lily diseases. Acta Hortic 1027:213–230

Decraemer W, Geraert E (2006) Ectoparasitic nematodes. In: Perry RN, Moens M (eds) Plant nematology. CABI, Oxfordshire, pp 154–181

Dekker EL, Derks AFLM, Asjes CJ, Lemmers MEC, Bol JF, Langeveld SA (1993) Characterization of potyviruses from tulip and lily which cause flower-breaking. J Gen Virol 74:881–887

Deligoz I, Sevik MA (2019) The occurrence of the Tulip breaking virus in tulips in the northern part of Turkey. Folia Hortic 31(2):263–268

Derks AFLM (1988) Symptomen van tulpemozaiekvirus. Vakwerk 62(16):13–15

Derks AFLM (1995) Lily. In: Loebenstein G, Lawson RH, Brunt AA (eds) Virus and virus-like diseases of bulb and flower crops. John Wiley & Sons, Chichester, pp 313–321

Derks AFLM, Asjes CJ (1975) Lily symptomless virus in tulip. Acta Hortic 47:67–68

Derks AFLM, Vink-van den Abeele JL, Van Schadewijk AR (1982) Purification of tulip breaking virus and production of anti-sera for use in ELISA. Eur J Plant Pathol 88:87–98

Derks AFLM, Lemmers MEC, van Gemen BA (1994) Lily mottle virus in lilies: characterization of strains and its differentiation from tulip breaking virus in tulips. Acta Hortic 377:281–288

Derks AFLM, Lemmers MEC, Hollinger TC (1997) Detectability in lily bulbs depends on virus, host and storage conditions. Acta Hortic 430:633–640

Dey KK, Melzer MJ, Li C, Sun X, Adkins S (2017) First report of tuberose mild mottle virus infecting tuberose (*Polianthes tuberosa*) in the United States. Plant Dis 102(2):461

Dubey VK, Aminuddin, Singh VP (2008) First report of a subgroup IA Cucumber mosaic virus isolate from gladiolus in India. Aust Plant Dis Notes 3:35–37

Dubey VK, Aminuddin A, Singh VP (2009) First report of Bean yellow mosaic virus in gladiolus in India. J Plant Pathol 91(4):787–791

Dubey VK, Aminuddin, Singh VP (2010) Molecular characterization of Cucumber mosaic virus infecting Gladiolus, revealing its phylogeny distinct from the Indian isolate and alike the Fny strain of CMV. Virus Genes 41:126–134

Dubey VK, Chandrasekhar K, Srivastava A, Aminuddin, Singh VP, Dhar K, Arora PK (2015) Expression of coat protein gene of *Cucumber mosaic virus* (CMV-subgroup IA) Gladiolus isolate in *Nicotiana tabacum*. J Plant Interact 10(1):296–304

Duraisamy GA, Pokorny R (2009) Survey of virus pathogens in gladiolus, iris, and tulip in the Czech Republic. Acta Universitatis Agriculturae Silviculturae Mendelianae Brunensis 57:79–86

Duraisamy GS, Pokorny R, Holkova L (2011) Possibility of Bean yellow mosaic virus detection in Gladiolus plants by different methods. J Plant Dis Prot 118:2–6

Engelmann J, Hamacher J (2008) Plant virus diseases: ornamental plants. In: Encyclopedia of virology, 3rd edn. Academic Press, New York, NY, pp 207–229

Fang L, Coutts RH (2013) Investigations on the tobacco necrosis virus D p60 replicase protein. PLoS One 8:e80912

He X, Xue F, Xu S, Wang W (2016) Rapid and sensitive detection of Lily symptomless virus by reverse transcription loop-mediated isothermal amplification. J Virol Methods 238:38–41

Hemachandra Reddy P, Venkataravanappa V, Swarnalatha P, Ashwathappa KV, Krishna Reddy M (2019) Characterization of recombinant bean yellow mosaic virus belongs to group-IV infecting gladiolus. Int J Curr Microbiol App Sci 8(9):1644–1657

Horner MB, Pearson MN (1988) Purification and electron microscopy studies of a probably potyvirus from *Polianthes tuberosa* L. J Phytopathol 122:261–266

Horst RK (2008) Westcott's plant disease handbook, 7th edn. Springer, Dortdrecht, p 1317

Jung HJ, Ueda S, Ryu KH, Lee SY, Choi JK (2000) A novel strain of cucumber mosaic virus isolated from Lilium longiflorum. Plant Pathol J 16:306–311

Kamo K, Gera A, Cohen J, Hammond J, Blowers A, Smith F, Van Eck J (2005) Transgenic Gladiolus plants transformed with the bean yellow mosaic virus coat-protein gene in either sense or antisense orientation. Plant Cell Rep 23:654–663

Kamo K, Jordan R, Guaragna MA, Hsu H-t, Ueng P (2010) Resistance to Cucumber mosaic virus in Gladiolus plants transformed with either a defective replicase or coat protein subgroup II gene from Cucumber mosaic virus. Plant Cell Rep 29:695–704

Katoch M, Abdin MZ, Ram R, Zaidi AA (2003) An overview of diagnostics for viruses infecting gladiolus. Crop Prot 22:153–156

Kaur C, Kumar S, Raj SK, Chauhan PS, Sharma N (2015) Characterization of a new isolate of Bean yellow mosaic virus Group-IV Associated with Mosaic disease of Gladiolus in India. J Plant Pathol Microb 6:10

Kaur C, Raj R, Susheel K, Purshottam DK, Agrawal L, Chauhan PS, Raj SK (2019) Elimination of Bean yellow mosaic virus from infected cormels of three cultivars of gladiolus using thermo-, electro- and chemotherapy. Biotech 9:154

Kehoe MA, Coutts BA, Buirchell BJ, Jones RAC (2014) Plant virology and next generation sequencing: experiences with a potyvirus. PLoS One 9(8):e104580

Khan JA, Dijkstra J (eds) (2002) Plant viruses as molecular pathogens. Food Products Press An imprint of The Haworth Press, New York, NY; London; Oxford

Kim JY, Shin HK, Lee SY, Lim JH, Ko JY, Chung SK, Choi JK (1995) Detection of lily symptomless virus, cucumber mosaic virus and tulip breaking virus according to the tissue positions and organs in lily plants by tissue blotting immunoassay. J Kor Soc Hortic Sci 36 (6):843–850

Kim JH, Yoo HN, Bae EH, Jung YT (2012) Development of an indirect ELISA and immunocapture RT-PCR for lily virus detection. J Microbiol Biotechnol 22(12):1776–1781

King AMQ, Lefkowitz E, Adams MJ, Carstens EB (2011) Virus taxonomy. In: Ninth Report of the International Committee on Taxonomy of Viruses. Elsevier, Amsterdam

Krishnareddy M, Smitha R, Devaraju, Jalali S (2007) Molecular characterization of a potyvirus infecting tuberose (*Polianthes tuberosa*) in southern India. Indian Phytopathol 60:251–258

Kulshrestha S, Mehra A, Hallan V, Raikhy G, Ram R, Zaidi AA (2005) Molecular evidence for occurrence of Tuberose mild mottle virus infecting tuberose (*Polianthes tuberosa*) in India. Curr Sci 89:870–872

Kwon JY, Ryu KH, Choi SH (2013) Reverse transcription polymerase chain reaction-based system for simultaneous detection of multiple lily-infecting viruses. Plant Pathol J 29:338–343

Langeveld SA, Dore JM, Memelink J, Derks AFLM, Van der Vlught CIM, Asjes CJ, Bol JF (1991) Identification of potyviruses using the polymerase chain reaction with degenerate primers. J Gen Virol 72:1531–1541

Lee KH, Choi HS, Choi GS, Kim JS (1996) Virus diseases of lilies in Korea. Acta Hortic 414:195–201

Lee JA, Choi SK, Yoon JY, Hong JS, Ryu KH, Lee SY, Choi JK (2007) Variation in the pathogenicity of lily isolates of Cucumber mosaic virus. Plant Pathol J 23(4):251–259

Lesnaw JA, Ghabrial SA (2000) Tulip breaking: past, present, and future. Plant Dis 84 (10):1052–1060

Lim MS, Kim SM, Choi SH (2016) Simultaneous detection of three lily-infecting viruses using a multiplex Luminex bead array. J Virol Methods 231:34–37

Lin L, Zheng H-Y, Chen J, Chen J-P, Zhang Q-Y, Zhao M-F, Antoniw JF, Adams MJ (2004) A new potyvirus from tuberose (*Polianthes tuberosa*) in China. Arch Virol 149:1107–1116

Makkouk KM, Hus HT, Kumari SG (1993) Detection of three plant viruses by dot-blot and tissue-blot immunoassays using chemiluminescence and chromogenic substrates. J Phytopathol 139:97–102

McWhorter FP (1957) A localized occurrence of Cucumber mosaic virus in Gladiolus. Plant Dis Rep 41:141–143

Mowat WP, Stefanac Z (1974) Aphid-transmitted viruses from lilies in Britain. Ann Appl Biol 76:281–287

Navalinskiene M, Samuitiene M (2001) Viral diseases of flower plants 15. Identification of viruses affecting gladiolus (Gladiolus L). Biogeosciences 1:31–35

Nesi B, Lazzereschi S, Pecchioli S, Grassotti A, Rizzo D, Stefani L, Paoli M, Della Bartola M, Materazzi A (2013) Detection of virus infecting Lilium spp. By RT-PCR and real-time PCR. Acta Hortic 1002:215–221

Newburn LR, Nicholson BL, Yosefi M, Cimino PA, White KA (2014) Translational readthrough in Tobacco necrosis virus-D. Virology 451:258–265

Nezamabad PS, Habibi MK, Dizadji A, Kalantari S (2015) Elimination of Bean yellow mosaic virus through thermotherapy combined with meristem-tip culture in gladiolus corms. J Crop Prot 4:533–543

Niimi Y, Han DS, Mori S, Kobayashi H (2003) Detection of cucumber mosaic virus, lily symptomless virus and lily mottle virus in Lilium species by RT-PCR technique. Sci Hortic 97:57–63
Owen J, Palukaitis P (1988) Characterization of cucumber mosaic virus. I. Molecular heterogeneity mapping of RNA 3 in eight CMV strains. Virology 166(2):495–502
Palukaitis P, Roossinck MJ, Dietzgen RG, Francki RIB (1992) Cucumber MOSAIC virus. Adv Virus Res 41:281–348
Park IS, Kim KW, Kyun H, Chang MU (1998) The viruses in gladiolus hybrids cultivated in Korea Bean yellow mosaic virus and Clover yellow vein virus. Korean J Plant Pathol 14:74–82
Pearson MN, Cohen D, Cowell SJ, Jones D, Blouin A, Lebas BSM, Shiller JB, Clover GRG (2009) A survey of viruses of flower bulbs in New Zealand. Australas. Plant Pathol 38:305–309
Pearson MN, Horner MB (1986) A potyvirus of *Polianthes tuberosa* in New Zealand. Australas Plant Pathol 15:39
Polder G, van der Heijden GWAM, van Doorn J, Clevers JGPW, van der Schoor R, Baltissen AHMC (2010) Detection of the tulip breaking virus (TBV) in tulips using optical sensors. Precis Agric 11:397–412
Raj SK, Srivastava A, Chandra G, Singh BP (2002) Characterization of cucumber mosaic virus isolate infecting Gladiolus cultivars and comparative evaluation of serological and molecular methods for sensitive diagnosis. Curr Sci 83(9):1132–1137
Ram R, Sharma A, Singh RK, Chauhan D, Zaidi AA (1999) Cucumber mosaic virus on Asiatic Hybrid lilies in India. Plant Dis 83:78
Romanow LR, Van Eijk JP, Eikelboom W, Van Schadewijk AR, Peters D (1990) Determining levels of resistance to tulip breaking virus (TBV) in tulip (*Tulipa* L.) cultivars. Euphytica 51 (3):273–280
Roossinck MJ (2002) Evolutionary history of Cucumber mosaic virus deduced by phylogenetic analyses. J Virol 76(7):3382–3387
Rosner A, Stein A, Levy S, Lilien-Kipnis H (1994) Evaluation of linked PCR-transcription amplification procedure for bean yellow mosaic virus detection in gladioli. J Virol Methods 47:227–236
Rubino L, Martelli GP (2010) Necrovirus. In: Mahy BWJ, van Regenmortel MHV (eds) Desk encyclopedia of plant and fungal virology. Academic Press, New York, NY, pp 233–235
Ryu KH, Park HW, Choi JK (2002) Characterization and sequence analysis of a lily isolate of cucumber mosaic virus from Lilium tsingtauense. Plant Pathol J 18:85–92
Selvarajan R, Gupta MD (1996) Characterization of a virus causing mosaic on gladiolus. Int J Trop Plant Dis 14:217–222
Selvarajan R, Gupta M, Datta Misra RL (1999) Production of gladiolus mosaic virus-free plants through meristem tip culture and in combination with thermotherapy and chemotherapy. J Orna Hortic 2(2):101–106
Sharma A, Mahinghara BK, Singh AK, Kulshrestha S, Raikhy G, Singh L, Verma N, Hallan V, Ram R, Zaidi AA (2005) Identification, detection and frequency of lily viruses in northern India. Sci Hortic 106:213–227
Singh BR, Dubey VK, Aminuddin (2007) Inhibition of mosaic disease of Gladiolus caused by Bean yellow mosaic- and Cucumber mosaic viruses by virazole. Sci Hortic 114:54–58
Sit TL, Lommel SA (2010) Tombusviridae, encyclopedia of life sciences (ELS). Wiley, Chichester
Sochacki D (2013) The occurrence of the viruses in tulip crops in Poland. J Hortic Res 21(1):5–9
Srivastava KM, Raizada RK, Singh BP (1983) Bean yellow mosaic virus on Gladiolus psittacinus in India. Indian J Plant Pathol 1:83–88
Stein A (1995) Gladiolus. In: Loebenstein G, Lawson RH, Brunt AA (eds) Virus and virus-like diseases of bulb and flower crops. Wiley, New York, NY, pp 281–292
Sun N, Deng CL, Zhao XL, Zhou Q, Ge GL, Liu Y, Yan WL, Xia Q (2014) Extraction of total nucleic acid based on silica-coated magnetic particles for RT-qPCR detection of plant RNA virus/viroid. J Virol Methods 196:204–211
Suzuki M, Kuwata S, Kataoka J, Masuta C, Nitta N, Takanami Y (1991) Functional analysis of deletion mutants of cucumber mosaic virus RNA3 using an *in vitro*. Virology 183:106–113

University of Illinois Extension (1983) RPD no. 612. http://ipm.illinois.edu/diseases/rpds/612.pdf

Van der vlugt CIM, Linthorst HJM, Asjes CJ, Van schadewijk AR, Bol JE (1988) Detection of tobacco rattle virus in different parts of tulip by ELISA and cDNA hybridisation assays. Neth J Pl Path 94:149–160

Wade GC (1984) Gladiolus diseases. Tasman J Agric 19:36–40

Wahyuni W, Dietzgen R, Hanada K, Francki R (1992) Serological and biological variation between and within subgroup I and II strains of Cucumber mosaic virus. Plant Pathol 41:282–297

Wang RY, Wang JH, Wang Y, Xie ZK, An LZ (2007) Comparison of two gel filtration chromatographic methods for the purification of lily symptomless virus. J Virol Methods 139:125–131

Wei T, Lebas BSM, Shiller JB, Quinn BD, Clover GRG (2012) Detection of five viruses infecting dormant bulbs by TaqMan-based real-time RT-PCR. Australas Plant Pathol 41:93–98

Yamamoto M (1971) Experiments on tulip bulb production in relation to cucumber mosaic virus. Acta Hortic 23:267–272

Zaidi AA, Ram R, Zaidi SNH, Mukherjee D (1993) Diagnosis of viruses in some ornamental plants with special reference to serological methods: new developments. Ind Rev Life Sci 13:157–174

Zhang Y, Wang Y, Meng J, Xiea Z Wang R, Kutcher R, Guo Z (2015) Development of an immunochromatographic strip test for rapid detection of lily symptomless virus. J Virol Methods 220:13–17

Zhang Y, Wang Y, Xie Z, Yang G, Guo Z, Wang L (2017) Simultaneous detection of three lily viruses using Triplex IC-RT-PCR. J Virol Methods 249:69–75

Zhang Y, Wang Y, Xie Z, Wang R, Guo Z, He Y (2020) Rapid detection of Lily mottle virus and Arabis mosaic virus infecting lily (Lilium spp.) using reverse transcription loop-mediated isothermal amplification. Plant Pathol J 36(2):170–178

Zhao B, Yang D, Zhang Y, Xu Y, Zhao X, Liang J, Fan X, Du Y, Zhu Z, Shi B, Zhang Q, Zhang X, Cai Y, Zhao K (2018) Rapid visual detection of Lily mottle virus using a loop-mediated isothermal amplification method. Arch Virol 163:545–548

Zheng HY, Chen J, Zhao MF, Lin L, Chen JP, Antoniw JF, Adams MJ (2003) Occurrence and sequences of Lily mottle virus and Lily symptomless virus in plants grown from imported bulbs in Zhejiang province, China. Arch Virol 148:2419–2428

Zitter TA, Murphy JF (2009) Cucumber mosaic. In: The Plant Health Instructor. https://doi.org/10.1094/PHI-I-2009-0518-01

Zoon FC, van Bruggen AS, de Heij A, Asjes CJ, van den Ende JE (2002) Effect of green manure crops and organic amendments on incidence of nematode-borne tobacco rattle virus. Acta Hortic 570:287–292

Zou MQ, Zhang F, Li JF, Wang N (2011) Rapid detection of lily symptomless virus with CdTe quantum dots by flow cytometry. J Immunoass Immunochem 32:259–268

Molecular Characterization of Cucumber Mosaic Virus Strains Infecting Gerbera Cultivars and Its Disease Management

16

Karmveer Kumar Gautam, Susheel Kumar, and Shri Krishna Raj

Abstract

Gerbera (*Gerbera jamesonii*) is cultivated all over the world and is widely popular as an ornamental plant. It is grown in beds or pots and used as cut flower, in making bouquets, and for decoration in ceremonial functions. It has been commercially cultivated by a large number of growers in India as a primary source of income. The gerbera cultivation area in India is increasing day by day due to its increasing uses and market demands. Gerbera production has been hampered by numerous viruses that affect its flower quality and quantity. Among them, *Cucumber mosaic virus* (CMV) causes considerable loss in gerbera industry. In this chapter, we describe about detection methods, molecular characterization of CMV strains infecting gerbera, and their management strategies for its quality production.

Keywords

Gerbera jamesonii · *Cucumber mosaic virus* · Molecular identification · Virus-free plants · Virus-resistant transgenic plants

16.1 Introduction

Gerbera (*Gerbera jamesonii*) of family Asteraceae is a popular ornamental plant and commonly known as African daisy. *G. jamesonii* is under cultivation for cut flower production in subtropical and Mediterranean climates of Israel, Italy, Spain, Portugal, Morocco, Colombia, Japan, South Africa, Australia, and Southern India.

K. K. Gautam · S. Kumar · S. K. Raj (✉)
Plant Molecular Virology Laboratory, CSIR-National Botanical Research Institute, Lucknow, Uttar Pradesh, India

S. K. Raj et al. (eds.), *Virus Diseases of Ornamental Plants*,
https://doi.org/10.1007/978-981-16-3919-7_16

The gerbera flowers are daisy-like and are of wide range of colors, viz., yellow, orange, cream, white, pink, red, and various other intermediate shades. It is considered as the fifth most used cut flower in the world (http://en.wikipedia.org/wiki/Gerbera). Gerbera has a great ornamental value due to its typical capitulum inflorescence which is highly valued by consumers as individual vase decorations and bouquet (Mata et al. 2009). Gerbera cultivars of commercial importance throughout the world are Zingaro (red), Silvester (white), Salvadore (yellow), Rosaline (pink), Flamingo (pale rose), Vista (red), Terra queen (pink), Valentine (pink), Fredaisy (pink), Fredorella (red), etc. The cultivars Cream Clementine, Maroon Clementine, Delphi, Vista, Uranus, Terraqueen, Dusty, Valentine, Winter Queen, Rosetta, Gloria, Pricilla, Zingaro, Dune, and Monique are commercially cultivated in Uttarakhand, Himachal Pradesh, Karnataka, Jammu and Kashmir, Assam, Telangana, Meghalaya, Maharashtra, Tamil Nadu, Nagaland, and Mizoram states of India (https://vikaspedia.in/agriculture/cropproduction/package-of-practices/flowers/gerbera%20cultivation). Tropical Floritech Pvt. Ltd. in Bangalore is the leading company for tissue culture production and commercial cultivation of gerbera in India (Choudhary and Prasad 2000).

However, the gerbera production is challenged by numerous viral pathogens such as *Impatiens necrotic spot virus* (INSV), *Tobacco mosaic virus* (TMV), *Tomato black ring virus* (TBRV), *Cucumber mosaic virus* (CMV), *Tobacco rattle virus* (TRV), and *Tomato spotted wilt virus* (TSWV) that affect its flower quality and quantity. Among them, CMV is considered as the most important due to its infectivity and transmission in a variety of host species worldwide (Edwardson and Christie 1987). CMV causes significant losses to most of the major crops around the world; therefore, it is a bottleneck to the crop production (Hull and Davies 1992; Raj et al. 2008) including gerbera (Gautam et al. 2014). Hence, in this chapter, we describe about disease symptom, detection methods, and biological and molecular characterization of CMV strains infecting gerbera and their management strategies for enhancing the quality and production of gerbera in India.

16.2 Biological and Molecular Characterization of *Cucumber Mosaic Virus* Strains Infecting Gerbera Cultivars

Finlay in 1975 (Finlay 1975), for the first time, reported CMV on gerbera (*Gerbera jamesonii* Bolus) plants from two nurseries in Australia based on mottled leaf and distorted petals with color break symptoms on flower, virus transmission mechanically and by aphid *Myzus persicae*, and transmission electron microscopic (TEM) and serological study (Finlay 1975).

Verma and co-worker, in 2004 (Verma et al. 2004), reported CMV on gerbera (*G. jamesonii*) for the first time from India based on virus transmission, enzyme-linked immunosorbent assay (ELISA), and TEM study of virus particle and its molecular characteristics. The CMV was isolated from gerbera expressing color break on the petals, asymmetrical ray florets, and deformed flower symptoms on gerbera growing in floriculture fields at Palampur and nearby nurseries. The virus was also transmitted in a nonpersistent manner by *Myzus persicae* and *Aphis gossypii* and was identified as CMV using ELISA with CMV-specific antibodies. Polyhedral particles approximately 29 nm were observed with electron microscopy of leaf dips from symptomatic gerbera leaves (Verma et al. 2004).

For molecular characterization, the total RNA was isolated from leaf tissues of infected gerbera and *N. glutinosa*. Reverse transcription-polymerase chain reaction (RT-PCR) using CMV-specific primers produced an amplicon predicted size of 540 bp. Sequence analysis of the amplicons (GenBank accession no. AJ634532) resulted in 91–99% homology with the partial inter cistronic region and partial coat protein gene (1042–1574 bp) of CMV subgroup I, hence identified as CMV subgroup I. This was the only report from India describing the CMV by ELISA and analysis of a small sequence amplified by RT-PCR from gerbera exhibiting color break symptoms on petals, asymmetrical ray florets, and deformed flowers (Verma et al. 2004).

Gautam and co-worker, in 2014 (Gautam et al. 2014), reported CMV on gerbera (*G. jamesonii*) based on complete RNA3 genome sequences associated with severe chlorotic mosaic and flower deformation disease in two cultivars (Zingaro and Silvester) growing in a polyhouse at Lucknow, India. The naturally infected gerbera plants exhibited severe chlorotic mosaic, greening of vein symptoms, severe color breaking, and flower deformation symptoms (Fig. 16.1).

Fig. 16.1 A view of naturally infected gerbera plants exhibiting chlorotic mosaic symptom in polyhouse conditions (**a**), a close view of infected leaf showing severe chlorotic mosaic and greening of vein symptoms (**b**), and severe color breaking and flower deformation symptoms (**c**) as compared to healthy flower (**d**) (Gautam et al. 2014)

The transmission of causal virus was attempted using the leaf sap of naturally infected gerbera (*G. jamesonii*) plants of cultivars (cvs.) Zingaro and Silvester, separately on some recipient host seedlings. During sap transmission tests, the virus was successfully transmitted from naturally infected gerbera to healthy gerbera seedlings which developed similar chlorotic mosaic symptoms at 40–45 dpi, suggesting Koch's postulates. The inoculations of sap taken from cultivars Zingaro and Silvester also induced more or less similar local and necrotic lesions and systemic mosaic symptoms on *C. sativus*, *C. annuum*, *P. hybrida*, *N. glutinosa*, *N. tabaccum* cv. White Burley, and *N. rustica* at 30–35 dpi.

For molecular detection of the virus, the total RNA was extracted from 100 mg leaf tissue of symptomatic plants of gerbera cv. Zingaro and Silvester, and reverse transcription-polymerase chain reaction (RT-PCR) was performed using CMV-CP gene-specific primers. Electrophoresis of RT-PCR products resulted in amplification of expected size ~650 bp bands in naturally infected gerbera samples of cv. Zingaro and Silvester, similar to as in CMV-Banana strain taken as positive control, confirming the presence of CMV.

Further, the complete RNA3 genome of CMV was amplified by RT-PCR using CMV-RNA3 primer from three infected gerbera leaf samples. The amplicons obtained were cloned, sequenced, and deposited in GenBank under the accessions JN692495, JX913531 (from cv. Zingaro), and JX888093 (from cv. Silvester). These sequences shared 98–99% identities to each other and with a strain of CMV-Banana reported from India, and 90–95% identities with various strains of CMV reported worldwide (Gautam et al. 2014).

16.2.1 Genome Organization of CMV Infecting Gerbera and Phylogenetic Affinities

The analysis of sequence data of the three isolates of gerbera under study, Zing 1 (JN692495), Zing 2 (JX913531), and Silves 1 (JX888093), revealed the presence of full-length RNA3 genome of 2219 nucleotides consisted of two ORFs: movement protein (MP) gene of 840 nucleotides (ATG and TAG at 123 and 962 nucleotide positions, respectively) putatively translating 279 amino acids and coat protein (CP) gene of 657 nucleotides (ATG and TAG at 1261 and 1917 nucleotide positions, respectively) translating 218 amino acid residues (Fig. 16.2). Both ORFs were separated by 300-nucleotide-long intergenic region (IR) and flanked by 5′un-translated region (UTR) and 3′UTR of 122 and 302 nucleotides, respectively (Gautam 2015).

Phylogenetic relationships of virus isolates, JN692495 (Zing 1) and JX913531 (Zing 2) and JX888093 (Silves 1), under study were inferred using complete RNA3 genome sequences along with the selected strains of CMV of subgroups IA, IB, and II. Phylogenetic analysis clearly grouped subgroup IA, IB, and II members to their respective clusters (Fig. 16.3). The virus isolates under study clustered together and showed close relationship with CMV-Banana (EF178298), CMV-Eggplant (HQ343232), and CMV-Petunia (JN642676) strains reported from India

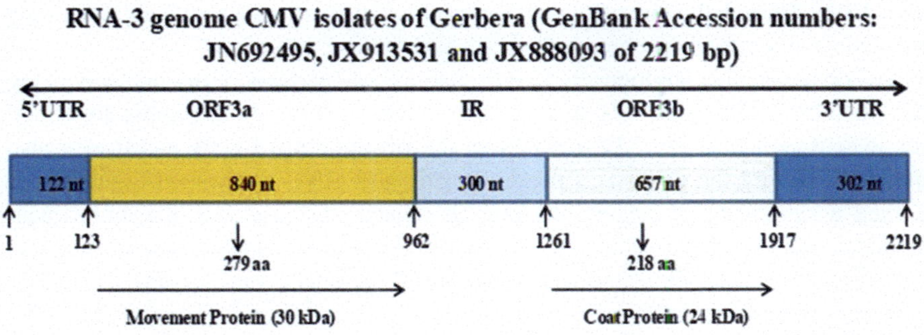

Fig. 16.2 Diagrammatic representation of RNA3 genome showing two ORFs: moment protein (MP of 279 amino acids) and coat protein (CP of 218 amino acid) and IR (300 nt) and 5′un-translated region (UTR) and 3′UTR (Gautam 2015)

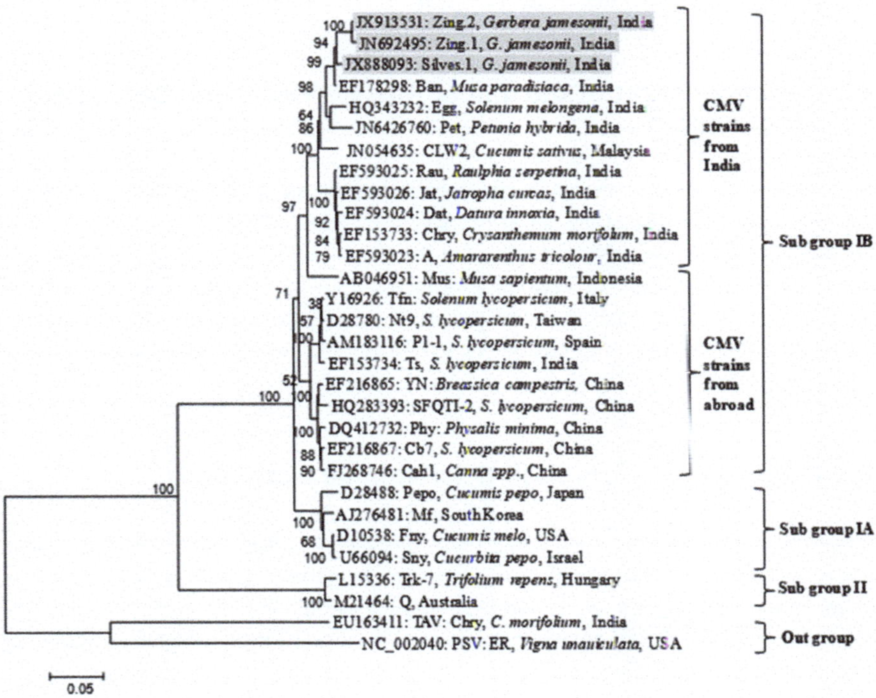

Fig. 16.3 Phylogenetic analysis of RNA3 genome sequences of gerbera virus isolates under study (JN692495, JX913531, and JX888093) with known CMV strains of subgroups IA, IB, and II. The isolates under study showed close relationships with Indian strains of CMV of subgroup IB (Gautam 2015)

(Fig. 16.3). Though the virus isolates under study clustered together, Zing 1 and Zing 2 isolates were closest as compared to Silves 1 isolate.

To observe the phylogenetic relationships of gerbera virus isolates Zing 1 (JN692495), JX913531 (Zing 2), and JX888093 (Silves 1) at protein level, CP amino acid sequences of CMV of all the strains of subgroups IA, IB, and II were also analyzed. During the analysis, Zing 1, Zing 2, and Silves 1 grouped together and formed a sub-cluster lying in subgroup IB cluster. As expected, the amino acid sequence analysis also revealed similar results as obtained with RNA3 genome, and CMV gerbera isolates clustered with subgroup IB members, more closely with CMV strains reported from India.

Both phylogenetic analyses revealed the closest affinity of gerbera virus isolates Zing 1, Zing 2 and Silves 1 under study with CMV-Banana strain, and close relationships with several other strains of CMV of subgroup IB. This study provided evidence that the gerbera virus isolates (Zing 1, Zing 2, and Silves 1) causing severe chlorosis and flower deformation in two cultivars (Zingaro and Silvester) of *G. jamesonii* in India belong to subgroup IB of CMV (Gautam 2015).

16.3 Management of Cucumber Mosaic Virus Infecting Gerber Cultivars

Management of viral diseases is much more difficult than that of diseases caused by other pathogens (Varma et al. 2002) because they have complex disease cycle and efficient vector transmission and there are no effective virucides available that can destroy them. Integration of various approaches like the avoidance of sources of infection, control of vectors, cultural practices (conventional), and use of resistant host plants (nonconventional) has been used for the management of diseases caused by plant viruses.

16.3.1 Management Through Conventional Methods

Prevention is the key for managing virus diseases because virus-infected plants cannot be cured. If viral infection is suspected in gerbera plants, samples should be sent to testing facilities to confirm the presence of the virus. Once the disease has been identified, the only management option is to discard infected plants (Whipker 2014). However, managing the vector of the virus can minimize the disease.

It is a well-known fact that sanitation of the cultivation fields enhances crop production by many folds. Remove all plant debris as well as weeds and flowering plants growing nearby production areas as these can be sources of new infections and infestations. It was suggested that soil sterilization can also eliminate the developmental stages of vector species (Elizabeth 2016).

Biological control of aphid vectors of CMV by ladybird (*Coccinella transversalis*) has also been studied by Kumar (2009). The biological control of aphid vector population (capable of transmitting CMV and TAV and potyviruses in several plant species) has been attempted by ladybird (*C. transversalis*), a predator of aphids. The feeding behavior of *C. transversalis* has been observed on chrysanthemums. Different larval stages, as well as adult ladybird predators, have been explored for minimizing aphid population (Kumar 2009).

16.3.2 Management of CMV by Development of Quick and Reliable Virus Diagnostic Protocols for Gerbera Viruses

The development of quick and reliable virus diagnostic protocols for detection of viruses in gerbera was a prerequisite for indexing of gerbera materials in bulk and for identifying virus-/disease-free materials to be used for large-scale gerbera propagation and its mass multiplication through tissue culture industry. In this direction, Gautam (2015) attempted for standardization and development of two protocols: Western blot immunoassay and RT-PCR using antiserum of CMV and CMV-specific primers, respectively, for successful detection of *Cucumber mosaic virus* (CMV) in two varieties of gerbera being cultivated in India.

Western blot immunoassay for CMV detection in gerbera was developed by Gautam (2015). During serological detection by Western blot immunoassay, the crude protein preparations from naturally infected gerbera cvs. Zingaro and Silvester samples reacted positively with the antiserum of CMV (PVAS242a, ATCC, USA), raised against the capsid protein, and showed two bands of 26 and 52 kDa, similar as in case of CMV-Banana taken as positive control; however, no such band was observed in a sample of healthy gerbera. The 26 kDa band was of coat protein of CMV, and 52 kDa band seems to be dimmer than 26 kDa protein as reported in case of many CMV strains. These results of Western blot immunoassay indicated the CMV infection in gerbera plants (Gautam 2015).

Molecular detection of *Cucumber mosaic virus* isolates in gerbera by RT-PCR was also developed by Gautam (2015). The RT-PCR performed with CMV-CP gene-specific primers resulted in amplification of expected size ~650 bp bands in naturally infected gerbera samples of cvs. Zingaro and Silvester, which was similar as in CMV-Banana-infected sample taken as positive control. The sap-inoculated gerbera (cvs. Zingaro and Silvester), *C. sativus* and *N. tabaccum* cv. White Burley, plants also showed ~650 bp amplicon when tested by RT-PCR (Fig. 16.4), confirming the presence of CMV in sap-inoculated plants (Gautam 2015).

Zang and co-workers, in 2009 (Zhang 2009), developed multiplex RT-PCR for detecting *Tomato black ring virus* (TBRV), *Tobacco mosaic virus* (TMV), and *Cucumber mosaic virus* (CMV), simultaneously, and for detecting *Tomato spotted wilt virus* (TSWV), *Tobacco rattle virus* (TRV), and *Tobacco mosaic virus* (TMV) and real-time RT-PCR for detecting four viruses, TBRV, TMV, CMV, and TSWV, respectively, from *G. jamesonii* Bolus. The multiplex RT-PCR for detecting TBRV, TMV, and CMV could detect as low as 1 μg of the three leaf tissues. While the multiplex RT-PCR for detecting TSWV, TRV, and TMV could detect as low as 1 mg of the three leaf tissues. The real-time RT-PCR could detect as low as 1 ng,

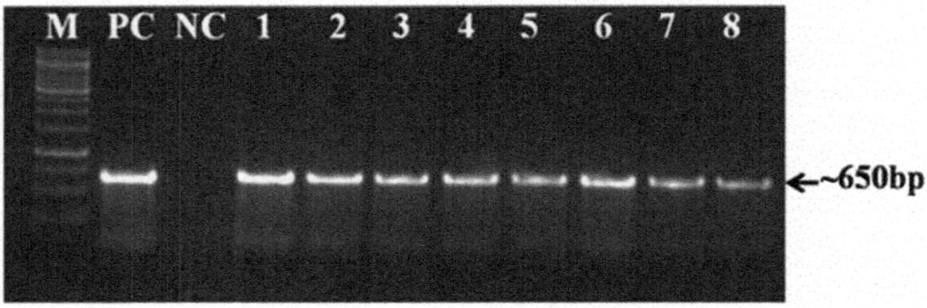

Fig. 16.4 Detection of CMV by RT-PCR from naturally infected and sap-inoculated plants using CMV-CP gene-specific primers. Lanes: PC = CMV-Banana-infected tobacco plant (as positive control); NC = healthy gerbera plants (as negative control); 1–4 = infected gerbera samples; 5–8 sap-inoculated test plants of *N. tabacum* cv. White Burley, *N. glutinosa*, *N. rustica*, and *C. sativus*; M = DNA marker (Gautam 2015)

1 ng, 100 pg, and 1 μg of the leaf tissues with TBRV, TMV, CMV, and TSWV, respectively (Zhang 2009).

16.3.3 Management Through In Vitro Chemotherapy and Development of CMV-Free Gerbera Plants

Viruses spread from mother plant to their progenies through infected cuttings, tubers, and other vegetative plant materials that have great possibility of virus transmission. Use of virus-free planting material and their transplantation in greenhouses that isolates crop from other plants which harbor viral diseases has been suggested for better crop production yield (Agrios 2005).

Literature survey revealed the only record of a postgraduate dissertation by Zhang (2009), who attempted development of virus-free plants of *G. jamesonii* cv. Bolus for management of four viruses: TBRV, TMV, CMV, and TRV. He inoculated these viruses on *G. jamesonii* and used its three different explants (tip, leaf, and torus) for their elimination through heat treatment followed by in vitro tissue culture. He found that tip and torus culturing was the best method for obtaining virus-free gerbera plants. He verified the virus-free plants by multiplex RT-PCR and real-time RT-PCR and found that the rates of virus-free by tip cultures were 75.0% and 66.7%, respectively, while rates were 52.0% and 54.5%, respectively, by torus cultures. In this study, the highest 75.0% and 54.5% virus-free gerbera plants were obtained by tip culture and torus culture, respectively, in combination with thermotherapy. The success of virus-free for *G. jamesonii* by tip culture and torus culture was the first report from China (Zhang 2009).

Then after, Gautam and co-workers in 2017 (Gautam et al. 2017) attempted successful elimination of CMV through in vitro chemotherapy (using 30 mg/L virazole) of ~4 × 8 mm^2 capitulum explants of infected gerbera cv. Zingaro for its quality improvement. A total of 38 plants were developed from 57 explants on Murashige and Skoog (MS) medium supplemented with 1 mg/L 6-benzylaminopurine (BAP), 0.5 mg/L indole-3-acetic acid (IAA), and 0.5 mg/L

Fig. 16.5 Different regeneration stages in the development of virus-free plants. Floral bud showing callus initiation after 1 month (**a**, **b**), shoot proliferation (**c**, **d**) after 60 days on MS medium supplemented with BA (1.0 mg/L), IAA (0.5 mg/L), and Ads (0.5 mg/L) hormones, rooting of shootlets in gerbera rooting media after 20 days (**e**), hardening of plants at culture room conditions for 1 week (**f**) and in glasshouse condition (**g**). A panoramic view of glasshouse-grown virus-free gerbera cv. Zingaro plants in comparison with CMV-infected gerbera plants at blooming stage (**h**) (Gautam et al. 2017)

adenine sulfate. Various regeneration stages in the development of virus-free plants have been showed in Fig. 16.5. Floral bud showed callus initiation after 1-month incubation. The shoot proliferation was achieved after 60 days on MS medium supplemented with BA (1.0 mg/L), IAA (0.5 mg/L) and hormones (0.5 mg/L). The rooting of shootlets developed after 20 days in rooting media. The rooted shootlets were hardened at culture room conditions for 1 week and then transplanted in pots and maintained in glasshouse condition till flowering stage. The glasshouse-grown virus-free gerbera cv. Zingaro plants were compared with CMV-infected gerbera plants at blooming stage (Fig. 16.5) (Gautam et al. 2017).

The developed plants showed absence of CMV in 81.6% (31/38) of plants when screened by RT-PCR using coat protein-specific primers of CMV. The CMV-free plants showed better plant growth, increase of 53.7% in length of leaf lamina and 59.2% in leaf width, as well as better blooming performance, increase of 62.6% in flower size (diameter in cm) and 69.1% in number of flowers per pot having intense red flower color as compared to the control ones. Elimination of CMV by in vitro chemotherapy (using virazole) of capitulum explants of gerbera cv. Zingaro is being reported for the first time from India (Gautam et al. 2017).

16.3.4 Management Through Genetic Transformation of Gerbera and Development of CMV-Resistant Transgenic Gerbera Plants

Pathogen-derived resistance has been observed to be mediated either by the protein encoded by the transgene (protein-mediated) or by the transcript produced from the transgene (RNA-mediated) also known as posttranscriptional gene silencing (PTGS) or both (Varma et al. 2002). In literature, several virus-resistant transgenic plants have been develop in various plants but in gerbera first reported from Poland for TSWV (Korbin et al. 2002) and then after from India for CMV (Gautam 2015).

Keeping in view of trait improvement and development of in-built resistance against viruses in gerbera, (Gautam 2015) attempted the genetic transformation using CMV-CP gene and leaf, petiole, and petiole base explants of gerbera. When different hormone combinations were added to MS basal medium, base petiole explant has shown higher regeneration efficiency as compared to petiole and leaf explants. The 7–25 mg/L concentrations of kanamycin were used in this study for *Agrobacterium*-mediated transformation of base petiole explants of gerbera. The use of 15 mg/L concentration of kanamycin was found optimum during transformation experiments.

Conclusively, a total of 310 base petiole explants of gerbera cv. Zingaro were co-cultivated for transformation. Out of them, 97 explants were successfully regenerated on selection medium. All shoots were transferred to rooting medium where 52% plants had developed a well-organized branched root system. The developed transgenic plants were placed under greenhouse conditions for acclimatization from where the survived 88% plants were chosen for molecular validation. The results of PCR, Southern, and Northern blot analyses confirmed that a total of 89% of plants had CP gene integration in their genome. The transgenic plants when challenged with mechnical inoculation of CMV. The results showed virus resistance in 53% and virus tolerance (delayed and mild symptoms) in 33% plants, while the rest of the plants showed severe disease symptoms for virus infection after challenged with mechanical inoculation of CMV. The developed protocol may be adopted for transferring any other gene of agronomic or economic interest in gerbera plants (Gautam 2015).

16.4 Conclusion

G. jamesonii is an important commercial ornamental plant. Its popularity increases day by day worldwide. In India, gerbera industry is supposed to be the sunrise industry in export point of view. It takes an important role in India as well as the world economy. However, several factors are responsible for bottleneck of gerbera production, and among them, viruses have an important role. Control of insect and fungal, bacterial, and phytoplasmal pathogens can be done using insecticide, fungicide, antibiotic, etc. The control of gerbera plant viruses is slightly difficult because of nonavailability of effective viricide for viruses. However, several conventional and nonconventional methods, viz., sanitation, biocontrol, and development of virus-free plants and transgenic gerbera plants, are available for effective management of gerbera viruses.

The present chapter particularly focuses on disease symptoms of infected gerbera as well as biological, serological, and molecular detection of viruses infecting gerbera worldwide. This review indicated that viruses until reported on gerbera have single-stranded RNA genome. Therefore, RNA-based pathogen-derived resistance against gerbera viruses would be useful for gerbera virus management. The protocol developed for elimination of CMV and production of virus-free elite varieties of gerbera cv. Zingaro plants may be utilized to save the germplasm from virus infection. The technique may also be used for the mass propagation of virus-free elite varieties of gerbera, which may be of importance to the floriculture industry.

This information summarized in this chapter will be useful for gerbera growers worldwide that ultimately would benefit in uplifting the economic and social status of the gerbera-related farmers. Moreover, an eco-friendly approach like biological control of virus-transmitting vectors in nature has also been suggested for virus-disease management which neither has adverse effect on human health nor possesses hazards to the environment. The developed diagnostic protocols may be used for quick and reliable detection of viruses in gerbera and for indexing of gerbera materials in bulk, and for identifying virus-/disease-free materials to be used for large-scale gerbera propagation for farmers and its mass multiplication through tissue culture industry.

References

Agrios GN (2005) Plant pathology, 5th edn. Elsevier Academic Press, Burlington, MA, p 952

Choudhary ML, Prasad KV (2000) Protected cultivation of ornamental crops - an insight. Indian Hortic 45(1):49–53

Edwardson JR, Christie RG (1987) Cucumoviruses. Viruses Infect Forage Legum Gainesville 1:143–1214

EI B-MC, Hausbeck MK (2016) Diseases of Gerbera. In: McGovern RJ, Elmer WH (eds) Handbook of florists' crops diseases, Handbook of plant disease management. Springer, New York, NY. https://doi.org/10.1007/978-3-319-32374-9_18-1

Finlay JR (1975) Cucumber mosaic virus in gerbera. Aust Plant Pathol Soc Newsl 4(2):14

Gautam KK (2015) Molecular detection and characterization of a strain of *Cucumber mosaic virus* infecting Petunia and Gerbera and their disease management. PhD thesis, Lucknow University, Lucknow

Gautam KK, Raj R, Kumar S, Raj SK, Roy RK, Katiyar R (2014) Complete sequence of RNA3 of *Cucumber mosaic virus* isolates infecting *Gerbera jamesonii* suggested its grouping under IB subgroup. Virus Dis 25:398–340

Gautam KK, Kaur C, Raj R, Kumar S, Raj SK, Purshottam DK, Roy RK (2017) Elimination of *Cucumber mosaic virus* from gerbera (*Gerbera jamesonii*) cv. Zingaro through *in vitro* chemotherapy of capitulum explants. Indian J Biotechnol 16:641–647

Hull R, Davies JW (1992) Approaches to non-conventional control of plant virus diseases. Crit Rev Plant Sci 11:17–33

Korbin M, Podwyszyńska M, Komorowska B, Wawrzyńczak D (2002) Transformation of gerbera plants with tomato spotted wilt virus (TSWV) nucleoprotein gene. Acta Hortic 572:149–157

Kumar S (2009) Molecular characterization of Cucumoviruses causing severe mosaic and ringspot diseases in chrysanthemum and development of their management strategies. PhD thesis, University of Lucknow, Lucknow

Mata TLD, Segeren MI, Fonseca AS, Colombo CA (2009) Genetic divergence among gerbera accessions evaluated by RAPD. Sci Hortic 121:92–96

Raj SK, Khan MS, Kumar S, Pratap D, Vishnoi R (2008) *Cucumber mosaic virus* infecting vegetable and pulse crops in India. In: Rao GP, Kumar PL, Pena RJH (eds) Characterization, diagnosis & management of plant viruses, Vol. 3: Vegetable and pulse crops. Studium Press LLC, Houston, TX, pp 39–62

Varma A, Jain RK, Bhat AI (2002) Virus resistance transgenic plant for environmentally safe management of virus diseases. Indian J Biotechnol 1:73–76

Verma N, Singh AK, Singh L, Kulshreshtha S, Raikhy G, Hallan V, Ram R, Zaidi AA (2004) Occurrence of *Cucumber mosaic virus* in *Gerbera jamesonii* in India. Plant Dis 88 (10):1161–1161

Whipker BE (2014) Gerbera: mottling and necrotic spotting. E-Gro Alert 3(4)

Zhang LL (2009) Study on main viruses of *Gerbera jamesonii* bolus with virus-free by tissue culture and its rapid detection. Master's thesis, Northwest University of Science and Technology, China. http://www.dissertationtopic.net/down/274429

Viruses Infecting *Narcissus tazetta* and Their Possible Management 17

Rashmi Raj, Susheel Kumar, P. S. Chauhan, and S. K. Raj

Abstract

Narcissus tazetta L. is a bulbous ornamental plant which belongs to the family Amaryllidaceae. It is popular for its sweet-smelling beautiful flowers and is also used for the production of fragrant oil and perfumes in India. Narcissus is also the source of alkaloids like lycorine, homolycorine, and galantamine used as traditional medicines for a variety of complaints. Narcissus is a commercially important ornamental plant; therefore, it is important to maintain its quality and quantity. Various groups of phytopathogens (fungi, bacteria, viruses, nematodes, and mites) affecting narcissus are of great concern as they caused loss of economic value at commercial cultivation. These pathogens may be controlled by chemicals except viruses. Viruses cause economic losses of narcissus in terms of reduction in bulb size, the number of bulbs, and the quality of narcissus flowers. In this review, we will describe about narcissus plant, type of virus disease, virus detection, characterization, and their possible management.

R. Raj
Plant Molecular Virology Laboratory, CSIR-National Botanical Research Institute, Lucknow, Uttar Pradesh, India

Academy of Scientific and Innovative Research (AcSIR), Ghaziabad, UP, India

S. Kumar · S. K. Raj (✉)
Plant Molecular Virology Laboratory, CSIR-National Botanical Research Institute, Lucknow, Uttar Pradesh, India

P. S. Chauhan
Academy of Scientific and Innovative Research (AcSIR), Ghaziabad, UP, India

Microbial Technology Division, Council of Scientific and Industrial Research National Botanical Research Institute, Lucknow, UP, India

S. K. Raj et al. (eds.), *Virus Diseases of Ornamental Plants*,
https://doi.org/10.1007/978-981-16-3919-7_17

Keywords

Narcissus tazetta · Alkaloid · Virus disease · Disease management

17.1 Introduction

Narcissus (*Narcissus tazetta* L.), also known as "daffodils," is a vegetatively propagated bulbous ornamental plant. It is worldwide accepted as an ornamental plant for cut flower in the floriculture industry (Hanks 2002). Most of the *Narcissus* genus blooms in spring while some in early winter, and some species are flowered in late autumn (Graham and Barrett 2004). Narcissus plants naturally grow in meadows, in woodlands, along watercourses, and in rocky outcroppings up to subalpine altitudes. The Mediterranean region and the central region of Europe (Spain, Portugal, and the Iberian Peninsula) are the native place of the narcissus species (Hanks 2002; Thakur 2017). Some narcissus species are also extending from Southern France, Italy, the Balkans, North Africa, etc. (Thakur 2017). Narcissus crop has also been cultivated from the earliest sixteenth century in the Netherlands (Hanks 2002). *N. tazetta* is a common species worldwide; however, they are found in Central Asia, Nepal, India, China, Japan, etc. with increasing international trading (Harvey and Selby 1997; Mathew 2002; Hanks 2002). In India, *N. tazetta* most probably was brought by the Mughals. Hundreds of years ago, *N. tazetta* was cultivated in Kashmir valley as the most scented flower (Willis 2012).

Different species of narcissus are present all over the world and are produced for commercial uses. All *Narcissus* spp. are perennial geophytes, are naturally occurring in a wide range of habitats, and flower in the spring and autumn (Thakur 2017). *Narcissus* sp. is classified on the basis of different flower forms or by botanical name into 13 divisions (Kington 1998): trumpet daffodils, large-cupped daffodils, small-cupped daffodils, double daffodils, triandrus daffodils, cyclamineus daffodils, jonquilla daffodils, tazetta (poetaz or bunch-flowered) daffodils, poeticus (poet's) daffodils, bulbocodium daffodils, split-corona daffodils, miscellaneous daffodils, wild variants, and wild hybrids. Narcissus is also the source of alkaloids like lycorine, homolycorine, and galantamine used as traditional medicines for a variety of disorders like cancer (Duke and Duke 1983). The Bible also provides many references about Mediterranean narcissus which have been used against cancer as they contain anticancer isocarbostyril constituents such as narciclasine and pancratistatin (Alexander and Antonio 2008). The galantamine alkaloid has been used for management of Alzheimer's disease (Bastida et al. 2006; Takos and Fred 2013).

Narcissus is a genus of herbaceous perennial bulbous plant which dies back after flowering to an underground storage bulb. Narcissus bulb consists of a disc-shaped stem plate and below the bulb having roots. The growing bulbs of narcissus typically produce four to eight small flowers with six tepals and cup-shaped corona with three types of trumpet, bowl and disc-shaped depending on the type of species. Generally, narcissus flowers are white or yellow (orange or pink is also found) in color, having a uniform or contrasting colored petals and corona. The fragrance of narcissus flower remains for 7–10 days after blooming. It usually goes ~5–80 cm above ground, depending on the species, and a single bulb may produce four to five bulbs and the life of each bulb is about 3–4 years (Hanks 2002).

17.2 Viruses Affecting Narcissus

Pests and diseases of narcissus plants are of great concern because they cause loss of economic value at commercial cultivation. Different pests and diseases are reported on narcissus plants which include insects, flies, snails, nematodes, fungi, bacteria, and viruses (Hanks 2002; Hanks et al. 2003; Hanks and Chastagner 2018). The bacterial diseases are not common in narcissus plants (Hanks 2002); however, some bacterial infections caused by *Pseudomonas* and *Pectobacterium carotovorum* sp. are reported by Kamenetsky and Okubo in 2012.

Narcissus is affected by a large number of RNA viruses. Under natural conditions, various viruses infect narcissus which include *Potyvirus* (*Narcissus yellow stripe virus*, *Narcissus latent virus*, *Narcissus tip necrosis virus*, *Narcissus late season yellows virus*, *Narcissus mosaic virus*, *Cyrtanthus elatus virus-A*), *Carlavirus* (*Narcissus common latent virus* and *Narcissus symptomless virus*), *Nepovirus* (*Tobacco ring spot virus*), *Carmovirus* (*Narcissus tip necrosis virus*) and *Cucumovirus* (*Cucumber mosaic virus*) (Brunt 1995, 1977, 1980; Chen et al. 2003; Wylie and Jones 2012). Among them, potyviruses are found to be the most prevalent which cause severe mosaic, yellow stripe symptoms, and reduced plant growth, flower quality, and production of bulbs (Brunt 1995; Aminuddin et al. 1999; Yadav and Khan 2008; Wylie and Jones 2012).

17.3 Disease Caused by Carlaviruses

Narcissus common latent virus and *Narcissus symptomless virus* belong to the genus *Carlavirus* of the family *Betaflexiviridae*. This family of viruses is non-enveloped with flexuous filamentous particles having a size of 12–13 nm. Virus genomes are linear of about 5.5–9.0 kb in length that contains 5′ terminus with m^7G and 3′ terminus polyadenylated tail. The size of viral capsid protein has ~18–44 kDa. Family *Betaflexiviridae* currently had 108 species, divided into 13 genera in two subfamilies. The two subfamilies include *Quinvirinae* that contains three genera (*Carlavirus*, *Fovevirus*, and *Robigovirus*) and *Trivirinae* that contains ten genera (*Capillovirus*, *Chordovirus*, *Citrivirus*, *Divavirus*, *Prunevirus*, *Ravavirus*, *Tepovirus*, *Trichovirus*, *Vitivirus*, and *Wamavirus*).

This family of virus(es) is mainly transmitted by mechanical inoculation and by vectors. In nature, this virus is transmitted by some vectors which include mealybugs, scale insect, and aphids in a non-persistent manner. *Narcissus common latent virus* and *Narcissus symptomless virus* cause mild mosaic symptoms on leaves and symptomless symptoms (Chen et al. 2006, 2003; Ward et al. 2009; Yan et al. 2008).

17.4 Disease Caused by Nepoviruses

Arabis mosaic virus and *Tomato black ring virus* belong to the genus *Nepovirus* of the family *Secoviridae*. This family of viruses is single-stranded positive-sense RNA virus. Genomes of *Secoviridae* are linear bipartite, which are around 24–27 kb in length, and they are non-enveloped with icosahedral particles and their diameter is around 25–30 nm. This family of virus(es) is mainly transmitted by mechanical inoculation and by vectors. Naturally, these viruses are transmitted by nematodes, mites, thrips, leafhoppers, and aphid vector in a non-persistent manner. *Arabis mosaic virus* (ArMV) and *Tomato black ring virus* (TBRV) cause yellow stripe symptoms on leaves and ring spot on leaves and lettuce ring spot (Kamenetsky and Okubo 2012).

ArMV infection in plants of *Narcissus* spp. was reported from the UK, Japan, Lithuania, and the Netherlands (Broadbent et al. 1962; Asjes 1971, 1974; Iwaki and Komuro 1974; Brunt 1995; Samuitiene et al. 2008). The infected daffodil plants exhibit symptoms of twisting and pale chlorosis of leaves (Asjes 1974). This virus is transmitted by nematode vectors (*Xiphinema* spp.) in a non-persistent manner and also by mechanical sap inoculation (Sastry et al. 2019).

17.5 Disease Caused by Carmovirus

Narcissus tip necrosis virus belongs to the genus *Carmovirus* of the family *Tombusviridae*. Viruses of this family have single-stranded positive-sense RNA viruses, which contains 76 species and is divided into three subfamilies. Genomes of *Tombusviridae* are monopartite except *Dianthoviruses* that contains a bipartite genome. The size of the genome is ~4.0–5.4 kb in length with 5′ cap and a poly-A tail which encoded four to six ORFs (open reading frames). The size of viral capsid protein was ~28–35 K. *Narcissus tip necrosis virus* causes necrotic lesion symptoms on the tip of the leaf.

17.6 Disease Caused by Cucumber Mosaic Cucumovirus

Cucumber mosaic virus belongs to the genus *Cucumovirus* of the family *Bromoviridae*. The family currently has 36 species, divided into six genera. This family of plant viruses contains tri-segmented, single-stranded positive-sense RNA viruses. Virion morphology of virus is variable (spherical or bacilliform). This family of viruses is mainly transmitted by pollen and insect vectors and by mechanical inoculation in a non-persistent manner. The diameter of the virions is ~26–35 nm of genera *Anulavirus*, *Bromovirus*, and *Cucumovirus* and 18–26 nm of genera *Alfamovirus*, *Ilarvirus*, and *Oleavirus* depending on the virus

The occurrence and prevalence of *Cucumber mosaic virus* and four nematode-transmitted viruses, Tobacco rattle virus (TRV), Tomato black ring virus (TBRV), Arabis mosaic virus (ArMV), and Strawberry latent ring spot virus (ALRSV) in British narcissus crops, was published for the first time in 1966 (Brunt 1966b). *Narcissus* spp., showing virus-like symptoms from Otago Province, New Zealand, were surveyed for viruses using ELISA and mechanical transmission tests. High incidence of virus infection was detected at five sites. *Cucumber mosaic virus*, *Narcissus latent virus*, *Narcissus mosaic virus*, *Narcissus tip necrosis virus*, and *Narcissus yellow stripe virus* (NYSV) were detected from *Narcissus* spp. in New Zealand (Clark and Guy 2000).

17.7 Narcissus Mosaic Potexvirus

Narcissus mosaic virus (NMV) is a plant pathogenic virus in the family *Alphaflexiviridae*, which infects narcissus. NMV was isolated in the Netherlands and the UK from *Narcissus pseudonarcissus* in 1946, and it was generally symptomless (van Slogteren and de Bruyn Ouboter 1946). Narcissus mosaic virus (NMV) is widespread in British crops of trumpet, large-cupped, and double daffodils. Many commercial daffodil cultivars seem to be totally infected and rouging or selection is therefore impracticable. Strict precautions by breeders and raisers to prevent infection of new cultivars are recommended. Healthy daffodil seedlings were readily infected with NMV by mechanical inoculation, but the virus was not detected in them until 17 months after inoculation, when a mild mosaic appeared. Unlike some viruses with elongated particles, NMV precipitates with antiserum in agar gel. Purified preparations reacted with antiserum to a Dutch isolate of NMV but not with antisera to seven other viruses having similar particles and in vitro properties, or to narcissus yellow stripe virus (Brunt 1966a).

Recently, a multiplex TaqMan RT-qPCR assay was developed by Jin et al. (2017) to simultaneously detect Narcissus yellow stripe virus (NYSV) and Narcissus mosaic virus (NMV), frequently causing mixed narcissus infection. Feasibility verification was confirmed in natural samples. Primers and probes were designed based on the conserved CP gene regions of NYSV or NMV and their suitability for singleplex and multiplex TaqMan RT-qPCR assays as well as for conventional RT-PCR. Conventional RT-PCR, singleplex, and multiplex TaqMan RT-qPCR assays proved to be NYSV and NMV specific. *P*-values and coefficients of variation of TaqMan RT-qPCR assays indicated high reproducibility. Significantly increased sensitivity was achieved compared to conventional RT-PCR. The detection limit of both viruses was 103 copies with superior correlation coefficients and linear standard curve responses between plasmid concentrations and Ct values. NYSV and NMV infection of narcissus leaves, petals, and bulbs could successfully be detected via multiplex RT-qPCR method at 1.25 mg. The multiplex TaqMan RT-qPCR assay provides rapid, specific, sensitive, and reliable testing to simultaneously detect

NYSV and NMV, supplying useful routine monitoring for different narcissus samples (Jin et al. 2017).

17.8 Disease Caused by Potyviruses

The *Potyvirus* genus belongs to the family *Potyviridae*. *Potyviridae* family was divided into six genera which include *Potyvirus*, *Bymovirus*, *Rymovirus*, *Maculavirus*, *Tritimovirus*, and *Ipomovirus* (Berger et al. 2005). Now according to the new ICTV report, the *Potyviridae* family has been divided into eight genera including six previous genera and two new genera *Brambyvirus* and *Poacevirus* (Wylie et al. 2017). All genera contain a monopartite genome having single RNA except one genus (Bymovirus), which contains a bipartite genome with two RNA molecules, i.e., RNA-1 and RNA-2 (Shukla et al. 1998; Wylie et al. 2017). The viruses of this family are transmitted in nature by various vectors such as aphids, mites, whiteflies, fungus, and nematodes in nature.

Genus potyvirus-type species *Potato virus* Y is the largest group of this family infecting a number of plant species in nature worldwide (Shukla et al. 1994). About 160 species were assigned into potyvirus which account for ~30% of reorganized plant viruses (Gallo et al. 2018). They have flexuous, single open reading frame (ORF) positive-sense RNA with ~9.4–11 kb in size, characterized by a 5′ VPg covalently linked untranslated region (5′ UTR) and a 3′ UTR region containing poly-A tail. Potyvirus genome contains a single long open reading frame which is translated into a large polyprotein having 340–370 kDa, and this polyprotein further yields the predicted ten mature proteins, a characteristic of the *Potyvirus* genus (Urcuqui-Inchima et al. 2001; Adams et al. 2005). These polypeptide chains cleaved into ten functional proteins were identified as P1 protein, HC-Pro, P3, 6K1, CI protein, 6K2, VPg, Nla-Pro, Nlb, and CP proteins (Shukla et al. 1998; Sharma et al. 2014).

Potyviruses induce a different type of symptoms accordingly to the species of the host plant, the cultivar, and the virus strain. Symptoms showing in infected host plants like necrotic or chlorotic lesions, mosaic, stunting, stripe, vein banding, mottling, vein clearing, flower breaking, and wilting most commonly lead to stunting (Sharma et al. 2014). A number of potyviruses infected narcissus which include *Narcissus yellow strip virus* that causes mosaic and chlorotic yellow stripe symptoms on leaves and reduces the size of bulbs (Caldwell and Kissick 1950; Rees 1966; Brunt 1995; Chen et al. 2006; Raj et al. 2019); *Narcissus latent virus* which causes mosaic chlorotic spots on leaves and transient leaf tip chlorosis symptoms (Brunt 1966a, b; Wylie and Jones 2012; Berniak et al. 2013); *Narcissus mosaic virus* which causes yellow mosaic, leaf distortion and chlorosis, and flower color breaking symptoms (Brunt 1995; Miglino et al. 2005; Hunter et al. 2011); *Narcissus late season yellows virus* which causes leaf streaking and stunting symptoms and reduces bulb size (Wylie and Jones 2012); *Cyrtanthus elatus virus-A* which causes chlorotic stripe symptoms along with leaf distortion and stunting (Chen et al. 2003; Kumar et al. 2015; Ohshima et al. 2016; Raj et al. 2018);

Ornithogalum mosaic virus which causes mosaic or chlorotic striping symptoms on leaves (Hong Ying et al. 2009); *Narcissus white streak virus* which showed greenish-purple streak symptoms on leaves that turn white to yellow, reducing flower stalk, bulb size, and yields (Hanks 1993; Yan et al. 2008); and *Narcissus degeneration virus* which causes mosaic or chlorotic striping symptoms on leaves (Chen et al. 2006; Ward et al. 2009; Ohshima et al. 2016).

Narcissus latent virus (NLV) is common in many cultivars of narcissus and bulbous iris, but was detected in only 1 of 19 cultivars of nerine. It induced symptoms in some narcissus cultivars, but inconspicuous infection in bulbous iris and nerine. NLV was not seed-borne in narcissus or *Nicotiana clevelandii* but was transmitted readily by aphids (*Acyrthosiphon pisum*, *Aphis gossypii*, and *Myzus persicae*) in a non-persistent manner and by sap inoculation to 12 of 53 species from 3 of 16 families. Purified virus preparations were obtained from infected *N. clevelandii* by clarification of buffered leaf extracts with diethyl ether and carbon tetrachloride, followed by one or two cycles of differential centrifugation and molecular permeation chromatography. NLV has filamentous particles, i.e., three times of 650 nm. They contain 5% nucleic acid and a single polypeptide of mol. wt 32.6 × 103. The biological and physical properties of NLV place it in the carlavirus group; it is serologically related to Lily symptomless virus, but not to other 14 authentic carlaviruses (Brunt 2008).

17.9 Diagnosis of Virus(es)

It is very critical to diagnose diseases caused by unknown pathogens and viruses. Several strategies can be used for diagnosis of potyviruses such as electron microscopy and biological, biochemical, and genomic (nucleic acid) properties.

17.10 Electron Microscopy

Electron microscopy is a valuable tool in plant virology. Particularly, transmission electron microscopy (TEM) is a very useful tool for virus detection and for basic and applied research of plant viruses. Through TEM, potyvirus particles and their in situ location in ultrathin sections of plant tissue are viewed. In purified or crude preparations, TEM is used to (1) detect the presence of virions, (2) provide information of morphology and size, and (3) monitor the purity and relative amount of virus during purification (Baker et al. 1985). After partial purification, transmission electron microscopy was done where typical flexuous rod-shaped virus particles were observed about 680 nm in length and 11 nm width (in diameter) at 80,000 magnification that indicated the presence of potyvirus.

17.11 Nucleic Acid-Based Techniques

17.11.1 Reverse Transcriptase-Polymerase Chain Reaction (RT-PCR)

RT-PCR is a molecular-based and most effective technique for the diagnosis of plant RNA viruses. In this technique, virus gene-specific primers are used during PCR for amplification of viral genomic fragment. For potyvirus detection, different primers of conserved region were used in the past. Since CP is the most conserved region of potyvirus, conserved sequence like WCIEN box or QMIKAA motif universal primers were designed to detect and identify potyviruses (Pappu et al. 1993; Colinet and Kummert 1993; Bateson and Dale 1995; Langeveld et al. 1991). The most conserved motif (GNNSGQPSTVVDN) of the NIb region (Gibbs et al. 2003) and forward primer of this conserved motif are used for detection of specific numerous members of *Potyviridae* family (Chen and Adams 2001). Szemes et al. in 2001 also used the CP primer of N-terminal region or 3′-noncoding region for detection of a potyvirus. Then after HC-Pro, C-terminal region of NIb and C-terminal of CP were also used for detection of potyvirus (Gadiou et al. 2009; Kajic et al. 2008; James et al. 2003; Raj et al. 2018).

17.11.2 Extraction and Quantification of Total RNA

For the molecular characterization of potyvirus, the leaf samples of narcissus plants showing mosaic and yellow stripe symptoms were collected. The total RNA was isolated from 100 mg infected leaf samples as well as healthy narcissus leaf and suspended in 30 μL nuclease-free water. The concentration of RNA was quantified through a spectrometer, and their OD at the ratio of absorbance at 260/280 nm was used and the ratio was ~160–190 μg/mL, suggesting the quality of RNA was appropriate.

17.11.3 Detection of Potyvirus by RT-PCR

For detection of potyvirus, RT-PCR was performed using two sets of degenerate primer pairs: the first is Pot 9502/CPUP that amplify complete CP region and the second PotI/PotII spanning 3′-UTR and partial NIb region of potyvirus.

The total RNA of eight infected narcissus leaf samples was used as a template and oligo-dt primer was used for cDNA synthesis. PCR reaction was performed in a 25 μL reaction volume which contained Taq buffer, $MgCl_2$, dNTPs, Taq DNA polymerase, primers (Pot9502/CPUP), nuclease-free water, and cDNA as template. Amplified products were electrophoresed in 1% agarose gel with 1 kb DNA ladder used as size marker. Results revealed the expected size ~750 bp amplification with CP gene-specific primers similar to that of the positive control indicated the presence of potyvirus infection in narcissus plants.

For characterization of potyvirus in narcissus, the obtained positive amplicons of RT-PCR of ~1.5 kb size (3′UTR to partial NIb region) were gel eluted, ligated into pGEM-T easy vector, and screened on selection medium after mobilization in competent *E. coli* DH5α cells. The plasmid was isolated from positive clones and digested with *EcoR*I to check the presence of insert. The digested products were separated by 1% agarose gel electrophoresis. The result showed two expected size bands of ~3.0 kb of vector and ~1.5 kb band of DNA inset. The positive clones were sequenced, and the obtained sequence data were assembled and analyzed to eliminate any sequence ambiguity. The obtained sequence data were analyzed using published sequences available in GenBank through BLASTn and *Genomatix DiAlign* tools for obtaining the global and specific sequence identity of the virus isolates under study which were identified as two potyvirus isolates *Cyrtanthus elatus virus* (CEVA) and *Narcissus yellow stripe virus* (NYSV) in narcissus samples.

Based on the criteria, distinct species must have <76% nucleotide identity and <82 amino acid sequence identity at the full-length genome and large polypeptide, respectively (Adams et al. 2005) and recommendations of ICTV (Wylie et al. 2017). BLASTn analysis NYSV isolates shared 100% nucleotide sequence identity with each other and 97–75% identity with other NYSV isolates which was reported worldwide during BLASTn analysis. BLASTn analysis of CyEVA isolates under study revealed their close homology to several *Vallota speciosa virus* isolates (revised as *Cyrtanthus elatus virus* A (CyEVA) by ICTV). CyEVA shared 98–80% sequence identities with CyEVA isolates reported worldwide.

17.12 Management Strategies of Potyviruses in Narcissus

Potyviruses cause significant losses of narcissus crops; therefore, their management is important for the improvement of the crop production. The genus *Potyvirus* is one of the most devastating groups of the plant pathogens worldwide, and yield losses due to infection of this virus commonly range from 8% to 35%, although losses as high as 94% have also been recorded. Narcissus is propagated by bulbs, and if mother stocks are infected with this virus then they can act as a source of virus spread from one generation to another and continue to many generations. Therefore, their management is essential by utilizing various management strategies (Verma et al. 2004).

17.13 Sanitation of Cropping Field and Control of Vectors

For the management of potyviruses, sanitation of field and planting practices are used. Vectors could be also avoided by early sowing of seeds, growing of plants in isolated or remote areas, and rotation of crops to minimize specifically soilborne virus diseases. Use of non-host plants in and around the cultivated area may be a successful means of reducing the virus infection as suggested by Matsuura et al.

(2006). Vectors are the main source of virus spread from infected to healthy plants. Mehrotra in 1991 suggested the use of some barriers like cages toward the onset of insect vectors (Mehrotra 1991). Khan et al. (2006) used Malathion insecticide for the control of whitefly in chili crop (Khan et al. 2006). In 2016, Shukla et al. used insecticidal fern protein in cotton for protection against whitefly (Shukla et al. 2016). Therefore, the virus disease management from mother stock (bulbs) of infected narcissus plants is essential. In natural conditions, potyviruses are mainly transmitted through aphid vectors and nematodes (Poutaraud et al. 2004; Adams et al. 2005) from infected plants to healthy plants growing nearby. Potyviruses can also be transmitted through seeds and vegetatively propagated tubers (Aminuddin and Singh 1993; Johansen et al. 1994; Shukla et al. 1994; Sastry 2013).

17.14 Developing Virus-Free Narcissus Plants

In general, virus accumulates in different plant parts including bulbs and survives in the growing tissues of mother bulbs of narcissus and therefore is a major reason for virus dissemination to their subsequent generation and from one place to another cultivation area (Aminuddin and Singh 1993; Johansen et al. 1994). Therefore, management of potyviruses was considered as highly required for quality and for quantity improvement of narcissus bulblets.

Different therapies were used for the management of potyvirus disease in narcissus plants. The success rate was varied from different protocols due to the relationship between host, virus, and treatment (Panattoni and Triolo 2010). Three different therapies are used for the elimination of potyvirus: electrotherapy, chemotherapy, and combination of electrotherapy with chemotherapy.

17.15 Chemotherapy

In this therapy, ribavirin, which is an antiviral chemical, is used which has the capability of eliminating the virus from infected plants by restricting their replication (De Fazio et al. 1978; Verma et al. 2004). The antiviral effect of ribavirin is that it forced RNA viruses into error catastrophe (Crotty et al. 2001). Ribavirin (1-D-ribofuranosyl-1,2,4-triazole3-carboxamide) is a base analog of adenine or guanine. The different concentration of ribavirin in MS medium and its concentration and duration of treatment should be optimized for different hosts and viruses (Hu et al. 2012). Due to the phytotoxicity of ribavirin, it can affect the growth of in vitro plants (Hansen and Lane 1985; Paunovic et al. 2007). Ribavirin induces broad-spectrum activity against several RNA and DNA viruses due to its guanosine analog base (Leyssen et al. 2005). In the process of transcription of triphosphate ribonucleotides, IMPDH inhibitors are directly involved and therefore assist in inhibition of viral nucleic acid replication (Crotty et al. 2001). The main target of antiviral activity of these molecules is the inosine monophosphate dehydrogenase, which is an enzyme that catalyzes the conversion of inosine 5′-monophosphate (IMP) in xanthosine

5′-monophosphate and is able to change the pathway for the production of guanosine mono-, di-, and triphosphate. SAH is another inhibitor which is involved in virus replication. In transmethylation reactions, *S*-adenosyl methionine (SAM) is used and donates methyl groups to a large range of acceptors together with nucleic acids, viral proteins, and phospholipids and then converted them to *S*-adenosylhomocysteine. The methylation process regulated them negatively by an increase in SAH and a decrease in SAM or SAM/SAH ratio. Removal of SAH plays an important role, and it is initiated by SAH hydrolase, which has the ability of converting this molecule into homocysteine and adenosine. Thus, the collection of SAH makes their conversion and resulting blockage of the maturation of viral RNA impossible, especially without stopping the formation of "cap" (De Clercq 2005).

In vitro chemotherapy was done for elimination of virus from infected narcissus bulbs using virazole (chemical name ribavirin). For the standardization of in vitro chemotherapy, use three different concentrations of ribavirin, 30 mg/L, 40 mg/L, and 50 mg/L, in MSg3 media. Result showed that at high concentrations, 50 mg/L ribavirin and 40 mg/L ribavirin, the germination rates were poor, i.e., 16% and 36%, respectively, because they are toxic to plants, but at low concentration, 30 mg/L ribavirin, the germination rate was high, i.e., 56%, because this level was less toxic. The virus-free plants were obtained 13%, 26%, and 46% in ribavirin 30 mg/L, 40 mg/L, and 50 mg/L, respectively.

17.16 Electrotherapy

In this technique, plant tissue is treated with electric current for the elimination of virulence activity of viruses by means of disrupting or degrading viral nucleoproteins. It is suggested as a presumption that when plant tissue is treated with electric current, then inside the tissue, nucleoproteins of viruses may be denatured because of increased temperature (Lozoya-Saldaña et al. 1996; Black et al. 1971). A research work showed that when nucleoproteins are present within the cells, thermal inactivation may not work because cells are thermally isolated by the cell wall. In 1974, Blanchard obtained virus-free plants using direct current (Blanchard 1974). This study established a basis for electric current mediated for management of plant RNA virus.

Gonzalez et al. (2006) proposed that viral particles may be denatured in apoplastic space during transportation through the plasmodesmata. Hormozi-Nejad et al. (2010) proposed that due to denaturation of a nucleoprotein, the virus cannot move from cell to cell, therefore preventing virus movement into healthy cells. These studies reveal a basis for electric current-induced removal of plant RNA viruses. Since the 1990s, it appeared that electric current is used in a number of crops for the production of virus-free plants. This technique was first utilized on potato plants for the elimination of *Potato virus* X (PVX) on a large scale. In this technique, use of an electric current of 15 mA for 5 min leads to 60–100% elimination of PVX in various plant cultivars.

In in vitro electrotherapy, three different electric currents were used for elimination of potyvirus in infected narcissus bulb explants. Infected bulbs were sterilized and kept for therapy of electric current in electrophoresis tank in 1× TAE buffer and a medium placed for germination under aseptic condition. Electrotherapy was used with different electric currents (10, 20, and 30 mA for 20 min). After 30 days, the germination rate was recorded, the rate of germination was higher (i.e., 63%) in case of 10 mA for 20 min electric current, and the percentage of virus-free plants (i.e., 40%) was obtained from 20 mA/20 min electric current. The combination of electrotherapy with chemotherapy was also used for elimination of potyvirus from narcissus bulb explants using a different combination of ribavirin (30–50 mg/L) + electric current (10–30 mA/20 min). The percentage of virus-free plants was found highest in the combination of chemotherapy and electrotherapy (30 mg/L ribavirin + 20 mA/20 min electric current) (i.e., 50%) as compared with other therapies; hence, the therapy efficiency rate was also the highest.

17.17 Combination of Chemotherapy with Electrotherapy

Previous studies showed that the use of a combination of chemotherapy with electrotherapy was found to have the best result in case of PVY elimination from potato and BYMV from gladiolus (Kaur et al. 2019).

The combination of electrotherapy with chemotherapy was also attempted for elimination of potyvirus from infected narcissus bulbs. Three combinations, ribavirin (30.0 mg/L) + electric current (10 mA/20 min), ribavirin (30 mg/L) + electric current (20 mA/20 min), and ribavirin (30 mg/L) + electric current (30 mA/20 min), were used. After 30 days, regeneration responses were recorded. It was found that 60% explants were germinated in 30.0 mg/L ribavirin with 10 mA/20 min electric current, 53% explants germinated in 30.0 mg/L ribavirin with 20 mA/20 min electric current, while only 33% explants germinated in 30.0 mg/L ribavirin with 30 mA/20 min electric current.

17.18 Conclusion

The nucleic acid spot hybridization test and RT-PCR utilizing NYSV and CyEVA probes and potyvirus-specific primers, respectively, can help in the diagnosis of NYSV and CyEVA potyviruses in narcissus bulbs to check their dissemination from one field/country to another field and for the purpose of quarantines. Moreover, the present study will help to design the efficient management strategy utilizing its genomic sequences through genetic engineering methods.

The method for elimination of virus from infected narcissus explants provides knowledge about the regeneration of narcissus and optimum conditions for chemotherapy, electrotherapy, and combination of chemo- with electrotherapy that would be used for the elimination of virus from other ornamental plants. The combination of chemotherapy (30 mg/L ribavirin) with electrotherapy (20 mA/20 min electric current) was found the best therapy among them. They were also achieved for the first time in India. This finding would improve the quality and productivity of bulbs and flowers and ultimately enhance the commercial production of narcissus and enhance the economic value of narcissus growers.

Acknowledgments Authors are thankful to the Director of CSIR-NBRI, Lucknow, India, for providing the facilities. RR is also thankful to the University Grant Commission for Rajiv Gandhi National fellowships and AcSIR for Ph.D. registration. This study was funded by the In-house project OLP105.

References

Adams MJ, Antoniw JF, Fauquet CM (2005) Molecular criteria for genus and species discrimination within the family Potyviridae. Arch Virol 150(3):459–479

Alexander K, Antonio (2008) Chemistry, biology and medicinal potential of narciclasine and its congeners evidente. Chem Rev 108(6):1982–2014

Aminuddin, Singh BP (1993) Multiplication of Gladiolus cultivars for producing virus free propagules. Indian J Virol 9:74–77

Aminuddin, Khan JA, Raj SK (1999) Association of an unknown potyvirus isolate with severe mosaic of Narcissus tazetta L. Indian J Exp Biol 137:1034–1036

Asjes CJ (1971) Virus diseases in narcissus in the Netherlands. Daffodil J 8:3–11

Asjes CJ (1974) Soil-borne virus diseases in ornamental bulb crops and their control in The Netherlands. Agric Environ 1:303–315

Baker KK, Donald CR, Giilet J (1985) Electron microscopy: current applications to plant virology. Plant Dis 69:85–90

Bastida J, Lavilla R, Viladomat F (2006) Chemical and biological aspects of "Narcissus" alkaloids. In: Cordell GA (ed) The alkaloids: chemistry and biology, vol 63. Elsevier Inc, Amsterdam, pp 87–179

Bateson MF, Dale JL (1995) Banana bract mosaic virus: characterization using potyvirus specific degenerate PCR primers. Arch Virol 140:515–527

Berger PH, Adams MJ, Barnett OW, Brunt AA, Hammond J, Hill JH, Jordan RL, Kashiwazaki S, Rybicki E, Spence N, Stenger DC, Ohki ST, Uyeda I, van Jaayen A, Valkonen J, Vetten HJ (2005) Potyviridae. In: Fauquet CM, Mayo MA, Maniloff J, Desselberger U, Ball LA (eds) Virus taxonomy. VIIIth Report of the International Committee on Taxonomy of Viruses. Elsevier/Academic Press, London

Berniak H, Komorowska B, Sochacki D (2013) Detection of narcissus latent virus isolates using one-step RT-PCR assay. J Hortic Res 21(1):11–14

Black JD, Forsyth FR, Fensom DS, Ross RB (1971) Electrical stimulation and its effects on growth and ion accumulation in tomato plants. Can J Bot 49(10):1809–1815

Blanchard RO (1974) Electrotherapy a new approach to wound healing. Proc Am Phytopathol Soc 1:133–134

Broadbent L, Green DE, Walker P (1962) Narcissus virus diseases. Daffodil Tulip Yb 28:1–7

Brunt AA (1966a) Narcissus mosaic virus. Ann Appl Biol 58:13–23

Brunt AA (1966b) The occurrence of Cucumber mosaic virus and four nematode transmitted viruses in British narcissus crops. Plant Pathol 15:157–160

Brunt AA (1977) Some hosts and properties of Narcissus latent virus, a carlavirus commonly infecting narcissus and bulbous iris. Ann Appl Biol 87:355–364

Brunt AA (1980) A review of problems and progress in research on viruses and virus diseases of narcissus in Britain. Acta Hortic 110:23–30

Brunt AA (1995) Bulb and corm crops. Narcissus. In: Loebenstein G, Lawson RH, Brunt AA (eds) Virus and virus-like diseases of bulb and flower crops. John Wiley and Sons, Chichester, pp 322–334

Brunt AA (2008) Some hosts and properties of narcissus latent virus, a carlavirus commonly infecting narcissus and bulbous iris. Ann Appl Biol 87:355–364

Caldwell J, Kissick E (1950) Varietal response to narcissus stripe virus. Daffodil Tulip Yb 16:63–74

Chen J, Adams MJ (2001) A universal PCR primer to detect members of the Potyviridae and its use to examine the taxonomic status of several members of the family. Arch Virol 146:757–766

Chen J, Chen JP, Langeveld SA, Derks AFLM, Adams MJ (2003) Molecular characterization of carla- and potyviruses from Narcissus in China. J Phytopathol 151:26–29

Chen J, Lu YW, Shi YH, Adams MJ, Chen JP (2006) Complete nucleotide sequence of the genomic RNA of *Narcissus yellow stripe virus* from Chinese narcissus in Zhangzhou city, China. Archiv Virol 15:1673–1677

Clark VR, Guy PL (2000) Five viruses in Narcissus spp. from New Zealand. Australas Plant Pathol 29:227–229

Colinet D, Kummert J (1993) Identification of a sweet potato feathery mottle virus isolate from China (SPFMV-Ch) by the polymerase chain-reaction with degenerate primers. J Virol Methods 45:149–159

Crotty S, Cameron CE, Andino R (2001) RNA virus error catastrophe: direct molecular test by using ribavirin. Proc Natl Acad Sci 98(12):6895–6900

De Clercq E (2005) Antiviral drug discovery and development: where chemistry meets with biomedicine. Antivir Res 67:66–75

De Fazio G, Caner J, Vicente M (1978) Inhibitory effect of virazole (ribavirin) on the replication of tomato white necrosis virus (VNBT). Arch Virol 58:153–156

Duke JA, Duke PK (1983) Medicinal plants of the bible, vol 1. Trad-Medic Books, New York, NY, p 98

Gadiou S, Šafárová D, Navratil M (2009) Differentiation of Plum pox virus isolates by single-strand conformation polymorphism and low-stringency single specific primer PCR analysis of HC-Pro genome region. Acta Virol 53(1):53

Gallo A, Valli A, Calvo M, García JA (2018) A functional link between RNA replication and virion assembly in the potyvirus plum pox virus. J Virol 92(9):e02179–e02117

Gibbs A, Mackenzie AM, Gibbs MJ (2003) The 'potyvirid primers' will probably provide phylogenetically informative DNA fragments from all species of Potyviridae. J Virol Methods 112(1–2):41–44

Gonzalez JE, Sanchez R, Sanchez A (2006) Biophysical analysis of electric current mediated nucleoprotein inactivation process. In: First International Workshop on Bioinformatics, Santa Clara, Cuba

Graham SW, Barrett SCH (2004) Phylogenetic reconstruction of the evolution of stylar polymorphisms in Narcissus (Amaryllidaceae). Am J Bot 91(7):1007–1021

Hanks GR (1993) Narcissus. In: De Hertogh AA, Le Nard M (eds) The physiology of flower bulbs. Elsevier, Amsterdam, pp 463–558

Hanks GR (2002) Narcissus and daffodil. In: Hanks GR (ed) The genus narcissus. Horticulture Research International, Taylor and Francis, Kirton, London. ISBN 978-0-415-27344-2

Hanks GR, Chastagner GA (2018) Diseases of Daffodil (Narcissus). In: Handbook of florists' crops diseases, pp 1129–1228

Hanks GR, Kennedy R, Millar M (2003) Narcissus leaf diseases: forecasting and control of white mould and smoulder. Final report on Project BOF 041. HDC, East Malling

Hansen AJ, Lane WD (1985) Elimination of apple chlorotic leafspot virus from apple shoot cultures by ribavirin. Plant Dis 69:134–135

Harvey BMR, Selby C (1997) Micropropagation of Narcissus (Daffodils) in biotechnology in agriculture and forestry. In: Bajaj YPS (ed) High-tech and micropropagation. VI, vol 40, pp 225–251

Hong Ying Z, YuWen L, JiaLe S, Lin L, Jiong C, JianPing C (2009) Prokaryotic expression of coat protein gene of *Ornithogalum mosaic virus* and its use in the antiserum preparation (in Chinese). Acta Agric Zhejiang 21:198–201

Hormozi-Nejad MH, Mozafari J, Rakhshandehroo F (2010) Elimination of Bean common mosaic virus using an electrotherapy technique. J Plant Dis Protect 5:201–205

Hu GJ, Hong N, Wang LP, Hu HJ, Wang GP (2012) Efficacy of virus elimination from in vitro-cultured sand pear (Pyrus pyrifolia) by chemotherapy combined with thermotherapy. Crop Prot 37:20–25

Hunter DA, Fletcher JD, Davies KM, Zhang H (2011) Colour break in reverse bicolour daffodils is associated with the presence of Narcissus mosaic virus. Virol J 8:412

Iwaki M, Komuro Y (1974) Viruses isolated from Narcissus (*Narcissus* spp.) in Japan. V. Arabis mosaic virus. Ann Phytopathol Soc Jpn 40:344–353

James D, Varga A, Thompson D, Hayes S (2003) Detection of a new and unusual isolate of Plum pox virus in Plum (Prunus domestica). Plant Dis 87:1119–1124

Jin J, Shen JG, Cai W, Xie GH, Liao FR, Gao FL, Ma JF, Chen XH, Wu ZJ (2017) Narcissus yellow stripe virus and Narcissus mosaic virus detection in Narcissus via multiplex TaqMan-based reverse transcription-PCR assay. J Appl Microbiol 122:1299–1309. https://doi.org/10.1111/jam.13422)

Johansen E, Edwards MC, Hampton RO (1994) Seed transmission of viruses: current perspectives. Annu Rev Phytopathol 32:363–386

Kajic V, Cerni S, Krajacic M, Mikec I, Škoric D (2008) Molecular typing of Plum pox virus isolates in Croatia. J Plant Pathol 90:S9–S13

Kamenetsky R, Okubo H (2012) Ornamental geophytes: from basic science to sustainable production. CRC Press, Boca Raton, FL. ISBN-9780429061776

Kaur C, Raj R, Agrawal L, Kumar S, Purshottam DK, Chauhan PS, Raj SK (2019) Elimination of Bean yellow mosaic virus from infected cormels of three cultivars of gladiolus using thermo, electro and chemotherapy. 3 Biotech 9(4):154

Khan MS, Raj SK, Bano T, Garg VK (2006) Incidence and management of mosaic and leaf curl disease in cultivars of chilli (Capsicum annuum). J Sci Food Agric 44:99–102

Kington S (1998) The international Daffodil register and classified list. Royal Horticultural Society, London. 978-1-874431-69-5

Kumar S, Raj R, Kaur C, Raj SK (2015) First report of Cyrtanthus elatus virus A in Narcissus tazetta in India. Plant Dis 99(11):1658

Langeveld SA, Dore JM, Memelink J, Derks A, Vandervlugt CIM, Asjes CJ, Bol JF (1991) Identification of potyviruses using the polymerase chain-reaction with degenerate primers. J Gen Virol 72:1531–1541

Leyssen P, Balzarini J, De Clrecq E, Neyts J (2005) The predominant mechanism by which ribavirin exerts its antiviral activity in vitro against flavivirus and apramyxoviruses is mediated by inhibition of IMP dehydrogenase. J Virol 79:1943–1947

Lozoya-Saldaña H, Abelló FJ, García GR (1996) Electrotherapy and shoot tip culture eliminate potato virus X in potatoes. Am J Potato Res 73(4):149–154

Mathew B (2002) Classification of the genus narcissus. In: Hanks GR (ed) Narcissus and daffodil, the genus Narcissus. Taylor & Francis, New York, NY, pp 31–52

Matsuura S, Hoshino S, Koga H (2006) Verbena as a trap crop to suppress thrips-transmitted Tomato spotted wilt virus in chrysanthemum. J Gen Plant Pathol 72:180–185

Mehrotra RS (1991) Plant pathology, 8th edn. Tata McGraw Hill-publishing, New Delhi

Miglino R, Jodlowska A, Van Schadewijk AR (2005) First report of Narcissus mosaic virus infecting Crocus spp. cultivars in the Netherlands. Plant Dis 89(3):342

Ohshima K, Nomiyama R, Mitoma S, Honda Y, Yasaka R, Tomimura K (2016) Evolutionary rates and genetic diversities of mixed potyviruses in Narcissus Infection. Genet Evol 45:213–223

Panattoni A, Triolo E (2010) Susceptibility of grapevine viruses to thermotherapy on in vitro collection of Kober 5BB. Sci Hortic 125:63–67

Pappu SS, Brand R, Pappu HR, Rybicki EP, Gough KH, Frenkel MJ, Niblett CL (1993) A polymerase chain reaction method adapted for selective amplification and cloning of 3' sequences of potyviral genomes - application to Dasheen mosaic virus. J Virol Methods 41:9–20

Paunovic S, Ruzic D, Vujovic T, Milenkovic S, Jevremovic D (2007) In vitro production of Plum pox virus-Free plums by chemotherapy with ribavirin. Biotechnol Biotechnol Equip 21 (4):417–421

Poutaraud A, Desbiez C, Lemaire O, Lecoq H, Herrbach E (2004) Characterisation of a new potyvirus species infecting meadow saffron Colchicum autumnale. Arch Virol 149 (7):1267–1277

Raj R, Kaur C, Agrawal L, Chauhan PS, Kumar S, Raj SK (2018) Full-length genome sequence of *Cyrtanthus elatus virus-A* isolated from *Narcissus tazetta* in India. 3 Biotech 8(3):168

Raj R, Agrawal L, Kumar S, Chauhan PS, Raj SK (2019) Complete genome sequence analysis of Narcissus yellow stripe virus infecting *Narcissus tazetta* in India. 3 Biotech 9:409

Rees AR (1966) Effects of infection with narcissus mosaic, narcissus yellow stripe and tobacco rattle viruses on the growth and flowering of narcissus cv 'Minister Talma'. Ann Appl Biol 58: 25–30

Samuitiene M, Navalinskiene M, Jackeviciene E (2008) *Arabis mosaic virus* on ornamental plants. Biologia 54:264–268

Sastry KS (2013) Mechanism of seed transmission. In: Seed-borne plant virus diseases. Springer, New Delhi, pp 85–100

Sastry KS, Mandal B, Hammond J, Scott SW, Briddon RW (2019) *Narcissus* spp. (Daffodil). In: Encyclopedia of plant viruses and viroids. Springer, New Delhi. https://doi.org/10.1007/978-81-322-3912-3_608

Sharma P, Sahu AK, Verma RK, Mishra R, Choudharya DK, Gaur RK (2014) Current status of Potyvirus in India. Arch Phytopathol Plant Protect 47:906–918

Shukla DD, Ward CW, Brunt AA (1994) The potyviridae. CAB International, Wallingford

Shukla DD, Ward CW, Brunt AA, Berger PH (1998) Potyviridae family. In: Description of plant viruses, vol 366. Association of Applied Biologists, Warwickshire, p 12

Shukla AK, Upadhyay SK, Mishra M, Saurabh S, Singh R, Singh H, Thakur N, Rai P, Pandey P, Hans AL, Srivastava S (2016) Expression of an insecticidal fern protein in cotton protects against whitefly. Nat Biotechnol 34(10):1046

van Slogteren E, de Bruyn Ouboter MP (1946) Daffodil Tulip Yb 12:3

Takos AM, Fred R (2013) Towards a molecular understanding of the biosynthesis of *Amaryllidaceae* alkaloids in support of their expanding medical use. Int J Mol Sci 14 (6):11713–11741

Thakur P (2017) Narcissus. In: Misra RL, Misra S (eds) Commercial ornamental crops cut flowers. Kruger Brentt Publishers, London, pp 373–388

Urcuqui-Inchima S, Haenni AL, Bernardi F (2001) Potyvirus proteins: a wealth of functions. Virus Res 74:157–175

Verma N, Ram R, Hallan V, Kumar K, Zaidi AA (2004) Production of Cucumber mosaic virus-free chrysanthemums by meristem tip culture. Crop Prot 23:469–473

Ward LI, Veerakone S, Tang J, Clover GRG (2009) First report of *Narcissus degeneration virus*, *Narcissus late season yellows virus*, and *Narcissus symptomless virus* on Narcissus in New Zealand. Plant Dis 93:964

Willis D (2012) In: Rushton BS, Peter R, Jan D, Philip O (eds) 'Yellow fever', A prospect of the history and culture of daffodils

Wylie SJ, Jones MGK (2012) Complete genome sequences of seven carlavirus and potyvirus isolates from Narcissus and Hippeastrum plants in Australia, and proposals to clarify their naming. Arch Virol 157:1471–1480

Wylie SJ, Adams M, Chalam C, Kreuze J, López-Moya JJ, Ohshima K, Praveen S, Rabenstein F, Stenger D, Wang A, Zerbini FM (2017) Ictv Report Consortium. ICTV Virus taxonomy profile: Potyviridae. J Gen Virol 98(11):2893

Yadav N, Khan JA (2008) Identification of a potyvirus associated with mosaic disease of Narcissus sp. in India. New Dis Rep 14:56

Yan L, RongHao S, CuiYun Y, JinPing L, JunXi J, FuMing D (2008) Identification of viruses infecting Narcissus in Chongming island of Shanghai (in Chinese). Acta Phytotaxon Sinica 38: 456–461

Begomoviruses Infecting *Jatropha* Species Grown for Ornamental Values 18

Sunil Kumar Snehi and S. K. Raj

Abstract

Some *Jatropha* species such as *Jatropha podagrica*, *J. multifida*, and *J. integerrima* are grown in India as ornamental plants. The severe yellow mosaic disease symptoms were observed in these three species at CSIR-NBRI garden, Lucknow, in the year 2009–2011. Infection of begomovirus was suspected based on typical yellow mosaic and infestation of whiteflies. Association of begomovirus with mosaic disease was detected by polymerase chain reaction using begomovirus gene-specific primers which resulted in the ~1.2 kb amplicons in all these ornamental species of *Jatropha*. For molecular identification of associated begomovirus species, the ~1.2 kb amplicons were sequenced and sequence data were analyzed. The highest sequence identities of ~1.2 kb (partial DNA-A) genome that showed close phylogenetic relationships with *Jatropha mosaic India virus* in *J. podagrica*, *Tomato leaf curl Patna virus* in *J. multifida*, and *Papaya leaf curl virus* in *J. integerrima* were identified. We have summarized in this chapter worldwide distribution of mosaic disease associated with begomovirus species of ornamental *Jatropha* species including India based on their genome sequence analysis and genetic diversity.

Keywords

Begomovirus · Mosaic disease · Molecular identification · Sequence analysis · Genetic diversity analysis

S. K. Snehi (✉)
Department of Microbiology, Barkatullah University, Bhopal, Madhya Pradesh, India

S. K. Raj
Plant Molecular Virology Laboratory, CSIR-National Botanical Research Institute, Lucknow, Uttar Pradesh, India

S. K. Raj et al. (eds.), *Virus Diseases of Ornamental Plants*,
https://doi.org/10.1007/978-981-16-3919-7_18

18.1 Introduction

The genus *Jatropha* of the family Euphorbiaceae has 476 species and distributed throughout the world, and among them *Jatropha curcas*, *J. gossypifolia*, *J. integerrima*, *J. multifida*, *J. podagrica*, and *J. cuneata* are recorded in India. Many excellent characteristics, including high yield, resistance to drought, as well as content and good quality of the plant oil, have generated the interest of many researchers to *J. curcas*, while other species are of ornamental value or traditionally used for their medicinal values.

The natural infection of begomoviruses has been reported in *African cassava mosaic virus* on *J. multifida* in East and West Africa (Okoth 1991). Begomoviruses (family *Geminiviridae*) are transmitted by whitefly species which cause diseases of important crops in the tropics and subtropics (Stanley et al. 2005). Their genome consists of one or two circular single-stranded DNA components, referred to as DNA-A and DNA-B, each about 2.6–2.8 kb in size (Stanley et al. 2005; Fauquet et al. 2008). DNA-A encodes replication-associated protein (AC1) essential for viral replication; replication enhancer protein (AC3); transcriptional activator protein (AC2) that controls late gene expression and is involved in RNAi suppression; AC4 protein for host range determination, symptom severity, and virus movement; coat protein (AV1) for encapsidation and insect transmission; and pre-coat protein (AV2) for virus accumulation and symptom development. However, DNA-B encodes nuclear shuttle protein (BV1) and movement protein (BC1), both of which are vital for efficient systemic spread and symptom expression (Hanley-Bowdoin et al. 1999; Briddon et al. 2001). A number of begomoviruses occurring in the Old World (Eastern Hemisphere, Europe, Africa, Asia) are monopartite and have only a single component equivalent to DNA-A. The cloned genomic component of some of these monopartite begomoviruses has been shown to produce typical symptoms, confirming that single genomic component is solely responsible for the disease (Stanley et al. 2005; Pant et al. 2001).

18.2 Survey and Symptomatology on Ornamental Species of Jatropha

During surveys in 3 subsequent years (2009–2011), severe mosaic disease-like symptoms were also observed on other species of *Jatropha*, viz., *J. podagrica*, *J. integerrima*, and *J. multifida* grown for their ornamental values in CSIR-NBRI, garden Lucknow, India. The naturally infected *J. podagrica* exhibited severe yellow mosaic and vein yellowing symptoms (Fig. 18.1a), *J. multifida* showed mosaic and leaf curl symptoms (Fig. 18.1b), and *J. integerrima* showed mosaic and yellow mosaic symptoms (Fig. 18.1c). The disease incidence was about ~30–35% in *Jatropha* species growing field.

Fig. 18.1 Naturally infected *J. podagrica* (**a**), *J. multifida* (**b**), and *J. integerrima* (**c**) showing mosaic symptoms

18.3 Molecular Detection of Begomovirus in Jatropha Species

To find out how many begomoviruses/species are associated with the mosaic disease of these *Jatropha* species, the total nucleic acid DNA were isolated from newly emerging infected leaves of *J. podagrica*, *J. multifida*, and *J. integerrima* by Dellaporta et al. (1983) method, and PCRs were performed using a pair of begomovirus-specific primers PALIv 722/PALIc 1960 (Rojas et al. 1993). PCRs were set up in a 50 μL reaction mixture containing template DNA (100 ng), dNTPs (10 mM each), primers (each 25 pM), $MgCl_2$ (25 mM), *Taq* DNA polymerase (3.0 U, Bangalore Genei Pvt. Ltd), and *Taq* buffer (1×, Bangalore Genei Pvt. Ltd) in a Peltier thermal cycler PTC200 engine (MJ Research, Waltham, MA, USA) with the following PCR conditions: initial denaturation at 94 °C for 5 min, followed by 35 cycles of denaturation at 94 °C for 1 min, specific annealing at 52 °C temperatures for 1 min, and extension at 72 °C for 1.5 min. The final extension cycle was for 5 min at 72 °C.

An expected size amplicons of ~1.2 kb were observed from 3/3 symptomatic samples but not from healthy one collected from the same location on 1% agarose gel electrophoresis.

To detect DNA-B genome, the PCR was also performed using DNA-B-specific primers with the following conditions: initial denaturation at 94 °C for 5 min, followed by 30 cycles of denaturation at 94 °C for 1 min, annealing at 52 °C for 1 min, extension at 72 °C for 1.5 min, and a final extension at 72 °C for 5 min. However, our several attempts failed to amplify DNA-B genome in *J. podagrica*, *J. multifida*, and *J. integerrima* samples.

Attempts were also made to detect association of DNA-β using DNA-β-specific primers (Briddon et al. 2003) with the following PCR conditions: initial denaturation at 94 °C for 5 min, followed by 30 cycles of denaturation at 94 °C for 1 min, annealing at 55 °C for 1 min, extension at 72 °C for 1.5 min, and final extension at 72 °C for 5 min. With all affords, DNA-β genome could not be amplified during the PCR in *J. podagrica* and *J. multifida*, but in *J. integerrima* ~1.3 kb amplicon of DNA-β genome was amplified by PCR.

18.3.1 Cloning and Sequencing

For molecular identification of begomovirus associated with the mosaic disease of *J. podagrica*, *J. multifida*, and *J. integerrima*, the ~1.2 kb amplicons of DNA-A and ~1.3 kb DNA-β obtained were cloned into the pGEM-T Easy Vector System-1 and *E. coli* strain DH5α were transformed. Three independent clones of each amplicon were sequenced in both orientations. The consensus sequence data of three identical sequences were combined to partial DNA-A genome. The sequence resulted in the presence of 1177 bp (*J. podagrica*), 1288 bp (*J. multifida*) and 1201 bp (*J. integerrima*) partial DNA-A of begomovirus species, which were deposited in the GenBank database under the following accession numbers: *J. podagrica*, HQ848382; *J. multifida*, JQ043440; and *J. integerrima*, HQ848381.

18.4 Identification of Begomoviruses Infecting *Jatropha* Species

18.4.1 Sequence Analysis of *J. podagrica* Begomovirus Isolate

The analysis of partial DNA-A (1177 nt) of *J. podagrica* virus isolate (HQ848382) contained four ORFs, encoding partial AV1 gene (1–357 nt, coat protein), in the virion sens, complete AC3 (354–758 nt, replication enhancer protein, REn) and AC2 (499–906 nt, transcriptional activator protein, TrAP) and partial AC1 (836–1177 nt, replication-associated protein) genes which were located in the complementary sense.

BLASTn analysis of *J. podagrica* virus isolate under study (HQ848382) showed highest sequence identity (96–99%) with several isolates of Jatropha mosaic India virus (JMIV: JN807768, GU906292, GU574210, JN807767, FJ346232, JN698951, HQ910408, GQ847545, and JN698953) of *J. curcas* and *Withania somnifera* from India, 95% with Papaver enation virus (PaEV: HM149260) of *Papaver somniferum* from India, 91% with Tomato leaf curl Lucknow virus (JN135234) of tomato from India, 87% with *Sri Lankan cassava mosaic virus* (AJ579307, AJ890225, AJ607394, AJ890229, andAJ314737) in cassava from India, 86% with *Indian cassava mosaic virus* (ICMV: AY738105, AY730035, AJ314739) in cassava from India, and 85% with Jatropha curcas mosaic virus (JCMV: GQ924760) in *J. curcas* from India and other begomovirus isolates reported worldwide.

The partial DNA-A nucleotide sequence of *J. podagrica* virus isolate (HQ848382) was also analyzed by Genomatix DiAlign program with the respective sequences of selected begomoviruses. The under study virus isolate revealed highest sequence similarities (99%) with Jatropha mosaic India virus (JN807768, GU906292 and GU574210) and 94–98% similarities with other Jatropha mosaic India virus isolates (JN807767, FJ346232, and HM149260), and the virus isolate shared less than 87% sequence similarities with ToLCLuV (JN135234), SrLCMV (AJ579307 and AJ890225), ICMV (AY738105 and AY730035), JCMV (GQ924760), and other begomovirus isolates. The multiple sequence alignment of

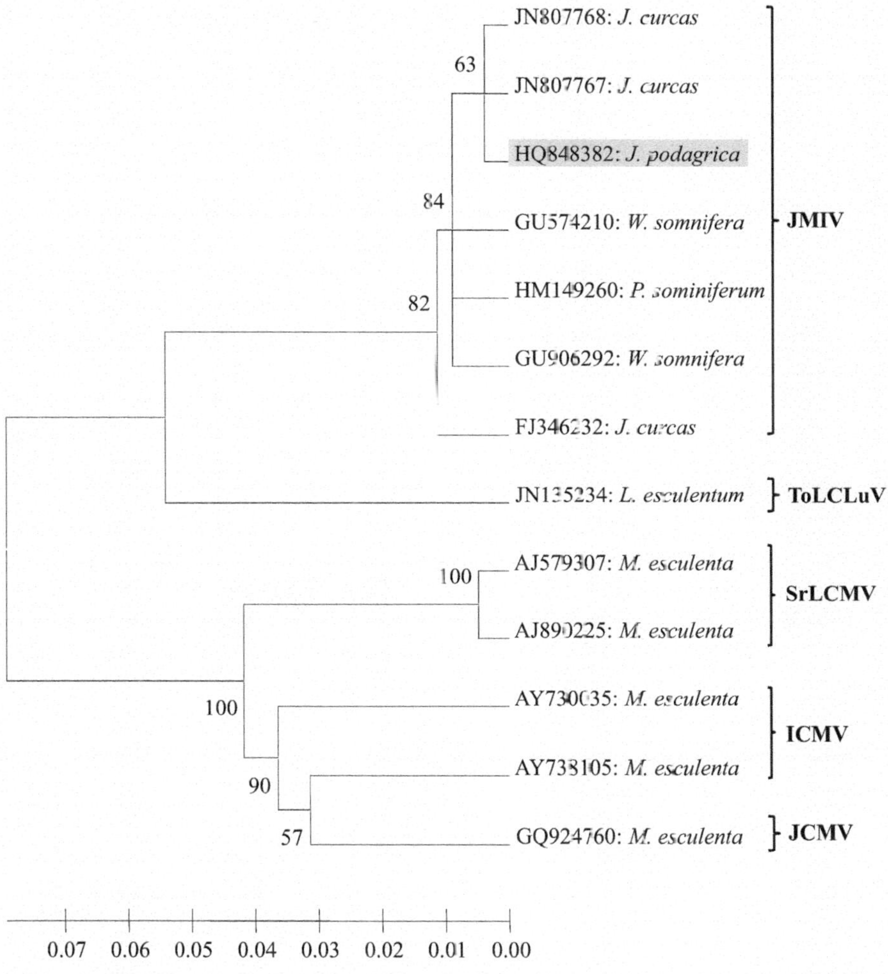

Fig. 18.2 Phylogenetic analysis of the partial DNA-A genome of *J. podagrica* virus isolate under study (HQ848382) showing close relationships with JMIV isolates

J. podagrica virus isolate (HQ848382) under study did not have any substitution with the JMIV partial DNA-A genome.

Phylogenetic analysis of *J. podagrica* virus isolate (HQ848382) shared close relationships with several isolates of JMIV (JN807767, JN807768, GU574210, HM149260, GU906292, FJ346232) in *J. curcas*, *W. somnifera*, and *P. somniferum* from India. The virus isolate also showed distinct relationships with ToLCLuV (JN135234), SrLCMV (AJ579307, AJ890225), ICMV (AY738105, AY730035), and JCMV (GQ924760) isolates reported from India (Fig. 18.2).

18.4.2 Sequence Analysis of *J. multifida* Begomovirus Isolate

Based on highest sequence identities of partial DNA-A genome and its closest phylogenetic relationships with *Jatropha mosaic India virus*, the virus isolates associated with mosaic disease of *J. multifida* were identified as isolates of *Jatropha mosaic India virus*.

The sequence analysis of partial DNA-A (1288 nt) of *J. multifida virus* isolate (HQ848381) contained four ORFs, encoding partial AV1 gene (1–423 nt, coat protein), in the virion sens, complete AC3 (420–824 nt, replication enhancer protein, REn) and AC2 (565–972 nt, transcriptional activator protein, TrAP), and partial AC1 (878–1288 nt, replication-associated protein) genes which were located in the complementary sense (Table 18.1).

BLASTn analysis of *J. multifida* virus isolate (HQ848381) shared 94% sequence identities with *Tomato leaf curl Patna virus* (ToLCPatV: EU862323, GU253915) in tomato and tobacco from India, 94% with *Tobacco leaf curl Pusa virus* (TobLCPuV: HQ180391) in tobacco from India, and 93% with *Cotton leaf curl virus* (CoLCV: GU440580) in cotton from India. Less than 89% sequence identity with *Tomato leaf curl Laos virus* (ToLCLaV: AF195782), *Tomato leaf curl Malaysia virus* (ToLCMalV: AF327436), *Tomato yellow leaf curl Vietnam virus* (ToLCMinV: EU189150), *Tomato leaf curl Mindanao virus* (ToYLCVeiV: EU487046), Ageratum yellow vein virus (AgYVV: JN809821), and other begomovirus isolates is also reported worldwide.

Table 18.1 Sequence identities of the begomovirus isolated from *J. podagrica* (HQ848382) under study with other related begomoviruses

Accession number	Abbreviation used for begomovirus	Host	Location in India	% identities by BLAST	% similarities by Genomatix
JN807768	JMIV	*J. curcas*	Gujarat	99	99
GU906292	JMIV	*W. somnifera*	Rajasthan	99	99
GU574210	JMIV	*W. somnifera*	Lucknow	99	99
JN807767	JMIV	*J. curcas*	Gujarat	99	98
FJ346232	JMIV	*J. curcas*	Lucknow	98	98
HM149260	PaEV/JMIV	*P. somniferum*	Lucknow	95	94
JN135234	ToLCLuV	*L. esculentum*	Lucknow	91	86
AJ579307	SrLCMV	*M. esculenta*	Kerala	87	83
AJ890225	SrLCMV	*M. esculenta*	Kerala	87	84
AY738105	ICMV	*M. esculenta*	Kerala	86	82
AY730035	ICMV	*M. esculenta*	Maharashtra	86	82
GQ924760	JCMV	*J. curcas*	Dharwad	85	82

JMIV jatropha mosaic India virus, PaEV papaver enation virus, ToLCLuV tomato leaf curl Lucknow virus, SrLCMV Sri Lankan cassava mosaic virus, ICMV India cassava mosaic virus, JCMV jatropha curcas mosaic virus

Table 18.2 Identities of the begomovirus alignment of under study isolate *J. multifida* (HQ84381) with other related begomoviruses

Accession number	Abbreviation used	Host	Country	% identities (BLAST)	% similarity (Genomatix)
EU862323	ToLCPatV	Tomato	India	94	92
GU253915	ToLCPatV	Tobacco	India	94	92
HQ180391	ToLCPuV	Tobacco	India	94	92
GU440580	CoLCV	Cotton	India	93	92
AF195782	ToLCLaV	Tomato	Taiwan	88	85
AF327436	ToLCMalV	Tomato	Malaysia	87	85
EU487046	ToLCMinV	Tomato	Philippines	85	82
EU189150	ToYLCVeiV	Tomato	Germany	85	83
JN809821	AgYVV	*Sauropus* sp.	Thailand	86	81

ToLCPatV tomato leaf curl Patna, ToLCPuV tobacco leaf curl Pusa virus, CoLCV Cotton leaf curl virus, ToLCLaV tomato leaf curl Laos virus, ToLCMalV tomato leaf curl Malaysia virus, ToYLCVeiV tomato yellow leaf curl Vietnam virus, ToLCMinV tomato leaf curl Mindanao virus, AgYVV Ageratum yellow vein virus

The percent nucleotide pairwise identity at partial DNA-A *J. multifida* virus isolate (HQ848381) through the Genomatix DiAlign revealed the highest sequence similarity (92%) with *Tomato leaf curl Patna* (EU862323, GU253915), *Tobacco leaf curl Pusa virus* (HQ180391), and *Cotton leaf curl virus* (GU440580). Less than 85% sequence similarities achieved with *Tomato leaf curl Laos virus* (AF195782), *Tomato leaf curl Malaysia virus* (AF327436), *Tomato yellow leaf curl Vietnam virus* (EU189150), *Tomato leaf curl Mindanao virus* (EU487046), and *Ageratum yellow vein virus* (JN809821) isolates were reported worldwide (Table 18.2).

Phylogenetic analyses of *J. multifida* isolates (HQ848381) were perused using the MEGA v. 4.0 program with selected begomovirus isolates. During phylogenetic analysis, the virus isolate under study shared closest relationship with *Tobacco leaf curl Pusa virus* (HQ180391) reported in tobacco from Bihar, India. The virus isolate under study also showed close relationships with *Tomato leaf curl Patna* (GU253915, EU862323) and *Cotton leaf curl virus* (GU440580) and distinct relationships with *Tomato leaf curl Laos virus* (AF195782), *Tomato leaf curl Malaysia virus* (AF327436), *Tomato leaf curl Mindanao virus* (EU487046), *Ageratum yellow vein virus* (JN809821), and *Tomato yellow leaf curl Vietnam virus* (EU189150) (Fig. 18.3).

18.4.3 Sequence Analysis of *J. integerrima* Virus Isolate

The analysis of partial DNA-A (1201 nt) of *J. integerrima virus* isolate (JQ043440) contained four ORFs, encoding partial AV1 gene (1–381 nt, coat protein), in the virion sens, complete AC3 (378–782 nt, replication enhancer protein, REn) and AC2 (523–927 nt, transcriptional activator protein, TrAP), and partial AC1 (830–1201 nt,

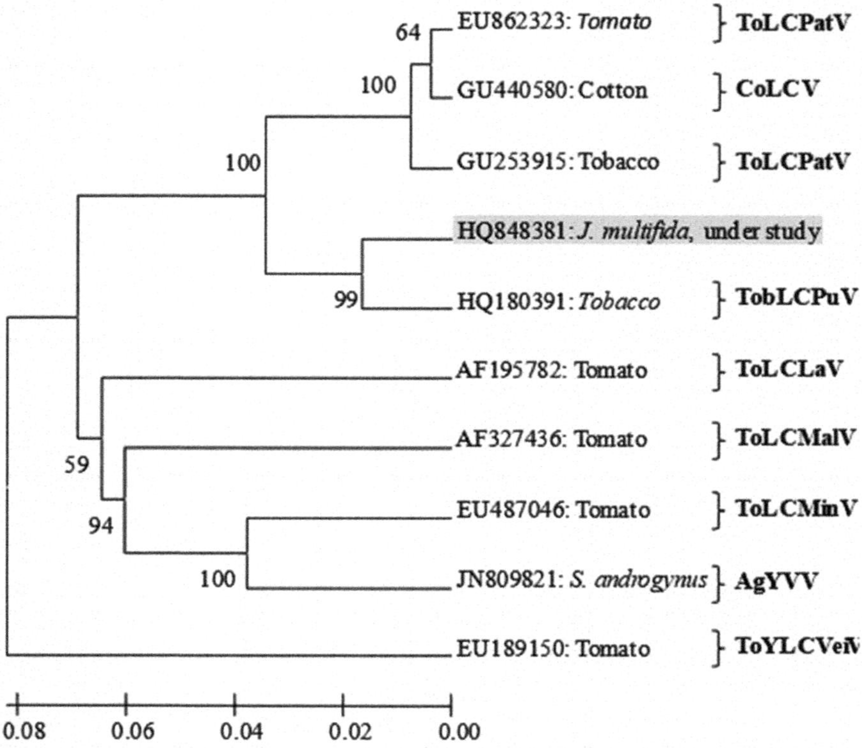

Fig. 18.3 Phylogenetic analysis of the partial DNA-A genome of virus isolate under study (HQ848381) associated with mosaic disease of *J. multifida* showing relationships with isolates of TobLCPuV and other begomoviruses

replication-associated protein) genes which were located in the complementary sense.

BLASTn analysis of partial DNA-A of the virus isolate *J. integerrima* (JQ043440) under study showed highest sequence identity (98%) with Papaya leaf curl virus (PLCV: HM143914); 93% with *Papaya leaf curl virus* (Y15934), *Tomato leaf curl Karnataka virus* (ToLCKV: AY754812), and *Radish leaf curl virus* (RaLCV: EU194914); and 92% with *Tomato leaf curl Karnataka virus* (ToLCKV: HM851186, HM803118, AY738101), *Tomato leaf curl virus-Bangalore II* (ToLCV-BII: U38239), *Tomato leaf curl New Delhi virus* (ToLCNDV: DQ629102), and *Papaya leaf curl virus* (AY738092, JN807765).

The partial DNA-A nucleotide sequence of *J. integerrima* (JQ043440) was analyzed by Genomatix DiAlign program with the respective sequences of selected begomoviruses. The under study virus isolate revealed highest sequence similarities (97%) with PLCV (HM143914); 92% with ToLCKV (AY754812); 91% with RaLCV (EU194914) and ToLCKV (HM851186, HM803118, U38239); 90% with

Table 18.3 Identities of the begomovirus alignment of under study isolate *J. integerrima* (JQ043440) with other related begomoviruses

Accession number	Abbreviation	Host	Location in India	% identity (BLAST)	% similarity (Genomatix)
HM143914	PLCV	Tobacco	Lucknow	98	97
AY754812	ToLCKV	Tomato	Jhansi	93	92
Y15934	PLCV	Papaya	Lucknow	93	90
EU194914	RaLCV	Tobacco	Bihar	93	91
AY738092	PLCV	Cotton	Patna	92	90
JN807765	PLCV	Soybean	Lucknow	92	89
HM851186	ToLCKV	Tomato	New Delhi	92	91
HM803118	ToLCKV	Tomato	New Delhi	92	91
U38239	ToLC-BII	Tomato	Bangalore	92	91
AY738101	ToLCKV	Tobacco	Shimoga	92	90
DQ629102	ToLCNDV	Tomato	New Delhi	92	90

PLCV papaya leaf curl virus, RaLCV radish leaf curl virus, ToLCKV tomato leaf curl Karnataka virus, ToLCV-BII tomato leaf curl virus-Bangalore II, ToLCNDV tomato leaf curl New Delhi virus

PLCV (Y15934, AY738092) ToLCKV (AY738101), and ToLCNDV (DQ629102); and 89% with PLCV (JN807765) (Table 18.3).

BLASTn analysis of the isolates with maximum homology was considered for multiple sequence alignment to know the nucleotide difference in under study virus isolate and other begomovirus isolates. The multiple sequence alignment of *J. integerrima* virus isolate (JQ043440) under study contained eight nucleotide unique substitutions at the following positions: 33 (A/T), 178 (G/C), 205 (A/G), 345 (A/G), 421 (C/A), 429 (A/T), 488 (T/C), and 1169 (T/C).

Phylogenetic analyses of partial DNA-A genome of *J. integerrima* virus isolate (JQ043440) under study were perused using the MEGA v. 4.0 program with selected begomovirus isolates. During phylogenetic analysis, the virus isolate (JQ043440) under study shared close relationship with several isolates of PLCV (HM143914, Y15934, EU194914, AY738092, JN807765). The virus isolate also shared distinct relationships with ToLCNDV (DQ629102) and ToLCKV (AY754812, HM803118, HM851186, AY738101, U38239) begomovirus isolates (Fig. 18.4).

18.5 Genetic Diversity Among Begomoviruses Infecting *Jatropha* Species

During surveys in 4 subsequent years (2008–2011), mosaic disease-like symptoms were observed on *J. podagrica*, *J. multifida*, and *J. integerrima* grown as ornamental at CSIR-NBRI, Lucknow. To find out how many begomoviruses/species are associated with the mosaic disease of these *Jatropha species*, PCRs were performed using begomovirus-specific primers (PALIv 722/PALIc 1960) and RCA and the expected sizes ~1.2 kb and ~2.7 kb amplicons were obtained and sequenced. The

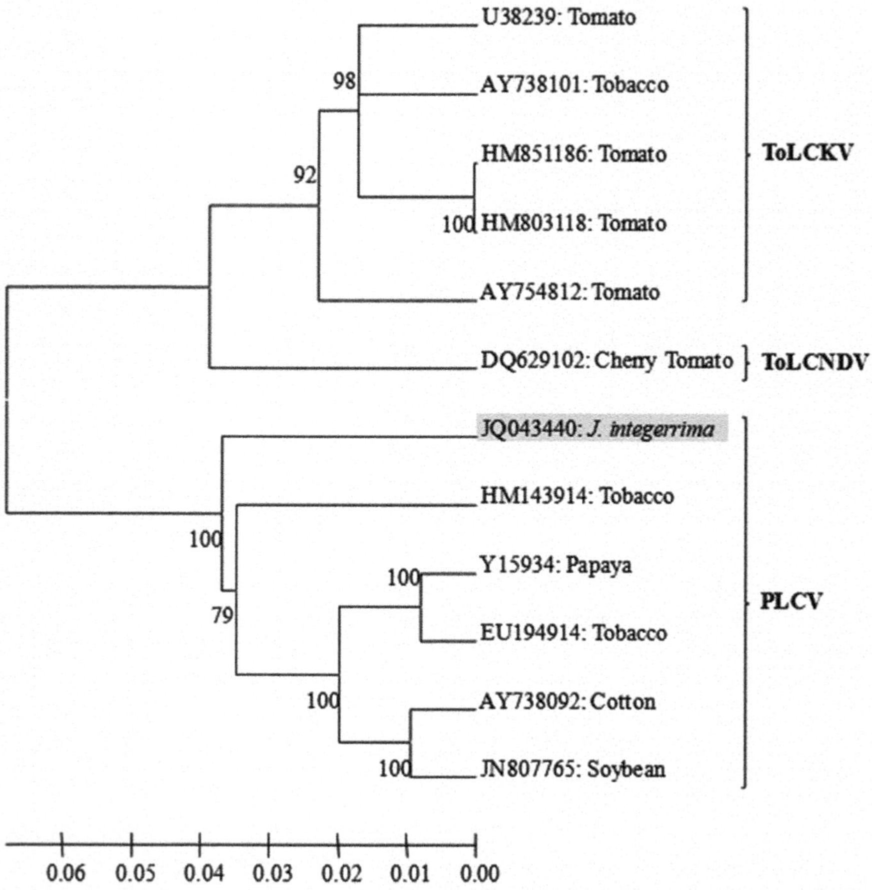

Fig. 18.4 Phylogenetic analysis of the partial DNA-A of *J. integerrima* (JQ043440) (highlighted in gray) showing relationships with PLCV and other begomovirus strains

sequence data were analyzed for sequence identities and phylogenetic relationships of these begomovirus isolates.

The sequence analysis of ~1.2 kb partial DNA-A amplicons also revealed presence of three begomovirus species: *Jatropha mosaic India virus* (HQ848382) on *J. podagrica*, *Tomato leaf curl Patna virus* (JN381198) on *J. multifida*, and *Papaya leaf curl virus* (JQ043440) on *J. integerrima*. Phylogenetic analysis of complete DNA-A of virus isolates under study (JMIV: HM230683), (JCMV: JN692494), (CYVMV: EU727086) and (JYMIV: FJ177030) from *J. curcas* and *J. gossypifolia* with other *Jatropha* begomoviruses (JCMV, GQ924760 and JF496657; JLCV, EU798996 and GU451249; ACMV, JN053421 and JN053447) reported worldwide formed separate clusters; therefore, they were considered as four distinct begomovirus species in India and abroad (Fig. 18.5).

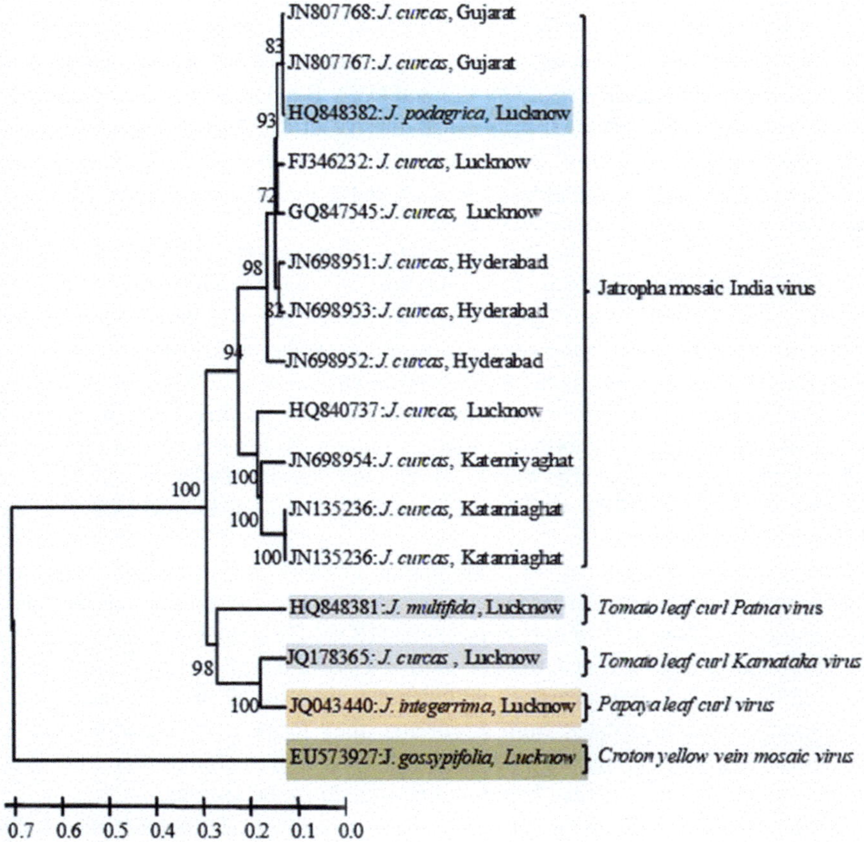

Fig. 18.5 Phylogenetic analysis of partial DNA-A genome of begomoviruses infecting ornamental *Jatropha* species in India

During the study, a total three begomovirus species, viz., Jatropha mosaic India virus, Jatropha curcas mosaic virus, and *Jatropha mosaic India virus* on *J. podagrica*, *Tomato leaf curl Patna virus* on *J. multifida*, and *Papaya leaf curl virus* on *J. integerrima* have been identified. These results indicated that genetic diversity exists among the begomoviruses infecting ornamental species of *Jatropha* grown in India.

18.6 Conclusion

Aswatha Narayana et al. (2006, 2007) reported the natural occurrence of Jatropha mosaic virus disease for the first time in southern India and they identified the causal organism as a distinct begomovirus closely related to Indian cassava mosaic viruses. Then after, Raj et al. (2008) reported the association of a begomovirus with Jatropha mosaic disease in north India which possessed highest identities and closest relationships with *Indian* and *Sri Lankan cassava mosaic virus* isolates. Gao et al. (2010) reported a new strain of *Indian cassava mosaic virus* based on analysis of complete nucleotide sequences of DNA-A and DNA-B bipartite genome of *Jatropha curcas* isolated from Dharwad, southern India. Recently, *African cassava mosaic virus* (Ramkat et al. 2011a, b) is reported on *J. curcas* from Kenya; however, the isolate under study is a monopartite genome since no DNA-B genome has been detected by PCR using its specific primers.

The natural infection of begomoviruses has been reported in *African cassava mosaic virus* on *J. multifida* in East and West Africa (Okoth 1991). Molecular detection and identification of begomovirus isolates are associated with mosaic disease of ornamental *Jatropha* species from India (Snehi et al. 2016).

Association of the new monopartite begomovirus species (JMIV) with mosaic disease of *J. curcas* is a new report from northern India. In this study, JCMV also identified from *J. curcas* in northern India is similar to earlier reports from Dharwad (GQ924760) and Jalgaon (JF496657). Both virus species JMIV and JCMV are monopartite begomoviruses which may be considered as a serious threat to the cultivation of *J. curcas*.

The sequence analysis of ~1.2 kb partial DNA-A amplicons also revealed presence of four begomovirus species: Jatropha mosaic India virus (HQ848382) on *J. podagrica*, *Tomato leaf curl Patna virus* (HQ848381) on *J. multifida*, *Papaya leaf curl virus* (JQ04340) on *J. integerrima*, and *Tomato leaf curl Karnataka virus* (JQ178365) on *J. curcas*. During phylogenetic analysis of these isolates (from ornamentals and one isolated from *J. curcas*), they formed five separate clusters with other begomovirus reported on *Jatropha* species from India; therefore, they were considered as five distinct begomoviruses in *Jatropha* species.

During this study, a total three begomovirus species, viz., Jatropha mosaic India virus on J. podagrica, Tomato leaf curl Patna virus on *J. multifida*, and Papaya leaf curl virus on *J. integerrima*, have been identified. These results indicated that genetic diversity exists among the begomoviruses infecting various species of Jatropha ornamental species grown in India.

The recent developments in biotechnology and emergence of genetic engineering have offered the powerful techniques for molecular breeding and incorporation of genes from taxonomically unrelated species in developing varieties of plant species with novel, useful agronomical and value-added traits. In recent years, genetic engineering technique using virus gene elements has been employed as an alternative strategy to produce virus-resistant plants. Therefore, the coat protein and replicase protein genes of begomovirus characterized in this study may be utilized in the future for the development of transgenic Jatropha ornamental species for the development of inbuilt resistance against begomovirus.

References

Aswatha Narayana DS, Shankarappa KS, Govindrappa MR, Prameela HA, Gururaj Rao MR, Rangaswamy KT (2006) Natural occurrence of Jatropha mosaic virus disease in India. Curr Sci 91:584–586

Aswatha Narayana DS, Rangaswamy KT, Shankarappa KS, Maruthi MN, Reddy-Laksminarayana CN, Rekha AR, Murthy Keahava KV (2007) District begomoviruses closely related to cassava mosaic viruses cause Indian Jatropha mosaic disease. Int J Virol 3:1–11

Briddon R, Mansoor S, Bedford I, Pinner M, Saunders K, Stanley J, Zafar Y, Malik K, Markham P (2001) Identification of DNA components required for induction of cotton leaf curl disease. Virology 285:234–243

Briddon RW, Bull SE, Amin I (2003) Diversity of DNA β: a satellite molecule associated with some monopartite begomoviruses. Virology 312:106–121

Dellaporta SL, Wood JJ, Hicks JB (1983) A plant DNA mini-preparation: version II. Plant Mol Biol Report 1:19–21

Fauquet CM, Briddon RW, Brown JK, Moriones E, Stanley J, Zerbini M, Zhou X (2008) Geminivirus strain demarcation and nomenclature. Arch Virol 153:783–821

Gao S, Qu J, Chua NH, Ye J (2010) A new strain of Indian cassava mosaic virus causes a mosaic disease in the biodiesel crop Jatropha curcas. Arch Virol 155:607–612

Hanley-Bowdoin L, Settlage SB, Orozco BM, Nagar S, Robertson D (1999) Geminiviruses: models for plant DNA replication, transcription and cell cycle regulation. Crit Rev Plant Sci 18:71–106

Okoth VAG (1991) Whitefly as a vector of *African cassava mosaic virus* (ACMV). Namulonge research Station, Ministry of Agriculture, Animal Industry and Fisheries, Kampala. Plant Dis 77:340–347

Pant V, Gupta D, Choudhury NR, Malathi VG, Varma A, Mukherjee SK (2001) Molecular characterisation of the Rep protein of the blackgram isolate of Indian mungbean yellow mosaic virus. J Gen Virol 82:2559–2567

Raj SK, Snehi SK, Kumar S, Khan MS, Pathre U (2008) First molecular identification of a begomovirus in India that is closely related to Cassava mosaic virus and causes mosaic and stunting of Jatropha curcas L. Aust Plant Dis Notes 3:69–71

Ramkat RC, Alberto C, Fatemeh M, Margit L (2011a) Biotechnological approaches to determine the impact of viruses in the energy crop plant Jatropha curcas. Virol J 8:386

Ramkat RC, Alberto C, Fatemeh M, Margit L (2011b) Occurrence of African cassava mosaic virus (ACMV) and East African cassava mosaic virus–Uganda (EACMV-UG) in *Jatropha curcas*. BMC Proc 5:P93

Rojas MR, Gilbertson RL, Russel DR, Maxwell DP (1993) Use of degenerate primers in the polymerase chain reaction to detect whitefly transmitted geminiviruses. Plant Dis 77:340–347

Snehi SK, Raj SK, Prasad V, Singh S (2016) Molecular detection and identification of begomovirus isolates associated with mosaic disease of ornamental Jatropha species from India. Int J Bact Virol 1(1):1–004

Stanley J, Bisaro DM, Briddon RW, Brown JK, Fauquet CM, Harrison BD, Rybicki EP, Stenger DC (2005) Geminiviridae. In: Fauquet CM, Mayo MA, Maniloff J, Desselberger U, Ball LA (eds) Virus taxonomy. VIIIth report of the ICTV. Elsevier/Academic Press, London, pp 301–326

Begomovirus Interaction with Ornamental Plants: Recent Advances and Methods for Engineering Resistance

19

Rachana Singh and S. K. Raj

Abstract

Ornamental plants are cultivated for decorative and utility purposes in urban landscapes throughout the world. They are much valued for their aesthetic properties and constitute an important part of the global horticulture industry. Begomovirus infections are increasing day by day as these are causing significant losses and raising danger to a large number of cultivated crops and also to ornamental plants. They are posing a significant negative impact on ornamentals provoking a wide range of symptoms, reducing both decorative value and quality of propagated material, and causing large economic damage. Significant growth of the ornamental plant market in recent years promotes the spread of viral diseases. Among the most popular ornamentals are chrysanthemum, rose, clematis, canna, and lavender. In the presented chapter, the various methods used for the engineering of ornamental plants to guard them against begomovirus pathogenesis are described.

Keywords

Ornamental plants · Begomovirus · Resistance · Pathogen-derived resistance · AmiRNA · CRISPR-Cas9

R. Singh
Amity Institute of Biotechnology, Amity University Uttar Pradesh, Lucknow Campus, Lucknow, UP, India

S. K. Raj (✉)
Plant Molecular Virology Laboratory, CSIR-National Botanical Research Institute, Lucknow, Uttar Pradesh, India

S. K. Raj et al. (eds.), *Virus Diseases of Ornamental Plants*,
https://doi.org/10.1007/978-981-16-3919-7_19

19.1 Introduction

Floriculture generally refers to flower farming and involves the farming of ornamental and flowering plants. India has a large market for floriculture. Flowers are an integral part of our society and around the world, and they are cultivated and used for enormous purposes like social, religious, and aesthetic purposes. The high demand for cut and loose flowers has made floriculture an important industry. The commercialization of ornamental emerged as a new industry of floriculture. Apart from that, the flowers and ornamentals are also much in demand for industrial applications because of flavor, fragrances, medicine, natural color, etc. Due to the huge demand of floriculture industry, it is necessary to ensure the quality and quantity of ornamental plants which are affected by biotic agents and majorly by plant viruses. Plant viruses are identified as a key factor affecting production and also the economy (Marwal et al. 2014).

Plant viruses are organisms consisting of RNA or DNA as genetic material and coat protein. They are transmitted through vectors, vegetative propagation, or seed dissemination. Some viruses are also transmitted through mechanical contact. The viruses cause diseases that led to economic losses. The symptoms of viruses include mosaic, mottle, ringspot, yellow vein, necrosis, curling, flower/foliage variegation, and overall plant stunting. Among plant viruses, *Geminiviridae* is the largest family infecting several crops, weeds, and ornamental plants (Marwal et al. 2014; Prajapat et al. 2011). In this family, the genus *Begomovirus* causes the most devastating effects on plants. Begomoviruses are identified, confirmed, and reported as the most detrimental pathogens that cause epidemics in economically important crops especially in tropical and subtropical regions.

19.2 Interaction of Begomoviruses with Some Ornamental Plants

19.2.1 Begomovirus Interaction with *Rosa chinensis*

Begomovirus infection and its list are increasing day by day. In ornamentals, rose plants' (*Rosa chinensis*) highly stunted growth and leaf curling were reported from Faisalabad, Pakistan. Based on symptoms, rose plants were checked for begomovirus infection through PCR. The method of rolling circle amplification was used for the complete genome sequence. Based on sequence homologies with other begomoviruses, the *Rose leaf curl virus* (RoLCuV) was identified as a new virus, and the name is given according to the symptom and sequence similarity. RoLCuV showed close identity (83%) with *Tomato leaf curl Pakistan virus*, while associated beta-satellite showed 96% identity with *Digera arvensis* yellow vein betasatellite (DiAYVB), justifying a new isolate for the beta-satellite. Recombination analysis of newly identified begomovirus revealed it as a recombinant of Tomato leaf curl Pakistan virus from its coat protein region.

The infectious molecules for virus/satellite were prepared and inoculated from *Agrobacterium tumefaciens* to *Nicotiana benthamiana* plants. RoLCuV alone was unable to induce any level of symptoms on *N. benthamiana* plants, but co-inoculation with cognate beta-satellite produced infection symptoms. Further investigation to understand the trans-replication ability of beta-satellites revealed their flexibility to interact with *Rose leaf curl virus* (Khatri et al. 2014).

19.2.2 Begomovirus Interaction with *Calendula officinalis*

Calendula officinalis L. (Asteraceae) is an important annual ornamental plant grown in gardens during the winter season and has an aesthetic beauty of bright yellow-colored flower. It belongs to the family Asteraceae and is commonly known as pot marigold. The plant is native to Central and Southern Europe, Western Asia, and the USA. Besides, it has considerable importance to the cosmetic/pharmaceutical industry because it's used in the manufacture of antiseptic creams. Plant viruses affect the aesthetic value of ornamental *Calendula* by reducing its rate of growth as well as the quality and quantity of its flowers. *Calendula officinalis* plants are affected by the *Cucumber mosaic virus* (Lisa and Lecoq 1984; Naqvi and Samad 1985), *Turnip mosaic virus* (Lisa et al. 1981), and *Tobacco mosaic virus* (Hristova et al. 1994). A rosette disease transmitted by whiteflies and grafting was recorded on *C. officinalis* by (Gupta and Verma 1983). Singh et al., in 2018 (Singh et al. 2018), also worked on *Calendula* samples showing symptoms of yellow vein disease and identified as *Tomato leaf curl New Delhi virus* (ToLCNDV). The findings suggested that *Calendula* spp. plants infected with ToLCNDV may act as an alternate host (reservoir) for other economically important plants (Singh et al. 2018).

In the year 2014, yellow vein net disease was noticed on *C. officinalis* plants in Lucknow, Uttar Pradesh, India. The severe infection was observed on *C. officinalis* plants with yellow vein net symptoms on leaves and also deformation in leaves. Based on the symptoms, PCR amplification, and DNA sequence analysis, the disease was identified as Ageratum enation virus (AEV) and Ageratum leaf curl-beta satellite (ALCuB), further confirmed by phylogenetic analysis and pathogenicity tests (Jaidi et al. 2015).

The calendula plants with yellow vein disease symptoms were observed in Madikeri dist. of Karnataka state, India. The PCR-based diagnostic with begomovirus-specific primer and whole-genome sequencing revealed the association of *Papaya leaf curl virus* (PaLCuV) with the symptomatic calendula plants with maximum nucleotide (nt) identity of 91.5–94.9% with begomovirus (PaLCuV) strains infecting *Croton bonplandianum*, *Acalypha* sp., and *Capsicum annuum* in India (Venkataravanappa et al. 2020).

19.2.3 Begomovirus Interaction with *Passion Fruit* (*Passiflora edulis*)

Passion fruit (*Passiflora edulis*; syn. *P. adults* f. *flavicarpa*) is a prominent tropical fruit of Brazil. Brazil is the world's largest producer of passion fruit, with ca. 825,000 tons annually. In the orchards in Southwestern Bahia, Brazil, severe mosaic symptoms, followed by yellow spots, unusual small leaves, fruit malformation and cracking, stunted plant growth, and large population of whitefly (*Bemisia tabaci*) infestation, were observed in passionflower (*Passiflora edulis*). The disease was observed in various orchards of different countries. Leaf samples were collected from 57 orchards located in ten counties and evaluated by PCR followed by sequencing. The complete nucleotide sequences of DNA-A for two isolates showed 97% identity with *Passionfruit severe leaf distortion virus* (PSLDV). A very huge study was performed with approximately 235,000 visually assessed plants that showed symptoms characteristic of begomovirus infection.

A similar outbreak, also in the state of Bahia, was first reported by Novaes et al. (2003). High incidence of a begomovirus (family *Geminiviridae*, genus *Begomovirus*) is tentatively named as *Passionflower little leaf mosaic virus* (PLLMV). The infected plants were also heavily colonized by whitefly identified as *Bemisia tabaci*. But the attempts to transmit the virus from infected to healthy passionflowers through *B. tabaci* biotype B reared on soybean failed, indicating that the biotype of whitefly might be different which is colonizing the passionflower.

19.2.4 *Begomovirus* Infecting Wild Ornamentals

In the year 2010, yellowing of leaf vein disease was observed on Spanish flag (*Lantana camara*) in Sirsa, Haryana province, India. Earlier, no report was found to show an association of begomovirus and DNA satellites with *Lantana camara*. Therefore, molecular characterization and understanding of the genomic analysis of begomovirus infecting *Lantana camara* was imperative for pathogen diagnosis and disease management. This was the first report and molecular characterization of a begomovirus associated with its two satellites infecting a new host *Lantana camara* in India (Marwal et al. 2013).

Vinca minor L., an ornamental plant with typical symptoms of a begomovirus, was reported from Lahore, Pakistan. The PCR with symptomatic and healthy leaf samples, using universal primers conserved for the coat protein region of begomoviruses, shows positive amplification of ~0.78 kb from symptomatic leaf samples, while healthy plant samples gave no amplification. Sequencing followed by BLASTn analysis confirmed 93% nucleotide sequence identity with *Pedilanthus leaf curl virus* originating from Pakistan that confirms the begomovirus infection on *V. minor* in Pakistan (Saleem et al. 2008).

19.2.5 Role of Recombination in Evolution of New Begomovirus Species

Recombination plays a key role in the evolution of *Begomovirus* and may be contributing to the emergence of new species. With the development of computational recombination detection tools and an increasing number of available genome sequences, many studies have reported evidence of recombination. Begomovirus-associated symptoms were observed in *Jasminum sambac* and *Millingtonia hortensis* plants growing in crop fields of Lakshmangarh, Rajasthan (India). Amplification of a PCR product was found up to the expected size (~550 bp). The PCR product was cloned and partially sequenced and it was utilized for in silico characterization. The in silico analysis suggested that interspecific recombination has resulted in significant diversity among *Begomovirus* (Marwal et al. 2012).

19.3 Management of Begomoviruses Through Genetic Engineering

Genetic management of begomoviruses is very important to reduce viral diseases and their impact upon the economy because viruses tend to recombine, reorganize, and rapidly emerge again and again. Most of the countries were not able to develop the long-lasting resistance in genotypes by breeders against the begomoviruses successfully. Genetic engineering offers alternative tools that emerged as the best management strategy against the begomoviruses. The newest techniques of modern breeding, hybrid seeds, automation, and machine-based farming and pest management can also be used to enhance the situation (Snehi et al. 2015).

19.3.1 Defense Through Selective Proteins

The *Agrobacterium tumefaciens*-mediated transformation of begomovirus-resistant genes for the production of a transgenic plant led toward the management of begomoviruses. Tobacco plants were the first to be engineered successfully by incorporating the resistant genes against viruses (Abel et al. 1986). Protoplast culture was also used for receiving transformational changes in early experiments (Horsch et al. 1985), but with the establishment of transformation techniques, every part of the plant (leaves, shoots, and roots) has been utilized for genetic transformation in dicotyledons (Snehi et al. 2015).

19.3.2 Resistance Through Pathogen Genes/Pathogen-Derived Resistance (PDR)

Hamilton, in 1980, was the first who developed the concept of resistance against the begomoviruses. He has proposed the idea of the transformation of genes derived from the pathogen itself (Hamilton 1980). This strategy emerges from the cross-protection concept that evolved from the cross-protection phenomenon. In this non-symptomatic virus, the strain is inoculated into the host to develop a defense in the host against the pathogenic virus. The virus strain inserted into the host pathogenic strain interrupts the genetic expression of the pathogen by disturbing the normal life cycle of the pathogen (Reimann-Philipp 1998).

This phenomenon was verified by various scientists on different begomoviruses by transforming the non-symptomatic virus and also checking the expression of the virus that is inoculated (Beachy 1993; Wilson 1993; Baulcombe 1994; Hull 1994; Lomonossoff 1995). Various genes were used for the confirmation of this concept of pathogen-derived resistance using viral particles, for example, coat protein (CP) genes, movement protein (MP), replicase gene (RG), antisense RNA (As-RNA), satellite RNA (S-RNA), and defective interfering genes (DI).

19.3.2.1 PDR Through Coat Proteins

The transgenic plant was developed by expressing the CP gene to acquire the virus resistance (Beachy 1993). In the first experiment, the CP gene was isolated from the *Tobacco mosaic virus* (TMV) and incorporated in tobacco plants to achieve resistance. Confirmation of transgenic plants was done by checking the presence of foreign DNA in primary and secondary transformants (Wilson 1993). The gene induced in transgenic plants can be isolated in progenies by antibodies raised by utilization of CP (Beachy 1993). It was found that transgenics show differences in coat protein gene expression with different hosts. The difference is based on viral transgenes or virus groups (Lomonossoff 1995). The coat protein-mediated resistance which is protein-mediated is achieved when a single copy of transgene is inserted, and the protein level enhances as transcription and translation takes place (Varma et al. 2002). Multiple copies of gene-inserted transgenes are dependent upon RNA, and such type of acquired resistance is due to high strain-specific RNA-mediated resistance (Lomonossoff 1995). The expression of the transgene is only up to mRNA level and shows a very low level of protein accumulation. But when the accumulation levels of mRNA exceed a designated threshold level, the gene silencing mechanism gets initiated which directly shows its effects upon viral multiplication and transgenic expression (Varma et al. 2002).

For imitation of the viral suppression phenomenon, an identical virus genome sequence to transgene is necessary (Varma et al. 2002). The particular phenomenon is named posttranscriptional gene silencing (PTGS) (English et al. 1996). Such kind of resistance is referred to as homology dependent as it reflects the homology-dependent silencing relationships (Mueller et al. 1995). This defense mechanism involves mRNA degradation by the invading viruses as well as the transgene (Fig. 19.1) (Waterhouse et al. 2001).

19.3.2.2 Coat Protein Gene-Based Resistance in Tomato for *Tomato Leaf Curl Virus*

Coat protein (CP) gene of *Tomato leaf curl virus* (TLCV) was cloned into an expression vector and mobilized to *Agrobacterium tumefaciens* through triparental

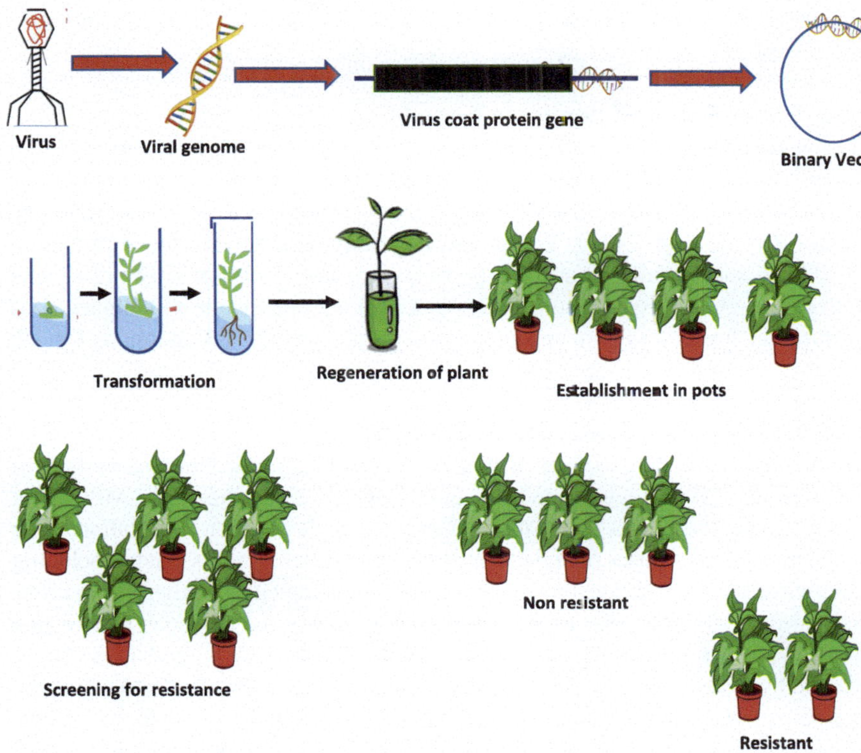

Fig. 19.1 Diagrammatic representation of genetic defence mechanism for Begomoviruses

mating. Cotyledon leaf explants of Pusa Ruby tomato were transformed by co-cultivation with *Agrobacterium* containing TLCV-CP construct. Kanamycin-resistant transformants were regenerated and established in glasshouse. T0-generation putative transgenic plants obtained were screened by PCR, Southern and Northern hybridization tests, and Western blot assay, which confirmed the incorporation and expression of the CP gene. CP expressing transgenic plants were self-pollinated. T1-generation transgenic plants were challenged by TLCV through whiteflies which showed variable degrees of disease resistance/tolerance compared to the untransformed control. This was the first demonstration of coat protein gene-based resistance in tomato against *Tomato leaf curl virus* in India (Raj et al. 2005).

These findings were useful to strengthen the understanding on the phenomenon of virus encapsidation by the capsid protein and its movement in infected plants. This information on CP-mediated resistance in case of geminiviruses was an additional support to the earlier work (Kunik et al. 1994) who first extended, reported, and applied this concept earlier only for RNA viruses. Moreover, these efforts were done to achieve resistance against TLCV infection in transgenic tomato plants. The protocols developed for tomato transformation had a wide scope in generation of transgenic plants using other genes of interest (Raj et al. 2005).

19.3.2.3 PDR Through Replicase Gene (RG)

After the coat protein-mediated resistance to raise the virus resistance transgenics, the replicase gene is the second to be considered as important for begomovirus management. The RG has used either functional or truncated gene to achieve the resistance (Snehi et al. 2015). Varma and the group worked to achieve resistance with RG and successfully reported to be worked against 16 viruses. The resistance acquired through RG expression involves inoculation of the plants with higher virus concentration levels and is considered to be strain-specific (Wintermantel et al. 1997).

19.3.2.4 PDR Through Movement Proteins

Movement proteins are associated with the viral encoded intracellular movement of plant viruses (Muskhogean and Koonin 1993). Interaction of MPs with plasmodesmata correlates the modification and facilitation of intracellular movements of plant viruses (Cooper et al. 1995). Six different virus genera including begomoviruses have displayed longer intracellular movements through MP interference (Varma et al. 2002). Resistance expressed by transgenic tobacco via defective TMV-MP against various viruses (CMV, TRSV, TRV, AIMV, and PCSV) was explained by Cooper et al. (1995). In a similar context, Tacke et al. (1996) generated a transgenic potato that expressed PLRV-MP and was found to be resistant against PVX and PVY. Further efforts in this line of action of production of transgenics lead toward the production of TSWV mutants which shown resistance toward various TSWV strains (Varma et al. 2002).

19.3.2.5 PDR Through RNA Silencing

The RNA silencing mechanism is characterized by the suppression of gene expression by sequence specified mRNA degradation. This mechanism was initially known as posttranscriptional gene silencing (PTGS) in plants (Han et al. 2000) and as RNA interference in animals (Cogoni and Macino 1997). The ribonuclease dicer (RNA-D) and argonaute (AGO) are the key molecules of this mechanism. Overall, there are three specific pathways in this mechanism. It is based on three specific pathways: endogenous mRNA silencing through miRNAs, cytoplasmic interfering (Vanitharani et al. 2004), and DNA methylation. The mechanism was proved successful in the plants, for example, *Arabidopsis thaliana* (used as a model plant) encoded ten argonaute proteins, six RNA-dependent RNA polymerase, and four dicer enzymes.

19.3.2.6 MicroRNA-Mediated Resistance

MicroRNAs (miRNAs) are small noncoding RNA molecules (22–24 nt) present in and regulate the expression of genes in plants and animals. Virologists have employed them to control begomovirus diseases. In a study, cotton leaf curl Multan begomovirus (CLCuMV) causes devastating cotton leaf curl disease (CLCuD) in cotton plants. Baig and Khan (2014) used in silico approaches in identifying the cotton miRNA targets in the genomes of CLCuMV and beta-satellite. A total of 18 nt sequences representing full-length DNA-A of CLCuMV and 58 nt sequences of full-length beta-satellite were screened against a set of 69 mature miRNAs of *G. hirsutum* (Dhir et al. 2019).

Small RNA (siRNA) regulates the expression of several genes in all plants and constitutes a natural immunity against viruses (Blevins et al. 2011). siRNA-based genetic engineering (SRGE) technology had been explored for crop protection against viruses for nearly 30 years. Viral resistance has been developed in diverse crops with SRGE technology, and a few viral-resistant crops have been approved for commercial release. Here, the efforts generating viral resistance with SRGE in different crops were summarized and the evolution of the technology, its efficacy in different crops for different viruses, and its application status in different crops were analyzed.

At this moment, siRNAs levels are negatively correlated with the level of infection severity (Akbergenov et al. 2006). Thus, to deactivate or defeat the host defenses, viruses have developed RNA silencing suppressors (Chellappan et al. 2004). Here the βC1 proteins are likely to play their part as suppressors. These proteins are CLCuMuB-βC1 (Amin et al. 2011), TYLCCNB-βC1 (Azhar et al. 2010), BYVMV-βC1 (*Bhendi yellow vein mosaic virus*) (Cui et al. 2005), ToLCCNV-βC1 (*Tomato leaf curl China virus*) (Sharma et al. 2010), and ToLCJAV-βC1 (*Tomato leaf curl Java virus*) (Gopal et al. 2007).

19.3.2.7 AmiRNA Mediated

AmiRNAs play an important role in plant development, signal transduction, and response to biotic and abiotic stress. The miRNAs can be engineered to alter their target specificity and such artificial miRNAs (amiRNAs) that have been shown to provide resistance against many begomovirus infections. Begomovirus-linked cotton leaf curl disease is a major constraint to cotton cultivation across Pakistan and northwestern India. The amiRNA-based studies revealed that the two amiRNA constructs, related to cotton miRNA169a sequences, were produced that contained 21 nt of the V2 gene sequence of *cotton leaf curl Burewala virus* (CLCuBuV), and Nicotiana benthamiana were transformed. The first amiRNA construct (P1C) maintained the miR169a sequence except for the replaced 21 nt, whereas in the second (P1D), the sequence of the miRNA169a backbone was altered to restore some of the hydrogen bonding of the mature miRNA duplex (Ali et al. 2013).

19.3.3 Limitations of PDR and Their Possible Solutions

Though PDR has been reported with successful implementation to produce tolerant cultivars against begomoviruses, there are certain limitations. For example, PRD exhibits suppression of targeted genes reducing its ability to express. But still enough amounts of functional proteins can be generated from that particularly reduced transcript. This may not be enough to produce any phenotypic variation in the silenced plant. Thus, PDR does not guarantee complete functionality of the silenced genes.

19.3.4 Ribozyme Mediated

Cleavage of target RNA was affected by antisense RNA ribozymes which have intrinsic endonucleolytic activity. However, double-standard RNA mechanism or antisense dominance controlled the endonucleolytic activity in vitro. According to Mishra et al. (2014), hammerhead ribozyme was designed to target rep-mRNA of MYMIV and was developed as an antiviral agent, and during his study, it was also found that RNA silencing is induced on the introduction of catalytically active as well as inactive ribozymes. It was also demonstrated that endonucleolytic activity of ribozymes is a true phenomenon, while muted version may have similar downregulation of target RNA by using RNA silencing suppressors (Mishra et al. 2014).

19.3.5 CRISPR-Cas9-Mediated Resistance

Conventional strategies can fail to control rapidly evolving and emerging plant viruses. Genome engineering strategies have recently emerged as promising tools to introduce desirable traits in many eukaryotic species, including plants. Among these genome engineering technologies, the CRISPR (clustered regularly interspaced palindromic repeats)/CRISPR-associated 9 (CRISPR/Cas9) systems have evolved recently, because of their simplicity, efficiency, and reproducibility. CRISPR/Cas9 technique is being used to engineer virus resistance in plants, either by directly targeting and cleaving the infecting viral genome or by modifying the host plant genome to introduce viral immunity. To date, the CRISPR/Cas9 technique has been used to develop resistance against many begomoviruses such as *Tomato yellow leaf curl virus* (TYLCV), *Bean yellow dwarf virus* (BeYDV), and *Cotton leaf curl Kokhran virus* (CLKCoV). These resistant plants were developed by targeting the virus genome using this technique. It was concluded by pinpointing the gaps in knowledge and the outstanding questions regarding CRISPR-/Cas9-mediated viral immunity (Zaidi et al. 2016; Islam and Jan 2017).

References

Abel PP, Nelson RS, De B et al (1986) Delay of disease development in transgenic plants that express the tobacco mosaic virus coat protein gene. Science 232:738–744

Akbergenov R, Si-Ammour A, Blevins T et al (2006) Molecular characterization of geminivirus-derived small RNAs in different plant species. Nucleic Acids Res 34(2):462–471

Ali I, Amin I, Briddon RW, Mansoor S (2013) Artificial microRNA-mediated resistance against the monopartite begomovirus Cotton leaf curl Burewala virus. Virol J 10:231

Amin I, Hussain K, Akbergenov R et al (2011) Suppressors of RNA silencing encoded by the components of the cotton leaf curl begomovirus-betasatellite complex. Mol Plant-Microbe Interact 24(8):973–983

Azhar MT, Amin I, Anjum ZI et al (2010) Both malvaceous and non-malvaceous betasatellites are associated with two wild cotton species grown under field conditions in Pakistan. Virus Genes 41(3):417–424

Baig MS, Khan J (2014) Identification of Gossypium hirsutum miRNA targets in the genome of Cotton leaf curl Multan virus and Betasatellite. Indian J Biotechnol 12:336–342

Baulcombe D (1994) Novel strategies for engineering virus resistance in plants. Curr Opin Biotechnol 5(2):117–124

Beachy RN (1993) Transgenic resistance to plant viruses. Semin Virol 4:327–416

Blevins T, Rajeswaran R, Aregger M, Borah BK, Schepetilnikov M, Baerlocher L, Farinelli L, Meins M, Hohn T, Pooggin MM (2011) Massive production of small RNAs from a non-coding region of cauliflower mosaic virus in plant defense and viral counter-defense. Nucleic Acids Res 39(12):5003–5014

Chellappan P, Vanitharani R, Fauquet CM (2004) Short interfering RNA accumulation correlates with host recovery in DNA virus-infected hosts, and gene silencing targets specific viral sequences. J Virol 78(14):7465–7477

Cogoni C, Macino G (1997) Isolation of quelling-defective (qde) mutants impaired in posttranscriptional transgene-induced gene silencing in *Neurospora crassa*. Proc Natl Acad Sci 94 (19):10233–10238

Cooper B, Lapidot M, Heick JA et al (1995) A defective movement protein of TMV in transgenic plants confers resistance to multiple viruses whereas the functional analog increases susceptibility. Virology 206(1):307–313

Cui X, Li G, Wang D et al (2005) A begomovirus DNA β -encoded protein binds DNA, functions as a suppressor of RNA silencing, and targets the cell nucleus. J Virol 79(16):10764–10775

Dhir S, Srivastava A, Yoshikawa N, Khurana SMP (2019) Plant viruses as virus induced gene silencing (VIGS) vectors. In: Khurana S, Gaur R (eds) Plant biotechnology: progress in genomic era. Springer, Singapore. https://doi.org/10.1007/978-981-13-8499-8_22

English JJ, Mueller E, Baulcombe DC (1996) Suppression of virus accumulation in transgenic plants exhibiting silencing of nuclear genes. Plant Cell 8(2):179–188

Gopal P, Kumar PP, Sinilal B (2007) Differential roles of C4 and β C1 in mediating suppression of post-transcriptional gene silencing: evidence for transactivation by the C2 of Bhendi yellow vein mosaic virus, a monopartite begomovirus. Virus Res 123(1):9–18

Gupta M, Verma VS (1983) Calendula rosette disease. Gartenbauwissenschaft 48:106–107

Hamilton RI (1980) Viruses. Plant disease: an advanced treatise, vol 5. Academic Press, New York, NY, p 279

Han S, Wu Z, Yang H et al (2000) Ribozyme-mediated resistance to rice dwarf virus and the transgene silencing in the progeny of transgenic rice plants. Transgenic Res 9(3):195–203

Horsch RB, Fry JE, Hoffman NL et al (1985) A simple and general method for transferring genes into plants. Science 227:1229–1232

Hristova D, Barkerdzhieva N, Svrakov K (1994) *Tomato mosaic virus* isolated from *Calendula officinalis*. Plant Protection Institute, 2230 Kostinbrod, Bulgaria. In: Int Conf Pl Virol. Apriltsi, Troyon, Bulgaria, vol 32, p 153

Hull R (1994) Resistance to plant viruses: obtaining genes by non-conventional approaches. Euphytica 75(3):195–205

Islam WJ, Jan WZ (2017) Genetic defense approaches against begomoviruses. J Appl Virol 6 (3):2306–6210

Jaidi M, Kumar S, Srivastava A, Raj SK (2015) First report of Ageratum enation virus and Ageratum leaf curl betasatellite infecting *Calendula officinalis* in India. New Dis Rep 32:6. https://doi.org/10.5197/j.2044-0588.2015.032.006

Khatri S, Nazia N, Fauquet CM, Muhammad M, Muhammad SNR (2014) A beta satellite-dependent begomovirus infects ornamental rose: characterization of begomovirus infecting rose in Pakistan. Virus Genes 49(1):124–131. https://doi.org/10.1007/s11262-014-1076-6
Kunik T, Salomon R, Zamir D, Navot N, Zeidan M, Michelson I, Gafni Y, Czosnek H (1994) Transgenic tomato plants expressing the tomato yellow leaf curl virus capsid protein are resistant to the virus. Bio/Technology 12:500–504
Lisa V, Lecoq H (1984) Zucchini yellow mosaic virus. In: Description of plant viruses, vol 282. Commonwealth Mycological Institute/Association of Applied Biologists, Kew
Lisa V, Boccardo G, D'Agostino G, Dellavalle G, D'Aquilio M (1981) Characterization of a potyvirus that causes zucchini yellow mosaic. Phytopathology 71:667–672
Lomonossoff GP (1995) Pathogen-derived resistance to plant viruses. Annu Rev Phytopathol 33 (1):323–343
Marwal A, Prajapat R, Sahu A, Gaur RK (2012) Computational characterization of begomovirus infecting two ornamental plants: Jasminum sambac and Millingtonia hortensis. Asian J Biol Sci 5:240–249. https://doi.org/10.3923/ajbs.2012.240.249
Marwal A, Sahu AK, Gaur RK (2013) Molecular Characterization of Begomoviruses and DNA satellites associated with a new host Spanish Flag (*Lantana camara*) in India. Int Scholar Res Not 2013:915703. https://doi.org/10.5402/2013/915703
Marwal A, Sahu AK, Gaur R (2014) Transmission and host interaction of Geminivirus in weeds. In: Plant virus-host interaction: molecular approaches and viral evolution, vol 1. Academic Press, Elsevier, New York, NY. https://doi.org/10.1016/B978-0-12-411584-2.00007-X
Mishra SK, Chilakamarthi U, Deb JK, Mukherjee SK (2014) Unfolding of in planta activity of anti-rep ribozyme in presence of a RNA silencing suppressor. FEBS Lett 588(10):1967–1972
Mueller E, Gilbert J, Davenport G et al (1995) Homology - dependent resistance: transgenic virus resistance in plants related to homology - dependent gene silencing. Plant J 7(6):1001–1013
Muskhogean AR, Koonin EV (1993) Cell-to-cell movement of plant viruses. Arch Virol 133 (3):239–257
Naqvi QA, Samad A (1985) Purification and properties of Calendula yellow net virus. Indian J Virol 1:143–146
Novaes QS, Astua JF, Yuki VA, Kitajima EW, Camargo LEA, Rezende JAM (2003) Partial characterization of a bipartite begomovirus infecting yellow passionflower in Brazil. Plant Pathol 52:648–654
Prajapat R, Marwal A, Bajpai V, Gaur RK (2011) Genomics and proteomics characterization of alphasatellite in weed associated with begomovirus. Int J Plant Pathol 2(1):1–14
Raj SK, Rachana S, Pandey SK, Singh BP (2005) Agrobacterium-mediated tomato transformation and regeneration of transgenic lines expressing tomato leaf curl virus coat protein gene for resistance against TLCV infection. Curr Sci 88(10):1674–1679
Reimann-Philipp U (1998) Mechanisms of resistance: expression of coat protein. In: Plant virology protocols: from virus isolation to transgenic resistance. Humana Press, Totowa, NJ, pp 521–532
Saleem M, Haider A, Tahir A, Saeed A, Ahmed B, Parveen R, Rashid N (2008) First report of a begomovirus infecting the ornamental plant Vinca minor L. Aust Plant Dis Notes 3:150–151
Sharma P, Ikegami M, Kon T (2010) Identification of the virulence factors and suppressors of posttranscriptional gene silencing encoded by Ageratum yellow vein virus, a monopartite begomovirus. Virus Res 149(1):19–27
Singh J, Khanna S, Ranjan K, Pant RP, Kumar P, Sirohi A, Baranwal VK (2018) Evidence of Association of begomovirus with the yellow vein disease of an ornamental plant pot marigold (*Calendula officinalis*) from Western Uttar Pradesh. J Pure Appl Microbiol 11(3). https://doi.org/10.22207/JPAM.11.3.48
Snehi SK, Raj SK, Prasad V et al (2015) Recent research findings related to management strategies of Begomoviruses. J Plant Pathol Microbiol 6(6):273
Tacke E, Salamini F, Rohde W (1996) Genetic engineering of potato for broad—spectrum protection. Nat Biotechnol 14:1597–1601

Vanitharani R, Chellappan P, Pita JS et al (2004) Differential roles of AC2 and AC4 of cassava geminiviruses in mediating synergism and suppression of posttranscriptional gene silencing. J Virol 78(17):9487–9498

Varma A, Jain RK, Bhat AI (2002) Virus resistant transgenic plants for environmentally safe management of viral diseases. Indian J Biotechnol 1:73–86

Venkataravanappa V, Sonavane P, Reddy CNL, Reddy MK (2020) Associations of papaya leaf curl virus and betasatellite with yellow vein disease of pot marigold (*Calendula officinalis* L.). Arch Phytopathol Plant Protect 52:1289. https://doi.org/10.1080/03235408.2019.1704350

Waterhouse PM, Ming-Bo W, Lough T (2001) Gene silencing as an adaptive defense against viruses. Nature 411(6839):834

Wilson TM (1993) Strategies to protect crop plants against viruses: pathogen-derived resistance blossoms. Proc Natl Acad Sci 90(8):3134–3141

Wintermantel WM, Banerjee N, Oliver JC et al (1997) Cucumber mosaic virus is restricted from entering minor veins in transgenic tobacco exhibiting replicase-mediated resistance. Virology 231(2):248–257

Zaidi SSA, Tashkandi M, Mansoor S, Mahfouz MM (2016) Engineering plant immunity: using CRISPR/Cas9 to generate virus resistance. Front Plant Sci 7:1673

Begomovirus on Ornamental Plants: Diversity and Management

20

Seweta Srivastava, N. N. Tiwari, Malyaj R. Prajapati, R. K. Jain, Jitender Singh, and A. K. Tiwari

Abstract

Ornamental plants are extensively disseminated in India and throughout the globe because they have extreme environmental adaptability. The infection of begomoviruses in many plants has been extensively observed and testified worldwide as crops stay in the field for a specific period, while different ornamental plants grow in or nearby these agricultural fields throughout the year. Geminiviruses have been more of a concern, specifically diseases caused by begomoviruses, because they have now touched the epidemic magnitudes. Ornamental plants serve as a substitute host for begomovirus in gardens and may permit their transmission to other economically significant crops and medicinal plants, thus boosting the host range of this phytovirus in varied regions of India. Whitefly (*Bemisia tabaci*) is the vector that efficiently transmits begomoviral infections in ornamentals, crops, and weeds and is predominant in the tropical and subtropical regions of the world. Hence, cumulative information about its epidemiology, sequence diversity, and biodiversity are extremely significant in order to implement preventive tactics for their proper management.

S. Srivastava
Plant Protection Sciences, Lovely Professional University, Jalandhar, Punjab, India

N. N. Tiwari · R. K. Jain
Department of Biotechnology, Agra Engineering College, Agra, Uttar Pradesh, India

M. R. Prajapati · J. Singh (✉)
College of Biotechnology, Sardar Vallabhbhai Patel University of Agriculture & Technology, Meerut, Uttar Pradesh, India

A. K. Tiwari (✉)
Central Lab, Sugarcane Research Institute, Shahjahnapur, Uttar Pradesh, India

S. K. Raj et al. (eds.), *Virus Diseases of Ornamental Plants*,
https://doi.org/10.1007/978-981-16-3919-7_20

Keywords

Begomovirus · Whitefly · Ornamental · Epidemiology · Diversity

20.1 Introduction

In India, numerous plant species are grown as ornamentals for their aesthetic values (Kumar et al. 2005). India is elegantly flagging its track to arise as a substantial performer of the world floriculture craft and a new floral super-influencer of the upcoming future trade carrying marvelous and booming potential. Variation in climatic conditions offers opportunity of growing all the most important cut flower species of the world, from tropical, subtropical, or temperate climatic zones (Zaidi et al. 2008; Marwal et al. 2020). Globally, there are adequate reports of begomovirus infection in ornamental plants (He et al. 2009; Ilyas et al. 2013). With regard to the Indian context, published reports on begomovirus infection in weeds and crops are available (Chatterjee et al. 2007; Das et al. 2008; Ghosh et al. 2007; Jyothsna et al. 2011; Khan et al. 2002; Tiwari et al. 2012a, b; Saxena and Tiwari 2017), but there is considerably fewer scientific data published till date with respect to begomoviral infections in plants having ornamental importance (Marwal et al. 2013a, b, c, d, e). In developing countries like India, farmers are mainly concerned on those crops which have some edible values or serve as a staple food. That is why we neglected the ornamental plants or are not taken into consideration while accomplishing the investigational studies on begomoviruses.

A large number of virus species have been observed causing diseases in numerous ornamental plants, thereby affecting their quality as well as production values (Marwal et al. 2014a, b). Diverse ornamental plant species growing in diversified gardens of India were found to be displaying begomovirus-like symptoms, viz., leaf curling and yellowing. Begomoviruses are categorized as massive group of Geminiviridae family of plant viruses (Mansoor et al. 2003) and are able to infect an extensive array of plants such as main crops, ornamentals, and vegetables along with fruits and cause a remarkable loss in agricultural and horticultural production universally (Lima et al. 2013). Ornamental plants are broadly dispersed with elevated environmental acquiescence (Raj et al. 2007; Urbino et al. 2013). In the absence of seasonal crops or main host, ornamental plants are the main abode and prevailing grounds for begomovirus endurance (Ilyas et al. 2013). Recently, Marwal et al. (2012a, b) described a high rate of recombination and pseudo-recombination events that contribute in the progression of new viral species of begomoviruses isolated from ornamental plants in India. Thus, there is extensive demand for comprehensive knowledge on the miscellany, diversity, and allocation of begomovirus in ornamental plants. Roughly, the recombinants showed an advanced pathogenic phenotype which is habitually more virulent as compared to crossed parents (Zhou et al. 1997; Harrison and Robinson 1999). The principal carrier of begomovirus virus is the whitefly (*Bemisia tabaci*). To date, there has been no record of the begomovirus infection in trees as the infestation of the whitefly at tree level is unlikely because whitefly transmits the begomoviral particles that cause plant infection at low height only by residing under the leaves (Marwal et al. 2014a, b).

Begomoviruses have circular ssDNA genomes, with all the functions mandatory for virus replication, regulation of gene expressions, and DNA encapsidation by coat protein (CP) encoded on the DNA-A section, and the proteins required for virus movement in the plant on the DNA-B component (Hanley-bowden et al. 1999). The CP provides a protective coat for the viral DNA and is also vital for virus transmission from diseased to healthy plants by the whitefly vector, *Bemisia tabaci*. Thus, the CP gene is highly conserved among begomoviruses originating from the same geographical region, and modified to transmission by local vector populations (McGrath and Harrison 1995; Maruthi et al. 2002). The CP gene is consequently a crucial constituent for endurance of the begomoviruses and has been broadly used to illustrate and create the relationships between many begomoviruses (Harrison et al. 2002). The amplicon of begomovirus was efficaciously amplified in naturally infected ornamental plant species by polymerase chain reaction (PCR)/rolling circle amplification (RCA) using begomovirus-specific primers and RCA kit (Snehi 2014).

Plant viruses have revealed to be the cause of the yellow vein symptoms on the foliage of several ornamental plants. Ornamental plants, like oxalis (*Oxalis debilis* var. *corymbosa*) cv. Golden Veined Oxalis, geranium (*Pelargonium peltatum*) cv. Crocodile, pseuderanthemum (*Pseuderanthemum carruthersii*) cv. Golden Net Bush (narrow leaf and broad leaf variants), honeysuckle (*Lonicera japonica*) cv. Yellow Net Honeysuckle, and coleus (*Solenostemon scutellarioides*) cv. Electric Lime, showing foliar yellow vein were used. Graft transmissions indicated that, except for coleus, the foliar yellow vein of all other plants was caused by infectious agents, likely plant viruses. With the exception of Golden Veined Oxalis, attempts to identify a virus as the causal agent of the foliar yellow vein in geranium, *Pseuderanthemum*, and honeysuckle failed.

Successful whitefly (*Bemisia tabaci* biotype B) transmission of the foliar yellow vein of Golden Veined Oxalis suggested that a begomovirus was the causal agent. Sequence information and phylogenetic analyses revealed that the Golden Veined Oxalis begomovirus was closely related to *Tomato yellow spot virus* and *Sida mottle virus*, two begomoviruses reported to infect solanaceous and legume crops. The host range of the virus isolated from Golden Veined Oxalis begomovirus, designated as Oxalis yellow vein virus, is not known. This finding illustrates the potential spread of plant viruses to different geographical areas through the commercialization of virus-infected ornamental plants.

Many new viruses can flourish upon these ornamental plants; therefore, they are contemplated as reservoirs of economically important plant viruses which are often ignored during virus diversity studies (Urbino et al. 2013). Many scientific reports were published to ensure that these ornamental plants act as an alternative host and became a reservoir for the survival of begomoviruses (Raj et al. 2007) and could also be disseminated without the presence of the main crop (Ilyas et al. 2013). Thus, there is a pressing need for further investigation on the distribution and diversity of begomovirus infection in the ornamental plants. The focus in this chapter is mainly on the two distinct symptoms produced by begomoviruses, viz., leaf curling and yellow vein of various ornamental plants of India.

20.2 Ornamental Plants with Leaf Curl Symptoms

The symptoms of leaf curl disease are very complex, and the typical symptoms include leaf curling, puckering of leaves, vein yellowing, stunting, excessive branching, from pale yellowing to deep yellowing, and little leaves (Vasudeva and Sam 1948). Apart from this, it also causes extreme distortion of leaves, stunting of plants, and premature drop of flower and fruits. In some genotypes, it causes green vein banding, twisting, and green enation on the undersurface of the leaf, upward rolling of margin, and islands of golden colors scattered amidst the normal green tissue (Singh and Lal 1964). The type of symptom produced is dependent on the genotype cultivated and the developmental stage at which the infection occurs (Kumar and Kumar 2018). At cellular level, structural changes have been observed like hypertrophy of nucleus and accumulation of dark granules and the aggregate of viruslike particles in the cytoplasm (Saikia and Muniyappa 1989).

20.2.1 Rose

Rosa indica is a woody perennial plant that is cultivated mostly in gardens throughout India. Rose plants showing symptoms of leaf curl disease showed involvement of distinct begomovirus, namely, *Rose leaf curl virus* (RoLCuV). The virus is a monopartite belonging to the Old World begomovirus (Marwal et al. 2013a, b, c, d, e). The associated betasatellite showed identity to Digera leaf curl betasatellite. The rolling circle amplification was used to obtain the whole genomic sequence of begomovirus from rose (*Rosa indica*) which is related with betasatellite revealing leaf distortion and leaf curling along with dwarfing symptoms reported from India (Sahu et al. 2014).

The sequence of the begomovirus was found to be 2741 bp and exhibited the typical genome pattern of a monopartite Old World begomovirus. The sequence correlated with the highest nucleotide sequence identity (98%) to the isolate of Rose leaf curl virus (RoLCuV), confirming it as an isolate of RoLCuV (Sahu et al. 2014). The discovery of betasatellite (ssDNA satellites previously known as DNA-β) associated with monopartite members of the genus has increased in the last decades worldwide (Briddon et al. 2008). A satellite molecule termed "betasatellite" (previously known as DNA-β) has been associated with monopartite and bipartite begomoviruses and is required for systemic infection and symptom development (Briddon and Stanley 2006; Srivastava et al. 2014). Betasatellites are half the size of their helper viruses (approximately 1350 nucleotides in length) with a highly conserved structure gene encoded in the complementary sense known as βC1 and depend entirely on the helper virus for replication, movement, and transmission (Shafiq et al. 2020). The satellite region also contains a hairpin-loop structure

TAA/GTATTAC similar to the origin of replication of geminiviruses, and an adenine-rich region (A-rich) (Briddon et al. 2003). Leaf curl disease of rose is a relatively recent phenomenon, and Sahu et al. (2014) for the first time identified RoLCuV and associated betasatellite in India. RoLCuV has so far only been identified from *Rosa chinensis* (Chinese rose) in Faisalabad, Pakistan (Khatri et al. 2014). Further study that is still required to understand the trans-replication capability of betasatellites revealed their flexibility to interact with Rose leaf curl virus.

20.2.2 Shoeblack Plant

Hibiscus rosa-sinensis, known colloquially as Chinese hibiscus, China rose, Hawaiian hibiscus, rose mallow, and shoeblack plant, is a species of tropical hibiscus, a flowering plant in the Hibisceae tribe of the family Malvaceae (Chin 1986; Pfeil et al. 2002). China rose is esteemed for its aesthetic value (Rajeshwari et al. 2005). It is widely cultivated in tropical and subtropical regions, but is not known in the wild, so that its native distribution is uncertain. An origin in some part of tropical Asia is likely. It is widely grown as an ornamental and hedge plant (Parrilla et al. 2012) in the tropics and subtropics (Wee 2003; Akhtar et al. 2014).

Hibiscus leaf curl disease (HLCuD) occurs extensively in India. Infected hibiscus plants show symptoms of upward curling of leaves and enations on the abaxial leaf surface and reduction in leaf size along with vein thickening and stunting (Rajeshwari et al. 2005). A new begomovirus, tentatively named hibiscus yellow vein leaf curl virus (HYVLCV), was identified in *Hibiscus rosa-sinensis* plants showing symptoms of leaf curl, yellow vein, and vein enation on the undersides of the leaf in Taiwan (Huang et al. 2020). The disease occurs in Africa, China, and Pakistan/northwestern India, where its causative agent has been successfully characterized and found to consist of a complex of monopartite begomoviruses and a small symptom-modulating, single-stranded DNA betasatellite transmitted by whitefly *Bemisia tabaci* (Azhar et al. 2010; Sattar et al. 2013).

Bemisia tabaci, the vector of this viral disease, can feed and multiply on many diverse cultivated and noncultivated plant species (Ling et al. 2011), including China rose (Attique et al. 2003). Thus, once China rose gets infected with *Begomovirus*, it could serve as a potential reservoir (Akhtar et al. 2014).

20.2.3 Trumpet Bush

Tecoma stans L. is commonly known as a yellow trumpetbush and belongs to family Bignoniaceae. It is widely cultivated in Indian gardens as an ornamental plant native to tropical and subtropical regions of Central and South America (Gentry 1992). *Tecoma stans* is a small perennial tree or shrub with a natural distribution from the southern USA to Argentina (Mabberley 2008; Martínez and Ramos 2012) and is extensively used in the Mexican traditional medicines (Torres and Lopez 2011; Singh et al. 2013a, b; Anburaj et al. 2016). Yellow trumpetbush showing begomovirus-specific leaf curl symptoms was observed in the gardens of Lakshmangarh (Rajasthan). The begomovirus component isolated from *Tecoma stans* revealed highest nucleotide sequence identity of 100% with Lantana yellow vein virus (Marwal et al. 2013a, b, c, d, e).

20.2.4 Common Zinnia

Zinnia elegans belonging to the Asteraceae family of Kingdom Plantae is an annual summer flowering plant and is native to Mexico (Von Linnaeus 1759). Due to its good medicinal properties, this plant is used for different purposes. *Z. elegans* is reported to have some antifungal properties and found to inhibit the growth of *Fusarium moniliforme* (Hafiza et al. 2002).

Presence of Tobacco leaf curl virus that is associated with leaf curling symptoms of *Zinnia* was examined earlier from Tanzania (Storey 1931). After sometime, Padma et al. (1974) did serological assay and proved the presence of Zinnia mild mosaic mottle virus on *Zinnia* plants from India. Moreover, leaf curling and mosaic were found as key symptoms on *Zinnia* associated with Zinnia leaf curl virus (NCBI) from Pakistan. As per the published reports, Zinnia leaf curl virus was mainly reported in Asian countries (NCBI), while little information is available on the occurrence of Zinnia leaf curl virus from other parts of the world. Since the last two decades, the number of begomoviruses has significantly increased from different provinces of the world. In India, this phytopathogenic problem has been given more attention on the cultivated and noncultivated crop species, viz., *Cucurbita maxima* (Singh et al. 2001), *Dimorphotheca sinuata* (Raj et al. 2007), *Amaranthus cruentus* (Raj et al. 2008), *Momordica charantia* (Tiwari et al. 2010a; Raj et al. 2010), *Cucurbita pepo* (Tiwari et al. 2010b), *Cleome gynandra* (Raj et al. 2010), *T. dioica* (Raj et al. 2011), *Ageratum conyzoides* (Pandey et al. 2011), *Luffa cylindrica* (Tiwari et al. 2012a), and *Luffa acutangula* (Tiwari et al. 2012b). Several plant species, namely, *Acalypha indica*, *Ageratum conyzoides*, *Croton bonplandianum*, *Coccinia grandis*, *Corchorus olitorius*, *Eclipta alba*, *Malvastrum coromandelianum*, *Nicotiana plumbaginifolia*, *Parthenium hysterophorus*, *Solanum nigrum*, *Sonchus oleraceus*, and *Trigonella corniculata*, grow naturally or in neighboring fields of cultivated crops in many places of our country which are recognized as alternate hosts or also act as reservoir hosts of one or many begomoviruses (Raj et al. 2011). The virus diseases cannot be managed by any chemical treatment in the field (Valkonen 1998). However, they can be managed based on strategies that prevent infection, but we can manage the population of whiteflies by using the appropriate insecticides. Consequently, an immediate attention is required to check the further spread of this begomovirus in nature (Pandey and Tiwari 2012).

Zinnia elegans plants with excessive leaf curling, little leaf, stunting, and witches' broom appearance were observed from many parts of India. Rihne et al. (2020) reported mixed infection of both phytoplasma and virus. The presence of "*Candidatus* Phytoplasma asteris" was confirmed by using sequence comparison and phylogenetic and virtual RFLP analysis of *16S rRNA* gene sequences of zinnia isolate (16SrI-B subgroup). The association of begomovirus with symptomatic samples of *Zinnia elegans* was identified as a strain of Tomato leaf curl virus, which later on confirmed by using coat protein sequence comparison and phylogenetic tree analysis of coat protein gene.

20.2.5 Croton

Croton (*Codiaeum variegatum* L.) is admired as both an indoor and outdoor ornamental pot plants grown in tropical gardens of India mostly as a hedge. Numerous croton species are well-known for their bright and shiny colored vegetation ranging from yellow to orange, pink to purplish, and red and light to dark green, with all aggregation of multicolored mixtures. Crotons are evergreen perennial shrubs; thus, they boost the aesthetic value of gardens year after year (Mahesh et al. 2010). *C. variegatum* was earlier tested to be infected with a begomovirus by using nucleic acid hybridization tests (Raj et al. 1996). Another species of croton, *Codiaeum bonplandianum*, was also examined to be infected by a begomovirus (Mandal 1989).

A decade ago, *Codiaeum variegatum* plants were reported with typical begomovirus symptoms including reduction in leaf size, severe twisting and curling of the leaves, and stunted growth of the affected plant from the southern part of India (Mahesh et al. 2008), consequently compressing its aesthetic worth. Nevertheless, the begomovirus nature, causing leaf curl disease of croton species in India, is still unknown. The transmission of leaf curl virus disease was reported through grafting of infected scions, and through the whitefly vector, *Bemisia tabaci*, thereby confirming that the particular viral disease was caused by a begomovirus (Mahesh et al. 2010).

20.2.6 Chinaberry Tree

The fast-growing tree *Melia azedarach* belongs to the Meliaceae family. It is commonly known as Pride of India or the Chinaberry tree. The tree grows up to 15 m in height and is utilized for timber purpose in the Indian subcontinent (Langeland and Burks 2005). The double-colored leaves and beautifully patterned purple flowers were shown to have an aesthetic value. It is also extensively grown as an ornamental tree to make eye-catching boundary around gardens. *Melia azedarach* trees were found with severe crinkled and curled leaves showing the typical symptoms of a begomovirus infection confirmed by PCR (Marwal et al. 2014a, b).

Zhang et al. (2000) showed sequence analysis of begomovirus infecting *M. azedarach*, which had 97% identity with the Vinca yellow vein virus isolated from *Vinca rosea* growing at ground level at a distance of 30 m from the *M. azedarach* tree. As we all know, whitefly-mediated viral transmission has long been recognized as the major way of begomoviral transmission in plants (Mansoor et al. 2003). We can easily observe the presence of whiteflies on the lower leaf surface of the begomovirus-infected plant during early morning hours (Markham et al. 1994). Notwithstanding, we found whiteflies present under the leaf surface of short-heighted infected plants because they were totally absent in *M. azedarach* at 6–7 m in height just above the ground surface. Marwal et al. (2014a, b) firstly reported the aerial transmission of the begomovirus causing leaf curl disease in an ornamental tree *Melia azedarach* from the Indian subcontinent. Thus, identification of transmission of the begomovirus other than whiteflies represents a new possibility, and as such, the begomovirus poses a serious threat to economically important ornamental and crop plants.

20.2.7 Madagascar Periwinkle

Vinca rosea is an herbaceous shrub with glossy dark leaves commonly known as Madagascar periwinkle belonging to the Apocynaceae family (Vadeyar et al. 2010). *Vinca rosea* is mostly grown as an ornamental plant in perennial tropical regions but also cultivated for its alkaloids having anticancer activities (Jaleel et al. 2006). Typical symptoms of begomovirus infection, i.e., leaf curling and stunted growth, were observed in *V. rosea* growing in gardens and fields of Punjab (India). The vector is as usual whitefly (*Bemisia tabaci*) that transmits begomoviral infections in ornamentals, crops, and weeds, and the infection is prevalent in the tropical as well as subtropical regions of the world (Marwal et al. 2012b).

Sequence analysis of the begomovirus infecting *V. rosea* showed 74% nucleotide sequence identity with the MS-6 coat protein (AV1) gene (complete cds FJ002571) of Tomato leaf curl virus, and the Lucknow coat protein (CP) gene (complete cds HQ630856) of the Papaya leaf curl virus (Marwal et al. 2013b). β satellites were also identified in the begomovirus-infected plant samples of *V. rosea*, and this β satellite (JQ693151) showed nearly 83% identity both with Cotton leaf curl virus-associated DNA β satellite from India (complete sequence DQ364230) and Cotton leaf curl virus-associated DNA β satellite (complete sequence, clone Lu6 AM774309) (Marwal et al. 2013b).

20.3 Ornamental Plants with Yellow Vein Symptoms

There are several ornamental plants commercially utilized around the world having yellow vein symptoms. As cited earlier, a bit of these symptoms is of genetic nature, but some others may be produced by plant viruses or by physiological disorders. In general, the ornamental plants with infrequent characteristics like variegations and variations in leaf morphology are highly sought by the horticulturists and ornamental plant enthusiasts (Valverde et al. 2012; Singh et al. 2017).

20.3.1 Pseuderanthemum

Pseuderanthemum (*Pseuderanthemum carruthersii*) belongs to Acanthaceae family, and it is categorized as a famous ornamental plant of Hawaii and Puerto Rico, predominantly showing foliar vein yellowing in different cultivars. *Pseuderanthemum palatiferum* and *P. carruthersii* var. *atropurpureum* used as a panacea in Vietnam and some lignans (estrogen-like compounds in plants which act as antioxidants) and triterpenes (strong smelling compounds produced to protect plants against parasites) have been isolated (Vo et al. 2012). Several previous studies showed that the probable cause of foliar vein yellowing of a *Pseuderanthemum* sp. is begomoviral particles. However, the virus was not transmitted by *B. tabaci* using several biotypes of whiteflies from different locations. Nevertheless, positive results were obtained with begomovirus-specific probe, and the name *Pseuderanthemum* yellow vein virus was proposed as the causal agent of this viral disease associated with *Pseuderanthemum* spp. (Bedford et al. 1994). When rub-inoculated or sprayed, extracts of *P. bicolor* have been reported to reduce virus incidence in *Crotalaria juncea* and *Cucumis melo* var. *momordica* (Verma and Khan 1984).

20.3.2 Honeysuckle

Honeysuckle (*Lonicera japonica*) belonging to the Caprifoliaceae family is a common wild and cultivated perennial plant grown throughout the world. It is propagated vegetatively through cuttings. It has been reported as a plant which grows in forest openings but can occupy intensely shaded areas where it advances gradually until the canopy is open (Randall 1996). It is considered as an aggressive invasive weed in the USA due to its morphological plasticity in comparison to *Lonicera sempervirens* (Schweitzer and Larson 1999). The allelopathic compounds produced in the root system have been demonstrated to compete with natural pine regeneration in forests (Skulman et al. 2004). In China, honeysuckle has been utilized to make medicine for the treatment of epidemic febrile diseases, wind-heat, sores, carbuncles, and other infectious diseases. *Lonicera japonica* is also used as food, to make cosmetics, and as ornamental groundcover (Shang et al. 2011). There are at least 12 cultivars available for sale through the Internet. Some of them are the Elegant Creeper, Hall's prolific, Cream Cascade, Mint crisp, and Interold Dart's World. The most popular and widely available cultivar is Halliana also called Hall's honeysuckle (Schierenbeck 2004).

Japanese honeysuckle is an important larval host of the tobacco budworm (*Heliothis virescens*) and the corn earworm (*Helicoverpa zea*) due to the volatile chemical constituents in its flowers (Schlotzhauer et al. 1996). Some other pests reported are aphids and whiteflies. These aforesaid pests are generally observed in the warmer regions of the USA, but they do not cause any serious damage to honeysuckle plants (Williams et al. 2001). The whitefly *B. tabaci* B biotype is able to produce phytotoxic disorders in honeysuckle adjacent to the non-“B” biotype (Bedford et al. 1994). Several viruses other than begomoviruses belonging to different families have also been reported to infect this ornamental plant (Gulati-Sakhuja et al. 2011). Some of the cultivars exhibited a yellow vein mosaic symptom sometimes followed by enations along the veins' abaxial surface (Fig. 20.1). It has been shown that these symptoms can be produced by three whitefly-transmitted begomoviruses. Two of these viruses are *Honeysuckle yellow vein mosaic virus* (HYVMV) (Kitamura et al. 2004) and *Tobacco leaf curl Japan virus* (TbLCJV). Ueda et al. (2008), Ogawa et al. (2008), and Park et al. (2011) have described HYVMV and TbLCJV are being able to infect tomato (*Solanum lycopersicum*) plant also. Wang et al. (2011) described a third virus named *Honeysuckle yellow vein virus* which shared high nucleotide identity with HYVMV and TbLCJV, and it can also cause infection in tomato as well.

20.3.3 Pot Marigold

Calendula officinalis (family Asteraceae) is an annual ornamental plant, and in India, it is mainly grown for its beautiful flowers. Origin of this plant is from Central and Southern Europe, Western Asia, and USA. Additionally, it has a substantial importance to the cosmetic/pharmaceutical trade because it is utilized to manufacture antiseptic creams (Singh et al. 2017). Yellow vein net disease of *Calendula* was observed on several plants growing in gardens of India with similar disease symptoms consisting of vein yellowing, shortening of leaves and petioles, and stunting, and the infecting virus isolate was identified as a *Begomovirus* (Khan et al. 2005). Characteristic symptoms of the disease on *Calendula* consist of vein yellowing, shortening of leaves and petioles, stunting of plants, and reduction in growth, number, and size of flowers. With the help of experimentation, it was proved that the respective virus was transmitted from naturally infected *Calendula* to healthy seedlings by whiteflies (*Bemisia tabaci*), but not by mechanical or aphid transmission (Khan et al. 2005).

Samples of calendula with yellow vein disease were collected from Meerut, Uttar Pradesh, India, and the total genomic DNA was isolated from the symptomatic and asymptomatic leaf samples and subjected to PCR using cp gene-specific primer of begomovirus. The PCR amplification of ~770 bp was obtained, and the resultant sequence was submitted to GenBank, with accession number KT833850. The sequence data revealed close similarity of sequences with coat protein gene (AV1) components of other potato begomoviruses, which are all tentative strains of Tomato leaf curl New Delhi virus (ToLCNDV). The result demonstrated that *Calendula* plants infected with Tomato leaf curl New Delhi virus may act as an alternate host (reservoir) for other economically important plants (Singh et al. 2017).

To identify the begomovirus associated with the disease, PCR was done by using the total DNA extracted from *Calendula* leaves with and without symptoms, with a pair of primers specific to the coat protein (CP) gene of the genus *Begomovirus* (Singh et al. 2013c). Sequence analysis of virus infecting *Calendula* revealed the highest nucleotide sequence identities (95%, 94% and 93%) of the virus particles

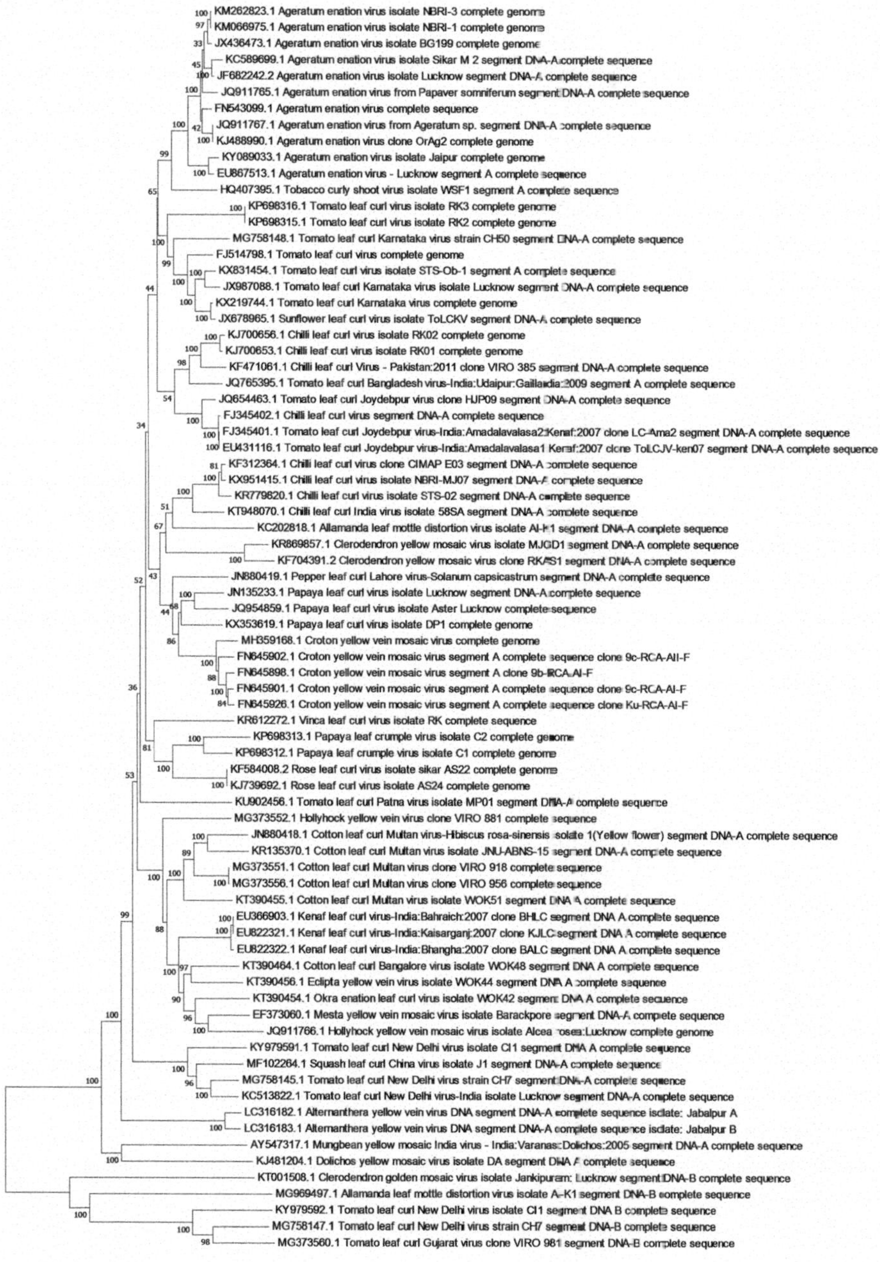

Fig. 20.1 Phylogenetic tree of begomoviruses; DNA-A component reported from India on different hosts. The tree was generated with 1000 bootstrap value represented along each root

with *Tobacco curly shoot virus* (AF240675), *Ageratum enation virus* (AJ437618), and *Tomato leaf curl Bangladesh virus* (AF188481), respectively (Hallan 1998).

It was also identified on the basis of whitefly transmission that the isolated virus is begomoviral particles, and this was confirmed earlier by using amplification of an expected size DNA band with begomovirus detecting primers by PCR and positive hybridization of PCR products with a known begomovirus-specific probe and its high sequence identity to known begomoviruses. *Calendula* leaf showing yellow vein net symptoms also associated with *Cucumber mosaic virus* has been reported previously (Naqvi and Samad 1985). However, Khan et al. (2005) firstly recorded the natural begomoviral infection of *Calendula*.

20.3.4 Indian Chrysanthemum

A begomovirus is also associated with a disease of ornamental plant *Chrysanthemum indicum*, a flowering plant belonging to Asteraceae family of Kingdom Plantae. It is native to Asia and North-East Europe. Most species originate from East Asia but the center of diversity is in China (Liu et al. 2012). *Chrysanthemum indicum* is a perennial, aromatic, medicinal plant used to make traditional folk medicine in several countries of Asia and Europe, against many women's diseases, nephritis, and headache and also in the neurological treatments (Chang et al. 2010).

Chrysanthemum indicum also has antimicrobial, antioxidant and anti-inflammatory properties (Park et al. 2012). Infected leaf samples of *Chrysanthemum indicum* plant show vein yellowing symptoms of leaves. The vector of the virus is the whitefly (*Bemisia tabaci*) that causes begomoviral infections. Therefore, enhanced knowledge about its epidemiological factors and sequence diversity along with biodiversity is vastly significant for executing preventive strategies (Marwal et al. 2013b).

Ashwathappa et al. (2020) also examined the leaf samples of chrysanthemum plants collected from farmer's field of Karnataka (India) showing characteristic discoloration in the form of mosaic, mottling, and downward leaf curling symptoms of begomoviral infection. These symptoms of begomovirus-infected chrysanthemum are confirmed by PCR using specific primers in samples. Based on begomovirus species demarcation criteria (91% nucleotide sequence identity), the virus responsible for the infection of chrysanthemum is acknowledged as a strain of ToLCNDV (Tomato leaf curl New Delhi virus). *Chrysanthemum* infected by betasatellite amplified from the begomovirus revealed maximum identity of 89% with Tomato leaf curl Bangladesh betasatellite (ToLCBDB) infecting papaya fruit in India (Varun and Saxena 2018). Additionally, break point recombination analysis revealed that the genome and betasatellite of ToLCNDV causing infection in chrysanthemum are recombinant with previously known begomovirus. Cryptic whitefly species principally present in the chrysanthemum fields having infection of begomovirus was identified as Asia-II-5 group.

20.3.5 Golden Dewdrop

Golden dewdrop (*Duranta erecta*) is an ornamental plant of the family Verbenaceae often grown as garden hedge in India. During the survey, yellow mosaic disease was examined on *Duranta erecta* growing in many parks/gardens in India with disease incidence of 80–90% (Marwal et al. 2013c). Severe yellowing mosaic is observed on infected plants followed by upward leaf curling symptoms and stunting that are more commonly seen in severely infected plants than in plants without infection. These symptoms of a begomoviral infection were firstly reported on *Duranta erecta* in Pakistan (Iram et al. 2004), and hence a similar infection was assumed (Jaidi et al. 2015). PCR amplifications of total DNA were performed with begomovirus-specific degenerate primers (Rojas et al. 1993) resulting in the generation of amplicons of an anticipated size of ~1.2 kb in all samples from leaves showing disease symptoms and no amplicons from any of the four healthy samples suggestive of begomovirus infection in *Duranta erecta*. Earlier, begomovirus infection was also found associated with Tomato leaf curl New Delhi virus (Tahir et al. 2006) and *Catharanthus* yellow mosaic virus in *Duranta repens* (Mustujab et al. 2015) from Pakistan. *Catharanthus* yellow mosaic virus has also been reported on *Clerodendrum inerme* (Sivalingam et al. 2011) and *Borrichia peruviana* (Nehra et al. 2014) in India (Table 20.1).

20.4 Conclusion

Thus, due to begomoviruses, there is leeway of a serious menace to most of the economically important ornamental as well as crop plants, and there is a pressing need of further inclusive study mainly emphasizing on the amplification of the whole viral genome and their recombination study that bring to light on its origin and to recognize conceivable begomovirus infection in the country in order to evaluate the involvement each makes to losses with a view to emerging management strategies. Outcomes of numerous molecular and unconventional methods are efficiently applied for disease management, crop protection, and development of quarantine strategies at national and international level in India. Moreover, advanced molecular techniques such as RNAi will be used for the development of transgenic plant resistance to begomovirus. This will form the sounding base of our forthcoming investigations.

Table 20.1 Distribution of begomoviruses, with accession numbers and locations in South India

Name of virus	Accession number	Location
Cotton leaf curl Bangalore virus	KT390464	Varanasi
Ageratum enation virus	KC589699	Lakshmangarh
Mesta yellow vein mosaic virus	EF373060	Barrackpore
Cotton leaf curl Multan virus	KT390455	Mirzapur
Cotton leaf curl Multan virus	JN880418	India
Papaya leaf crumple virus	KP698313	India
Papaya leaf crumple virus	KP698312	India
Mungbean yellow mosaic India virus	AY547317	Varanasi
Tomato leaf curl New Delhi virus	MG758145	Hessaraghatta
Cotton leaf curl Multan virus	MG373551	New Delhi
Hollyhock yellow vein virus	MG373552	New Delhi
Cotton leaf curl Multan virus	MG373556	New Delhi
Tomato leaf curl New Delhi virus	KC513822	Lucknow
Kenaf leaf curl virus	EU822322	Bhangha
Pepper leaf curl Lahore virus	JN880419	India
Okra enation leaf curl virus	KT390454	Varanasi
Kenaf leaf curl virus	EU366903	Bahraich
Squash leaf curl China virus	MF102264	Tamil Nadu
Kenaf leaf curl virus	EU822321	Kaisarganj
Rose leaf curl virus	KF584008	Sikar
Rose leaf curl virus	KJ739692	Sikar
Alternanthera yellow vein virus	LC316182	Jabalpur
Alternanthera yellow vein virus	LC316183	Jabalpur
Papaya leaf curl virus	JN135233	Lucknow
Papaya leaf curl virus	JQ954859	Lucknow
Ageratum enation virus	KY089033	Jaipur
Tomato leaf curl Patna virus	KU902456	India
Cotton leaf curl Multan virus	KR135370	India
Hollyhock yellow vein mosaic virus	JQ911766	Lucknow
Croton yellow vein mosaic virus	FN645902	Haryana
Croton yellow vein mosaic virus	FN645901	Haryana
Chilli leaf curl virus	KF312364	India
Chilli leaf curl virus	KR779820	India
Papaya leaf curl virus	KX353619	India
Eclipta yellow vein virus	KT390456	Mirzapur
Ageratum enation virus	JQ911765	Lucknow
Ageratum enation virus	EU867513	Lucknow
Papaya leaf curl virus	JN135233	Lucknow
Ageratum enation virus	KY089033	Jaipur
Papaya leaf curl virus	JQ954859	Lucknow
Tomato leaf curl Karnataka virus	MG758148	Kolar
Ageratum enation virus	KM262823	Lucknow
Chilli leaf curl Virus	KF471061	Banswara

(continued)

Table 20.1 (continued)

Name of virus	Accession number	Location
Tomato leaf curl virus	KX831454	Lucknow
Chilli leaf curl virus	KX951415	India
Ageratum enation virus	FN543099	Kangra
Ageratum enation virus	JF682242	Lucknow
Ageratum enation virus	JQ911767	India
Ageratum enation virus	JX436473	Pantnagar
Chilli leaf curl India virus	KT948070	Bhera
Ageratum enation virus	KM066975	Lucknow
Ageratum enation virus	KJ488990	Lucknow
Vinca leaf curl virus	KR612272	Sikar
Tomato leaf curl New Delhi virus	KY979591	Tamil Nadu
Croton yellow vein mosaic virus	MH359168	India
Tomato leaf curl Joydebpur virus	JQ654463	India
Tobacco curly shoot virus	HQ407395	India
Allamanda leaf mottle distortion virus	KC202818	Kalyani
Chilli leaf curl virus	KJ700656	Sikar
Chilli leaf curl virus	KJ700653	Sikar
Chilli leaf curl virus	FJ345402	Ponduru
Tomato leaf curl Joydebpur virus	FJ345401	Amadalavalasa
Tomato leaf curl Joydebpur virus	EU431116	Amadalavalasa
Croton yellow vein mosaic virus	FN645926	Punjab
Croton yellow vein mosaic virus	FN645898	Haryana
Tomato leaf curl Karnataka virus	KX219744	Nandyal
Tomato leaf curl virus	KP698316	Sikar
Tomato leaf curl virus	KP698315	Sikar
Dolichos yellow mosaic virus	KJ481204	India
Sunflower leaf curl virus	JX678965	Karnataka
Tomato leaf curl Bangladesh virus	JQ765395	Fatehnagar
Clerodendron yellow mosaic virus	KR869857	Jankipuram
Clerodendron yellow mosaic virus	KF704391	Sikar
Tomato leaf curl Karnataka virus	JX987088	Lucknow
Tomato leaf curl virus	FJ514798	Punjab
Tomato leaf curl New Delhi virus	MG758147	Hessaraghatta
Allamanda leaf mottle distortion virus	MG969497	Kalyani
Tomato leaf curl Gujarat virus	MG373560	New Delhi
Clerodendron golden mosaic virus	KT001508	Jankipuram
Tomato leaf curl New Delhi virus	KY979592	Tamil Nadu

References

Akhtar KP, Ullah R, Saeed M, Sarwar N, Mansoor S (2014) China rose (*Hibiscus rosa-sinensis*): a new natural host of Cotton leaf curl Burewala virus in Pakistan. J Plant Pathol 96(2):385–389

Anburaj G, Marimuthu M, Sobiyana P, Manikandan R (2016) A review on Tecoma stans. Int J Eng Res Modern Educ 1:43–49

Ashwathappa KV, Venkataravanappa V, Reddy CNL et al (2020) Association of Tomato leaf curl New Delhi virus with mosaic and leaf curl disease of Chrysanthemum and its whitefly cryptic species. Indian Phytopathol 73:533–542

Attique MR, Rafiq M, Ghaffar A, Ahmad Z, Mohyuddin AI (2003) Hosts of Bemisia tabaci (Genn.) (Homoptera: Aleyrodidae) in cotton areas of Punjab, Pakistan. Crop Prot 22:715–720

Azhar MT, Amin I, Anjum ZI, Arshad M, Briddon RW (2010) Both malvaceous and non-malvaceous betasatellites are associated with two wild cotton species grown under field conditions in Pakistan. Virus Genes 41:417–424

Bedford ID, Briddon RW, Brown JK, Rosell RC, Markham PG (1994) Geminivirus transmission and biological characterisation of Bemisia tabaci (Gennadius) biotypes from different geographic regions. Ann Appl Biol 125(2):311–325

Briddon RW, Stanley J (2006) Subviral agents associated with plant single-stranded DNA viruses. Virology 344(1):198–210

Briddon RW, Bull SE, Amin I, Idris AM, Mansoor S, Bedford ID, Dhawan P, Rishi N, Siwatch SS, Abdel-Salam AM, Brown JK, Zafar Y, Markham PG (2003) Diversity of DNA β, a satellite molecule associated with some monopartite begomoviruses. Virology 312:106–121

Briddon RW, Brown JK, Moriones E, Stanley J, Zerbini M, Zhou X, Fauquet CM (2008) Recommendations for the classification and nomenclature of the DNA-β satellites of begomoviruses. Arch Virol 153:763–781

Chang K, Choi EM, Kim GH (2010) Chemical Constituents of *Chrysanthemum indicum* L. flower oil and effect on osteoblastic MC3T3-E1 cells. Food Sci Biotechnol 19:815–819

Chatterjee A, Roy A, Ghosh SK (2007) First record of a begomovirus associated with yellow vein mosaic disease of Urena lobata in India. Aust Plant Dis Notes 2:27–28

Chin HF (1986) Introduction. In: The hibiscus: queen of tropical flowers. Tropical Press, Kuala Lumpur, pp 2–3

Das S, Ghosh R, Paul S, Roy A, Ghosh SK (2008) Complete nucleotide sequence of a monopartite begomovirus associated with yellow vein mosaic disease of mesta from north India. Arch Virol 153:1791–1796

Gentry AH (1992) Bignoniaceae Part II (Tribe Tecomeae). In: Flora neotropica. New York Botanical Garden, New York, NY, pp 285–290

Ghosh R, Paul S, Roy A, Mir JI, Ghosh SK, Srivastava RK, Yadav US (2007) Occurrence of begomovirus associated with yellow vein mosaic disease of kenaf (Hibiscus cannabinus) in northern India. Plant Health Prog 8:1

Gulati-Sakhuja A, Rains L, Tian T, Liu HY (2011) The complete nucleotide sequence and genome organization of a novel carmovirus-honeysuckle ringspot virus isolated from honeysuckle. Arch Virol 156(9):1635

Hafiza MA, Parveen B, Ahmad R, Hameid K (2002) Phytochemical and antifungal screening of Medicago sativa and Zinnia elegans v. J Biol Sci 2(2):130–132

Hallan V (1998) Genomic organization of geminivirus causing leaf curl in tomato (Lycopersicon esculentum). PhD thesis, University of Lucknow, Lucknow

Hanley-bowden I, Settlage SB, Orozco BM, Nagar S, Robertson D (1999) Geminiviruses: models for plant DNA replication, transcription and cell cycle regulation. Crit Rev Plant Sci 18:71–106

Harrison BD, Robinson DJ (1999) Natural genomic and antigenic variation in whitefly-transmitted geminiviruses (begomoviruses). Annu Rev Phytopathol 37:369–398

Harrison BD, Swanson MM, Fargette D (2002) Begomovirus coat protein: serology, variation and functions. Physiol Mol Plant Pathol 60:257–271

He ZF, Mao MJ, Yu H, Li HP, Chen X (2009) Molecular characterization of a distinct begomovirus infecting Allamanda cathartica in Guangdong, China. Arch Virol 54:1199–1202

Huang CH, Tai CH, Sharma N, Chao CH, Chang CJ, Jan FJ (2020) Characterization of a new monopartite begomovirus with a betasatellite associated with leaf curl, yellow vein, and vein enation in *Hibiscus rosa-sinensis*. Plant Dis 104(5):1318–1327

Ilyas M, Nawaz K, Shafiq M, Haider MS, Shahid AA (2013) Complete nucleotide sequences of two begomoviruses infecting Madagascar periwinkle (Catharanthus roseus) from Pakistan. Arch Virol 158:505–510

Iram S, Amrao L, Mansoor MS, Malik AH, Briddon RW, Zafar Y (2004) First report of a begomovirus associated with leaf curl disease of *Durcnta erecta* in Pakistan. Plant Pathol 54:260

Jaidi M, Kumar S, Srivastava A, Raj SK (2015) First report of Clerondendron yellow mosaic virus on golden dewdrop (*Duranta erecta*) in India. New Dis Rep 32:12

Jaleel CA, Gopi R, Lakshmanan GMA, Panneerselvam R (2006) Triadimefon induced changes in the antioxidant metabolism and ajmalicine production in *Catharanthus roseus* (L.) G. Don. Plant Sci 171(2):71–276

Jyothsna P, Rawat R, Malathi VG (2011) Molecular characterization of a new begomovirus infecting leguminous weed Rhynchosia minima in India. Virus Genes 42:407–414

Khan JA, Siddiqui MK, Singh BP (2002) The natural occurrence of a begomovirus in sunn hemp (Crotalaria juncea) in India. Plant Pathol 51:398

Khan AA, Naqvi QA, Khan MS, Singh R, Raj SK (2005) First report of a begomovirus infecting Calendula in India. New Dis Rep 11:6

Khatri S, Nahid N, Fauquet C, Mubin M, Nawaz-ul-Rehman MS (2014) A betasatellite-dependent begomovirus infects ornamental rose: characterization of begomovirus infecting rose in Pakistan. Virus Genes 49(1):124. https://doi.org/10.1007/SII262-014-1076-6

Kitamura K, Murayama A, Ikegami M (2004) Evidence for recombination among isolates of Tobacco leaf curl Japan virus and Honeysuckle yellow vein mosaic virus. Arch Virol 149 (6):1221–1229

Kumar P, Kumar M (2018) Chapter 5: Leaf curl disease: a significant constraint in the production of tomato in India. In: Kimatu JN (ed) Advances in plant pathology. IntechOpen, London, pp 93–101

Kumar JIN, Soni H, Kumar RN (2005) Aesthetic values of selected floral elements of Khatana and Waghai forests of Dangs, western Ghats. Indian J Tradit Knowl 4(3):275–286

Langeland KA, Burks KC (2005) *Melia azedarach*. In: Identification and biology of non-native plants in florida's natural areas. University of Florida, Gainesville, FL, pp 96–97

Lima AT, Sobrinho RR, González-Aguilera J, Rocha CS, Silva SJ, Xavier CA, Silva FN, Duffy S, Zerbini FM (2013) Synonymous site variation due to recombination explains higher genetic variability in begomovirus populations infecting non-cultivated hosts. J Gen Virol 94:418–431

Ling K-S, Harrison HF, Simmons AM, Zhang SC, Jackson DM (2011) Experimental host range and natural reservoir of Sweet potato leaf curl virus in the United States. Crop Prot 30:1055–1062

Liu PL, Wan Q, Guo YP, Yang J, Rao GY (2012) Phylogeny of the genus Chrysanthemum L.: evidence from single-copy nuclear gene and chloroplast DNA sequences. PLoS One 7(11): e48970

Mabberley DJ (2008) Mabberley's plant book. Cambridge University Press, Cambridge

Mahesh YS, Shankarappa KS, Rangaswamy KT, Prameela HA, Aswathanarayana DS, Maruthi MN (2008) Detection of a begomovirus associated with leaf curl disease of ornamental croton (*Codiaeum variegatum*) in southern India. J Plant Pathol 90:143–149

Mahesh YS, Shankarappa KS, Rangaswamy KT, Prameela HA, Aswathanarayana DS, Divya BL, Nagesha N, Maruthi MN (2010) Detection and characterisation of a begomovirus associated with leaf curl disease of ornamental croton (*Codiaeum variegatum*). J Hortic Sci Biotechnol 85 (2):101–105

Mandal B (1989) Studies on croton yellow vein mosaic virus. MSc thesis, University of Agricultural Sciences, Bangalore, India, 132 p

Mansoor S, Briddon RW, Zafar Y, Stanley J (2003) Geminivirus disease complexes: an emerging threat. Trends Plant Sci 8:128–134
Markham PG, Bedford SL, Pinner MS (1994) The transmission of geminivirus by Bemisia tabaci. Pestic Sci 42:123–128
Martínez E, Ramos CH (2012) Flora del valle de Techuacán-Cuicatlán. Departamento de Botánica, Instituto de Biología. Universidad Autónoma de México, México. 58 p
Maruthi MN, Colvin J, Seal S, Gibson G, Cooper J (2002) Co-adaptation between cassava mosaic Geminiviruses and their local vector populations. Virus Res 86:71–85
Marwal A, Prajapat R, Sahu A, Gaur RK (2012a) Computational characterization of Begomovirus infecting two ornamental plants: Jasminum sambac and Millingtonia hortensis. Asian J Biol Sci 5:240–249
Marwal A, Sahu A, Prajapat R, Choudhary DK, Gaur RK (2012b) First report of association of a begomovirus with the leaf curl disease of a common weed, Datura inoxia. Indian J Virol 23 (1):83–84
Marwal A, Sahu A, Prajapat R, Gaur RK (2013a) First report of begomovirus infecting two ornamental plants: Jasminum sambac and Millingtonia hortensis. Indian Phytopathol 66 (1):115–116
Marwal A, Sahu AK, Gaur RK (2013b) First report of molecular characterization of begomovirus infecting an ornamental plant *Tecoma stans* and a medicinal plant *Justicia adhatoda*. Sci Int (Lahore) 25(4):837–839
Marwal A, Sahu AK, Gaur RK (2013c) First report on the association of a begomovirus with *Chrysanthemum indicum* exhibiting yellowing of leaf vein disease characterized by molecular studies. J Hortic Res 21(2):17–21
Marwal A, Sahu A, Sharma P, Gaur RK (2013d) Molecular characterizations of two begomoviruses infecting *Vinca rosea* and *Raphanus sativus* in India. Virol Sin 28:53–56
Marwal A, Sahu AK, Gaur RK (2013e) Molecular characterization of a begomovirus infecting a new host Golden Duranta (*Duranta erecta*) in India. Int J Curr Microbiol App Sci 2(9):45–48
Marwal A, Sahu AK, Gaur RK (2014a) First report of airborne begomovirus infection in *Melia azedarach* (Pride of India), an ornamental tree in India. Aerobiologia 30:211–215
Marwal A, Sahu AK, Gaur RK (2014b) Association of begomovirus & an alphasatellite with leaf curl disease of ornamental plant, *Vinca alba* in Punjab, India. J Agric Res 52(3):339–356
Marwal A, Srivastava AK, Gaur RK (2020) Improved plant tolerance to biotic stress for agronomic management. Agrica 9(2):84–100. https://doi.org/10.5958/2394-448X.2020.00013.9
McGrath PF, Harrison BD (1995) Transmission of tomato leaf curl geminiviruses by *Bemisia tabaci*: effects of virus isolate and vector biotype. Ann Appl Biol 126:307–316
Mustujab A, Briddon RW, Tahir M (2015) First report of Catharanthus yellow mosaic virus infecting Duranta repens in Pakistan. J Plant Pathol 97:209–220
Naqvi QA, Samad A (1985) Purification and properties of Calendula yellow net virus. Indian J Virol 1:143–146
Nehra C, Sahu AK, Marwal A, Gaur RK (2014) Natural occurrence of Clerodendron yellow mosaic virus on Bougainvillea in India. New Dis Rep 30:19
Ogawa T, Sharma P, Ikegami M (2008) The begomoviruses Honeysuckle yellow vein mosaic virus and Tobacco leaf curl Japan virus with DNAβ satellites cause yellow dwarf disease of tomato. Virus Res 137(2):235–244
Padma R, Phatak HC, Singh S, Verma VS (1974) Zinnia mild mottle virus affecting Zinnia elegans in India. Hortic Res 14:55
Pandey N, Tiwari AK (2012) Identification of Zinnia leaf curl virus infecting Zinnia elegans in India. ISABB J Biotechnol Bioinformatics 2(1):6–10
Pandey N, Tiwari AK, Rao GP, Shukla K (2011) Detection and identification of Ageratum enation virus infecting Ageratum conyzoides in India. Acta Phytopathol Entomol Hungrica 46:203
Park J, Lee H, Kim MK, Kwak HR, Auh CK, Lee KY et al (2011) Phylogenetic lineage of Tobacco leaf curl virus in Korea and estimation of recombination events implicated in their sequence variation. Virus Res 159(2):124–131

Park S, Lee JB, Kang S (2012) Topical application of *Chrysanthemum indicum* L. attenuates the development of atopic dermatitis-like skin lesions by suppressing serum IgE levels, IFN-γ, and IL-4 in Nc/NgaMice. Evid Based Complement Alternat Med 2012:821967. 8 p

Parrilla G, Fiallo-Olivé E, Navas-Castillo J (2012) First report of China rose (*Hibiscus rosa-sinensis*) as a host of Alfalfa mosaic virus in Spain. Plant Dis 96:462

Pfeil BE, Brubaker CL, Craven LA, Crisp MD (2002) Phylogeny of hibiscus and the tribe hibisceae (malvaceae) using chloroplast DNA sequences of ndhF and the rpl16 Intron. Syst Bot 27 (2):333–350

Raj SK, Srivastava KM, Singh BE (1996) Evidence from nucleic acid hybridization tests for geminivirus infection of ornamental crotons in India. Eur J Plant Pathol 102:201–203

Raj SK, Khan MS, Snehi SK, Kumar S, Khan AA (2007) Natural occurrence of a Begomovirus on Dimorphotheca sinuata in India. Aust Plant Dis Notes 2:25–26

Raj SK, Snehi SK, Khan MS, Chandra P, Pandey RM (2008) First molecular identification of a begomovirus associated with yellow vein net disease in grain amaranth (Amaranthus cruentus L.) in India. Aust Plant Dis Notes 3:129–131

Raj SK, Snehi SK, Khan MS, Tiwari AK, Rao GP (2010) Detection of Ageratum enation virus from cat's whiskers (*Cleome gynandra* L.) with leaf curl symptoms in India. J Gen Plant Pathol 76 (4):292–294

Raj SK, Snehi SK, Tiwari AK, Rao GP (2011) Molecular identification of an Ageratum enation virus isolate associated with mosaic disease of pointed gourd (Trichosanthes dioica) in India. Phytoparasitica 39:497. https://doi.org/10.1007/s12600-011-0182-4

Rajeshwari R, Reddy RVC, Maruthi MN, Colvin J, Seal SE, Muniyappa V (2005) Host range, vector relationships and sequence comparison of a begomovirus infecting hibiscus in India. Ann Appl Biol 147(1):15–25

Randall JM (1996) Weed control for the preservation of biological diversity. Weed Technol 10 (2):370–383

Rihne T, Mitra S, Bahadur A, Panda P, Banyal N, Rao GP (2020) Mixed infection of phytoplasma and begomovirus associated with leaf curling and witches' broom disease of *Zinnia elegans* in India. Indian Phytopathol 73:527–532

Rojas MR, Gilbertson RL, Russell DR, Maxwell DP (1993) Use of degenerate primers in the polymerase chain reaction to detect whitefly transmitted geminiviruses. Plant Dis 77:340–347

Sahu AK, Marwal A, Shahid MS, Nehra C, Gaur RK (2014) First report of a begomovirus and associated beta-satellite in Rosa indica and in India. Aust Plant Dis Notes 9(1):147

Saikia AK, Muniyappa V (1989) Epidemiology and control of tomato leaf curl virus in Southern India. Trop Agric 66:350–354

Sattar MN, Kvarnheden A, Saeed M, Briddon RW (2013) Cotton leaf curls disease – an emerging threat to cotton production worldwide. J Gen Virol 94:695–710

Saxena S, Tiwari AK (eds) (2017) Begomoviruses: occurrence and management in Asia and Africa. Springer, Singapore

Schierenbeck KA (2004) Japanese honeysuckle (Lonicera japonica) as an invasive species; history, ecology, and context. Crit Rev Plant Sci 23(5):391–400

Schlotzhauer WS, Pair SD, Horvat RJ (1996) Volatile constituents from the flowers of Japanese honeysuckle (Lonicera japonica). J Agric Food Chem 44(1):206–209

Schweitzer JA, Larson KC (1999) Greater morphological plasticity of exotic honeysuckle species may make them better invaders than native species. J Torrey Bot Soc:15–23

Shafiq M, Qurashi F, Mushtaq S, Hussain M, Hameed A, Haider MS (2020) Chapter 13 - DNA plant viruses: biochemistry, replication, and molecular genetics. In: Awasthi LP (ed) Applied plant virology: advances, detection, and antiviral strategies. Academic Press, New York, NY, pp 169–182

Shang X, Pan H, Li M, Miao X, Ding H (2011) Lonicera japonica Thunb.: ethnopharmacology, phytochemistry and pharmacology of an important traditional Chinese medicine. J Ethnopharmacol 138(1):1–21

Singh DV, Lal SB (1964) Occurrence of viruses of tomato and their probabales strains at Kanpur. Proc Natl Acad Sci India 34:179–187

Singh R, Raj SK, Chandra G (2001) Association of a monopartite begomovirus with yellow mosaic disease of pumpkin (Cucurbita maxima) in India. Plant Dis 85(9):1029–1029

Singh A, Nagori BP, Mathur K (2013a) Tecoma stans: an important medicinal plant. Int J Pharmaceut Erud 3:13–21

Singh J, Singh A, Kumar P, Rani A, Barnwal VK, Sirohi A, Bhatnagar SK, Singh D (2013b) Evidence of association of a tomato leaf curl New Delhi virus with chilli leaf curl disease in western Uttar Pradesh, India. Vegetos 26(2):203–211

Singh S, Singh SK, Yadav A (2013c) A review on cassia species: pharmacological, traditional and medicinal aspects in various countries. Am J Phytomed Clin Ther 1(3):291–312

Singh J, Khanna S, Ranjan K, Pant RP, Kumar P, Sirohi A, Barnwal VK (2017) Evidence of association of begomovirus with the yellow vein disease of an ornamental plant pot marigold (*Calendula officinalis*) from Western Uttar Pradesh. J Pure Appl Microbiol 11(3):1609–1615

Sivalingam PN, Satheesh V, John P, Chandramohan S, Malathi VG (2011) Complete genome sequence of an isolate of Clerodendron yellow mosaic virus - a new begomovirus from India. Acta Virol 55:357–360

Skulman BW, Mattice JD, Cain MD, Gbur EE (2004) Evidence for allelopathic interference of Japanese honeysuckle (Lonicera japonica) to loblolly and shortleaf pine regeneration. Weed Sci 52(3):433–439

Snehi SK (2014) Molecular identification of begomoviruses affecting important ornamental plants in India. In: 4th World Congress on Virology, 06–08 October, 2014 Hilton San Antonio Airport, TX, USA

Srivastava A, Kumar S, Raj SK (2014) First report of Ageratum enation virus, betasatellite and alphasatellite causing leaf curl and enation disease of Amaranthus hypochondriacus in India. Plant Dis 98(9):1285–1285

Storey HH (1931) Tobacco leaf curl virus on Zinnia elegans. Nature 128:187

Tahir M, Haider MS, Shah AH, Rashid N, Saleem F (2006) First report of bipartite begomovirus associated with leaf curl disease of Duranta repens in Pakistan. J Plant Pathol 88:339

Tiwari AK, Sharma PK, Khan MS, Snehi SK, Raj SK, Rao GP (2010a) Molecular detection and identification of Tomato leaf curl New Delhi virus isolate causing yellow mosaic disease in Bitter gourd (Momordica charantia), a medicinally important plant in India. Med Plants 2 (2):117–123

Tiwari AK, Mall S, Khan MS, Snehi SK, Sharma PK, Rao GP, Raj SK (2010b) Detection and identification of Tomato leaf curl Palampur virus infecting Cucurbita pepo in India. Guanaxi Agric Sci 41:1291–1.295

Tiwari AK, Snehi SK, Khan MS, Raj SK, Sharma PK, Rao GP (2012a) Molecular detection and identification of Tomato leaf curl New Delhi virus associated with yellow mosaic and leaf curling disease of Luffa cylindrica crops in India. Indian Phytopathol 65(1):80–84

Tiwari AK, Snehi SK, Raj SK, Rao GP, Sharma PK (2012b) Molecular identification and genetic diversity among six begomovirus isolates affecting cultivation of cucurbitaceous crops in Uttar Pradesh, Archives Phytopathol. Plant Prot 45(1):62–72

Torres AP, Lopez RG (2011) Photosynthetic daily light integral during propagation of tecoma stans influences seedling rooting and growth. HortScience 46:282–286

Ueda S, Onuki M, Hanada K, Takanami Y (2008) Unique grouping of the Far East Asian begomovirus complex based on sequence analyses of the DNA-A genome and associated DNAβ satellite molecules isolated from tomato, honeysuckle and Eupatorium plants in Japan. Arch Virol 153(3):417–426

Urbino C, Gutiérrez S, Antolik A, Bouazza N, Doumayrou J, Granier M, Martin DP, Peterschmitt M (2013) Within-host dynamics of the emergence of tomato yellow leaf curl virus recombinants. PLoS One 8:e58375. https://doi.org/10.1371/journal.pone.0058375

Vadeyar LK, Hooli AA, Holihosur SN, Kallapur SV (2010) Bioefficacy of *Vinca rosea* leaf powder on Aedes aegypti L. J Biopest 3(1):89–191

Valkonen J (1998) Virus disease control in plants using natural and engineered resistance and some considerations regarding biosafety. Currents 17:51–55

Valverde RA, Sabanadzovic S, Hammond J (2012) Viruses that enhance the aesthetics of some ornamental plants: beauty or beast? Plant Dis 96(5):600–611

Varun P, Saxena S (2018) Association of tomato leaf curl Gujarat virus and tomato leaf curl Bangladesh betasatellite on papaya showing typical leaf curl symptoms in North India. 3 Biotech 8(5):243

Vasudeva RS, Sam RJ (1948) A leaf curl disease of tomato. Phytopathology 38:364–369

Verma HN, Khan MAA (1984) Management of plant virus diseases by Pseuderanthemum bicolor leaf extract/Beeinflussung von Viruskrankheiten an Pflanzen durch Blattextrakte von Pseuderanthemum bicolor. Zeitschrift für Pflanzenkrankheiten und Pflanzenschutz/J Plant Dis Prot:266–272

Vo TN, Nguyen PL, Tuong LT, Pratt LM, Vo PN, Nguyen KPP, Nguyen NS (2012) Lignans and triterpenes from the root of Pseuderanthemum carruthersii var. atropurpureum. Chem Pharm Bull 60(9):1125–1133

Von Linnaeus C (1759) Systema Naturae, vol 2, 10th edn, p 1189. 1221, 1377

Wang Y, Ji J, Tae-Kyun O, Sung O, Kim SH, Lee HJ, Shim MY, Choi CW, Kim SH, Kim I-S, Kim YS (2011) Occurrence of Honeysuckle Yellow Vein Virus (HYVV) containing a monopartite DNA-A genome in Korea. Eur J Plant Pathol 129(3):361–370

Wee YC (2003) Tropical trees and shrubs: a selection for urban planting. Sun Tree Publishing, Singapore, p 65

Williams PA, Timmins SM, Smith JM, Downey PO (2001) The biology of Australian weeds 38. Lonicera japonica Thunb. Plant Prot Q 16(3):90–100

Zaidi AA, Hallan V, Raikhy G, Singh AK, Ram R (2008) Viruses of ornamental plants in India - current research status and future prospects. In: XII International Symposium on Virus Diseases of Ornamental Plants, Haarlem (Netherlands). https://doi.org/10.17660/ActaHortic.2011.901.7

Zhang Z, Schwartz S, Wagner L, Miller W (2000) A greedy algorithm for aligning DNA sequences. J Comput Biol 7(1–2):203–214

Zhou X, Liu Y, Calvert L, Munoz C, Otim-Nape GW (1997) Evidence that DNA-A of a geminivirus associated with severe cassava mosaic disease in Uganda has arisen by interspecific recombination. J Gen Virol 78:2101–2111

Begomovirus Diseases of Ornamental and Fruit Plants: Discoveries and Management Approaches

21

Muhammad Shafiq Shahid and Abdullah Mohammed Al-Sadi

Abstract

Begomoviruses (family *Geminiviridae*) consist of single-stranded (ss) and circular genome particles which are transmitted by a complex of whitefly (*Bemisia tabaci—Gennadius*) biotype vectors and are also infecting fruits and ornamental plants all over the world. There have been few reports about the association of DNA satellite (alphasatellite and betasatellite) molecules with begomoviruses infecting fruits and ornamental plants. The introduction of different whitefly biotypes into diverse geographical areas and polyphagous nature of insect vector is responsible for transmitting the begomoviral diseases to different fruits and ornamental plants. Advancement in virus detection technology as high-throughput sequencing also helps in the identification of different begomoviruses from fruit and ornamental plants. The begomovirus diseases in fruits and ornamental plants mainly focussed on the molecular characterization and phylogenetic and recombination analysis followed by different management approaches using CRISPER/Cas technologies. This book chapter highlights the begomovirus diseases of fruits and ornamental plants, their discoveries using different virus detection technologies including high-throughput DNA sequencing and application of their management approaches.

Keywords

Begomovirus · High-throughput sequencing · Whitefly · Ornamental plants · Fruit plants

M. S. Shahid · A. M. Al-Sadi (✉)
Department of Plant Sciences, College of Agricultural and Marine Sciences, Sultan Qaboos University, Al-Khod, Oman
e-mail: alsadi@squ.edu.om

S. K. Raj et al. (eds.), *Virus Diseases of Ornamental Plants*,
https://doi.org/10.1007/978-981-16-3919-7_21

21.1 Introduction

Plant-infecting begomoviruses (family *Geminiviridae*) are accountable for the huge crop losses globally which are responsible for infecting mono- and dicotyledon plant species. Geminiviruses consist of either a single or double genomic molecule of circular single-stranded DNA (ssDNA) of ~2.8 kb genomic DNA particles (Zhang et al. 2001). The recent criterion of International Committee on Taxonomy of Viruses has divided family *Geminiviridae* into nine genera depending on their genome size and organization, nature and nature of its vector and host range (Zerbini et al. 2017). Among them, *Begomovirus* (genus) is of unique importance which has more than 300 species in geminiviruses (Brown et al. 2015) having a large range of vegetables, ornamentals and fruit plants. Additionally, the begomovirus disease occurrences and their tendency to contaminate economically valuable crops are rising day by day. Previous studies have revealed that various begomovirus species have evolved frequently (Nawaz-ul-Rehman and Fauquet 2009). This alliance of virus complex, consequently, augments the opportunities of nucleotide modification, genome recombination and genomic component rearrangement in viral components which further enhance viral heterogeneity (Duffy and Holmes 2008; Harkins et al. 2009; Lima et al. 2013). The genome of begomovirus acquire either bipartite genome which is comprising of DNA-A and DNA-B components or monopartite genome which is only having a single DNA-A which is complimentary to the DNA-A genome of the bipartite begomovirus (Fig. 21.1). The monopartite begomovirus genome consists of six open reading frames (ORFs), where AC1, transcribed into replication-associated protein (Rep), AC2 into transcriptional activator protein (TrAP), AC3 into replication enhancer (REn) and AC4 into RNA silencing suppressor protein in complementary sense, V1 and V2 transcribed, respectively, into coat and pre-coat proteins from the virion sense strand. Majority of the monopartite begomoviruses identified from the Old World are also coupled with DNA molecules known as alphasatellites and/or betasatellites. Alphasatellites rely on its helper virus for its replication, whereas betasatellites require its helper virus for their encapsidation and movement (Srivastava et al. 2013, 2014). The betasatellite molecules are known as symptom-controlling molecule and referred to aggravate the symptom manifestation by controlling the host gene silencing (Kumar et al. 2014). Nevertheless, the purpose of alphasatellite in disease development is not clear yet. Besides alphasatellite and betasatellites, a newly tiny subviral particle together with dissimilar sweepoviruses has been detected named as 'deltasatellites'. These small molecules reveal to have structural resemblances in the conserved zone of betasatellites and contain adenine-rich region, coupled with a nonanucleotide TAATATTAC sequence as were identified in the first ToLCV-sat (later known as betasatellite molecule) (Lozano et al. 2016). However, the DNA-B of bipartite begomovirus encodes BC1 ORF (in complementary sense) and BV1 s (in the virion sense) strand. The BV1 is transcribed for nuclear shuttle protein (NSP), which is essential for passage from nucleus to cytoplasm, whereas BC1 is transcribed for movement protein (MP), which is responsible for cell-to-cell movement. The region between AC1 and AC4 contains stem loop structure called intergenic

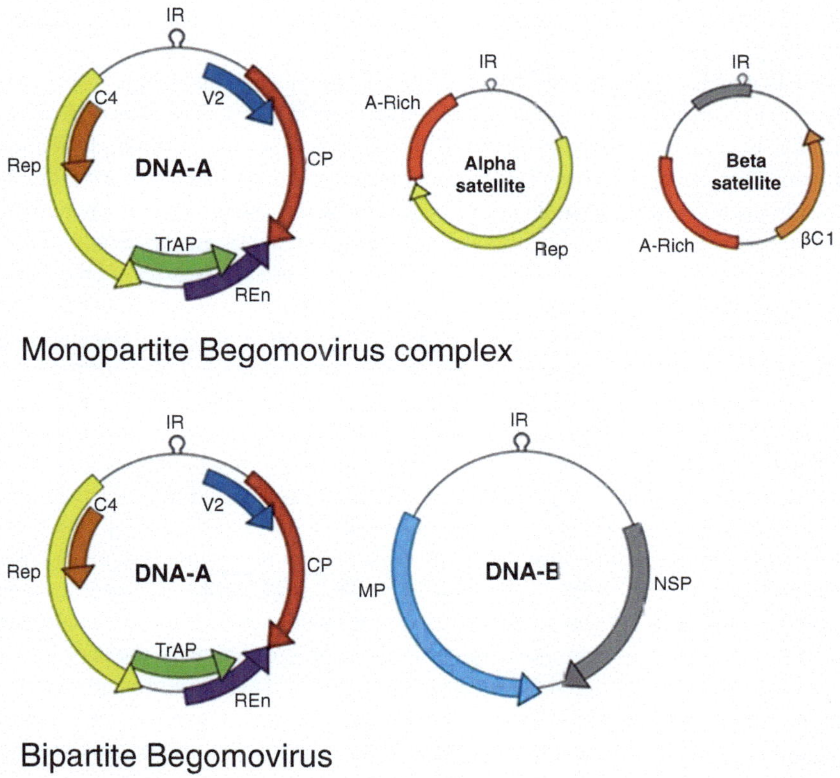

Fig. 21.1 Schematic presentation of monopartite and bipartite begomovirus complexes

region (IR) which consists of an origin of replication (Brown et al. 2012). In Old World (OW) (Africa, Asia, Australasia and Europe) countries, both monopartite and bipartite begomoviruses are frequently reported, whereas in New World (NW, the Americas), bipartite begomoviruses are indigenous with the exception of few newly reported monopartites (Brown et al. 2015).

21.2 Begomoviruses Infecting Fruits and Ornamental Plants

Fruit production is the key industry among agricultural crops throughout the world. China, India and Brazil are the largest and top fruit producer countries in the world. Overall, global fruit production has grown, and China is the main fruit producer after contributing 83.24 million tonnes (mt) in the total global fruit productions which is 548 mt. Begomoviruses are largely infecting economically important agricultural crops (fruits and ornamentals). However, recently, several studies have identified different begomoviruses infecting significantly ornamental and fruit plants resulting in a huge loss to horticulture commodities worldwide (Lima et al. 2013; Mansoor et al. 2003) (Table 21.1).

Due to the characteristics of vector (*B. tabaci*) biotype and the texture of fruit tree, there is less possibility of transmitting begomovirus disease in fruit plants. However, several fruit trees have been invaded by different begomoviruses. Among fruit crops, papaya (*Carica papaya* L.) plants have extensively been infected with diverse begomoviruses in different geographical locations. *C. papaya* has great nutritive and commercial value, largely cultivated in tropics and subtropics. Papaya-infecting begomoviruses have been reported from various geographical regions such as India, Nepal, Pakistan, China, Oman and Iran (Raj et al. 2008a; Shahid et al. 2013; Nadeem et al. 1997; Wang et al. 2004; Chang et al. 2003; Khan et al. 2012; Bananej et al. 2021), respectively. Ornamentals are also extensively variegated all over the globe which demonstrate a huge environmental adaptability. However, such ornamentals are most likely containing several unexplored begomoviruses which are often thought to be insignificant and never gain proper attention as achieving begomovirus diversity studies (Urbino et al. 2013). And to date, several begomoviruses (ageratum yellow vein virus, Mesta yellow vein mosaic virus and ageratum enation virus) have respectively been reported from *Althea rosea*, *Hibiscus cannabinus* and *Zinnia elegans* (Briddon et al. 2003; Celli et al. 2014; Das et al. 2008) ornamental plants. Numerous studies have shown that several ornamental plants act as an alternate source for replication of begomoviruses which help to proliferate for the period of no crops in the field (Kumar et al. 2010; Ilyas et al. 2013). Therefore, it is important to explore the spread and diversity of these viruses among ornamental plant species using rolling circle amplification and high-throughput sequencing technologies.

21.3 The Dynamics of Whitefly Vector

Whitefly (*Bemisia tabaci*) is the vector which is responsible for transmitting the begomovirus diseases to different horticultural crops under favourable warmer regions throughout the world. The first reason of the occurrence of begomoviruses is the introduction of whitefly biotypes globally. Whitefly cannot travel long distance; however, the possibility of spreading throughout continents would mainly be through transport of ornamentals, fruits and other vegetable crops. The second reason of emergence of new and recombinant begomoviruses is the polyphagous nature of whitefly biotypes which transmit begomoviruses between indigenous flora, no host plants and susceptible cultivars. Moreover, the movement of *B. tabaci* which depends on the presence of the vulnerable host, acceptable host phase, environmental factor as well as climate condition of the host plants can also serve a crucial role in the distribution of begomoviruses. The recent example for such evolution of begomovirus causing cotton leaf curl disease which exclusively spread in India, Pakistan and neighbouring China causes huge losses of cotton crops largely due to movement of whitefly biotypes from neighbouring countries to the newly cotton growing areas (Raj et al. 2007).

Table 21.1 List of the major begomoviruses infecting different ornamental and fruit plants from all over the world

Begomovirus	Acronym	Genome	Host	Country	Reference
Clerodendron yellow mosaic virus	ClYMV	Monopartite	*Chrysanthemum indicum*	India	Xu et al. (2020)
Papaya leaf crumple virus	PLCuCrV	Monopartite	*Vinca alba*	India	Marwal et al. (2013a)
Catharanthus yellow mosaic virus	CYMV	Monopartite	Madagascar periwinkle	Pakistan	Kumar et al. (2010)
Croton leaf curl virus	CrLCuV	Monopartite	*Codiaeum variegatum*	India	Marwal et al. (2014)
Velvet bean severe mosaic virus	VbSMV	Monopartite	*Mucuna pruriens*		Mahesh et al. (2010)
Ageratum enation virus	AEV	Monopartite	*Tagetes patula*	India	Marwal et al. (2013b)
Tomato leaf curl New Delhi virus	ToLCNDV	Bipartite	*Papaver somniferum*	India	Zaim et al. (2011)
Rose leaf curl virus	RoLCuV	Monopartite	*Rosa chinensis*	Pakistan	Srivastava et al. (2016)
Euphorbia mosaic Venezuela virus	EuMVV	Bipartite	*Euphorbia heterophylla*	Venezuela	Khatri et al. (2014)
Sida mosaic Bolivia virus-2	SiMBV	Bipartite	*Salvia hispanica*	Argentine	Zambrano et al. (2012)
Tomato yellow spot virus	ToYSV	Bipartite	*Salvia hispanica*	Argentine	Zambrano et al. (2012)
Sida leaf curl virus	SiLCuV	Monopartite	*Sida cordifolia*	China	Guo and Zhou (2006)
Mungbean yellow mosaic virus	MYMV	Bipartite	*Salvia splendens*	Bangladesh	Ara et al. (2012)
Tomato yellow leaf curl virus	TYLCV	Monopartite	*Salvia splendens*	Bangladesh	Ara et al. (2012)
Clerodendron golden mosaic China virus	ClGMCNV	Bipartite	*Salvia splendens*	USA	Valverde et al. (2012)
Tomato leaf Curl New Delhi virus	ToLCNDV	Bipartite	*Calendula officinalis*	India	Singh et al. (2017)
Sonchus yellow net virus	SYNV	Monopartite	*Jasminum sambac*	India	Marwal et al. (2013c)
Rose leaf curl virus	RoLCuV	Monopartite	*Millingtonia hortensis*	India	Marwal et al. (2013d)
Lantana yellow vein virus	LYVV	Monopartite	*Tecoma stans*	India	Marwal et al. (2013e)
Pedilanthus leaf curl virus	PeLCV	Monopartite	*Vinca minor* L.	Pakistan	Haider et al. (2008)
Pseuderanthemum yellow vein virus	PYVV	Monopartite	*Pseuderanthemum carruthersii*	Yemen	Bedford et al. (1994)
Tomato leaf curl Guangdong virus	ToLCGuV	Monopartite	*Allamanda* plants	China	He et al. (2009)
Tomato leaf curl New Delhi virus	ToLCNDV	Bipartite	*Zinnia elegans*	India	Rihne et al. (2020)
Tomato leaf curl Bhatinda virus	ToLCBhV	Monopartite	*Lantana camara*	India	Marwal et al. (2013f)
Cotton leaf curl Multan virus	CLCuMV	Monopartite	Hollyhock	India	Ashwathappa et al. (2012)
Jatropha mosaic begomovirus	JMB	Monopartite	*J. multifida*	Puerto Rico	Bird (1957)

(continued)

Table 21.1 (continued)

Begomovirus	Acronym	Genome	Host	Country	Reference
Jatropha mosaic begomovirus	JMB	Monopartite	*J. gossypifolia*	Jamaica	Roye et al. (2006)
Croton yellow vein mosaic virus	CYVMV	Monopartite	*J. gossypifolia*	India	Roossinck (2011)
Jatropha yellow mosaic India virus	JYMIV	Monopartite	*J. gossypifolia*	India	Snehi et al. (2011b)
African cassava mosaic virus	ACMV	Monopartite	*J. curcas*	Kenya	Ramkat et al. (2011)
Jatropha mosaic Nigeria virus	JMNV	Monopartite	*J. curcas*	Nigeria	Kashina et al. (2013)
Jatropha mosaic begomovirus	JMB	Monopartite	*J. gossypifolia*	America	Simmonds-Gordon et al. (2014)
Jatropha leaf crumple India virus	JLCrIV	Monopartite	*J. curcas*	India	Srivastava et al. (2015)
Passion fruit leaf distortion virus	PLDV	Bipartite	*Passiflora edulis*	Colombia	Vaca-Vaca et al. (2017)
Sida golden mosaic virus	SiGMV	Bipartite	*Sida santaremensis*	USA	Durham et al. (2010)
Honeysuckle yellow vein mosaic virus	HYVMV	Monopartite	*Lonicera* spp.	South Korea	Lee et al. (2011)
Rhynchosia rugose golden mosaic virus	RhRGMV	Monopartite	*Jatropha multifida*	USA	Polston et al. (2014)
Grapevine begomovirus A	GBA	Monopartite	*Vitis vinifera*	Argentina	Debat et al. (2020)
Tomato leaf curl Kerala virus	ToLCKeV	Monopartite	*Malva parviflora*	India	Raj et al. (2008b)
Passion fruit severe leaf distortion virus	PSLDV	Monopartite	*Passiflora edulis* f. *flavicarpa*	Brazil	Ferreira et al. (2010)
Ageratum yellow vein virus	AYVV	Monopartite	*Carica papaya*	Nepal	Shahid et al. (2013)
Cotton leaf curl Gezira virus	CLCuGeV	Monopartite	*Carica papaya*	Oman	Khan et al. (2012)
Ageratum enation virus	AEV	Monopartite	*Morus alba*	Pakistan	Sattar et al. (2018)
Mulberry crinkle leaf virus	MCLV	Monopartite	*Morus alba*	China	Lu et al. (2015)
Grapevine geminivirus A	GGVA	Monopartite	Grapevine	South Korea	Al Rwahnih et al. (2017)
Grapevine geminivirus A	GGVA	Monopartite	Grapevine	China	Sun et al. (2020)
Apple geminivirus	AGV	Monopartite	Apple	China	Liang et al. (2015)
Euphorbia leaf curl virus	EuLCuV	Monopartite	*Passiflora edulis*	Taiwan	Li and Zhou (2010)
Papaya leaf curl Guangdong virus	PaLCGuV	Monopartite	*Passiflora edulis*	Taiwan	Li and Zhou (2010)
Clerodendron golden mosaic China virus	ClGMCNV	Bipartite	*Salvia splendens*	China	Cheng et al. (2014)

21.4 Evolution of Begomoviruses

The mutation through recombination and DNA reassortment is very common in single-stranded DNA viruses. The rate of mutation is believed to be influenced by nature of begomovirus genome, type of invaded host, maturity of the infected plants and virion harmony. The rate of mutation is highest in begomoviruses infecting either wild or cultivated plants (Amrao et al. 2010; Van Der Walt et al. 2008). The genomic components slouching due to recoupling among different begomovirus species (Arguello-Astorga et al. 2007) and their invading hosts, as well as incorporation with the host genetic material, support the advent of 'new' begomoviral disease complexes following the emergence of more lethal strains of present begomoviruses. Furthermore, begomoviruses have an incredible capability of acclimating exceptionally distinct and varying environmental nooks. The existence of recombinant types of begomoviral strains may possibly compete for their divergence and adjustment to non-hosts and ecosystem. This may perhaps be one of the explanations that geminiviruses consist of higher than 320 virus species depicted until now (Brown et al. 2015). Recombination incidents might happen primarily in plants infested by mixed infection where two or more than two begomoviruses are involved in causing infection in a single host. Additionally, intermolecular pseudo-recombination in two bipartite begomoviruses has been described in Tomato mottle virus (TMV) and Bean dwarf mosaic virus (BDMV), where DNA-A of TMV recombine with DNA-B of BDMV, resulting in the production of new DNA-B genomic component of BDMV (Revers et al. 1996). Such strains might have been evolved unexpectedly which are the sources of huge crop losses. Several studies have been highlighting the diversity contained by begomoviruses owing to repeated recombination (Gilbertson et al. 1993; Padidam et al. 1999). Once such recent example is the breakdown of resistance in cotton infected with multiple cotton leaf curl disease in Pakistan (Mansoor et al. 2003).

21.5 Major Environment Factors Contributing Begomoviral Diseases

The air temperature may affect the dissemination and occurrence of begomoviruses and whitefly vector. However, one cannot predict the appearance of entirely new challenges because of temperature alteration. However, there is a direct relationship between temperature and the whitefly (*Bemisia tabaci*) biotypes. *B. tabaci*-transmitted geminiviruses (begomoviruses) are foremost challenges mainly in tropical countries. Temperatures ranging from 20 to 30 °C favours larger number of *B. tabaci* which is associated with great fecundity and larger longevity (Seal et al. 2006). A similar positive relation concerning the occurrence and off-season rainfall impacts on the incidence of whitefly vectors and inoculum sources has been recognized with cotton and maize infecting begomoviruses (Morales 2006). Through heavy airflow, vectors (whiteflies) are blown away to faraway locations which are responsible for the begomovirus transmission to that new areas (Thresh 2003). Mainly the air facilitates in the flight of virus vectors and also in controlling the direction of virus spread. The route of air and velocity will also have impact on the flights of the vectors.

21.6 Role of High-Throughput Sequencing in Discovery of Fruits and Ornamental Infecting Begomoviruses

The detection of new and begomoviruses has been boosted with the introduction of high-throughput sequencing (HTS) technologies. HTS technologies present a wide range of crucial applications by acting as a very effective technology in virus discovery which are inexpensive and time efficient. Multiple samples obtained from different crop (fruits or ornamental) sources can be handled rapidly for the sequencing. Along with this, consequent evaluation of the information contributes toward the more rapid processing of information while searching for the viral pathogens which have not been reported in recent times. However, the conventional molecular procedures for coping up with plant pathogenic viruses are still in use. However, typically, the workflow of the conventional molecular procedures required some prior knowledge and tons of efforts, but to deal with these issues, HTS tools have proven their importance and are extremely helpful. For instance, the sequences of primer pair which are used in the polymerase chain reaction (PCR) must be designed on the genome sequences, which would be available only from previously identified begomovirus strains or the restriction of concatemers of RCA products which can be generated by the use of an available restriction enzymes which possibly lead to the missing of the undiscovered begomovirus strains. In contrast, there is no hard requirement for the generic approaches employing HTS technologies with respect to begomoviruses which facilitates the discovery of nonspecific new begomoviruses.

21.7 Management Approaches Against Begomoviruses

To discuss about the management strategies of begomoviruses, it is required to know first the transmission ways of begomoviruses. Begomoviruses are spreading from one place to another due to two main factors: first, migration of contaminated propagative plant materials of ornamental and fruit crops which is most likely a great factor, and second, dissemination of insect vector into new areas through agriculture produce and possibly with the human migration. Viruses are extremely niggling organisms which can infect a broad spectrum of living beings encompassing as small as bacteria to humans on a larger scale. One can appreciate the importance of understanding viral dynamics from the fact that they have got the potential of destroying economies, and a classic example of such scenario is the most recent outbreak and spread of *Coronavirus* (COVID-19) from Wuhan, China, which not only led to many casualties but also caused losses of trillions of dollars throughout the world. Although viruses seem quite simple particles with regard to their genetic makeup, containing single- or double-stranded nucleic acid which could be linear (potyviruses) or circular (begomoviruses) enclosed in a protein coat, they are extremely tricky when it comes to their ability to mutate and recombine their genome. Additional genomic components, viz., satellite molecules (α-satellite and β-satellite), are also present particularly in plant pathogenic begomoviruses which are considered as key pathogenicity factors (Strickland 1950; Mansoor et al. 1999). To control such viral complexes, a stable solution to overcome these viruses is required, which depends solely on inexpensive discovery approaches. As a tool for initial screening, different methodologies such as enzyme-linked immunosorbent assay (ELISA), restriction analysis and reverse transcriptase polymerase chain reaction (RT-PCR) have been in use for many decades (Briddon et al. 2001; Nicolaisen 2011) for viral diagnostic purposes. Further, different molecular techniques are available to identify the begomoviruses infecting different fruits and ornamental plants. Among them rolling circle amplification (RCA) is a powerful technique being used to amplify all circular DNA genome from the targeted host plants. The concatemers of RCA are used in restriction endonuclease digestion to produce the linear monomer molecules which play a vital role in begomovirus diversity. Nevertheless, these methods of detection are reliable but require a certain degree of specificity and prior knowledge related to the viral pathogens, and some novel viruses might be over-looked in this case. This problem of novel pathogens being overlooked has effectively been resolved because of the HTS technologies. With the development and the use of HTS technology, the highest diversity of begomoviruses in ornamental and fruit crops will be investigated (Fig. 21.2).

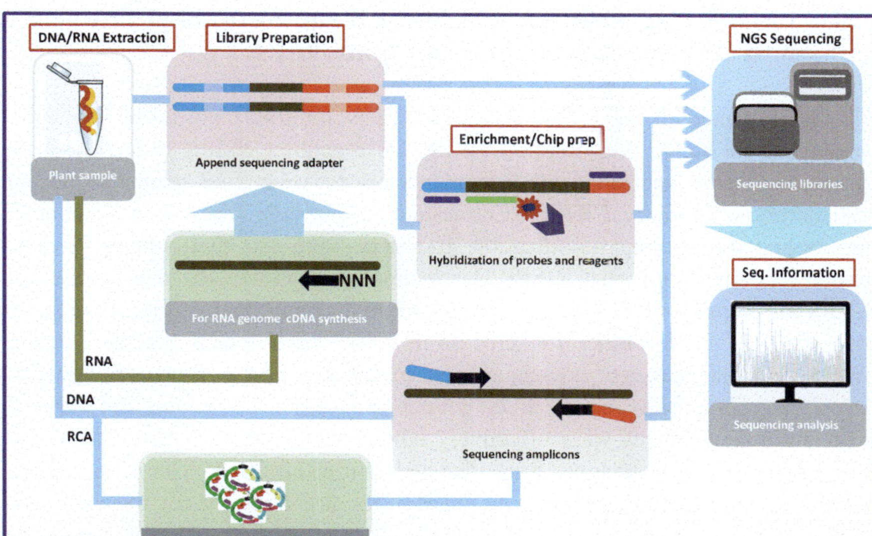

Fig. 21.2 Layout of high-throughput technologies to detect diverse begomoviruses infecting fruits and ornamental plants

Detection of begomovirus complexes is very crucial so that begomoviruses affecting ornamentals and fruits trees can be controlled effectively. Controlling of whitefly vector is also very important, which could be possible by selecting pesticide compounds; however, begomoviruses replicating in the ornamental and fruit plants cannot be killed by using any chemical spray, and the ornamental and fruit plants propagated vegetatively can also transmit virus, resulting in continued transmission of viral diseases to the next generations through propagations of infected materials. However, few reports claim the successful results of certain insecticide products which minimize the vector population on woody plants. Three rounds of 0.2% 'Malathion (a common foliar insecticide)' chemical claimed to be very effective in controlling virus vector (whitefly) transmitting begomoviruses to *J. curcas* crop at 21 days of intervals. The application of Malathion also has an impact on the plant growth development and significant increase in *J. curcas* fruit yields (Roossinck 2011). The developments in genetic engineering and recombinant DNA technology have provided an effective tool for plant molecular breeding to develop a variety of plant species by incorporation of genes from taxonomically unrelated species with novel and useful agronomical traits. Recently, transgenic woody plant *J. curcas* have been produced using a hairpin construct of double-stranded (ds) RNA in RNA interference containing gene sequences of cassava mosaic virus strains of Dharwad isolated from India, which suppresses viral gene sequence through genetic engineering (Snehi et al. 2011a). Therefore, it is possible to use other begomovirus, vector- or bacterial-based genes which can be identified and exploited in the future for the production of transgenic ornamental and fruit plants.

21.8 CRISPR/Cas: A Genome Editing Weapon Against Fruit- and Ornamental-Infecting Begomoviruses

Fruit and ornamental plants are cultivated largely for food and aesthetic values, respectively, for human consumption. CRISPR/Cas genome editing system has been proven a successful and reliable tool to mainly protect vegetables against biotic and abiotic stresses. Similarly, CRISPR/Cas system has played a vital role in ornamental plant architecture, colour modification, fragrance, size and shelf life of flowers. Disease resistance through CRISPR/Cas technology has developed a routine for the genome editing of different vegetables and ornamental plant species. Due to the successful genetic transformation process in ornamentals which enables creating novel colours and shapes of flowers and due to difficulty in agrobacterium transformation in ornamentals, mostly PDS gene is affected. The *Lilium* LpPDS and petunias gene sequences were effectively edited, where albino, pale yellow and albino–green chimeric mutants were detected (Ye et al. 2014; Yan et al. 2019). Flower induction and growth is a key phase in the flowering plant's life cycle. To enhance the flower characteristics, CRISPR/Cas9 system has distinct economic advantages (Zhang et al. 2016). Similarly, CRISPR/Cas9 system was also used to target various MADS genes which produced MADS—null mutants in *Phalaenopsis* orchid (Chandler and Sanchez 2012). MADS genes are extremely expressed in floral tissues and possibly important for flower induction and advancement. The life of flower is another essential feature of all ornamental plants. Most of the flowers are destroyed mainly due to increased ethylene production (Tong et al. 2020). Many chemicals have been exploited to improve longevity which are unable to fully stop the senescence activity (Pandey et al. 2000). However, ethylene creation may be reduced by focusing on the enzyme involved in the ethylene synthesis pathway (1-aminocyclopropane-1-carboxylate oxidase [ACO]) (Noordegraaf 1992). Recently, such modification had been done in petunia, wherePhACO1 gene was altered applying CRISPR/Cas9 system, and the modified lines enhanced flower durability compared with the control or the wild type (John 1997). Likewise, genes responsible for other characteristics of ornamental plants have also been modified.

21.9 Concluding Remarks

HTS technologies evolve rapidly over the last decade, which revolutionized the pathogen identification, particularly virus's detection. HTS provides extra benefits to detect the unknown viruses compared with earlier or conventional technologies, which exclusively rely on the available data set. HTS has significantly increased the discovery of novel fruit- or ornamental-infecting begomoviruses. Many known and novel viruses have successfully been identified using HTS, which was not possible with older techniques. Also, prior knowledge of virus genome is not required, which has made this technology robust, rapid, universal and accurate in virus diagnostics and discoveries. To confer resistance in fruit and ornamental plants against begomovirus viruses, CRISPR/Cas technology is playing a key role as this system has successfully been used to confer resistance against single-stranded DNA and (+) single-stranded RNA viruses in recent years. All these findings claim that this emerging genome editing technology has potential applications for both DNA and RNA genomes and, by linking with HTS, has a great potential to combat plant-infecting viruses at entirely different genomic level. Based on the discussions mentioned above, it is very likely that HTS and CRISPR/Cas have contributed a lot in plant virology to identify and control plant viral diseases of fruit and ornamental plants.

Acknowledgements *Author Contributions*: MSS wrote the initial draft, AMA edited it and both authors proofread and approved the final draft.

Funding: Thanks to Sultan Qaboos University for supporting this work under multiple grants awarded to both authors.

Conflict of Interest Statement No conflict of interest is involved by any of the author in this study.

References

Al Rwahnih M, Alabi OJ, Westrick NM, Golino D, Rowhani A (2017) Description of a novel monopartite geminivirus and its defective subviral genome in grapevine. Phytopathology 107 (2):240–251

Amrao L, Amin I, Shahid MS, Briddon RW, Mansoor S (2010) Cotton leaf curl disease in resistant cotton is associated with a single begomovirus that lacks an intact transcriptional activator protein. Virus Res 152(1–2):153–163

Ara MR, Masud MM, Akanda AM (2012) Detection of plant viruses in some ornamental plants that act as alternate hosts. Agriculturists 10(2):46–54

Arguello-Astorga G, Ascencio-Ibáñez JT, Dallas MB, Orozco BM, Hanley-Bowdoin L (2007) High-frequency reversion of geminivirus replication protein mutants during infection. J Virol 81 (20):11005–11015

Ashwathappa KV, Venkataravanappa V, Nandan M, Vinaykumar HD, Shankarappa KS, Reddy MK, Reddy CL (2012) Molecular characterization of Cotton leaf curl Multan virus and DNA-satellites complex associated enation leaf curl and yellow vein mosaic disease of hollyhock

Bananej K, Shafiq M, Shahid MS (2021) Association of cotton leaf curl Gezira virus with tomato leaf curl betasatellite infecting Carica papaya in Iran. Aust Plant Dis Notes 16(1):1–4

Bedford ID, Briddon RW, Brown JK, Rosell RC, Markham PG (1994) Geminivirus transmission and biological characterisation of Bemisia tabaci (Gennadius) biotypes from different geographic regions. Ann Appl Biol 125(2):311–325

Bird J (1957) A whitefly-transmitted mosaic of Jatropha gossypifolia. Agricultural Experiment Station, Río Piedras, PR

Briddon RW, Mansoor S, Bedford ID, Pinner MS, Saunders K, Stanley J, Zafar Y, Malik KA, Markham PG (2001) Identification of DNA components required for induction of cotton leaf curl disease. Virology 285(2):234–243

Briddon RW, Robertson I, Markham PG, Stanley J (2003) Occurrence of South African cassava mosaic virus (SACMV) in Zimbabwe. New Dis Rep 8:6. http://www.bspp.org.uk/ndr/

Brown JK, Fauquet CM, Briddon RW, Zerbini M, Moriones E, Navas-Castillo J, King A, Adams M, Carstens E, Lefkowitz E (2012) Virus taxonomy. In: King AMQ, Lefkowitz E, Adams MJ, Carstens EB (eds) Ninth report of the International Committee on Taxonomy of Viruses, p 351373

Brown JK, Zerbini FM, Navas-Castillo J, Moriones E, Ramos-Sobrinho R, Silva JC, Fiallo-Olivé E, Briddon RW, Hernández-Zepeda C, Idris A, Malathi VG (2015) Revision of Begomovirus taxonomy based on pairwise sequence comparisons. Arch Virol 160(6):1593–1619

Celli MG, Perotto MC, Martino JA, Flores CR, Conci VC, Pardina PR (2014) Detection and identification of the first viruses in chia (Salvia hispanica). Viruses 6(9):3450–3457

Chandler SF, Sanchez C (2012) Genetic modification; the development of transgenic ornamental plant varieties. Plant Biotechnol J 10(8):891–903

Chang LS, Lee YS, Su HJ, Hung TH (2003) First report of papaya leaf curl virus Infecting papaya plants in Taiwan. Plant Dis 87(2):204

Cheng YH, Deng TC, Chen CC, Chiang CH, Chang CA (2014) First report of Euphorbia leaf curl virus and Papaya leaf curl Guangdong virus on passion fruit in Taiwan. Plant Dis 98 (12):1746–1746

Das S, Roy A, Ghosh R, Paul S, Acharyya S, Ghosh SK (2008) Sequence variability and phylogenetic relationship of betasatellite isolates associated with yellow vein mosaic disease of mesta in India. Virus Genes 37(3):414–424

Debat H, Zavallo D, Moyano S, Luna F, Asurmendi S, Gomez-Talquenca S (2020) A new world begomovirus infecting grapevine. BioRxiv

Duffy S, Holmes EC (2008) Phylogenetic evidence for rapid rates of molecular evolution in the single-stranded DNA begomovirus tomato yellow leaf curl virus. J Virol 82(2):957–965

Durham TC, Baker C, Jones L, Snyder LU (2010) First report of Sida golden mosaic virus infecting snap bean (Phaseolus vulgaris) in Florida. Plant Dis 94(4):487–487

Ferreira SS, Barros DR, De Almeida MR, Zerbini FM (2010) Characterization of Passionfruit severe leaf distortion virus, a novel begomovirus infecting passionfruit in Brazil, reveals a close relationship with tomato-infecting begomoviruses. Plant Pathol 59(2):221–230

Gilbertson RL, Hidayat SH, Paplomatas EJ, Rojas MR, Hou YM, Maxwell DP (1993) Pseudorecombination between infectious cloned DNA components of tomato mottle and bean dwarf mosaic geminiviruses. J Gen Virol 74(1):23–31

Guo X, Zhou X (2006) Molecular characterization of a new begomovirus infecting Sida cordifolia and its associated satellite DNA molecules. Virus Genes 33(3):279–285

Haider MS, Tahir M, Saeed A, Ahmed S, Parveen R, Rashid N (2008) First report of a begomovirus infecting the ornamental plant Vinca minor L. Aust Plant Dis Notes 3(1):150–151

Harkins GW, Delport W, Duffy S, Wood N, Monjane AL, Owor BE, Donaldson L, Saumtally S, Triton G, Briddon RW, Shepherd DN (2009) Experimental evidence indicating that mastreviruses probably did not co-diverge with their hosts. Virol J 6(1):1–4

He ZF, Mao MJ, Yu H, Li HP, Chen X (2009) Molecular characterization of a distinct begomovirus infecting Allamanda cathartica in Guangdong, China. Archiv Virol 154(8):1199–1202

Ilyas M, Nawaz K, Shafiq M, Haider MS, Shahid AA (2013) Complete nucleotide sequences of two begomoviruses infecting Madagascar periwinkle (Catharanthus roseus) from Pakistan. Arch Virol 158(2):505–510

John P (1997) Ethylene biosynthesis: the role of 1-aminocyclopropane-1-carboxylate (ACC) oxidase, and its possible evolutionary origin. Physiol Plant 100(3):583–592

Kashina BD, Alegbejo MD, Banwo OO, Nielsen SL, Nicolaisen M (2013) Molecular identification of a new begomovirus associated with mosaic disease of Jatropha curcas L. in Nigeria. Arch Virol 158(2):511–514

Khan AJ, Akhtar S, Al-Shihi AA, Al-Hinai FM, Briddon RW (2012) Identification of Cotton leaf curl Gezira virus in papaya in Oman. Plant Dis 96(11):1704–1704

Khatri S, Nahid N, Fauquet CM, Mubin M, Nawaz-ul-Rehman MS (2014) A betasatellite-dependent begomovirus infects ornamental rose: characterization of begomovirus infecting rose in Pakistan. Virus Genes 49(1):124–131

Kumar Y, Bhardwaj P, Hallan V, Zaidi AA (2010) Detection and characterization of Ageratum enation virus and a nanovirus-like satellite DNA1 from zinnia causing leaf curl symptoms in India. J Gen Plant Pathol 76(6):395–398

Kumar J, Kumar J, Singh SP, Tuli R (2014) Association of satellites with a mastrevirus in natural infection: complexity of wheat dwarf India virus disease. J Virol 88(12):7093–7104

Lee G, Kim S, Jung J, Auh CK, Choi E, Chang M, Lee S (2011) Agroinoculation of Nicotiana benthamiana with cloned honeysuckle yellow vein virus isolated from Lonicera japonica. Arch Virol 156(5):785–791

Li J, Zhou X (2010) Molecular characterization and experimental host-range of two begomoviruses infecting Clerodendrum cyrtophyllum in China. Virus Genes 41(2):250–259

Liang P, Navarro B, Zhang Z, Wang H, Lu M, Xiao H, Wu Q, Zhou X, Di Serio F, Li S (2015) Identification and characterization of a novel geminivirus with a monopartite genome infecting apple trees. J Gen Virol 96(8):2411–2420

Lima AT, Sobrinho RR, Gonzalez-Aguilera J, Rocha CS, Silva SJ, Xavier CA, Silva FN, Duffy S, Zerbini FM (2013) Synonymous site variation due to recombination explains higher genetic variability in begomovirus populations infecting non-cultivated hosts. J Gen Virol 94 (2):418–431

Lozano G, Trenado HP, Fiallo-Olivé E, Chirinos D, Geraud-Pouey F, Briddon RW, Navas-Castillo J (2016) Characterization of non-coding DNA satellites associated with sweepoviruses (genus Begomovirus, Geminiviridae)–definition of a distinct class of begomovirus-associated satellites. Front Microbiol 7:162

Lu QY, Wu ZJ, Xia ZS, Xie LH (2015) Complete genome sequence of a novel monopartite geminivirus identified in mulberry (Morus alba L.). Arch Virol 160(8):2135–2138

Mahesh YS, Shankarappa KS, Rangaswamy KT, Prameela HA, Aswathanarayana DS, Divya BL, Nagesha N, Maruthi MN (2010) Detection and characterisation of a begomovirus associated with leaf curl disease of ornamental croton (Codiaeum variegatum). J Hortic Sci Biotechnol 85 (2):101–105

Mansoor S, Khan SH, Bashir A, Saeed M, Zafar Y, Malik KA, Briddon R, Stanley J, Markham PG (1999) Identification of a novel circular single-stranded DNA associated with cotton leaf curl disease in Pakistan. Virology 259(1):190–199

Mansoor S, Amin I, Iram S, Hussain M, Zafar Y, Malik KA, Briddon RW (2003) Breakdown of resistance in cotton to cotton leaf curl disease in Pakistan. Plant Pathol 52(6):784–784

Marwal A, Sahu AK, Gaur RK (2013a) First report on the association of a begomovirus with Chrysanthemum indicum exhibiting yellowing of leaf vein disease characterized by molecular studies. J Hortic Res 21(2):17–21

Marwal A, Sahu AK, Choudhary DK, Gaur RK (2013b) Complete nucleotide sequence of a begomovirus associated with satellites molecules infecting a new host Tagetes patula in India. Virus Genes 47(1):194–198

Marwal A, Sahu A, Prajapat R, Gaur RK (2013c) First report of Begomovirus infecting two ornamental plants: Jasminum sambac and Millingtonia hortensis. Indian Phytopathol 66 (1):115–116

Marwal A, Sahu AK, Gaur RK (2013d) First report of molecular characterization of begomovirus infecting an ornamental plant Tecoma stans and a medicinal plant Justicia adhatoda. Sci Int 25 (4):837

Marwal A, Sahu AK, Gaur RK (2013e) Molecular characterization of a begomovirus infecting a new host golden duranta (Duranta erecta) in India. Int J Curr Microbial Appl Sci 2:45–48

Marwal A, Kumar Sahu A, Gaur RK (2013f) Molecular characterization of begomoviruses and DNA satellites associated with a new host Spanish Flag (Lantana camara) in India. Int Scholar Res Not 2013:915703

Marwal A, Sahu AK, Gaur RK (2014) Association of begomovirus and an alphasatellite with leaf curl disease of ornamental plant, Vinca alba in Punjab, India. J Agric Res 52(3):339

Morales FJ (2006) History and current distribution of begomoviruses in Latin America. Adv Virus Res 67:127–162

Nadeem A, Mehmood T, Tahir M, Khalid S, Xiong Z (1997) First report of papaya leaf curl disease in Pakistan. Plant Dis 81(11):1333–1333

Nawaz-ul-Rehman MS, Fauquet CM (2009) Evolution of geminiviruses and their satellites. FEBS Lett 583(12):1825–1832

Nicolaisen M (2011) An oligonucleotide-based microarray for detection of plant RNA viruses. J Virol Methods 173(1):137–143

Noordegraaf C (1992) Production and marketing of high quality plants. In: International Workshop on Floriculture & Nursery Industries and Environment, vol 353, pp 134–148

Padidam M, Sawyer S, Fauquet CM (1999) Possible emergence of new geminiviruses by frequent recombination. Virology 265(2):218–225

Pandey S, Ranade SA, Nagar PK, Kumar N (2000) Role of polyamines and ethylene as modulators of plant senescence. J Biosci 25(3):291–299

Polston JE, Londoño MA, Capobianco H (2014) The complete genome sequence of New World jatropha mosaic virus. Arch Virol 159(11):3131–3136

Raj SK, Khan MS, Snehi SK, Kumar S, Khan AA (2007) Natural occurrence of a Begomovirus on Dimorphotheca sinuata in India. Aust Plant Dis Notes 2(1):25–26

Raj SK, Snehi SK, Khan MS, Singh R, Khan AA (2008a) Molecular evidence for association of Tomato leaf curl New Delhi virus with leaf curl disease of papaya (Carica papaya L.) in India. Aust Plant Dis Notes 3(1):152–155

Raj SK, Snehi SK, Khan MS, Chandra P, Pandey RM (2008b) First molecular identification of a begomovirus associated with yellow vein net disease in grain amaranth (Amaranthus cruentus L.) in India. Aust Plant Dis Notes 3(1):129–131

Ramkat R, Calari A, Maghuly F, Laimer M (2011) Occurrence of African cassava mosaic virus (ACMV) and East African cassava mosaic virus–Uganda (EACMV-UG) in Jatropha curcas. BMC Proc 5(7):1–2

Revers F, Le Gall O, Candresse T, Le Romancer M, Dunez J (1996) Frequent occurrence of recombinant potyvirus isolates. J Gen Virol 77(8):1953–1965

Rihne T, Mitra S, Bahadur A, Panda P, Banyal N, Rao GP (2020) Mixed infection of phytoplasma and begomovirus associated with leaf curling and witches' broom disease of Zinnia elegans in India. Indian Phytopathol 5:1–6

Roossinck MJ (2011) The big unknown: plant virus biodiversity. Curr Opin Virol 1(1):63–67

Roye ME, Collins AM, Maxwell DP (2006) First report of a begomovirus associated with the common weed Jatropha gossypiifolia in Jamaica. Plant Pathol 55(2):286

Sattar MN, Qurashi F, Iqbal Z, Saleem HM (2018) Molecular characterization of ageratum enation virus and DNA-satellites associated with yellowing and leaf curl symptoms on mulberry in Pakistan. Can J Plant Pathol 40(3):442–449

Seal SE, VandenBosch F, Jeger MJ (2006) Factors influencing begomovirus evolution and their increasing global significance: implications for sustainable control. Crit Rev Plant Sci 25 (1):23–46

Shahid MS, Yoshida S, Khatri-Chhetri GB, Briddon RW, Natsuaki KT (2013) Complete nucleotide sequence of a monopartite Begomovirus and associated satellites infecting Carica papaya in Nepal. Virus Genes 46(3):581–584

Simmonds-Gordon RN, Collins-Fairclough AM, Stewart CS, Roye ME (2014) First report of a complete genome sequence for a begomovirus infecting Jatropha gossypifolia in the Americas. Arch Virol 159(10):2815–2818

Singh J, Khanna S, Ranjan K, Pant RP, Kumar P, Sirohi A, Baranwal VK (2017) Evidence of association of begomovirus with the yellow vein disease of an ornamental plant pot marigold (Calendula officinalis) from Western Uttar Pradesh. J Pure Appl Microbiol 11(3):1609–1615

Snehi SK, Khan MS, Raj SK, Prasad V (2011a) Complete nucleotide sequence of Croton yellow vein mosaic virus and DNA-β associated with yellow vein mosaic disease of Jatropha gossypifolia in India. Virus Genes 43(1):93–101

Snehi SK, Raj SK, Khan MS, Prasad V (2011b) Molecular identification of a new begomovirus associated with yellow mosaic disease of Jatropha gossypifolia in India. Arch Virol 156 (12):2303–2307

Srivastava A, Raj SK, Kumar S, Snehi SK, Kulshreshtha A, Hallan V, Pande SS (2013) Molecular identification of Ageratum enation virus, betasatellite and alphasatellite molecules isolated from yellow vein diseased Amaranthus cruentus in India. Virus Genes 47(3):584–590

Srivastava A, Kumar S, Raj SK (2014) First report of Ageratum enation virus, betasatellite and alphasatellite causing leaf curl and enation disease of Amaranthus hypochondriacus in India. Plant Dis 98(9):1285–1285

Srivastava A, Kumar S, Jaidi M, Raj SK (2015) Molecular characterization of a new begomovirus associated with leaf yellow mosaic disease of Jatropha curcas in India. Arch Virol 160 (5):1359–1362

Srivastava A, Kumar S, Jaidi M, Raj SK, Shukla SK (2016) First report of tomato leaf curl New Delhi virus on opium poppy (Papaver somniferum) in India. Plant Dis 100(1):232

Strickland AH (1950) The dispersal of pseudoeoeeidae (hemiptera-homoptera) by air currents in the gold coast. Proc R Entomol Soc Lond 25(1–3):1

Sun S, Hu Y, Jiang G, Tian Y, Ding M, Yu C, Zhou X, Qian Y (2020) Molecular characterization and genomic function of grapevine Geminivirus A. Front Microbiol 11:2121

Thresh JM (2003) Control of plant virus diseases in sub-Saharan Africa: the possibility and feasibility of an integrated approach. Afr Crop Sci J 11(3):199–223

Tong CG, Wu FH, Yuan YH, Chen YR, Lin CS (2020) High-efficiency CRISPR/Cas-based editing of Phalaenopsis orchid MADS genes. Plant Biotechnol J 18(4):889

Urbino C, Gutiérrez S, Antolik A, Bouazza N, Doumayrou J, Granier M, Martin DP, Peterschmitt M (2013) Within-host dynamics of the emergence of tomato yellow leaf curl virus recombinants. PLoS One 8(3):e58375

Vaca-Vaca JC, Carrasco-Lozano EC, López-López K (2017) Molecular identification of a new begomovirus infecting yellow passion fruit (Passiflora edulis) in Colombia. Arch Virol 162 (2):573–576

Valverde RA, Singh R, Sabanadzovic S (2012) Detection and identification of Clerodendron golden mosaic China virus in Salvia splendens. Eur J Plant Pathol 133(3):499–503

Van Der Walt E, Martin DP, Varsani A, Polston JE, Rybicki EP (2008) Experimental observations of rapid Maize streak virus evolution reveal a strand-specific nucleotide substitution bias. Virol J 5(1):1–1

Wang X, Xie Y, Zhou X (2004) Molecular characterization of two distinct begomoviruses from papaya in China. Virus Genes 29(3):303–309

Xu J, Kang BC, Naing AH, Bae SJ, Kim JS, Kim H, Kim CK (2020) CRISPR/Cas9-mediated editing of 1-aminocyclopropane-1-carboxylate oxidase1 enhances Petunia flower longevity. Plant Biotechnol J 18(1):287–297

Yan R, Wang Z, Ren Y, Li H, Liu N, Sun H (2019) Establishment of efficient genetic transformation systems and application of CRISPR/Cas9 genome editing technology in Lilium pumilum DC, Fisch, and Lilium longiflorum White Heaven. Int J Mol Sci 20(12):2920

Ye J, Qu J, Mao HZ, Ma ZG, Rahman NE, Bai C, Chen W, Jiang SY, Ramachandran S, Chua NH (2014) Engineering geminivirus resistance in Jatropha curcas. Biotechnol Biofuels 7(1):1–1

Zaim M, Kumar Y, Hallan V, Zaidi AA (2011) Velvet bean severe mosaic virus: a distinct begomovirus species causing severe mosaic in Mucuna pruriens (L.) DC. Virus Genes 43 (1):138–146

Zambrano K, Fernández-Rodríguez T, Marys E (2012) Molecular characterization of a new begomovirus that infects Euphorbia heterophylla and Solanum lycopersicum in Venezuela. Arch Virol 157(2):379–382

Zerbini FM, Briddon RW, Idris A, Martin DP, Moriones E, Navas-Castillo J, Rivera-Bustamante R, Roumagnac P, Varsani A, Consortium IR (2017) ICTV virus taxonomy profile: Geminiviridae. J Gen Virol 98(2):131

Zhang W, Olson NH, Baker TS, Faulkner L, Agbandje-McKenna M, Boulton MI, Davies JW, McKenna R (2001) Structure of the Maize streak virus geminate particle. Virology 279 (2):471–477

Zhang B, Yang X, Yang C, Li M, Guo Y (2016) Exploiting the CRISPR/Cas9 system for targeted genome mutagenesis in petunia. Sci Rep 6(1):1–8

Next-Generation Molecular Diagnostics for Plant Viruses

22

Ashish Srivastava, Sonal Srivastava, Taruna Gupta, and Narayan Rishi

Abstract

Viral disease is a major constraint for the yield loss in many economically important crops and ultimately causes economic losses. Plant viruses are distributed from one infected crop to nearby and distant crops either due to mechanical injuries or by their specific insect vectors. From time to time, new plant viruses such as Papaya ringspot virus, Chili leaf curl virus, Citrus tristeza virus, and Tomato yellow leaf curl virus emerge and re-emerge and become the major reason of outbreak of important viral diseases of crops. Many times, such diseases paralyze the economy of many countries due to severe yield losses. In developing countries like India, the management of such diseases is not practicable; therefore, the only way to eradicate such viruses is early detection and surveillance of the disease. Field-based easy diagnostic methods such as RPA, LAMP, and CRISPR/Cas have drawn the attention of researchers to develop a point-of-care testing system. Especially, clustered regularly interspaced short palindromic repeats (CRISPR)/CRISPR-associated protein (Cas) is an emerging technology for gene editing and diagnostics development. Several rapid nucleic acid diagnostic kits have been developed and validated using Cas9, Cas12, and Cas13 proteins. This book chapter summarizes the CRISPR/Cas-based next-generation molecular diagnostic techniques and portability of devices for field-based utilization.

A. Srivastava · S. Srivastava · T. Gupta · N. Rishi (✉)
Amity Institute of Virology & Immunology, Amity University Uttar Pradesh, Noida, India
e-mail: nrishi@amity.edu

S. K. Raj et al. (eds.), *Virus Diseases of Ornamental Plants*,
https://doi.org/10.1007/978-981-16-3919-7_22

Keywords

Recombinase polymerase amplification (RPA) · Loop-mediated isothermal amplification (LAMP) · CRISPR/Cas · Cas9 · Cas13 · Cas12 · Collateral cleavage · Fluorescent probe · DNA virus · RNA virus

22.1 Introduction

Plant viruses are small obligate infectious agents, comprise either DNA or RNA as genetic material, and hijack host machinery for their replication and protein expression. Viral species emerge and re-emerge which results in a threat to crop production. Plant viruses are the reason for devastating yield losses of several economically important crops around the globe. RNA virus genera as *Potyvirus*, *Tobamovirus*, *Cucumovirus*, *Tombusvirus*, *Tospovirus*, and *Potexvirus* and DNA virus genera as *Begomovirus*, *Mastrevirus*, *Nanovirus*, etc. are well-known devastating plant viruses and reported to be major pathogens causing major economic losses worldwide (Scholthof et al. 2011). Papaya ringspot virus (Kumar et al. 2019), Soybean mosaic virus (Hajimorad et al. 2018), Cyrtanthus elatus virus A, Narcissus yellow stripe virus (Raj et al. 2020), Bean yellow mosaic virus (Kaur et al. 2018), etc. are reported to infect many important crops. Among DNA viruses, begomoviruses are responsible for many of these diseases of dicot plants. Tomato yellow leaf curl virus (Kil et al. 2016), Tomato leaf curl New Delhi virus (Srivastava et al. 2016a), Chilli leaf curl virus (Kumar et al. 2018), Cotton leaf curl Multan virus (Srivastava et al. 2016b), and Ageratum enation virus (Srivastava et al. 2012, 2013, 2014, 2015, 2017) which are begomoviruses of *Jatropha* (Srivastava et al. 2015a, b, c; Snehi et al. 2012) are some of the many highly infectious begomoviruses. Similarly, many other viruses also cause important diseases in crops. These viruses mainly infect the annual crops or perennial shrubs; therefore, early diagnosis and disposal of infected planting material may be proved most beneficial to check the virus spread (Fig. 22.1).

COVID-19 may prove as "blessing in disguise for the new research on diagnostic techniques." CRISPR/Cas system is rapidly becoming well-known for its powerful potential in therapeutics but also getting unique identity in the field of molecular diagnostics. There are many research groups that are actively working on efficient CRISPR/Cas-based diagnostics development. Recently, improved understanding of other diverse CRISPR/Cas systems has expanded CRISPR applications that include genome-wide screening, multiplex genome editing, transcriptional regulation, gene therapy, and antiviral defense. This book chapter will discuss the potential application of different Cas proteins to develop new tools that offer cost-effective, portable diagnostics through nucleic acid screening.

Fig. 22.1 Showing virus symptoms on different plant leaves

22.2 Conventional Diagnostic Techniques

22.2.1 Serological Diagnosis

Serological or immunological assays are the most convenient way to detect plant viruses. These may be agglutination tests, precipitation tests, enzyme-linked immunosorbent assays (ELISA), and immunoblotting assays. ELISA-based tests are available in either a solid microtiter plate format or a paper strip format. Rapid tests are based on the paper ELISA format and are discussed in detail in the section on rapid tests. These assays are based on the specific interaction of both antigen and antibody. ELISA technique was first applied for the detection of potyvirus of potato (Clark and Adams 1977). The serological detection methods have been used to detect several viruses of bananas, potatoes, garlic, grapes, papaya, cardamom, etc. earlier (Agarwal et al. 2009; Mandal et al. 2004; Pramesh et al. 2013; Vijayanandraj et al. 2013; Singh et al. 2014; Sharma et al. 2014; Rai et al. 2018). Serological techniques are mainly used to detect viruses in bulk samples of plants due to ease of handling.

22.2.2 Molecular Hybridization Technique for Diagnosis

The molecular hybridization methodologies such as dot-blot, gel-blot, and in situ hybridization can be used to detect plant viruses and viroids. Srivastava et al. (2015d) have reported the dot-blot assay for detection of Papaya leaf curl virus from amaranths. Nanomolar amounts of potyviruses have been standardized (Berger et al. 1985). Many other RNA and DNA viruses and viroid can be detected by this technique.

22.3 Modern Techniques for Virus Diagnosis

22.3.1 Artificial Intelligence for Virus Detection

Artificial intelligence is a very novel aspect of surveillance of viral diseases in very large fields. Machine learning and deep learning are major methods to detect virus symptoms on leaves. VirLeafNet is a recent method which uses three convolutional neural network architecture to detect virus infection in mung bean by deep learning-based plant disease automatic detection method (Joshi et al. 2021). We have also used the automatic artificial intelligence-based tool for early-stage diagnosis of viral growth, where the symptoms may be available in certain parts like leaves. We developed an automatic computer vision-based method for the identification of yellow mosaic disease in *Vigna mungo* (Pandey et al. 2021). Statistical- and texture-based transmission electron microscopic images were designed for maximum accuracy of contributed methodology (Sengar et al. 2017).

22.3.2 PCR-Based Diagnostic Techniques

Polymerase chain reaction (PCR) is a very popular molecular technique for the detection of plant viruses worldwide. There are several reports of PCR-based detection of viruses such as potyviruses (Kumar et al. 2019; Hajimorad et al. 2018; Raj et al. 2020; Kaur et al. 2018) and begomoviruses (Kil et al. 2016; Srivastava et al. 2016a, 2012, 2013, 2014, 2015a, b, c, 2016b, 2017; Snehi et al. 2012). The major advantage of PCR-based detection is that it is used further for sequencing which confirms the virus infection.

22.3.3 Isothermal Amplification Methods for Detection

The major limitation of PCR-based diagnostics is the requirement of costly instruments due to its requirement of cyclic temperature change. Therefore, PCR is not a feasible instrument to detect viruses in the field. The isothermal amplification methods are required for point-of-care testing of viruses. Reverse transcription loop-mediated isothermal amplification (RT-LAMP)-based assay is a loop-mediated isothermal nucleic acid amplification approach that has been evaluated for the detection of several viruses. An RT-LAMP assay was used to detect eight RNA viruses and one DNA virus from infected rice plants and their insect vector insects (Sasaya 2015). Zhao et al. (2012) used multiple primers for tobacco viruses, cucumber mosaic virus, potyvirus (PVY), tobacco etch virus, tobacco mosaic virus, and tobacco vein banding mosaic virus. Another isothermal amplification method, reverse transcription recombinase polymerase amplification (RT-RPA), is a sensitive and easy to design molecular diagnostic technology which is rapid, is portable, and can be multiplexed with several pathogens. RT-RPA can also be used as a reliable diagnostic tool platform for many plant viruses, including potyviruses and potexviruses (Euler et al. 2013; Silva et al. 2018). Kumar et al. (2021) recently demonstrated RT-RPA-based diagnosis of Onion yellow dwarf virus (OYDV). Both RT-RPA and RT-LAMP can also be combined with other nucleic acid diagnostic assays to enhance the performance of the kit.

22.4 CRISPR/Cas-Based Diagnostics

CRISPR sequence together with Cas enzyme forms the basis of a technology known as CRISPR/Cas that is used for genome editing. CRISPR (clustered regularly interspaced short palindromic repeats) is a family of DNA sequences found within the genome of prokaryotic organisms. Cas (CRISPR-associated protein), the most commonly used endonuclease, utilizes a guide RNA to bind to a complementary DNA sequence, which is subsequently cleaved through Cas endonuclease activity. Protospacer adjacent motif is a two to six base pair DNA sequence immediately following the DNA sequence targeted by the Cas nuclease in the CRISPR bacterial adaptive immune system. The most commonly studied endonuclease belongs to type II CRISPR system, i.e., Cas9 that targets DNA phages, while type V CRISPR system, i.e., Cas12 also targets DNA and type III and type VI groups, exhibits RNA targeting activity which includes Cas13.

Cas9 can recruit guide RNA (gRNA), can facilitate its specific binding to target DNA, and can make a blunt double-stranded DNA break, which can then be repaired by either nonhomologous end joining or homologous direct recombination with a donor template DNA to create site-specific edits. Another class of Cas effectors (Class 2 type V) is Cas12 (Cpf1), which is a proficient enzyme that creates staggered cuts in dsDNA (Chen et al. 2018). Cas12 develops its crRNAs which leads to increased multiplexing ability. Cas12 has brought a platform for epigenome editing and was recently discovered that Cas12a can slice single-stranded DNA once activated by a target DNA molecule identical to its spacer sequence (Chen et al. 2018). They recognize a T-rich protospacer adjacent motif (PAM) instead of a G-rich PAM and generate dsDNA breaks with staggered 5′ ends. Cas13 (C2c2) are unique CRISPR-associated effectors which have specific recognition and cleavage activity for complementary RNA and known for trans cleavage or collateral cleavage of nearby RNA (Abudayyeh et al. 2016). This feature is widely being utilized for the diagnosis of viruses and other mRNAs (Kellner et al. 2020).

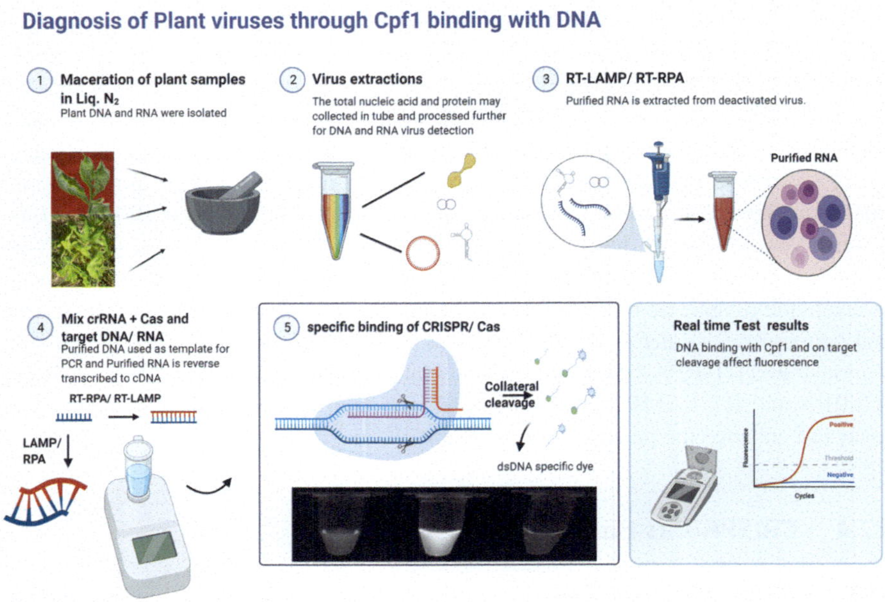

Fig. 22.2 Schematic representation of CRISPR/Cas-based diagnostics

CRISPR-based diagnostic tools may play an important role in developing such diagnostics. These tools were combined with various platforms of amplification like RPA and LAMP and also with various devices for detection such as plate readers, fluorimeter, and lateral flow devices to make them accessible and can be applied on point of operation (Fig. 22.2).

22.4.1 Cas9-Based Diagnostics

Pardee and coworkers first used nucleic acid sequence-based amplification (NASBA), an isothermal technique, with Cas9 to distinguish accurately between closely related Zika virus strains in vitro (Pardee et al. 2016). In this diagnostic test, they extracted RNA and amplified it via NASBA, and rehydrated the freeze-dried paper sensors, and the detection of the target RNA was indicated by a color change in the paper disc from yellow to purple. Later, Zhou et al. (2018) developed a CRISPR/Cas9-triggered nicking endonuclease-mediated strand displacement amplification method (namely CRISDA) for ultrasensitive nucleic acids in point-of-care diagnostics and field analyses. Wang et al. (2020) have come up with a unique diagnostic assay named clustered regularly interspaced short palindromic repeats/Cas9-mediated lateral flow nucleic acid assay (CASLFA) for diagnosis of African swine fever virus (ASFV) by lateral flow nucleic acid diagnosis kit and FnCas9 Editor Linked Uniform Detection Assay (FELUDA) (Azhar et al. 2020). Presently,

Cas9-based techniques were explored for human viruses only; however, these can be utilized successfully for plant viruses also.

22.4.2 Cas12-Based Diagnostics

Cas12 proteins around 130 kDa protein are highly used for in vitro diagnosis of several plant and human viruses. The Cas12a-based system utilizes an ssDNA probe in place of an RNA probe; therefore, it is more suitable for on-site viral detection. HOLMES (one-HOur Low-cost Multipurpose highly Efficient System) and DETECTR (DNA Endonuclease Targeted CRISPR Trans Reporter) are two major CRISPR/Cas12-based diagnostic systems which have been applied worldwide (Srivastava et al. 2020) coupled with isothermal amplification. Cas12-based diagnostics play an important role in the direct diagnosis of DNA viruses without undergoing RNA transcription, which makes it popular for the diagnosis of both DNA and RNA viruses. This technique showed its potential in detection of most prevalent RNA viruses/viroid in apple, Apple necrotic mosaic virus (ApNMV), Apple stem pitting virus (ASPV), Apple stem grooving virus (ASGV), Apple chlorotic leaf spot virus (ACLSV), and Apple scar skin viroid (ASSVd) (Jiao et al. 2021). The recent report of CRISPR/Cas12a system-based detection of two begomoviruses, tomato yellow leaf curl virus and tomato leaf curl New Delhi virus (Mahas et al. 2021). The potential application of the CRISPR/Cas13a system for viroid detection interference in plants is suggested by targeting specific motifs of three economically important viroids.

22.4.3 Cas13-Based Diagnostics

Cas13 is a unique RNA-guided protein that possesses RNase activity and, once activated with binding to target RNA, cleaves nearby RNAs in trans (Abudayyeh et al. 2016). A Cas13-based, efficient, and robust detection system for RNA viruses was developed for *Nicotiana benthamiana* (Mahas et al. 2019). The potential of Cas13-based detection in several RNA and DNA viruses makes it popular next-generation diagnostics.

22.5 Conclusions

Plant viruses are one of the major limiting factors for the growth and yield of plant products. Unlike animals, plants do not have any developed immune system and therefore, eradication of viruses is not possible. The only way to manage virus disease is early diagnosis and disposal. There is a need for cheap, quick, and reliable diagnostic techniques for surveillance of the plant viral disease in the interior and inaccessible villages. Based on the current development, CRISPR/Cas-based diagnostics can be considered as next-generation molecular diagnostics.

Acknowledgment The authors are thankful to the Department of Science and Technology, New Delhi (DST-INSPIRE-Faculty scheme, IFA-17-LSPA-100), for the grant to A. Srivastava; the Indian Council of Medical Research (ICMR—File No. 5/3/8/29/ITR-F/2020), for the grant to AS, TG; and the University Grant Commission (UGC) for Junior Research Fellowship to SS (795/CSIR—UGC NET JUNE 2017). Figure 22.2 is created with BioRender.com.

References

Abudayyeh OO, Gootenberg JS, Konermann S, Joung J, Slaymaker IM, Cox DB, Shmakov S et al (2016) C2c2 is a single-component programmable RNA-guided RNA-targeting CRISPR effector. Science 353:aaf5573

Agarwal S, Krishnareddy M, Jain RK (2009) Production of polyclonal antibodies using recombinant coat protein of Papaya ring spot virus and their use in immunodiagnosis. J Plant Biochem Biotechnol 18:109–111

Azhar M, Phutela R, Ansari AH, Sinha D, Sharma N, Kumar M, Aich M, Sharma S, Singhal K, Lad H, Patra PK, Makharia G, Chandak GR, Chakraborty D, Maiti S (2020) Rapid, field-deployable nucleobase detection and identification using FnCas9. BioRxiv. https://doi.org/10.1101/2020.04.07.028167

Berger PH, Thornbury DW, Pirone TP (1985) Detection of picogram quantities of potyviruses using a dot blot immunobinding assay. J Virol Methods 12(1–2):31–39

Chen JS, Ma E, Harrington LB, Da Costa M, Tian X, Palefsky JM, Doudna JA (2018) CRISPR-Cas12a target binding unleashes indiscriminate single-stranded DNase activity. Science 360 (6387):436–439

Clark MF, Adams AN (1977) Characteristics of the microplate method of enzyme linked immunosorbent assay for the detection of plant viruses. J Gen Virol 34:475–483

Euler M, Wang Y, Heidenreich D, Patel P, Strohmeier O, Hakenberg S, Niedrig M, Hufert FT, Weidmann M (2013) Development of a panel of recombinase polymerase amplification assays for detection of biothreat agents. J Clin Microbiol 51:1110–1117

Hajimorad MR, Domier LL, Tolin SA, Whitham SA, Saghai-Maroof MA (2018) Soybean mosaic virus: a successful potyvirus with a wide distribution but restricted natural host range. Mol Plant Pathol 19(7):1563–1579

Jiao J, Kong K, Han J, Song S, Bai T, Song C, Wang M, Yan Z, Zhang H, Zhang R, Feng J, Zheng X (2021) Field detection of multiple RNA viruses/viroids in apple using a CRISPR/Cas12a-based visual assay. Plant Biotechnol J 19:394. https://doi.org/10.1111/pbi.13474

Joshi RK, Kaushik M, Dutta MK, Srivastava A, Choudhary N (2021) VirLeafNet: automatic analysis and viral disease diagnosis using deep-learning in Vigna mungo plant. Ecol Informatics 61:101197

Kaur C, Raj R, Srivastava A, Kumar S, Raj SK (2018) Sequence analysis of six full-length bean yellow mosaic virus genomes reveals phylogenetic diversity in India strains, suggesting subdivision of phylogenetic group-IV. Achiv Virol 163:235–242

Kellner MJ, Koob JG, Gootenberg JS, Abudayyeh OO, Zhang F (2020) SHERLOCK: nucleic acid detection with CRISPR nucleases. Nat Protoc 15(3):1311

Kil EJ, Kim S, Lee YJ, Byun HS, Park J, Seo H, Kim CS, Shim JK, Lee JH, Kim JK, Lee KY, Choi HS, Lee S (2016) Tomato yellow leaf curl virus (TYLCV-IL): a seed-transmissible geminivirus in tomatoes. Sci Rep 6:19013

Kumar S, Raj R, Raj SK, Agrawal L, Chauhan PS, Srivastava A (2018) Study of biochemical and histopathological changes induced in the sweet pepper (Capsicum annuum L.) in response to Chilli leaf curl virus infection. Physiol Mol Plant Pathol 104:95–102

Kumar R, Srivastava A, Srivastava A, Srivastava S, Prasad V, Raj SK (2019) Studies on molecular variability of coat protein gene of Papaya ringspot virus-P isolates from India. Eur J Plant Pathol 155:369–376

Kumar R, Pant RP, Kapoor S, Khar A, Baranwal VK (2021) Development of a reverse transcription-recombinase polymerase amplification (RT-RPA) assay for the detection of onion yellow dwarf virus (OYDV) in onion cultivars. Indian Phytopathol 74:201–207

Mahas A, Aman R, Mahfouz M (2019) CRISPR-Cas13d mediates robust RNA virus interference in plants. Genome Biol 20:263

Mahas A, Hassan N, Aman R, Marsic T, Wang Q, Ali Z, Mahfouz MM (2021) LAMP-coupled CRISPR–Cas12a module for rapid and sensitive detection of plant DNA viruses. Viruses 13:466

Mandal B, Mandal S, Pun KB, Varma A (2004) First report of the association of a nanovirus with foorkey disease of large cardamom in India. Plant Dis 88:428

Pandey C, Baghel N, Dutta MK, Srivastava A, Choudhary N (2021) Machine learning approach for automatic diagnosis of Chlorosis in Vigna mungo leaves. Multimed Tools Appl 80:13407–13427

Pardee K, Green A, Takahashi MK, Braff D, Lambert G, Lee JW, Ferrante T, Ma D, Donghia N, Fan M, Daringer NM, Bosch I, Dudley DM, O'Connor DH, Gehrke L, Collins JJ (2016) Rapid, low-cost detection of zika virus using programmable biomolecular components. Cell 165 (5):1255–1266

Pramesh D, Singh P, Baranwal VK (2013) In vitro expression and purification of the coat protein gene of Garlic common latent virus and its application in ELISA-based diagnostics. J Plant Pathol 95(2):385–389

Rai R, Khurana SMP, Kumar S, Gupta N, Baranwal VK (2018) Serological detection of Grapevine leafroll-associated virus 4 in grapevine growing areas of India using polyclonal antiserum raised against the recombinant coat protein. Crop Prot 109:128–135

Raj R, Kaur C, Srivastava A, Kumar S, Raj SK (2020) Sequence analyses of RT-PCR products obtained from seven infected leaf samples revealed existence of three potyvirus species in Indian narcissus (Narcissus tazetta L.). 3 Biotech 10(10):428

Sasaya T (2015) Detection methods for rice viruses by a reverse-transcription loop-mediated isothermal amplification (RT-LAMP). Methods Mol Biol 1236:49–59

Scholthof KB, Adkins S, Czosnek H, Palukaitis P, Jacquot E, Hohn T, Hohn B et al (2011) Top 10 plant viruses in molecular plant pathology. Mol Plant Pathol 12(9):938–954

Sengar N, Srivastava A, Dutta MK (2017) Machine vision-based detection of ageratum enation virus infection using light microscopic images of poppy plants cells. In: International Conference on Emerging Trends in Computing and Communication Technologies (ICETCCT), pp 1–4

Sharma SK, Kumar PV, Baranwal VK (2014) Immunodiagnosis of episomal Banana streak MY virus using polyclonal antibodies to an expressed putative coat protein. J Virol Methods 207:86–94

Silva G, Oyekanmi J, Nkere CK, Bomer M, Kumar PL, Seal SE (2018) Rapid detection of potyviruses from crude plant extracts. Anal Biochem 546 17–22

Singh P, Prabha K, Jain RK, Baranwal VK (2014) N-terminal in coat protein of Garlic virus X is indispensable for its serological detection. Virus Genes 48(1):128–132

Snehi SK, Srivastava A, Raj SK (2012) Biological characterization and complete genome sequence of a possible strain of Indian cassava mosaic virus from Jatropha curcas in India. J Phytopathol 160:547–553

Srivastava S, Bisht H, Sidhu OP, Srivastava A, Singh PC, Pandey RM, Raj SK, Roy R, Nautiyal CS (2012) Changes in the metabolome and histopathology of Amaranthus hypochondriacus L. in response to Ageratum enation virus infection. Phytochemistry 80:8–16

Srivastava A, Raj SK, Kumar S, Snehi SK, Kulshreshtha A, Hallan V, Pande SS (2013) Molecular identification of Ageratum enation virus, betasatellite and alphasatellite molecules isolated from yellow vein diseased Amaranthus cruentus in India. Virus Genes 47(3):584–590

Srivastava A, Kumar S, Raj SK (2014) First report of Ageratum enation virus, betasatellite and alphasatellite causing leaf curl and enation disease of Amaranthus hypochondriacus in India. Plant Dis 98(9):1285

Srivastava A, Kumar S, Raj SK (2015) Molecular characterization of a begomovirus and betasatellite causing yellow vein net disease of Ageratum houstonianum. Plant Dis 99 (5):627–631

Srivastava A, Kumar S, Jaidi M, Raj SK (2015a) Molecular characterization of a new begomovirus associated with leaf yellow mosaic disease of Jatropha curcas in India. Arch Virol 160:1359–1362

Srivastava A, Jaidi M, Kumar S, Raj SK (2015b) Molecular identification of a new begomovirus associated with leaf crumple disease of Jatropha curcas L. in India. Arch Virol 160(2):617–619

Srivastava A, Kumar S, Jaidi M, Raj SK (2015c) Characterization of a novel begomovirus associated with yellow mosaic disease of three ornamental species of Jatropha grown in India. Virus Res 201:41–49

Srivastava A, Jaidi M, Kumar S, Raj SK, Shukla S (2015d) Association of Papaya leaf curl virus with the leaf curl disease of grain amaranth (Amaranthus cruentus L.) in India. Phytoparasitica 43:97–101

Srivastava A, Kumar S, Jaidi M, Raj SK, Shukla SK (2016a) First Report of Tomato leaf curl New Delhi virus on Opium Poppy (Papaver somniferum) in India. Plant Dis 100:232

Srivastava A, Kumar S, Jaidi M, Raj SK (2016b) Association of Cotton leaf curl Multan virus and its associated betasatellite with leaf curl disease of Hibiscus rosa-sinensis in India. New Dis Rep 33:4

Srivastava A, Agrawal L, Raj R, Jaidi M, Raj SK, Gupta S et al (2017) Ageratum enation virus infection induces programmed cell death and alters metabolite biosynthesis in Papaver somniferum. Front Plant Sci 8:1172

Srivastava S, Upadhyay DJ, Srivastava A. (2020) Next-generation molecular diagnostics development by CRISPR/Cas tool: rapid detection and surveillance of viral disease outbreaks. Front Mol Biosci 7:582499

Vijayanandraj S, Yogita M, Das A, Ghosh A, Mandal B (2013) Highly efficient immunodiagnosis of Large cardamom chirke virus using the polyclonal antiserum against Escherichia coli expressed recombinant coat protein. Indian J Virol 24(2):227–234

Wang X, Xiong E, Tian T, Cheng M, Lin W, Wang H, Zhang G, Sun J, Zhou X (2020) Clustered regularly interspaced short palindromic repeats/Cas9-mediated lateral flow nucleic acid assay. ACS Nano 14(2):2497–2508

Zhao L, Cheng J, Hao X, Tian X, Wu Y (2012) Rapid detection of tobacco viruses by reverse transcription loop-mediated isothermal amplification. Achiv Virol 157(12):2291–2298

Zhou W, Hu L, Ying L, Zhao Z, Chu PK, Yu XF (2018) A CRISPR–Cas9-triggered strand displacement amplification method for ultrasensitive DNA detection. Nat Commun 9:5012. https://doi.org/10.1038/s41467-018-07324-5

The manufacturer's authorised representative in the EU is Springer Nature Customer Service Centre GmbH, Europaplatz 3, 69115 Heidelberg, Germany. If you have any concerns regarding our products, please contact ProductSafety@springernature.com

Printed and bound by CPI Group (UK) Ltd, Croydon, CR0 4YY
30/09/2026
02225241-0011